Birkhäuser

Progress in Mathematics

Volume 328

More information about this series at http://www.springer.com/series/4848

Preface

This volume is based around the workshop and conference *Representations of p-adic groups* held at IISER Pune from July 10 to 19, 2017.

The first two chapters give expanded accounts of two of the workshop mini-courses. Each is concerned with aspects of the local Langlands conjecture or correspondence (LLC). This predicts a close relation between irreducible representations of reductive p-adic groups and suitable relatives of Galois representations, now known as Langlands parameters. Its influence pervades our subject and lies at the heart of applications to automorphic forms.

The chapter "Local Langlands and Springer Correspondences," by Anne-Marie Aubert, builds toward a precise formulation of the conjecture in a general setting, sketching in the considerable background that this requires. At the same time, it provides a valuable window on the author's work with Ahmed Moussaoui and Maarten Solleveld on links between the LLC and (generalized) Springer correspondences. It also explores a connection with the world of non-commutative geometry via the Aubert-Baum-Plymen-Solleveld conjecture. Among interrelated results and conjectures, the chapter includes, in particular, a conjectural description of the manner in which irreducible cuspidal (or supercuspidal in Harish-Chandra's terminology) representations fit into and generate the LLC.

The chapter "Arithmetic of Cuspidal Representations," by Colin Bushnell, looks at the Langlands correspondence for the family of groups $\mathrm{GL}_n(F)$ for F a non-Archimedean local field. In this case, via the Zelevinsky classification and properties of local constants, the correspondence can be viewed as a canonical bijection between equivalence classes of irreducible cuspidal representations of $\mathrm{GL}_n(F)$ and equivalence classes of irreducible n-dimensional representations of the Weil group of F. Its existence has been known for over 20 years. Many mysteries, however, remain. For example, can one describe the correspondence explicitly, or at least in some detail, in terms of the classification via types of the irreducible cuspidal representations of $\mathrm{GL}_n(F)$ due to Bushnell and Kutzko? Indeed, the linchpin of that classification—the notion of *simple character*—is unmistakably arithmetic in nature. It is natural then to attempt to understand the correspondence along such lines. Bushnell and Henniart, in a long series of papers, have made considerable progress in this direction. Bushnell's write-up provides a characteristically cogent account of their work.

With one exception, the remaining chapters present original papers. The exception is a pleasant survey by Chufeng Nien on local converse theorems. We give below a brief description of the original papers, ignoring restrictions on characteristic or residual characteristic. Please see the introductions to the papers for a more careful and detailed overview or—best of all—read the papers themselves!

- The contribution by Jeff Adler and Josh Lansky discusses group actions on root data. It fits into the authors' continuing study of liftings of representations of reductive groups over finite and local non-Archimedean fields.
- Moshe Adrian's paper gives a complete character formula, valid in all residual characteristics, for a simple supercuspidal representation of $\mathrm{SL}_2(F)$.
- Yeansu Kim and Ivan Matić classify strongly positive discrete series representations of even unitary and general unitary groups by exploiting and extending a formalism due to Tadić.
- Peter Latham and Monica Nevins look at questions concerning uniqueness of types for a broad class of supercuspidal representations of a semisimple and simply connected p-adic group.
- Maarten Solleveld looks at completions of Hecke algebras that arise naturally in studying the tempered representations of a reductive p-adic group. As an application, he computes the topological K-theory of the reduced C^*-algebra of many classical p-adic groups.
- Shuichiro Takeda's paper is concerned with p-adic symmetric spaces. It establishes an analogue in this setting of the classical result that irreducible tempered representations embed as direct summands of modules that are induced from discrete series representations of Levi subgroups.

We are enormously grateful to the authors for their labors. We would like to acknowledge also the work of the various referees who read carefully through the submissions and offered many detailed and constructive comments.

Funding for the workshop and conference was provided by the National Board for Higher Mathematics (NBHM) in India and by IISER Pune. We are thankful for their support. Finally, a special word of thanks to the participants for many stimulating conversations and discussions, mathematical and otherwise, over a memorable week and a half in Pune.

Paris, France — Anne-Marie Aubert
Pune, India — Manish Mishra
Norman, USA — Alan Roche
Pune, India — Steven Spallone

Participants and Speakers

∗ Jeff Adler, American University, USA
∗ Moshe Adrian, CUNY, USA
∗ U. K. Anandavardhanan, IIT Mumbai, India
Michael Arnold, University of East Anglia, UK
∗ Anne-Marie Aubert, IMJ-PRG, France
Mugur (Peter) Badea, Radboud Universiteit Nijmegen, The Netherlands
Kumar Balasubramanian, IISER Bhopal, India
Subham Bhakta, Chennai Mathematical Institute, India
Gautam Borisagar, IITRAM, India
∗ Colin Bushnell, King's College London, UK
Kolhatkar Sampada Chandrashekhar, IIS Bangalore, India
Jesua Epequin Chavez, IMJ-PRG, France
Jonathan Cohen, University of Maryland, USA
∗ Tyrone Crisp, Radboud Universiteit Nijmegen, The Netherlands
Peiyi Cui, Université Pierre et Marie Curie and Université de Rennes 1, France
Chandan S. Dalawat, HRI Allahabad, India
Bastian Drevon, Université de Versailles Saint-Quentin-en-Yvelines, France
Melissa Emory, University of Missouri, USA
Mikolaj Fraczyck, Université Paris-Sud, France
Venketasubramanian C. G., IISER Tirupati, India
Jessica Fintzen, University of Michigan, USA
Radhika Ganapathy, TIFR, Mumbai, India
Zahi Hazan, Tel Aviv University, Israel
Lu Hengfei, TIFR, Mumbai, India
Trias Justin, IMJ-PRG, France
∗ Yeansu Kim, Chonnam National University, South Korea
Neha Kwatra, IISER Mohali, India
Luis Lomeli, Instituto de Matemáticas PUCV, Chile
Sampath Lonka, University of Hyderabad, India
∗ Jia-jun Ma, Shanghai Jiao Tong University, China
Arnaud Mayeux, IMJ-PRG, France

Amiya Mondal, TIFR, Mumbai, India
* Ahmed Moussaoui, Université de Versailles Saint-Quentin-en-Yvelines, France
Santosh Nadimpalli, TIFR, Mumbai, India
* Monica Nevins, University of Ottawa, Canada
* Chufeng Nien, National Cheng Kung University, Taiwan
* Eric Opdam, University of Amsterdam, The Netherlands
* Dipendra Prasad, TIFR, Mumbai, India
* Alan Roche, University of Oklahoma, USA
Beth Romano, University of Cambridge, UK
Lauren C. Ruth, University of California, Riverside, USA
Claudia Schoemann, Mathematical Institute, University of Goettingen, Germany
* Vincent Sécherre, Université de Versailles Saint-Quentin-en-Yvelines, France
* Maarten Solleveld, Radboud Universiteit Nijmegen, The Netherlands
Anna Szumowicz, Université Pierre et Marie Curie, France, and Durham University, UK
* Shuichiro Takeda, University of Missouri, USA
Lanard Thomas, Université Pierre et Marie Curie, France
Sandeep Varma, TIFR, Mumbai, India
Tian An Wong, IISER Pune, India
Hongjie Yu, Université Paris 7, France

* = speaker

Contents

Local Langlands and Springer Correspondences 1
Anne-Marie Aubert

Arithmetic of Cuspidal Representations 39
Colin J. Bushnell

Root Data with Group Actions 127
Jeffrey D. Adler and Joshua M. Lansky

The Character of a Simple Supercuspidal Representation of $SL(2, F)$ 145
Moshe Adrian

Classification of Strongly Positive Representations of Even General Unitary Groups 161
Yeansu Kim and Ivan Matić

On the Unicity of Types for Toral Supercuspidal Representations 175
Peter Latham and Monica Nevins

Local Gamma Factors, Converse Theorems and Related Problems 191
Chufeng Nien

On Completions of Hecke Algebras 207
Maarten Solleveld

On Relatively Tempered Representations for p-adic Symmetric Spaces 263
Shuichiro Takeda

About the Editors

Anne-Marie Aubert is Research Director in Mathematics at the Centre National de la Recherche Scientifique (CNRS) at the Institut Mathématiques de Jussieu Paris Rive-Gauche, Paris, France. She is a member of the Comité National de la Recherche Scientifique, and has served on the editorial board of the Bulletin and the Memoirs of the French Mathematical Society.

Manish Mishra is Assistant Professor of Mathematics at the Indian Institute of Science Education and Research (IISER), Pune, India. He previously held post-doctoral positions at the Heidelberg University and the Hebrew University of Jerusalem. He completed his BTech at the Indian Institute of Technology Kanpur and his Ph.D. at Purdue University, USA.

Alan Roche is Associate Professor of Mathematics at the University of Oklahoma, USA. He previously held visiting positions at Purdue University and Oklahoma State University, USA.

Steven Spallone is Associate Professor of Mathematics at the Indian Institute of Science Education and Research (IISER), Pune, India. He graduated from the University of Pennsylvania and completed his Ph.D. at the University of Chicago, USA, in 1998 and 2004, respectively. His research interests include number theory and representation theory.

Local Langlands and Springer Correspondences

Anne-Marie Aubert

Contents

1 Introduction ... 2
2 Langlands Parameters ... 4
2.1 The Weil Group ... 4
2.2 The L-Group ... 6
2.3 Definitions for Langlands Parameters ... 8
2.4 L-Packets ... 10
3 Enhanced Langlands Parameters ... 11
3.1 The Group $\mathcal{S}_\phi$ ... 11
3.2 Definition of Enhanced Langlands Parameters ... 13
3.3 Desiderata for the Local Langlands Correspondence ... 14
3.4 Depth ... 17
4 Generalized Springer Correspondence ... 19
4.1 Cuspidal Enhanced Unipotent Classes ... 19
4.2 A Partition of the Enhancement of the Unipotent Variety of $\mathcal{G}^\circ$ ... 22
4.3 Twisted Group Algebras ... 23
4.4 A Partition of the Enhancement of the Unipotent Variety of $\mathcal{G}$ ... 23
5 Cuspidality for Enhanced L-Parameters: Definition and Conjecture ... 24
6 A Partition of the Set of Enhanced Langlands Parameters ... 25
6.1 Inertial Classes for Enhanced Langlands Parameters ... 25
6.2 The Cuspidal Support Map for Enhanced Langlands Parameters ... 27
6.3 A Generalized Springer Correspondence for Enhanced L-Parameters ... 28
7 A Galois Version of the ABPS Conjecture ... 29
7.1 Twisted Extended Quotients ... 29
7.2 The Bernstein Decomposition of the Category of Smooth Representations ... 30
7.3 The ABPS Conjecture ... 32
7.4 A Version for Enhanced L-Parameters ... 33
7.5 A Conjectural Diagram ... 34
References ... 35

It is a pleasure to acknowledge the excellent comments and questions of the workshop participants.

A.-M. Aubert (✉)
Institut de Mathématiques de Jussieu—Paris Rive Gauche, IMJ-PRG, C.N.R.S, Sorbonne Université, Université Paris Diderot, 75005 Paris, France
e-mail: anne-marie.aubert@imj-prg.fr

A.-M. Aubert et al. (eds.), *Representations of Reductive p-adic Groups*,
Progress in Mathematics 328, https://doi.org/10.1007/978-981-13-6628-4_1

Abstract These notes give an overview of results obtained jointly with Ahmed Moussaoui and Maarten Solleveld on the local Langlands correspondence, focusing on the links of the latter with both the generalized Springer correspondence and the geometric conjecture, the so-called ABPS Conjecture, introduced in collaboration with Paul Baum, Roger Plymen and Maarten Solleveld.

2010 Mathematics Subject Classification 20C08 · 14F43 · 20G20

1 Introduction

The local Langlands correspondence predicts a relation between two rather different kinds of objects: on the one side irreducible representations of reductive groups over a local field F, and on the other side certain analogs of Galois representations, called *Langlands parameters*.

Let G be the group of F-rational points of a connected reductive algebraic group $\mathbf{G}$ over a non-Archimedean local field F. Slightly more precisely, the local Langlands conjecture for G asserts the existence a finite-to-one surjection, subject to certain properties,

$$\mathrm{rec}_G \colon \mathrm{Irr}(G) \to \Phi(G),$$

from the set $\mathrm{Irr}(G)$ of (isomorphism classes of) irreducible smooth representations of G to the set $\Phi(G)$ of equivalence classes (for a certain equivalence relation) of Langlands parameters for G.

The (conjectural) fibers of rec_G are called L-packets for G. The irreducible representations π inside a given L-packet are said to be *L-indistinguishable*. To distinguish them, the idea (see Lusztig [44], Vogan [61], Arthur [2]) is to "enhance" ϕ by an additional datum ρ, which is an irreducible representation of a certain finite group $\mathcal{S}_\phi$ [defined by Eq. (2)]. The pair (ϕ, ρ) will be called an *enhanced Langlands parameter* for G. Let $\Phi_{\mathrm{e}}(G^\natural)$ be the collection of equivalence classes of such pairs (ϕ, ρ) with $\rho \in \mathrm{Irr}(\mathcal{S}_\phi)$. Then the LLC for G should be an injective map

$$\mathrm{Irr}(G) \to \Phi_{\mathrm{e}}(G), \tag{1}$$

which satisfies several natural properties. The map will almost never be surjective, but for every ϕ which is relevant for G the image should contain at least one pair (ϕ, ρ).

A remarkable aspect of the Langlands conjecture [61] is that it is better to consider not just one reductive group G at a time, but all inner forms (more precisely all inner twists) G_ϑ of G simultaneously. Inner twists share the same Langlands dual group. The hope is that one can turn (1) into a bijection by defining a suitable equivalence relation on the set of inner twists and taking the corresponding union of the sets $\mathrm{Irr}(G_\vartheta)$ on the left-hand side. Such a statement was proven for unipotent representations (also known as representations with unipotent reduction) of simple p-adic groups in [48].

The main goal of these notes is to provide an overview of the main results obtained jointly with Moussaoui and Solleveld in [15] on the local Langlands correspondence, focusing on the links of the latter with both the generalized Springer correspondence and the geometric conjecture, the so-called ABPS Conjecture, introduced in collaboration with Paul Baum, Roger Plymen and Maarten Solleveld and studied in several articles, notably [12]. We also give an account of the known results regarding the preservation (and non-preservation) of the depth by the LLC.

For maximal generality, we adhere to the setup for L-parameters used by Arthur in [2]. Let W_F be the Weil group of F and let ${}^LG = G^\vee \rtimes W_F$ be the L-group of G. We set $W_F' := W_F \times \mathrm{SL}_2(\mathbb{C})$. Let $G^\vee_{\mathrm{ad}}$ be the adjoint group of $G^\vee$ and $G^\vee_{\mathrm{sc}}$ be the simply connected cover of the derived group of $G^\vee$.

Let $\phi\colon W_F' \to {}^LG$ be an L-parameter. We denote by $\mathrm{Z}_{G^\vee_{\mathrm{ad}}}(\phi)$ the centralizer in $G^\vee_{\mathrm{ad}}$ of $\phi(W_F')$ and we define $\mathrm{Z}^1_{G^\vee_{\mathrm{sc}}}(\phi)$ to be its inverse image under $G^\vee_{\mathrm{sc}} \to G^\vee_{\mathrm{ad}}$. Let

$$\mathcal{S}_\phi := \pi_0(\mathrm{Z}^1_{G^\vee_{\mathrm{sc}}}(\phi)), \tag{2}$$

denote the component group of $\mathrm{Z}^1_{G^\vee_{\mathrm{sc}}}(\phi)$. It coincides with the group considered by both Arthur in [2] and Kaletha in [39, § 4.6].

We set

$$\mathcal{G}_\phi := \mathrm{Z}^1_{G^\vee_{\mathrm{sc}}}(\phi|_{W_F}). \tag{3}$$

A remarkable fact is that the finite group $\mathcal{S}_\phi$ is isomorphic to the component group

$$A_{\mathcal{G}_\phi}(u_\phi) := \pi_0(\mathrm{Z}_{\mathcal{G}_\phi}(u_\phi)), \tag{4}$$

of the centralizer in $\mathcal{G}_\phi$ of unipotent element $u_\phi := \phi(1, \left(\begin{smallmatrix}1&1\\0&1\end{smallmatrix}\right))$. It provides a way to plug the generalized Springer correspondences for the groups $\mathcal{G}_\phi$, where ϕ runs over $\Phi(G)$, in the study of the local Langlands correspondence for G.

In particular, it allows to transfer the Lusztig notion of cuspidality for representations of the groups $A_{\mathcal{G}_\phi}(u_\phi)$ into a notion of cuspidality for enhanced L-parameters for G: An enhanced L-parameter (ϕ, ρ) for G is called *cuspidal* if u_ϕ and ρ, considered as data for the complex reductive group $\mathcal{G}_\phi$, form a *cuspidal pair*. By definition, this means that the restriction of ρ from $A_{\mathcal{G}_\phi}(u_\phi)$ to $A_{\mathcal{G}_\phi^\circ}(u_\phi)$ is a direct sum of cuspidal representations in Lusztig's sense [45]. Intuitively, it says that ρ or $\rho|_{A_{\mathcal{G}_\phi^\circ}(u_\phi)}$ cannot be obtained (via an appropriate notion of parabolic induction) from any pair (u', ρ') that can arise from a proper Levi subgroup of $\mathcal{G}_\phi^\circ$. It is essential to use L-parameters enhanced with a representation of a suitable component group, for cuspidality cannot be detected from the L-parameter alone.

In particular, we conjecture in [15, § 6] (generalizing to G arbitrary the conjecture stated by Moussaoui when G is F-split: [52, Conjecture 1.2]) that the cuspidal enhanced Langlands parameters correspond by the LLC to the irreducible supercuspidal representations of G.

The validity of this conjecture is proved for representations with unipotent reduction of the group G of the F-rational points of any connected reductive algebraic group which splits over an unramified extension of F in [26, Theorem 2] (when G is simple of adjoint type, it is a special case of [47, 48]), for the Deligne-Lusztig depth-zero supercuspidal representations (as a consequence of [25]), and also for general linear groups and split classical p-adic groups (any representation) (see [52]), for inner forms of linear groups and of special linear groups, and for quasi-split unitary p-adic groups (any representation) (see [15, § 6]).

We thank Maarten Solleveld and the referee for helpful and constructive comments that greatly contributed to improving the final version of the paper.

2 Langlands Parameters

2.1 The Weil Group

Let F be a local non-Archimedean field with finite residual field $k_F = \mathbb{F}_q$. Let F_{sep} be a fixed separable closure of F, and let Γ_F denote the Galois group of F_{sep}/F. The field F admits a unique unramified extension F_m/F of degree m and contained in F_{sep}, for each integer $m \geq 1$. The composite of all the fields F_m is the unique maximal unramified extension of F contained in F_{sep} and will be denoted by F_{ur}. The extension F_{ur} will allow us to decompose the study of Γ_F in two steps of a different nature by considering separately the group $\text{Gal}(F_{\text{ur}}/F)$ and the group $I_F := \text{Gal}(F_{\text{sep}}/F_{\text{ur}})$, called the *inertia group* of F: We have an exact sequence of topological groups

$$1 \to I_F \to \Gamma_F \to \text{Gal}(F_{\text{ur}}/F) \to 0. \tag{5}$$

The extension F_m/F is Galois and the group $\text{Gal}(F_m/F)$ is cyclic. An F-automorphism of F_m is determined by its action on the residual field $k_{F_m} \simeq \mathbb{F}_{q^m}$ of F_m, and there is a unique element σ_m of $\text{Gal}(F_m/F)$ which acts on k_{F_m} by $\lambda \mapsto \lambda^q$. We set $\text{Fr}_m := \sigma_m^{-1}$. The map $\text{Fr}_m \mapsto 1 + m\mathbb{Z}$ gives a canonical isomorphism from $\text{Gal}(F_m/F)$ onto $\mathbb{Z}/m\mathbb{Z}$. Taking the inverse limit over m, we get a canonical isomorphism of topological groups between $\text{Gal}(F_{\text{ur}}/F)$ and

$$\widehat{\mathbb{Z}} := \varprojlim_{m \geq 1} \mathbb{Z}/m\mathbb{Z},$$

and a unique element $\text{Fr}_F \in \text{Gal}(F_{\text{ur}}/F)$ which acts on F_m as Fr_m, for all m. The element Fr_F is called the *geometric Frobenius substitution* on F_{ur} (its inverse σ_F is the *arithmetic* Frobenius substitution). An element of Γ_F is called a *geometric Frobenius element* (over F) if its image in $\text{Gal}(F_{\text{ur}}/F)$ is Fr_F. The Chinese Remainder Theorem gives a canonical isomorphism of topological groups $\widehat{\mathbb{Z}} \simeq \prod_\ell \mathbb{Z}_\ell$, where ℓ range over all prime numbers, and $\mathbb{Z}_\ell$ is the (additive) group of ℓ-adic integers.

We recall some properties of the *ramification groups* (with respect to the upper numbering) of Γ_F, as defined in [57, Remark IV.3.1]:

- $\Gamma_F^{-1} := \Gamma_F$ and $\Gamma_F^0 := I_F$, the inertia group.
- For every $r \in \mathbb{R}_{\geq 0}$, Γ_F^l is the compact subgroup of I_F that consists of all $\gamma \in \Gamma_F$ which, for every finite Galois extension E of F contained in F_{sep}, act trivially on the ring $\mathfrak{o}_E/\mathfrak{p}_E^{i(r,E)}$ (where $i(r, E) \in \mathbb{Z}_{\geq 0}$ can be found with [57, § IV.3]).
- $r \in \mathbb{R}_{\geq 0}$ is called a jump of the filtration if

$$\Gamma_F^{r+} := \bigcap_{t>r} \Gamma_F^t$$

does not equal Γ_F^r. The set of jumps of the filtration is countably infinite and need not consist of integers.

In order to formulate the Langlands correspondence, we need to introduce the *Weil group* W_F, a subgroup of Γ_F. Let ${}_{\mathrm{a}}W_F$ denote the inverse image in Γ_F of the cyclic subgroup of $\mathrm{Gal}(F_{\mathrm{ur}}/F)$ generated by Fr_F. Then ${}_{\mathrm{a}}W_F$ is the dense subgroup of Γ_F generated by the geometric Frobenius elements. It is normal in Γ_F and fits into an exact sequence (of abstract groups)

$$1 \to I_F \to {}_{\mathrm{a}}W_F \to \mathbb{Z} \to 0. \tag{6}$$

The Weil group W_F of F (relative to F_{sep}) is the topological group, with underlying abstract group ${}_{\mathrm{a}}W_F$, so that I_F is an open subgroup of W_F, and the topology on I_F, as subspace of W_F, coincides with its natural topology as $\mathrm{Gal}(F_{\text{sep}}/F_{\mathrm{ur}}) \subset \Gamma_F$. Thus W_F is locally profinite, and the inclusion $\iota_F : W_F \subset \Gamma_F$ is continuous.

The definition of W_F does depend on the choice of F_{sep}/F, but only up to inner automorphism of Γ_F.

We will need some basic aspects of the representation theory of W_F. The group Γ_F being profinite, its smooth representations are semisimple. On the contrary, W_F has smooth representations which are not semisimple. The irreducible representations of W_F are quite closely related to those of Γ_F (in particular, they have finite dimension):

(1) if τ is an irreducible representation of Γ_F, then $\tau \circ \iota_F$ is an irreducible smooth representation of W_F,
(2) for any irreducible smooth representation σ of W_F, there is an unramified character χ of W_F (i.e., $\ker(\chi)$ contains I_F) such that $\chi \otimes \sigma \simeq \tau \circ \iota_F$, for some irreducible smooth representations τ of Γ_F.

2.2 The L-Group

Let G be the group of F-points of a connected reductive algebraic group $\mathbf{G}$ defined over F. The group $\mathbf{G}$ is split over F_{sep} and is classified by its root data. Let $\mathbf{T}$ be a maximal torus of $\mathbf{G}$, defined and split over F_{sep}, and let $X(\mathbf{T}) := \text{Hom}_F(\mathbf{T}, \mathbb{G}_m)$ be the group of F_{sep}-rational characters of $\mathbf{T}$. We set $X^\vee(\mathbf{T}) := \text{Hom}_F(\mathbb{G}_m, \mathbf{T})$. Let $R(\mathbf{G}, \mathbf{T})$ and $R^\vee(\mathbf{G}, \mathbf{T})$ denote the sets of roots and coroots of $\mathbf{G}$ with respect to $\mathbf{T}$, respectively. The corresponding root datum for $\mathbf{G}$ over F_{sep} is

$$\mathcal{R}(\mathbf{G}) := (X(\mathbf{T}), R(\mathbf{G}, \mathbf{T}), X^\vee(\mathbf{T}), R(\mathbf{G}, \mathbf{T})^\vee).$$

Definition 2.1 The *Langlands dual group* of $\mathbf{G}$ is the the reductive connected algebraic group $\mathbf{G}^\vee$, defined over $\mathbb{C}$, whose roots (resp. coroots) are the coroots (resp. roots) of $\mathbf{G}$, so that a root datum for $\mathbf{G}^\vee$ is

$$\mathcal{R}(\mathbf{G}^\vee) := (X^\vee(\mathbf{T}), R^\vee(\mathbf{G}, \mathbf{T}), X(\mathbf{T}), R(\mathbf{G}, \mathbf{T})).$$

We denote by $G^\vee$ the $\mathbb{C}$-points of the group $\mathbf{G}^\vee$.

Examples 2.2 We have

- $\text{GL}_n(F)^\vee = \text{GL}_n(\mathbb{C})$, $\text{SL}_n(F)^\vee = \text{PGL}_n(\mathbb{C})$, and $\text{PGL}_n(F)^\vee = \text{SL}_n(\mathbb{C})$;
- $\text{Sp}_{2n}(F)^\vee = \text{SO}_{2n+1}(\mathbb{C})$, $\text{SO}_{2n+1}(F)^\vee = \text{Sp}_{2n}(\mathbb{C})$;
- $\text{SO}_{2n}(F)^\vee = \text{SO}_{2n}(\mathbb{C})$;
- if $\mathbf{G}$ a group of exceptional type, then $\mathbf{G}$ and $\mathbf{G}^\vee$ are of the same type (e.g. $\text{G}_2(F)^\vee = \text{G}_2(\mathbb{C})$).

Of special importance is the group $\mathbf{G} = \text{GL}_n$. Indeed, any complex reductive group $\mathbf{G}$ may be embedded into $\text{GL}_n(\mathbb{C})$ for some n.

We choose a Borel subgroup $\mathbf{B} \supset \mathbf{T}$ of $\mathbf{G}$ defined over F_{sep}, and let Δ and $\Delta^\vee$denote the corresponding basis of $R(\mathbf{G}, \mathbf{T})$ and $R^\vee(\mathbf{G}, \mathbf{T})$. A based root datum for $\mathbf{G}$ is $\mathcal{R}_0(\mathbf{G}) := (X(\mathbf{T}), \Delta, X^\vee(\mathbf{T}), \Delta^\vee)$. We choose a pinning $(G, B, T, \{x_\alpha\}_\alpha \in \Delta)$, which induces a splitting of the exact sequence

$$1 \longrightarrow \text{Int}(\mathbf{G}) \longrightarrow \text{Aut}(G) \longrightarrow \text{Aut}(\mathcal{R}_0(G)) \longrightarrow 1.$$

The F structure of $\mathbf{G}$ is given by a morphism of $\theta\colon \Gamma_F \longrightarrow \text{Aut}(\mathbf{G})$ which descends to $\theta\colon \Gamma_F \longrightarrow \text{Aut}(\mathcal{R}_0(\mathbf{G}))$.

To obtain the L-group as a group that actually sees $G = \mathbf{G}(F)$ and not just $\mathbf{G}$, we have to work in the Galois action of Γ_F on the group $\mathbf{G}^\vee$. Since $\text{Aut}(\mathcal{R}_0(\mathbf{G})) \simeq \text{Aut}(\mathcal{R}_0(\mathbf{G})^\vee) \simeq \text{Aut}(\mathcal{R}_0(\mathbf{G}^\vee))$, we have

$$\theta\colon \Gamma_F \longrightarrow \text{Aut}(\mathcal{R}_0(\mathbf{G})) \longrightarrow \text{Aut}(\mathbf{G}^\vee).$$

Definition 2.3 The (Weil form of the) L-group of G is the group

$$^{L}G := G^{\vee} \rtimes_{\theta} W_F.$$

Remark 2.4 In the case when G is F-split, we have a direct product, that is: $^{L}G = G^{\vee} \times W_F$.

Definition 2.5 Let $\mathbf{G}$ and $\mathbf{H}$ be two connected reductive algebraic groups over F. A homomorphism $\eta\colon {}^{L}H \to {}^{L}G$ is called an *L-homomorphism* if

- η is continuous;
- the diagram $\begin{array}{ccc} {}^{L}H & \xrightarrow{\eta} & {}^{L}G \\ & \searrow \quad \swarrow & \\ & W_F & \end{array}$ commutes;
- the restriction of η to $H^{\vee}$ is morphism of algebraic groups from $H^{\vee}$ to $G^{\vee}$.

General connected reductive F-groups need not be quasi-split, but they are always *forms* of split F-groups.

Definition 2.6 Two F-groups $G = \mathbf{G}(F)$ and $H = \mathbf{H}(F)$ are called *forms* of each other if $\mathbf{G}$ is isomorphic to $\mathbf{H}$ as algebraic groups, or equivalently if $\mathbf{G}(F_{\rm sep}) \cong \mathbf{H}(F_{\rm sep})$ as $F_{\rm sep}$-groups.

An isomorphism $\vartheta\colon \mathbf{H} \to \mathbf{G}$ determines a 1-cocycle

$$\gamma_\vartheta : \begin{array}{ccc} \Gamma_F & \to & \mathrm{Aut}(\mathbf{G}) \\ \sigma & \mapsto & \vartheta\sigma\vartheta^{-1}\sigma^{-1}. \end{array} \tag{7}$$

From γ_ϑ, one can recover H (up to isomorphism) as

$$H \cong \{g \in \mathbf{G}(F_{\rm sep}) \ : \ (\gamma_\vartheta(\sigma) \circ \sigma)g = g \quad \forall\sigma \in \mathrm{Gal}(F_{\rm sep}/F)\}.$$

Given another form $\vartheta'\colon \mathbf{H}' \to \mathbf{G}$, the groups H and H' are F-isomorphic if and only if the 1-cocycles γ_ϑ and $\gamma_{\vartheta'}$ are cohomologous, that is, if there exists a $f \in \mathrm{Aut}(\mathbf{G})$ such that

$$\gamma_\vartheta(\sigma) = f^{-1}\gamma_{\vartheta'}(\sigma)\,\sigma f\sigma^{-1} \quad \forall\sigma \in \mathrm{Gal}(F_{\rm sep}/F). \tag{8}$$

In this way, the isomorphism classes of forms of G are in bijection with the Galois cohomology group $H^1(F, \mathrm{Aut}(\mathbf{G}))$.

Definition 2.7 An F-group H is an *inner form* of G if the cocycle γ_ϑ takes values in the group of inner automorphisms $\mathrm{Inn}(\mathbf{G})$. On the other hand, if the values of γ_ϑ are not contained in $\mathrm{Inn}(\mathbf{G})$, then H is called an outer form of G.

Proposition 2.8 [59, § 16.4] *Every connected reductive F-group G is an inner form of a unique quasi-split F-group G^*.*

Example 2.9 Let D be a division algebra with center F, of dimension d^2 over F. Then $G = \mathrm{GL}_m(D)$ is an inner form of $\mathrm{GL}_n(F) = G^*$ where $n = dm$. There is a reduced norm map $\mathrm{Nrd}\colon \mathrm{GL}_m(D) \to F^\times$ and the derived group $\mathrm{SL}_m(D) := \ker(\mathrm{Nrd}\colon \mathrm{GL}_m(D) \to F^\times)$ is an inner form of $\mathrm{SL}_n(F)$. Every inner form of $\mathrm{GL}_n(F)$ or $\mathrm{SL}_n(F)$ is isomorphic to one of this kind.

When $n = 2$, the only possibilities for d are 1 or 2, and so the inner forms are, up to isomorphism, $\mathrm{GL}_2(F)$ and $D^\times$, and $\mathrm{SL}_2(F)$ and $\mathrm{SL}_1(D)$.

Definition 2.10 An *inner twist* of G consists of a pair (H, ϑ) as above, where $H = \mathbf{H}(F)$ and $\vartheta\colon \mathbf{H} \xrightarrow{\sim} \mathbf{G}$ are such that $\mathrm{im}(\gamma_\vartheta) \subset \mathbf{G}_{\mathrm{ad}}$.

Two inner twists of G are *equivalent* if (8) holds for some $f \in \mathrm{Inn}(\mathbf{G})$. Let $\mathrm{IT}(G)$ denote the set of equivalence classes of inner twists of G.

Remark 2.11 It is quite possible that two inequivalent inner twists (H, ϑ) and (H', ϑ') share the same group $H \cong H'$. This happens precisely when γ_ϑ and $\gamma_{\vartheta'}$ are in the same orbit of $\mathrm{Aut}(\mathbf{G})/\mathrm{Inn}(\mathbf{G})$ on $H^1(F, \mathbf{G}_{\mathrm{ad}})$.

Kottwitz has found an important alternative description of $H^1(F, \mathbf{G})$. Recall that the complex dual group $G^\vee = \mathbf{G}^\vee(\mathbb{C})$ is endowed with an action of $\mathrm{Gal}(F_{\mathrm{sep}}/F)$. There exists a natural isomorphism

$$\kappa_G\colon H^1(F, \mathbf{G}) \xrightarrow{\sim} \mathrm{Irr}\Big(\pi_0\big(\mathrm{Z}(G^\vee)^{W_F}\big)\Big),$$

see [41, Proposition 6.4]. This is particularly useful in the following way. An inner twist of G is the same thing as an inner twist of the unique quasi-split inner form $G^* = \mathbf{G}^*(F)$. Let $G^*_{\mathrm{ad}} = \mathbf{G}^*_{\mathrm{ad}}(F)$ be the adjoint group of G^*. Let $\mathbf{G}^\vee_{\mathrm{sc}}$ be the simply connected cover of the derived group $G^\vee_{\mathrm{der}}$ of $\mathbf{G}^\vee$. We have $G^\vee_{\mathrm{sc}} = (G_{\mathrm{ad}})^\vee$, and

$$\kappa_{G^*_{\mathrm{ad}}}\colon H^1(F, \mathbf{G}^*_{\mathrm{ad}}) \xrightarrow{\sim} \mathrm{Irr}\big(\mathrm{Z}(G^\vee_{\mathrm{sc}})^{W_F}\big). \tag{9}$$

The equivalence classes of inner twists of G are parametrized by the Galois cohomology group $H^1(F, \mathbf{G}^*_{\mathrm{ad}})$. For every $\vartheta \in H^1(F, \mathbf{G}^*_{\mathrm{ad}})$, we will denote by G_ϑ an inner twist of G^* which is parametrized by ϑ, and we will take G^* to be G_1.

All the inner twists of a given group G share the same L-group, because the action of W_F on $G^\vee$ is only uniquely defined up to inner automorphisms. This also works the other way round: From the Langlands dual group ${}^LG$, one can recover the inner form class of G. Hence, it is natural to consider all the inner twists of a given group simultaneously.

2.3 *Definitions for Langlands Parameters*

We write $W_F' := W_F \times \mathrm{SL}_2(\mathbb{C})$ (as version of the Weil-Deligne group of F).

Definition 2.12 A Langlands parameter (or L-parameter, for short) for ${}^LG$ is a smooth morphism $\phi\colon W_F' \to {}^LG$ such that

$$\begin{array}{ccc} W_F' & \xrightarrow{\quad\phi\quad} & {}^LG \\ & \searrow \quad \swarrow & \\ & W_F & \end{array}$$

commutes, $\phi(w)$ is semisimple for each $w \in W_F$ (i.e., $r(\phi(w))$ is semisimple for every finite dimensional representation r of ${}^LG$), and $\phi|_{\mathrm{SL}_2(\mathbb{C})}$ is a morphism of complex algebraic groups.

The group $G^\vee$ acts on the set $\underline{\Phi}({}^LG)$ of such ϕ's by conjugation. Let $\Phi({}^LG)$ denote the set of $G^\vee$-orbits in $\underline{\Phi}({}^LG)$

Definition 2.13 An L-parameter $\phi\colon W_F' \to {}^LG$ is said to be

- *unramified* if it has trivial restriction to I_F;
- *tame* if it has trivial restriction to the wild ramification group P_F of F;
- *essentially tame* if $\phi(P_F)$ lies in a maximal torus of $G^\vee$;
- *discrete* (or *elliptic*) if there is no proper W_F-stable Levi subgroup $L^\vee \subset G^\vee$ such that $\phi(W_F') \subset {}^LL$.
- *bounded* if $\phi'(W_F) \subset G^\vee$ is bounded, where $\phi(w) = (\phi'(w), w)$. (This is equivalent to $\phi'(\mathrm{Frob})$ being a compact element of $G^\vee$.

2.3.1 Relevance for Langlands Parameters

We will call a Levi factor of a parabolic subgroup of G simply a Levi subgroup of G. The bijection

$$R(\mathbf{G}, \mathbf{T}) \longleftrightarrow R^\vee(\mathbf{G}, \mathbf{T}) = R(G^\vee, T^\vee)$$

gives a basis $\Delta^\vee$ and provides a canonical bijection between the sets of conjugacy classes of parabolic subgroups of $\mathbf{G}$ and of $G^\vee$.

Definition 2.14 [21, § 3] A parabolic subgroup $P^\vee$ of $G^\vee$ is F-*relevant* if the corresponding class of parabolic subgroups of $\mathbf{G}$ contains an element $\mathbf{P}$ which is defined over F. Similarly, a Levi subgroup $L^\vee \subset G^\vee$ is $F - relevant$ if it is a Levi factor of a parabolic subgroup $P^\vee \subset G^\vee$ which is F-relevant.

Definition 2.15 A parabolic subgroup $P^\vee$ of $G^\vee$ is $W_F - quasi - stable$ if the projection $\mathrm{N}_{{}^LG}(P^\vee) \to W_F$ is surjective.

Remark 2.16 The W_F-quasi-stable parabolic subgroups of $G^\vee$ are precisely the neutral components of what Borel [21, § 3] calls parabolic subgroups of ${}^LG$.

Definition 2.17 Let $\phi \in \underline{\Phi}({}^LG))$ and let $P^\vee$ be a W_F-quasi-stable parabolic subgroup of $G^\vee$ with a Levi factor $L^\vee$ such that

- the image of ϕ is contained in $\mathrm{N}_{{}^LP}(L^\vee)$, where ${}^LP := P^\vee \rtimes W_F$;

- $P^\vee$ is a minimal for this property.

The L-parameter ϕ is G-*relevant* (or is an L-*parameter for* G) if $P^\vee$ is F-relevant.

Notation 2.18 We denote by $\Phi(G)$ the subset of $\Phi({}^L G)$ of the $\mathbf{G}^\vee$-conjugacy classes of L-parameters that are G-relevant.

A Langlands classification of L-parameters for G is obtained in [58].
We have

$$\Phi({}^L G) = \Phi(G^*),$$

where G^* is the quasi-split inner form of G as defined in Proposition 2.8, and it is expected that in general $\Pi_\phi(G)$ is nonempty if and only if ϕ is G-relevant.

Example 2.19 Let D be a division algebra with center F such that $\dim_F(D) = 4$. Then $G = D^\times$ is the unique non-split inner form of $G^* = \mathrm{GL}_2(F)$.

The only Levi subgroup of $D^\times$ defined over F is $D^\times$ itself, and it corresponds to the Levi subgroup $\mathrm{GL}_2(\mathbb{C})$ on the complex side.

Let $\phi \in \underline{\Phi}(\mathrm{GL}_2(F)) = \underline{\Phi}(D^\times)$ be the embedding $W_F' \hookrightarrow \mathrm{GL}_2(\mathbb{C}) \times W_F$. No proper parabolic subgroup of $\mathrm{GL}_2(\mathbb{C})$ contains $\phi(\mathrm{SL}_2(\mathbb{C})) = \mathrm{SL}_2(\mathbb{C})$, so ϕ is relevant for both $D^\times$ and $\mathrm{GL}_2(F)$.

2.4 L-Packets

2.4.1 For a Given Group G

The local Langlands conjecture (LLC) asserts that the set $\mathrm{Irr}(G)$ of isomorphism classes of irreducible smooth representations of G can be parametrized by the set $\Phi(G)$. This parametrization is usually not a bijection. In fact, it is conjectured that each conjugacy class $\phi \in \Phi(G)$ is associated with a finite set $\Pi_\phi(G)$ of isomorphism classes of irreducible smooth representations of G, and that they give a disjoint decomposition of $\mathrm{Irr}(G)$;

$$\mathrm{Irr}(G) = \bigsqcup_{\phi \in \Phi(G)} \Pi_\phi(G). \tag{10}$$

Such finite sets are called L-*packets for* G. This parametrization is based on the belief that there should be certain arithmetic invariants (e.g., L-factors) defined on both the representation side and the parameter side that are preserved by the correspondence. From this point of view, one can think that the L-packet Π_ϕ attached to some $\phi \in \Phi(G)$ consists of all irreducible smooth representations of G whose arithmetic invariants match those of ϕ.

2.4.2 For All the Inner Twists of G

The local Langlands correspondence predicts the existence of a partition of the set $\mathrm{Irr}(\mathrm{IT}(G)$ of equivalence classes of the irreducible smooth representations of all the groups G_i in $\mathrm{IT}(G)$ into finite subsets:

$$\mathrm{Irr}(\mathrm{IT}(G)) = \bigsqcup_{\phi \in \Phi({}^L G)} \Pi_\phi({}^L G), \tag{11}$$

where each *L-packet* $\Pi_\phi({}^L G)$ is the union of L-packets for the groups G_i.

Example 2.20 We keep the notation of Example 2.19. We have

$$\Pi_\phi(\mathrm{GL}_2(F)) = \{\mathrm{St}_{\mathrm{GL}_2(F)}\} \quad \text{and} \quad \Pi_\phi(D^\times) = \{\mathrm{St}_{D^\times} = \mathrm{triv}_{D^\times}\}.$$

Thus here (11) is

$$\Pi_\phi({}^L G) = \Pi_\phi(\mathrm{GL}_2(F)) \cup \Pi_\phi(D^\times) = \{\mathrm{St}_{\mathrm{GL}_2(F)}, \mathrm{St}_{D^\times}\}.$$

Let $\phi' \in \underline{\Phi}(\mathrm{GL}_2(\mathbb{C}))$ such that $\phi'(\mathrm{SL}_2(\mathbb{C})) = 1$ and $\phi'(W_F) \subset \mathrm{diag}(\mathrm{GL}_2(\mathbb{C})) \times W_F$. Then $L^\vee = \mathrm{diag}(\mathrm{GL}_2(\mathbb{C}))$ is the minimal Levi subgroup such that ${}^L L$ contains the image of ϕ_2. Thus the standard Borel subgroup $P^\vee$ of $\mathrm{GL}_2(\mathbb{C})$ satisfies the conditions in Definition 2.17. But its conjugacy class does not correspond to any parabolic subgroup of $D^\times$, so ϕ' is not relevant for $D^\times$.

Let $\mathbf{G} \subset \widetilde{\mathbf{G}}$ be a pair of quasi-split connected reductive groups, defined over F, with the same derived group. It is expected that the L-packets of $G = \mathbf{G}(F)$ are restrictions of L-packets of $\widetilde{G} := \widetilde{\mathbf{G}}(F)$ in the sense that for each L-packet $\Pi(G)$ for G there is a packet $\Pi(\widetilde{G})$ for $\widetilde{G}$ whose restriction to G is equal to Π. Typically, one uses this last expectation to construct L-packets for G from the knowledge of L-packets of $\widetilde{G}$, e.g., in the cases of $\mathrm{SL}_n \subset \mathrm{GL}_n$ and $\mathrm{Sp}_{2n} \subset \mathrm{GSp}_{2n}$. In some cases, however, we wish to move in the other direction and use the knowledge of L-packets of G to obtain structural information about the L-packets of $\widetilde{G}$, see [62].

3 Enhanced Langlands Parameters

3.1 The Group $\mathcal{S}_\phi$

To parametrize the irreducible representations in a given L-packet, we need more information than just the Langlands parameter itself. Let $\mathrm{Z}_{\mathbf{G}^\vee}(\phi)$ denote the centralizer of $\phi(W_F')$ in $\mathbf{G}^\vee$. This is a reductive algebraic group over $\mathbb{C}$, in general disconnected. We denote by $\mathrm{Z}_{\mathbf{G}^\vee}$ the center of $\mathbf{G}^\vee$ and we write

$$\mathcal{R}_\phi := \pi_0\big(\mathrm{Z}_{\mathbf{G}^\vee}(\phi)/\mathrm{Z}(\mathbf{G}^\vee)^{W_F}\big). \tag{12}$$

Remark 3.1 It is expected that $\Pi_\phi(G)$ is in bijection with $\mathrm{Irr}(\mathcal{R}_\phi)$ if G is quasi-split. However, for G not quasi-split, this is not always the case.

Let $\mathbf{G}^\vee_{\mathrm{ad}}$ be the quotient $\mathbf{G}^\vee/\mathrm{Z}_{\mathbf{G}^\vee}$ (i.e., the group $\mathbf{G}^\vee_{\mathrm{ad}}$ is the adjoint group of $\mathbf{G}^\vee$). Since $\mathrm{Z}_{\mathbf{G}^\vee}(\phi) \cap \mathrm{Z}(\mathbf{G}^\vee) = \mathrm{Z}(\mathbf{G}^\vee)^{W_F}$,

$$\mathrm{Z}_{\mathbf{G}^\vee}(\phi)/\mathrm{Z}(\mathbf{G}^\vee)^{W_F} \cong \mathrm{Z}_{\mathbf{G}^\vee}(\phi)\mathrm{Z}(\mathbf{G}^\vee)/\mathrm{Z}(\mathbf{G}^\vee). \tag{13}$$

The right-hand side can be considered as a subgroup of the adjoint group $\mathbf{G}^\vee_{\mathrm{ad}}$. Let $\mathrm{Z}^1_{\mathbf{G}^\vee_{\mathrm{sc}}}(\phi)$ be its inverse image under the quotient map $\mathbf{G}^\vee_{\mathrm{sc}} \to \mathbf{G}^\vee_{\mathrm{ad}}$. We can also characterize it as

$$\begin{aligned}\mathrm{Z}^1_{\mathbf{G}^\vee_{\mathrm{sc}}}(\phi) &= \big\{g \in \mathbf{G}^\vee_{\mathrm{sc}} \,:\, g\phi g^{-1} = \phi\, a_g \text{ for some } a_g \in B^1(W_F, \mathrm{Z}(\mathbf{G}^\vee))\big\} \quad (14)\\ &= \big\{g \in \mathrm{Z}_{\mathbf{G}^\vee_{\mathrm{sc}}}\big(\phi(\mathrm{SL}_2(\mathbb{C}))\big) \,:\, g\phi|_{W_F}g^{-1} = \phi|_{W_F}\, a_g \text{ for some } a_g \in B^1(W_F, \mathrm{Z}(\mathbf{G}^\vee))\big\}\\ &= \mathrm{Z}^1_{\mathbf{G}^\vee_{\mathrm{sc}}}(\phi|_{W_F}) \cap \mathrm{Z}_{\mathbf{G}^\vee_{\mathrm{sc}}}\big(\phi(\mathrm{SL}_2(\mathbb{C}))\big),\end{aligned}$$

where $B^1(W_F, \mathrm{Z}(\mathbf{G}^\vee))$ denotes the set of 1-coboundaries for group cohomology, that is, the maps $W_F \to \mathrm{Z}(\mathbf{G}^\vee)$ of the form $w \mapsto zwz^{-1}w^{-1}$ with $z \in \mathrm{Z}(\mathbf{G}^\vee)$. The neutral component of $\mathrm{Z}^1_{\mathbf{G}^\vee_{\mathrm{sc}}}(\phi)$ is $\mathrm{Z}_{\mathbf{G}^\vee_{\mathrm{sc}}}(\phi)^\circ$, so it is a complex reductive group.

Remark 3.2 We have $\mathrm{Z}^1_{\mathbf{G}^\vee_{\mathrm{sc}}}(\phi) = \mathrm{Z}_{\mathbf{G}^\vee_{\mathrm{sc}}}(\phi)$ whenever $\mathrm{Z}(\mathbf{G}^\vee_{\mathrm{sc}})^{W_F} = \mathrm{Z}(\mathbf{G}^\vee_{\mathrm{sc}})$, in particular if G is an inner twist of a split group .

Following Arthur [2], given ϕ, we define the group $\mathcal{S}_\phi$ as the component group of $\mathrm{Z}^1_{\mathbf{G}^\vee}(\phi)$:

$$\mathcal{S}_\phi := \pi_0\big(\mathrm{Z}^1_{\mathbf{G}^\vee_{\mathrm{sc}}}(\phi)\big). \tag{15}$$

Via (13), the map $\mathbf{G}^\vee_{\mathrm{sc}} \to \mathbf{G}^\vee_{\mathrm{ad}}$ induces a homomorphism $\mathcal{S}_\phi \to \mathcal{R}_\phi$. We set

$$\mathcal{Z}_\phi := \mathrm{Z}(\mathbf{G}^\vee_{\mathrm{sc}})/\mathrm{Z}(\mathbf{G}^\vee_{\mathrm{sc}}) \cap \mathrm{Z}_{\mathbf{G}^\vee_{\mathrm{sc}}}(\phi)^\circ. \tag{16}$$

Then (see [12, Lemma 1.7]) $\mathcal{S}_\phi$ is a central extension of $\mathcal{R}_\phi$ by $\mathcal{Z}_\phi$:

$$1 \to \mathcal{Z}_\phi \to \mathcal{S}_\phi \to \mathcal{R}_\phi \to 1. \tag{17}$$

Since $\mathbf{G}^\vee_{\mathrm{sc}}$ is a central extension of $\mathbf{G}^\vee_{\mathrm{ad}} = \mathbf{G}^\vee/\mathrm{Z}(\mathbf{G}^\vee)$, the conjugation action of $\mathbf{G}^\vee_{\mathrm{sc}}$ on itself and on $\mathcal{S}_\phi$ descends to an action of $\mathbf{G}^\vee_{\mathrm{ad}}$. Via the canonical quotient map, also $\mathbf{G}^\vee$ acts on $\mathcal{S}_\phi$ by conjugation.

We attach to a given L-parameter $\phi \in \underline{\Phi}(^L G)$ the following (possibly disconnected) complex reductive group:

$$\mathcal{G}_\phi := \mathrm{Z}^1_{\mathbf{G}^\vee_{\mathrm{sc}}}(\phi(W_F)). \tag{18}$$

The following proposition provides a way to link the local Langlands correspondence for G to the generalized Springer correspondence for the groups $\mathcal{G}_\phi$'s.

Proposition 3.3 [15, (92)] *The group $\mathcal{S}_\phi$ is isomorphic to the group*

$$A_{\mathcal{G}_\phi}(u_\phi) := \pi_0(Z_{\mathcal{G}_\phi}(u_\phi)),$$

where $u_\phi := \phi(1, \left(\begin{smallmatrix}1&1\\0&1\end{smallmatrix}\right))$.

3.2 Definition of Enhanced Langlands Parameters

Definition 3.4 An *enhanced* Langlands parameter (or enhanced L-parameter) for ${}^LG$ is a pair (ϕ, ρ), where $\phi \in \underline{\Phi}({}^LG)$ and ρ is an irreducible representation of the group $\mathcal{S}_\phi$ defined in (15).

Remark 3.5 The enhanced Langlands parameter (ϕ, ρ) is already determined by $\phi|_{W_F}$, u_ϕ and ρ. Sometimes, we will write $(\phi|_{W_F}, u_\phi, \rho)$ instead of (ϕ, ρ).

We let $\mathbf{G}^\vee$ act on the set of enhanced L-parameters for ${}^LG$ by

$$g \cdot (\phi, \rho) = (g\phi g^{-1}, g \cdot \rho) \quad \text{where} \quad (g \cdot \rho)(a) = \rho(g^{-1}ag). \tag{19}$$

Notation 3.6 Let $\Phi_{\mathrm{e}}({}^LG)$ denote the set of $\mathbf{G}^\vee$-conjugacy classes of enhanced Langlands parameters for ${}^LG$.

We set

$$\mathcal{Z}_\phi^{W_F} := \mathrm{Z}(G_{\mathrm{sc}}^\vee)^{W_F} / \big(\mathrm{Z}(G_{\mathrm{sc}}^\vee)^{W_F} \cap Z_{G_{\mathrm{sc}}^\vee}(\phi)^\circ\big). \tag{20}$$

According to [2, § 4]

$$\mathrm{Z}(G_{\mathrm{sc}}^\vee) \cap Z_{G_{\mathrm{sc}}^\vee}(\phi)^\circ \subset \mathrm{Z}(G_{\mathrm{sc}}^\vee)^{W_F}. \tag{21}$$

Hence, $\mathcal{Z}_\phi^{W_F}$ can be regarded as a subgroup of $\mathcal{Z}_\phi$ and

$$\mathcal{Z}_\phi / \mathcal{Z}_\phi^{W_F} \cong \mathrm{Z}(G_{\mathrm{sc}}^\vee)/\mathrm{Z}(G_{\mathrm{sc}}^\vee)^{W_F}. \tag{22}$$

By Schur's lemma, every enhanced Langlands parameter (ϕ, ρ) restricts to a character $\rho|_{\mathcal{Z}_\phi^{W_F}}$ of $\mathcal{Z}_\phi^{W_F}$. This can be inflated to a character ζ_ρ of $\mathrm{Z}(\mathbf{G}_{\mathrm{sc}}^\vee)^{W_F}$. With the Kottwitz isomorphism (9), we get an element $\vartheta := \kappa_{G_{\mathrm{ad}}^*}^{-1}(\zeta_\rho) \in H^1(F, \mathbf{G}_{\mathrm{ad}}^*)$. In this way, (ϕ, ρ) determines a unique inner twist G_ϑ of G. This can be regarded as an alternative way to specify for which inner twists of G an enhanced Langlands parameter is relevant. Fortunately, it turns out that it agrees with the earlier definition of relevance of Langlands parameters.

Indeed for $\phi \in \underline{\Phi}({}^LG)$, the following are equivalent (see [12, Prop. 1.8])

(1) ϕ is relevant for G_ϑ;
(2) $Z(G^\vee_{sc})^{W_F} \cap Z_{G^\vee_{sc}}(\phi)^\circ \subset \ker \zeta$;
(3) there exists a $\rho \in \mathrm{Irr}(\mathcal{S}_\phi)$ such that ζ is the lift of $\rho|_{\mathcal{Z}_\phi^{W_F}}$ to $Z(G^\vee_{sc})^{W_F}$,

where $\zeta \in \mathrm{Irr}(Z(G^\vee_{sc})^{W_F})$ and G_ϑ is the inner twist of G associated with $\vartheta = \kappa_{G^*_{ad}}^{-1}(\zeta)$ via (9).

Notation 3.7 We denote by $\Phi_e(G)$ the set of $\mathbf{G}^\vee$-equivalence classes of enhanced L-parameters for ${}^L G$ that are relevant for G.

Let $(\phi, \rho) \in \Phi_e({}^L G)$. If ρ is relevant for G, then the character $\zeta_\rho \in \mathrm{Irr}\big(Z(G^\vee_{sc})^{W_F}\big)$ can be extended in precisely $[Z(G^\vee_{sc}) : Z(G^\vee_{sc})^{W_F}]$ ways to a character of $Z(G^\vee_{sc})$. We choose such an extension and we denote it by ζ_G. Of course we pick $\zeta_G = \mathrm{triv}$ when $G = G^*$. Every $\phi \in \underline{\Phi}({}^L G)$ can be enhanced with a $\rho \in \mathrm{Irr}(\mathcal{S}_\phi)$ such that $\rho|_{\mathcal{Z}_\phi}$ inflates to ζ_G.

We define

$$\Phi_{e,\zeta_G}(G) := \{(\phi, \rho) \in \Phi_e(G) : \zeta_G \, \mathrm{id}_{V_\rho} = \rho \circ (Z(\mathbf{G}^\vee_{sc}) \to \mathcal{S}_\phi)\}, \tag{23}$$

where V_ρ is the vector space underlying the representation ρ.

Since $\mathcal{S}_\phi / \mathcal{Z}_\phi \cong \mathcal{R}_\phi$, we obtain

$$\Phi_{e,\mathrm{triv}}(G) = \{(\phi, \rho) : \phi \in \Phi(G), \rho \in \mathrm{Irr}(\mathcal{R}_\phi)\}, \qquad \text{if } G \text{ is quasi-split.}$$

3.3 Desiderata for the Local Langlands Correspondence

We are ready to formulate the version of the conjectural local Langlands correspondence stated in [12]. It is inspired by many sources, in particular [21, § 10], [61, § 4], [2, § 3] and [33, § 5.2].

Conjecture 3.8 *Let (G, ϑ) be an inner twist of a quasi-split F-group G^*. There exists a surjection*

$$\Phi_e({}^L G) \longrightarrow \mathrm{Irr}(G) : (\phi, \rho) \mapsto \pi_{\phi,\rho},$$

which becomes bijective when restricted to $\Phi_{e,\zeta_G}(G)$. We write its inverse as

$$\mathrm{Irr}(G) \longrightarrow \Phi_e(G) : \pi \mapsto (\phi_\pi, \rho_\pi).$$

Then the map $\mathrm{Irr}(G) \to \Phi(G) : \pi \mapsto \phi_\pi$ is canonical. These maps satisfy the properties (1)–(7) listed below.

Remark 3.9 The above bijection becomes more elegant if one considers the union over inner twists. It then says that there exists a surjection

$$\{(\phi,\rho) : \phi \in \Phi(G^*), \rho \in \mathrm{Irr}(\mathcal{S}_\phi)\} \to \{(G,\vartheta,\pi) : (G,\vartheta) \text{ inner twist of } G^*, \pi \in \mathrm{Irr}(G)\}$$

whose fibers have exactly $[\mathrm{Z}(G^\vee_{\mathrm{sc}}) : \mathrm{Z}(G^\vee_{\mathrm{sc}})^{W_F}]$ elements.

Properties:

(1) The central character of π equals the character of $\mathrm{Z}(G)$ constructed from ϕ_π in [21, § 10.1].
(2) Let $z \in H^1_c(W_F, \mathrm{Z}(\mathbf{G}^\vee))$ be a class in continuous group cohomology, and let $\chi_z \colon G \to \mathbb{C}^\times$ be the character associated with it in [21, § 10.2]. Thus $z\phi_\pi \in \tilde{\Phi}(G)$ and $\mathcal{S}_{z\phi_\pi} = \mathcal{S}_{\phi_\pi}$. Then the LLC should satisfy $(z\phi_\pi, \rho_\pi) = (\phi_{\chi_z\pi}, \rho_{\chi_z\pi})$.
(3) π is essentially square-integrable if and only if ϕ_π is discrete.
(4) π is tempered if and only if ϕ_π is bounded.
(5) Let P be a parabolic subgroup of G with Levi factor L. Suppose that $g \in \mathrm{N}_G(L)$ and $\breve{g} \in \mathrm{N}_{G^\vee}(L^\vee)$ are such that $\mathrm{Ad}(g) \colon L \to L$ and $\mathrm{Ad}(\breve{g}) \colon L^\vee \to L^\vee$ form a corresponding pair of homomorphisms, in the sense of [21, § 2]. Then

$$(\phi_{g\cdot\pi}, \rho_{g\cdot\pi}) = (\mathrm{Ad}(\breve{g})\phi_\pi, \breve{g}\cdot\rho_\pi) \quad \text{for all} \quad \pi \in \mathrm{Irr}(L).$$

(6) Suppose that $(\phi^L, \rho^L) \in \Phi_{\mathrm{e}}(L)$ is bounded. Then

$$\{\pi_{\phi,\rho} \colon \phi = \phi^L \text{ composed with } {}^L L \to {}^L G, \rho|_{\mathcal{S}^L_\phi} \text{ contains } \rho^L\} \tag{24}$$

equals the set of irreducible constituents of the parabolically induced representation $\mathrm{I}_P^G(\pi_{\phi^L,\rho^L})$.
(7) If ϕ^L is discrete but not necessarily bounded, then (24) is the set of Langlands constituents of $\mathrm{I}_P^G(\pi_{\phi^L,\rho^L})$, as in [6, p. 30].

3.3.1 Inner Forms of $\mathbf{GL}_n(F)$

The local Langlands correspondence for supercuspidal representations of $\mathrm{GL}_n(F)$ was established first for F of positive characteristic in [42], and later for F of characteristic zero in [34, 35, 56], independently.

Together with the Jacquet–Langlands correspondence, this provides the LLC for essentially square-integrable representations of inner forms $G = \mathrm{GL}_m(D)$ of $\mathrm{GL}_n(F)$. It is extended to all irreducible G-representations via the Zelevinsky classification [64], see [9, 36]. For these groups, every L-packet is a singleton and the LLC is a canonical bijective map

$$\mathrm{rec}_{\mathrm{GL}_m(D)} \colon \mathrm{Irr}(\mathrm{GL}_m(D)) \to \Phi(\mathrm{GL}_m(D)). \tag{25}$$

3.3.2 Inner Forms of $\mathrm{SL}_n(F)$

The local Langlands correspondence for $\mathrm{SL}_m(D)$ was established [36] (for F of characteristic zero) and [9] (for F of positive characteristic). It is derived from the LLC for $\mathrm{GL}_m(D)$, in the sense that every L-packet for $\mathrm{SL}_m(D)$ consists of the irreducible constituents of

$$\mathrm{Res}_{\mathrm{SL}_m(D)}^{\mathrm{GL}_m(D)}(\Pi_\phi(\mathrm{GL}_m(D))),$$

with $\phi \in \underline{\Phi}(\mathrm{GL}_m(D))$. Of course these L-packets have more than one element in general (as in Example 2.20).

3.3.3 The Local Langlands Conjecture for Tempered Representation

Let $\mathrm{Irr}^{\mathrm{t}}(G)$ denote the subset of $\mathrm{Irr}(G)$ consisting all irreducible smooth representations π whose matrix coefficients are $L^{2+\epsilon}$ on G modulo the center $\mathrm{Z}(G)$, and let $\Phi^{\mathrm{bd}}(G)$ be the subset of $\Phi(G)$ consisting $\mathbf{G}^\vee$-conjugacy classes of L-parameters ϕ such that the $\phi(W_F \times \mathrm{SU}_2)$ is bounded modulo the center.

The local Langlands conjecture for tempered representation asserts the existence of a surjective map

$$\mathrm{rec}_G^{\mathrm{t}} \colon \mathrm{Irr}^{\mathrm{t}}(G) \to \Phi^{\mathrm{bd}}(G), \tag{26}$$

whose fibers are the *tempered L-packets*. For each $\phi \in \Phi^{\mathrm{bd}}(G)$, the corresponding tempered L-packet $\Pi_\phi^{\mathrm{t}}(G)$ is expected to be parametrized by $\mathrm{Irr}(S_\phi)_{\zeta_G}$, and $\mathrm{Irr}^{\mathrm{t}}(G)$ to be a disjoint union of the $\Pi_\phi^{\mathrm{t}}(G)$. In the case when the characteristic of F is zero, this conjecture has been proved by Arthur in [3] for quasi-split orthogonal groups and symplectic groups, by Mok in [51] for quasi-split unitary groups, and by Kaletha, Minguez, Shin and White for non in [40]. Note that the unitary group $\mathrm{U}_{E/F}(n)$ admits a non-quasi-split inner form exactly when n is even.

The LLC was also proved for the groups $\mathrm{GSp}_4(F)$ [27, 30], its inner form [29], the group $\mathrm{Sp}_4(F)$ [28] and its inner form [23], and for the groups GSpin_4, GSpin_6, and their inner forms [4].

In [31], Ganapathy and Varma used Arthur's results to lift the LLC for split symplectic and special orthogonal groups on a non-Archimedean field of odd positive characteristic, but using a "Gan-Takeda type" characterization instead of the theory of endoscopy. They proved (see [31, Theorem 13.6.1]) that given a tempered representation π of G, there exists a unique bounded Langlands parameter ϕ_π, defined by suitable compatibility conditions on Langlands-Shahidi L-functions and γ-factors, and on Plancherel measures, together with the requirement that ϕ_π is discrete if π belongs to the discrete series.

For a pleasant and precise description of the state of the art with respect to the local Langlands correspondence, the reader should consult [37, 38, 43], the latter including global aspects and the link with the Ramanujan conjecture. A survey on other aspects of the Langlands correspondence may be found in [5] and the references therein.

3.4 Depth

Another invariant that makes sense on both sides of the LLC is the *depth*. The *depth* $d(\pi)$ of an irreducible smooth representation π of a reductive p-adic group G was defined by Moy and Prasad [54] in terms of filtrations $G_{x,r}$ (with x a point in the Bruhat-Tits building of G and $r \in \mathbb{R}_{\geq 0}$) of its parahoric subgroups $G_{x,0}$.

The depth of a Langlands parameter ϕ is defined to be the smallest number $d(\phi) \geq 0$ such that ϕ is trivial on Γ^r for all $r > d(\phi)$.

Yu [63, § 7.10] proved that the depth is preserved by the LLC for unramified tori. Recently, Mishra and Pattanayak proved that it is not preserved for wildly ramified tori [50].

3.4.1 Inner Forms of General Linear Groups

Let $\mathrm{GL}_m(D)$ an inner form of $\mathrm{GL}_n(F)$. Let $k_D = \mathfrak{o}_D/\mathfrak{p}_D$ be the residual field of D. Let $\mathfrak{A}$ be a hereditary $\mathfrak{o}_F$-order in $\mathrm{M}_m(D)$. The Jacobson radical of $\mathfrak{A}$ will be denoted by $\mathfrak{P}$. Let $r = e_D(\mathfrak{A})$ and $e = e_F(\mathfrak{A})$ denote the integers defined by $\mathfrak{p}_D\mathfrak{A} = \mathfrak{P}^r$ and $\mathfrak{p}_F\mathfrak{A} = \mathfrak{P}^e$, respectively. We have $e_F(\mathfrak{A}) = d\, e_D(\mathfrak{A})$. The normalizer in G of $\mathfrak{A}^\times$ will be denoted by

$$\mathfrak{K}(\mathfrak{A}) := \left\{g \in G \,:\, g^{-1}\mathfrak{A}^\times g = \mathfrak{A}^\times\right\}.$$

Define a sequence of compact open subgroups of $G = \mathrm{GL}_m(D)$ by

$$U^0(\mathfrak{A}) := \mathfrak{A}^\times, \quad \text{and} \quad U^j(\mathfrak{A}) := 1 + \mathfrak{P}^j, \;\; j \geq 1.$$

Then $\mathfrak{A}^\times$ is a parahoric subgroup of G and $U^1(\mathfrak{A})$ is its pro-unipotent radical. We define the *normalized level* of an irreducible representation π of G to be

$$\ell(\pi) := \min\{j/e_F(\mathfrak{A})\}, \tag{27}$$

where $(j, \mathfrak{A})$ ranges over all pairs consisting an integer $j \geq 0$ and a hereditary $\mathfrak{o}_F$-order $\mathfrak{A}$ in $\mathrm{M}_m(D)$ such that π contains the trivial character of $U^{j+1}(\mathfrak{A})$. Then (see, for instance, [8, Proposition 2.5]) the normalized level of $\pi \in \mathrm{Irr}(G)$ equals its Moy–Prasad depth:

$$\ell(\pi) = d(\pi).$$

Let ψ be a nontrivial character of F and let $c(\psi)$ be the largest integer c such that $\mathfrak{p}_F^{-c} \subset \ker\psi$. The ϵ factor of ϕ (and ψ) was defined in [60]. It takes the form

$$\epsilon(s, \phi, \psi) = \epsilon(0, \phi, \psi) q^{-(a(\phi)+nc(\psi))s} \text{ with } \epsilon(0, \phi, \psi) \in \mathbb{C}^\times. \tag{28}$$

Here $a(\phi) \in \mathbb{Z}_{\geq 0}$ is the Artin conductor of ϕ (called $f(\phi)$ in [57, § VI.2]). For any elliptic $\phi \in \Phi(GL_n(F))$, we have (see [8, Lemma 2.3])

$$d(\phi) = \begin{cases} 0 & \text{if } I_F \subset \ker(\phi), \\ \dfrac{a(\phi)}{n} - 1 & \text{otherwise.} \end{cases} \tag{29}$$

Let π be an irreducible representation of $\mathrm{GL}_m(D)$. Let $\epsilon(s, \pi, \psi)$ denote the Godement–Jacquet local constant [32]. It takes the form

$$\epsilon(s, \pi, \psi) = \epsilon(0, \pi, \psi)\, q^{-f(\pi,\psi)s}, \quad \text{where } \epsilon(0, \pi, \psi) \in \mathbb{C}^\times. \tag{30}$$

A representation of $D^\times$ is called *unramified* if it is trivial on $\mathfrak{o}_D^\times$. An unramified representation of $D^\times$ is a character and has depth zero.

Let π be a supercuspidal irreducible representation of G. We have (see [8, Proposition 2.6]:

$$f(\pi, \psi) = \begin{cases} n\,(c(\psi) + 1) - 1 & \text{if } m = 1 \text{ and } \pi \text{ is unramified,} \\ n\,(d(\pi) + 1 + c(\psi)) & \text{otherwise.} \end{cases} \tag{31}$$

We set

$$f(\pi) := f(\pi, \psi) - nc(\psi). \tag{32}$$

Theorem 3.10 *[8, Theorem 2.7]*
The depth $d(\pi)$ and the conductor $f(\pi)$ of each essentially square-integrable irreducible representation π of $\mathrm{GL}_m(D)$ are linked by the following relation:

$$d(\pi) = \begin{cases} 0 & \text{if } \pi \text{ is an unramified twist of } \mathrm{St}_{\mathrm{GL}_m(D)} \\ \dfrac{f(\pi) - n}{n} & \text{otherwise.} \end{cases}, \tag{33}$$

In particular

$$d(\pi) = \max\left\{\frac{f(\pi) - n}{n}, 0\right\}. \tag{34}$$

Theorem 3.10 is a key ingredient in the proof of the following result:

Theorem 3.11 [8, Theorem 2.9]
The LLC *for $G = \mathrm{GL}_m(D)$ preserves the depth, that is:*

$$d(\pi) = d(\phi_\pi), \quad \text{where } \phi_\pi = \mathrm{rec}_G(\pi).$$

3.4.2 Inner Forms of Special General Linear Groups

The situation is different for $\mathrm{SL}_m(D)$. All the irreducible representations in a given L-packet Π_ϕ have the same depth, so the depth is an invariant of the L-packet, say $d(\Pi_\phi)$. We have $d(\Pi_\phi) = d(\varphi)$ where φ is a lift of ϕ which has minimal depth among the lifts of ϕ, and the following holds:

$$d(\phi) \leq d(\Pi_\phi) \tag{35}$$

for any Langlands parameter ϕ for $\mathrm{SL}_m(D)$ [8, Proposition 3.4 and Corollary 3.4]. Moreover (35) is an equality *if* ϕ is essentially tame (in the terminology of Definition 2.13).

Remark 3.12 The notion of essentially tameness is consistent with the usual notion for Langlands parameters for $\mathrm{GL}_n(F)$. Indeed, any lift $\varphi\colon W_F \to \mathrm{GL}_n(\mathbb{C})$ of ϕ, is called essentially tame if its restriction to P_F is a direct sum of characters. We denote by $t(\varphi)$ the *torsion number* of φ, that is, the number of unramified characters χ of W_F such $\varphi\chi \cong \varphi$. Then ϕ and φ are essentially tame if and only if the residual characteristic p of F does not divide $n/t(\varphi)$ [22, Appendix].

However, let F be a local non-Archimedean field of characteristic 2, that is, F is of the form $F = \mathbb{F}_q((t))$, the field of Laurent series with coefficients in $\mathbb{F}_q$, with $q = 2^f$. This case is particularly interesting because there are countably many quadratic extensions of $\mathbb{F}_q((t))$. Then equality holds in (35) *only if* ϕ is essentially tame (*i.e.*, $t(\varphi) = 2$), as proved in [14].

In the case when G is a classical group, the characteristic of F is zero (i.e., F is a finite extension of $\mathbb{Q}_p$) and p is odd, Ganapathy and Varma proved in [31, Lemma 8.2.3] that the following inequality holds

$$d(\phi_\pi) \leq \lfloor d(\pi) \rfloor + 1, \tag{36}$$

where ϕ the Langlands parameter attached to π by Arthur. If moreover p is sufficiently large with respect to G, then it is shown in [31] that

$$d(\pi) \leq d(\phi_\pi). \tag{37}$$

Very recently, it was proved in [55], for a quasi-split classical group over F, with F of characteristic equal to zero and p sufficiently large, that the depth of representations in each L-packet equals the depth of the corresponding L-parameter, and that, for quasi-split unitary groups, the depth is constant in each L-packet.

4 Generalized Springer Correspondence

4.1 Cuspidal Enhanced Unipotent Classes

Let $\mathcal{G}$ be a complex (possibly disconnected) reductive group. Let $\mathcal{G}^\circ$ be its identity component. For $u \in \mathcal{G}$ unipotent, we denote by $A_{\mathcal{G}}(u)$ the component group of the centralizer of u in $\mathcal{G}$. We denote by $\mathrm{U}(\mathcal{G})$ the unipotent variety of $\mathcal{G}$.

Definition 4.1 The *enhancement* of $\mathrm{U}(\mathcal{G})$ is the set $\mathrm{U}_e(\mathcal{G})$ of $\mathcal{G}$-conjugacy classes of pairs (u, ρ), with $u \in \mathcal{G}$ unipotent and $\rho \in \mathrm{Irr}(A_{\mathcal{G}}(u))$. We call a pair (u, ρ) an *enhanced unipotent class*.

Let $D^b_{\mathcal{G}}(\mathrm{U}(\mathcal{G}))$ denote the constructible $\mathcal{G}$-equivariant-derived category on $\mathrm{U}(\mathcal{G})$ as in [19] and let $\mathrm{Perv}_{\mathcal{G}}(\mathrm{U}(\mathcal{G}))$ be its subcategory of $\mathcal{G}$-equivariant perverse sheaves.

By a $\mathcal{P}$-resolution of an algebraic variety X, we mean a variety Y endowed with a free $\mathcal{P}$-action and a smooth $\mathcal{P}$-equivariant morphism $Y \to X$.

Definition 4.2 The *integration functor* is the functor

$$\gamma^{\mathcal{G}}_{\mathcal{P}} \colon D^b_{\mathcal{P}}(\mathrm{U}(\mathcal{G})) \to D^b_{\mathcal{G}}(\mathrm{U}(\mathcal{G})),$$

defined by

$$(\gamma^{\mathcal{G}}_{\mathcal{P}} A)(Y) := (q_Y)_! A(Y)[2 \dim \mathcal{G}/\mathcal{P}],$$

for A any object of $D^b_{\mathcal{P}}(\mathrm{U}(\mathcal{G}))$ and Y a $\mathcal{G}$-resolution of $\mathrm{U}(\mathcal{G})$, where $q_Y : \mathcal{P}\backslash Y \to \mathcal{G}\backslash Y$ is the quotient functor and $A(Y)$ is defined by regarding Y as a $\mathcal{P}$-resolution of $\mathrm{U}(\mathcal{G})$.

Definition 4.3 Let $\mathcal{P}^\circ = \mathcal{L}^\circ\mathcal{U}$ be a parabolic subgroup of $\mathcal{G}^\circ$, with $\mathcal{L}^\circ$ a Levi component and $\mathcal{U}$ the unipotent radical. Let

$$m \colon \mathrm{U}(\mathcal{P}^\circ) \hookrightarrow \mathrm{U}(\mathcal{G}^\circ) \quad \text{and} \quad p \colon \mathrm{U}(\mathcal{P}^\circ) \to \mathrm{U}(\mathcal{L}^\circ)$$

denote inclusion and projection, respectively. The *parabolic induction functor* is the functor

$$\mathrm{i}^{\mathcal{G}^\circ}_{\mathcal{L}^\circ \subset \mathcal{P}^\circ} := \gamma^{\mathcal{G}^\circ}_{\mathcal{P}^\circ} \circ m_! \circ p^* : \mathrm{Perv}_{\mathcal{L}^\circ}(\mathrm{U}(\mathcal{L}^\circ)) \to \mathrm{Perv}_{\mathcal{G}^\circ}(\mathrm{U}(\mathcal{G}^\circ)).$$

Remarks 4.4 (1) The functor $\mathrm{i}^{\mathcal{G}^\circ}_{\mathcal{L}^\circ \subset \mathcal{P}^\circ}$ commutes with Verdier duality. It is left adjoint to $\mathrm{r}^{\mathcal{G}^\circ}_{\mathcal{L}^\circ \subset \mathcal{P}^\circ} := p_! \circ m^*$ and right adjoint to $'\mathrm{r}^{\mathcal{G}^\circ}_{\mathcal{L}^\circ \subset \mathcal{P}^\circ} := p_* \circ m^!$. These functors are exchanged by Verdier duality.
(2) If $\mathcal{F}_L$ is a simple object in $\mathrm{Perv}_{\mathcal{L}^\circ}(\mathrm{U}(\mathcal{L}^\circ))$, then $\mathrm{i}^{\mathcal{G}^\circ}_{\mathcal{L}^\circ \subset \mathcal{P}^\circ}(\mathcal{F}_{\mathcal{L}^\circ})$ is semisimple.

Definition 4.5 A simple object $\mathcal{F}$ in $\mathrm{Perv}_{\mathcal{G}^\circ}(\mathrm{U}(\mathcal{G}^\circ))$ is *cuspidal* if for any simple object $\mathcal{F}_{\mathcal{L}^\circ}$ in $\mathrm{Perv}_{\mathcal{L}^\circ}(\mathrm{U}(\mathcal{L}^\circ))$, $\mathcal{F}$ does not occur in $\mathrm{i}^{\mathcal{G}^\circ}_{\mathcal{L}^\circ \subset \mathcal{P}^\circ}(\mathcal{F}_{\mathcal{L}^\circ})$ (equivalently, if $\mathrm{r}^{\mathcal{G}^\circ}_{\mathcal{L}^\circ \subset \mathcal{P}^\circ}(\mathcal{F}) = 0$, resp. $'\mathrm{r}^{\mathcal{G}^\circ}_{\mathcal{L}^\circ \subset \mathcal{P}^\circ}(\mathcal{F}) = 0$) for any proper parabolic subgroup $\mathcal{P}^\circ$ of $\mathcal{G}^\circ$ with Levi factor $\mathcal{L}^\circ$.

Remarks 4.6 (1) Cuspidality is preserved by Verdier duality (see, for instance, [1, Remark 2.3]).
(2) The above definition of cuspidality is inspired by [46] (see also [1]). It is equivalent to the original definition given by Lusztig in [45] (as shown in [49, 23.2.(b)]).

For u a given unipotent element in $\mathcal{G}^\circ$, let $A_{\mathcal{G}^\circ}(u)$ denote the component group of the centralizer $\mathrm{Z}_{\mathcal{G}^\circ}(u)$ of u in $\mathcal{G}^\circ$. Let $\mathcal{O}_u = (u)_{\mathcal{G}^\circ}$ be the $\mathcal{G}^\circ$-conjugacy class of u.

With a slight abuse of notation, we will sometimes write $A_{\mathcal{G}^\circ}(\mathcal{O}_u) := A_{\mathcal{G}^\circ}(u)$. We will denote by

$$\rho \mapsto \mathcal{E}_\rho \tag{38}$$

the bijection between $\operatorname{Irr}(A_{\mathcal{G}^\circ}(u))$ and the irreducible $\mathcal{G}^\circ$-equivariant local systems $\mathcal{E}$ on $\mathcal{O}_u$. We denote by

$$\mathcal{E} \mapsto \rho_{\mathcal{E}} \tag{39}$$

the inverse bijection.

Definition 4.7 (1) A character $\rho \in \operatorname{Irr}(A_{\mathcal{G}^\circ}(u))$ is *cuspidal* if the perverse sheaf $\operatorname{IC}(\mathcal{O}, \mathcal{E}_\rho)$ is cuspidal.
(2) An *enhanced unipotent class* $(\mathcal{O}, \rho)$ in $\mathcal{G}^\circ$ is *cuspidal* if ρ is cuspidal.

Proposition 4.8 [45, Proposition 2.8] *If $(\mathcal{O}_u, \rho)$ is cuspidal, then u is a distinguished unipotent element in $\mathcal{G}^\circ$ (i.e., u does not meet* $\mathrm{U}(\mathcal{L}^\circ)$ *for any proper Levi subgroup $\mathcal{L}$ of $\mathcal{G}^\circ$).*

Remark 4.9 In general, not every distinguished unipotent element supports a cuspidal representation.

Example 4.10 For $\mathcal{G} = \mathrm{SL}_n(\mathbb{C})$, the unipotent classes in $\mathcal{G}$ are in bijection with the partitions $\lambda = (\lambda_1, \lambda_2, \dots, \lambda_r)$ of n: The corresponding $\mathcal{G}$-conjugacy class $\mathcal{O}_\lambda$ consists of unipotent matrices with Jordan blocks of sizes $\lambda_1, \lambda_2, \dots, \lambda_r$. We identify the center $\mathrm{Z}(\mathcal{G})$ with the group μ_n of complex n-th roots of unity. For $u \in \mathcal{O}_\lambda$, the natural homomorphism $\mathrm{Z}(\mathcal{G}) \to A_{\mathcal{G}}(u)$ is surjective with kernel $\mu_{n/\gcd(\lambda)}$, where $\gcd(\lambda) := \gcd(\lambda_1, \lambda_2, \dots, \lambda_r)$. Hence, the irreducible $\mathcal{G}$-equivariant local systems on $\mathcal{O}_\lambda$ all have rank one, and they are distinguished by their central characters, which range over those $\chi \in \operatorname{Irr}(\mu_n)$ such $\gcd(\lambda)$ is a multiple of the order of χ. We denote these local systems by $\mathcal{E}_{\lambda,\chi}$. The unique distinguished unipotent class in $\mathcal{G}$ is the regular unipotent class $\mathcal{O}_{(n)}$, consisting unipotent matrices with a single Jordan block. The cuspidal irreducible G-equivariant local systems are supported on $\mathcal{O}_{(n)}$ (by Proposition 4.8) and are of the form $\mathcal{E}_{(n),\chi}$, with $\chi \in \operatorname{Irr}(\mu_n)$ of order n (see [45, (10.3.2)]).

We will now extend the above notion of cuspidality from $\mathcal{G}^\circ$ to $\mathcal{G}$:

Definition 4.11 An enhanced unipotent class $(\mathcal{O}, \rho)$ with $\mathcal{O} = (u)_{\mathcal{G}}$ and $\rho \in \operatorname{Irr}(A_{\mathcal{G}}(u))$ is *cuspidal* if the restriction of ρ to $A_{\mathcal{G}^\circ}(u)$ is a direct sum of cuspidal irreducible representations of $A_{\mathcal{G}^\circ}(u)$.

Notation 4.12 We set

$$\operatorname{Irr}_{\mathrm{cusp}}(A_{\mathcal{G}}(u)) := \{\varepsilon \in \operatorname{Irr}(A_{\mathcal{G}}(u)) \text{ such that } \varepsilon \text{ is cuspidal}\}.$$

Definition 4.13 A *quasi-Levi subgroup* of $\mathcal{G}$ is a subgroup $\mathcal{M}$ of the form $\mathcal{M} = \mathrm{Z}_{\mathcal{G}}(\mathrm{Z}(\mathcal{L})^\circ)$, with $\mathcal{L}$ a Levi subgroup of $\mathcal{G}^\circ$. The group $\mathcal{M}$ is said to be *cuspidal* if there exists a cuspidal enhanced unipotent pair in $\mathcal{M}$.

Notation 4.14 Let $\mathfrak{B}(\mathrm{U_e}(\mathcal{G}))$ be the set of $\mathcal{G}$-conjugacy classes of pairs $(\mathcal{M}, (\mathcal{O}, \varepsilon))$, where $\mathcal{M}$ is a cuspidal quasi-Levi subgroup of $\mathcal{G}$, and $(\mathcal{O}, \varepsilon)$ is a cuspidal enhanced unipotent pair in $\mathcal{M}$. We will write the elements of $\mathfrak{B}(\mathrm{U_e}(\mathcal{G}))$ as $\mathfrak{t} := [\mathcal{M}, v, \varepsilon]_{\mathcal{G}}$.

4.2 *A Partition of the Enhancement of the Unipotent Variety of $\mathcal{G}^\circ$*

The purpose of this section is to describe a *theory à la Harish-Chandra for* $\mathrm{U_e}(\mathcal{G}^\circ)$, the first step being to define a *cuspidal support map* for enhanced unipotent classes of $\mathcal{G}$.

The enhancement $\mathrm{U_e}(\mathcal{G}^\circ)$ of $\mathrm{U}(\mathcal{G}^\circ)$ parametrizes the isomorphism classes of simple objects of $\mathrm{Perv}_{\mathcal{G}^\circ}(\mathrm{U}(\mathcal{G}^\circ))$. Indeed, the simple objects in $\mathrm{Perv}_{\mathcal{G}^\circ}(\mathrm{U}(\mathcal{G}^\circ))$ are the $\mathrm{IC}(\mathcal{O}, \mathcal{E})$, where $\mathcal{O}$ is a unipotent class in $\mathcal{G}^\circ$ and $\mathcal{E}$ is an irreducible $\mathcal{G}^\circ$-equivariant $\overline{\mathbb{Q}}_\ell$-local system on $\mathcal{O}$.

Let $(\mathcal{O}, \rho) \in \mathrm{U_e}(\mathcal{G}^\circ)$ be an arbitrary enhanced unipotent class, and set $\mathcal{F}_\rho := \mathrm{IC}(\mathcal{O}, \mathcal{E}_\rho)$. Then $\mathcal{F}_\rho$ occurs as a summand of $\mathrm{i}_{\mathcal{L}^\circ \subset \mathcal{P}^\circ}(\mathrm{IC}(\mathcal{O}_0, \mathcal{E}_0))$, for some quadruple $(\mathcal{P}^\circ, \mathcal{L}^\circ, \mathcal{O}_0, \mathcal{E}_0)$, where $\mathcal{P}^\circ$ is a parabolic subgroup of $\mathcal{G}^\circ$ with Levi subgroup $\mathcal{L}^\circ$ and $(\mathcal{O}_0, \mathcal{E}_0)$ is a cuspidal enhanced unipotent class in $\mathcal{L}^\circ$ (see [45, § 6.2] and [1, Cor. 2.7]) and, moreover, $(\mathcal{P}^\circ, \mathcal{L}^\circ, \mathcal{O}_0, \mathcal{E}_0)$ is unique up to $\mathcal{G}^\circ$-conjugation (see [45, Prop. 6.3]). We set $\varepsilon_0 := \rho_{\mathcal{E}_0}$ using the bijection (39), and we denote by $\mathfrak{t}^\circ := (\mathcal{L}^\circ, (\mathcal{O}_0, \varepsilon_0))_{\mathcal{G}^\circ}$, the $\mathcal{G}^\circ$-conjugacy class of $(\mathcal{L}^\circ, (\mathcal{O}_0, \varepsilon_0))$ and we call it the *cuspidal support* of the enhanced unipotent class $(\mathcal{O}, \rho)$.

The center $\mathrm{Z}(\mathcal{G}^\circ)$ of $\mathcal{G}^\circ$ maps naturally to $A_{\mathcal{G}^\circ}(\mathcal{O})$ and to $A_{\mathcal{L}^\circ}(\mathcal{O}_0)$. By construction [45, Theorem 6.5.a]

$$\rho \text{ and } \varepsilon \text{ have the same } \mathrm{Z}(\mathcal{G}^\circ)\text{-character.} \tag{40}$$

Definition 4.15 The *cuspidal support map* for $\mathrm{U_e}(\mathcal{G}^\circ)$ is the map

$$\Psi_{\mathcal{G}^\circ} \colon \mathrm{U_e}(\mathcal{G}^\circ) \to \mathfrak{B}(\mathrm{U_e}(\mathcal{G}^\circ)),$$

where $\mathfrak{B}(\mathrm{U_e}(\mathcal{G}^\circ))$ is as in Notation 4.14, which sends the $\mathcal{G}^\circ$-conjugacy class of $(\mathcal{O}, \rho)$ to its cuspidal support $\mathfrak{t}^\circ = (\mathcal{L}^\circ, (\mathcal{O}_0, \varepsilon_0))_{\mathcal{G}^\circ}$.

By (40), the map $\Psi_{\mathcal{G}^\circ}$ preserves the $\mathrm{Z}(\mathcal{G}^\circ)$-characters of the involved representations.

Notation 4.16 Let $\mathfrak{t}^\circ \in \mathfrak{B}(\mathrm{U_e}(\mathcal{G}^\circ))$.

(1) We denote by $\mathrm{U_e}(\mathcal{G}^\circ)^{\mathfrak{t}^\circ}$ the fiber of $\mathfrak{t}$ under the map $\Psi_{\mathcal{G}^\circ}$.
(2) Let $W_{\mathcal{L}^\circ} := \mathrm{N}_{\mathcal{G}^\circ}(\mathcal{L}^\circ)/\mathcal{L}^\circ$, and let $W_{\mathfrak{t}^\circ} := \mathrm{N}_{\mathcal{G}^\circ}(\mathfrak{t}^\circ)/\mathcal{L}^\circ$.

Theorem 4.17 [45]

(1) The group $W_{\mathcal{L}^\circ}$ *is a Weyl group and it coincides with* $W_{\mathfrak{t}^\circ}$ *for every* $\mathfrak{t}^\circ \in \mathfrak{B}(\mathrm{U_e}(\mathcal{G}^\circ))$.

(2) We have

$$\mathrm{U_e}(\mathcal{G}^\circ) = \bigsqcup_{\mathfrak{t}^\circ \in \mathfrak{B}(\mathrm{U_e}(\mathcal{G}^\circ))} \mathrm{U_e}(\mathcal{G}^\circ)^{\mathfrak{t}^\circ},$$

and $\mathrm{U_e}(\mathcal{G}^\circ)^{\mathfrak{t}^\circ}$ *is in bijection with* $\mathrm{Irr}(W_{\mathcal{L}^\circ})$.

The goal of the next two subsections is to generalize the previous notation and terminology from $\mathcal{G}^\circ$ to $\mathcal{G}$.

4.3 Twisted Group Algebras

Throughout this section, Γ is a finite group and K is an algebraically closed field whose characteristic does not divide the order of Γ. Suppose that $\kappa\colon \Gamma \times \Gamma \to K^\times$ is a 2-cocycle, that is,

$$\kappa(\gamma_1, \gamma_2\gamma_3)\kappa(\gamma_2, \gamma_3) = \kappa(\gamma_1, \gamma_2)\kappa(\gamma_1\gamma_2, \gamma_3) \quad \forall \gamma_1, \gamma_2, \gamma_3 \in \Gamma. \tag{41}$$

The κ-twisted group algebra of Γ is defined to be the K-vector space $K[\Gamma, \kappa]$ with basis $\{T_\gamma : \gamma \in \Gamma\}$ and multiplication rules

$$T_\gamma T_{\gamma'} = \kappa(\gamma, \gamma')T_{\gamma\gamma'} \quad \gamma, \gamma' \in \Gamma. \tag{42}$$

Its representations can be considered as projective Γ-representations. Schur showed (see [24, Theorem 53.7]) that there exists a finite central extension $\tilde{\Gamma}$ of Γ, such that

- $\mathrm{char}(K)$ does not divide $|\tilde{\Gamma}|$,
- every irreducible projective Γ-representation over K lifts to an irreducible K-linear representation of $\tilde{\Gamma}$.

Then $K[\Gamma, \kappa]$ is a direct summand of $K[\tilde{\Gamma}]$, namely the image of a minimal idempotent in $K[\ker(\tilde{\Gamma} \to \Gamma)]$. The condition on $\mathrm{char}(K)$ ensures that $K[\tilde{\Gamma}]$ is semisimple, so $K[\Gamma, \kappa]$ is also semisimple.

4.4 A Partition of the Enhancement of the Unipotent Variety of $\mathcal{G}$

Notation 4.18 Let $\mathfrak{t} = (\mathcal{M}, (\mathcal{O}_0, \varepsilon))_{\mathcal{G}} \in \mathfrak{B}(\mathrm{U_e}(\mathcal{G}))$. We set

$$W_{\mathfrak{t}} := \mathrm{N}_{\mathcal{G}}(\mathfrak{t})/\mathcal{M} \quad \text{and} \quad W_{\mathfrak{t}}^\circ := \mathrm{N}_{\mathcal{G}^\circ}(\mathcal{M}^\circ)/\mathcal{M}^\circ.$$

Theorem 4.19 [15, § 4] *Let* $(\mathcal{O}, \rho) \in \mathrm{U_e}(\mathcal{G})$. *There exists a 2-cocycle*

$$\kappa_{\mathfrak{t}} \colon W_{\mathfrak{t}}/W_{\mathfrak{t}}^{\circ} \times W_{\mathfrak{t}}/W_{\mathfrak{t}}^{\circ} \to \mathbb{C}^{\times}$$

and a map, called the cuspidal support map for $\mathrm{U_e}(\mathcal{G})$

$$\Psi_{\mathcal{G}} \colon \mathrm{U_e}(\mathcal{G}) \to \mathfrak{B}(\mathrm{U_e}(\mathcal{G}))$$

(which coincides with the map $\Psi_{\mathcal{G}^\circ}$ *defined above in the case when* $\mathcal{G}$ *is connected), such that*

$$\mathrm{U_e}(\mathcal{G}) = \bigsqcup_{\mathfrak{t} \in \mathfrak{B}(\mathrm{U_e}(\mathcal{G}))} \mathrm{U_e}(\mathcal{G})^{\mathfrak{t}}, \tag{43}$$

in which the fiber $\mathrm{U_e}(\mathcal{G})^{\mathfrak{t}}$ *of* $\mathfrak{t}$ *under the map* $\Psi_{\mathcal{G}}$ *is isomorphic to the set of isomorphism classes* $\mathrm{Irr}(\mathbb{C}[W_{\mathfrak{t}}, \kappa_{\mathfrak{t}}])$ *of irreducible representations of a* $\kappa_{\mathfrak{t}}$*-twisted version* $\mathbb{C}[W_{\mathfrak{t}}, \kappa_{\mathfrak{t}}]$ *of the group algebra* $\mathbb{C}[W_{\mathfrak{t}}]$ *of the finite group* $W_{\mathfrak{t}}$.

Notation 4.20 Let $\Sigma_{\mathfrak{t}}$ denote the bijection

$$\mathrm{U_e}(\mathcal{G})^{\mathfrak{t}} \longrightarrow \mathrm{Irr}(\mathbb{C}[W_{\mathfrak{t}}, \kappa_{\mathfrak{t}}])$$

mentioned in Theorem 4.19.

Remark 4.21 When $\mathcal{G}$ is connected, the cocycle $\kappa_{\mathfrak{t}}$ is trivial. When $\mathcal{G}$ is disconnected, the cocycle $\kappa_{\mathfrak{t}}$ is not always trivial.

An intuitive way of thinking of Theorem 4.19 is to view the set $\mathfrak{B}(\mathrm{U_e}(\mathcal{G}))$ as a "palette of colors" and the map $\Psi_{\mathcal{G}}$ as a way to paint the elements of the fiber $\mathrm{U_e}(\mathcal{G})^{\mathfrak{t}}$ of $\mathfrak{t}$ under $\Psi_{\mathcal{G}}$ in the same color as $\mathfrak{t}$. Moreover, for each color, the subset of elements of $\mathrm{U_e}(\mathcal{G})$ with that color has a nice structure (here, that of the isomorphism classes of irreducible representations of a twisted group algebra).

5 Cuspidality for Enhanced L-Parameters: Definition and Conjecture

Definition 5.1 An enhanced L-parameter $(\phi, \rho) \in \Phi(G)_{\mathrm{e}}$ is called *cuspidal* if ϕ is discrete and (u_ϕ, ρ) is a cuspidal enhanced unipotent class in $\mathcal{G}_{\lambda_\phi}$ in the terminology of Definition 4.11.

Recall that an irreducible smooth complex representation of the group G is called *supercuspidal* if it does not appear in any G-representation induced from a proper Levi subgroup of G. It was proved by Jacquet and by Harish-Chandra that an irreducible representation of G is supercuspidal if and only if its matrix coefficients have compact support modulo the center of G (see, for instance, [20, pp. 34–35]).

Cuspidality Conjecture 5.2 [15, § 6]
The cuspidal enhanced Langlands parameters correspond under the LLC *to the irreducible supercuspidal representations of* G.

Conjecture 5.2 is known to be true in the following cases:

- for general linear groups and split classical groups (any representation), with F of characteristic equal to 0, see [52],
- for inner forms of linear groups and of special linear groups, see [15, Example 6.11], quasi-split unitary groups (any representation) with F of characteristic equal to 0, see [15, Example 6.13], and for Deligne-Lusztig depth-zero representations, [15, Example 6.14],
- for the representations with unipotent reduction of the group G of the F-rational points of a connected reductive algebraic group which splits over an unramified extension of F, see [26, Theorem 2] (when G is simple of adjoint type it is a special case of [47, 48]).

6 A Partition of the Set of Enhanced Langlands Parameters

We shall define a similar partition of the set of enhanced Langlands parameters by plugging the above construction into the framework of the Langlands correspondence. The idea is very similar as above: Using the data defined in (47), we will construct a set $\mathfrak{B}^\vee({}^LG)$ which will play the role of the palette of colors for $\Phi_{\mathrm{e}}({}^LG)$, and we will define a "way to paint" the elements in $\Phi_{\mathrm{e}}({}^LG)$ with this set of colors (i.e., we will construct a cuspidal support map) such that there is a decomposition

$$\Phi_{\mathrm{e}}({}^LG) = \bigsqcup_{\mathfrak{s}^\vee \in \mathfrak{B}^\vee({}^LG)} \Phi_{\mathrm{e}}({}^LG)^{\mathfrak{s}^\vee}, \tag{44}$$

where, for each $\mathfrak{s}^\vee$ in $\mathfrak{B}^\vee({}^LG)$, the subset $\Phi_{\mathrm{e}}({}^LG)^{\mathfrak{s}^\vee}$ of elements with color $\mathfrak{s}^\vee$ is related to a generalized affine Hecke algebras with possibly unequal parameters.

6.1 *Inertial Classes for Enhanced Langlands Parameters*

Definition 6.1 Let $\mathbf{T}^\vee \subset \mathbf{G}^\vee$ be a torus such that the projection $\mathrm{Z}_{{}^LG}(\mathbf{T}^\vee) \to W_F$ is surjective, where $\mathrm{Z}_{{}^LG}(\mathbf{T}^\vee)$ is the centralizer of $\mathbf{T}^\vee$ in ${}^LG$. Then we call $\mathrm{Z}_{{}^LG}(\mathbf{T}^\vee)$ a Levi L-subgroup of ${}^LG$.

Remark 6.2 In [21], such groups are called Levi subgroups of ${}^LG$. However, we prefer to stick to the connectedness of Levi subgroups.

Remark 6.3 Choose a W_F-stable pinning for $\mathbf{G}^\vee$. This defines the notion of standard Levi subgroups of $\mathbf{G}^\vee$. An alternative characterization of the Levi L-subgroups of ${}^LG$ is as follows [15, Lemma 6.2]: Let $\mathrm{Z}_{{}^LG}(\mathbf{T}^\vee)$ be a Levi L-subgroup of ${}^LG$ as in Definition 6.1. There exists a W_F-stable standard Levi subgroup $\mathbf{L}^\vee$ of $\mathbf{G}^\vee$ such that $\mathrm{Z}_{{}^LG}(\mathbf{T}^\vee)$ is $\mathbf{G}^\vee$-conjugate to $\mathbf{L}^\vee \rtimes W_F =: {}^LL$ and $\mathrm{Z}_{{}^LG}(\mathbf{T}^\vee) \cap \mathbf{G}^\vee$ is conjugate to $\mathbf{L}^\vee$.

Conversely, every $\mathbf{G}^\vee$-conjugate of this ${}^LL$ is a Levi L-subgroup of ${}^LG$.

Notation 6.4 In the sequel, we will use the (slightly abusive) notation ${}^LL$ for an arbitrary L-Levi of G.

Definition 6.5 Let ${}^LL$ be a Levi L-subgroup of ${}^LG$.

(1) A *Langlands parameter for* ${}^LL$ is a group homomorphism $\varphi\colon W_F' \to {}^LL$ satisfying the requirements of Definition 2.12.
(2) An *enhancement of* φ is an irreducible representation ϱ of $\pi_0(\mathrm{Z}^1_{\mathbf{L}^\vee_{\mathrm{sc}}}(\varphi))$, where $\mathbf{L}^\vee_{\mathrm{sc}}$ is the simply connected cover of the derived group of $\mathbf{L}^\vee :== {}^LL \cap \mathbf{G}^\vee$.

Then (φ, ϱ) is called an *enhanced Langlands parameter for* ${}^LL$.

The group $\mathbf{L}^\vee$ acts on the collection of enhanced L-parameters for ${}^LL$ by (19) and we denote the set of $\mathbf{L}^\vee$-orbits of enhanced Langlands parameters for ${}^LL$ by $\Phi_{\mathrm{e}}({}^LL)$.

Definition 6.6 We say that an enhanced L-parameter (φ, ϱ) for ${}^LL$ is *cuspidal* if φ is discrete for ${}^LL$ and $\left(u_\varphi := \varphi\left(1, \left(\begin{smallmatrix}1&1\\0&1\end{smallmatrix}\right)\right), \varrho\right)$ is a cuspidal pair for $\mathrm{Z}^1_{\mathbf{L}^\vee_{\mathrm{sc}}}(\varphi|_{W_F})$.

Notation 6.7 If an enhanced Langlands parameter (φ, ϱ) for ${}^LL$ is cuspidal, then any element in its $\mathbf{L}^\vee$-orbit is cuspidal. We denote the subset of $\mathbf{L}^\vee$-orbits of cuspidal enhanced Langlands parameters for ${}^LL$ by $\Phi_{\mathrm{e,cusp}}({}^LL)$.

Let $\mathbf{L}^\vee$ be a Levi subgroup of $\mathbf{G}^\vee$, and let $\mathbf{L}^\vee_{\mathrm{c}}$ denote the pre-image of $L^\vee$ under $\mathbf{G}^\vee_{\mathrm{sc}} \to \mathbf{G}^\vee$. The derived group of $\mathbf{L}^\vee_{\mathrm{c}}$ is the simply connected cover of $\mathbf{L}^\vee_{\mathrm{der}}$, and we identify $\mathbf{L}^\vee_{\mathrm{sc}}$ with the inverse image of $\mathbf{L}^\vee_{\mathrm{der}}$ under $\mathbf{G}^\vee_{\mathrm{sc}} \to \mathbf{G}^\vee$.

Definition 6.8 A cuspidal datum for ${}^LG$ is a triple $({}^LL, \varphi, \varepsilon)$ where ${}^LL$ is a Levi L-subgroup of ${}^LG$, such that (φ, ε) is a cuspidal enhanced Langlands parameter for ${}^LL$. It is relevant for G if

- $\varepsilon = \zeta$ on $\mathbf{L}^\vee_{\mathrm{sc}} \cap \mathrm{Z}(\mathbf{G}^\vee_{\mathrm{sc}})^{W_F}$, where $\zeta \in \mathrm{Irr}(\mathrm{Z}(\mathbf{G}^\vee_{\mathrm{sc}})^{W_F})$ parametrizes the inner twist G of G^* via the Kottwitz isomorphism (9):
- $\varepsilon = 1$ on $\mathbf{L}^\vee_{\mathrm{sc}} \cap \mathrm{Z}(\mathbf{L}^\vee_{\mathrm{c}})^\circ$.

Recall from [33, § 3.3.1] that the group of unramified characters of a Levi subgroup L of G is naturally isomorphic to $\mathrm{Z}(\mathbf{L}^\vee \rtimes I_F)^\circ_{W_F}$. We consider this as an object on the Galois side of the local Langlands correspondence and we write

$$X_{\mathrm{nr}}({}^LL) := (\mathrm{Z}_{\mathbf{L}^\vee \rtimes I_F})^\circ_{W_F}. \tag{45}$$

Given $(\varphi, \varrho) \in \Phi_e({}^L L)$, and $\xi \in X_{nr}({}^L L)$, we define $(\xi \cdot \varphi, \varrho) \in \Phi_e({}^L L)$ by $\xi \cdot \varphi := \varphi$ on $I_F \times \mathrm{SL}_2(\mathbb{C})$ and $(\xi \cdot \varphi)(\mathrm{Fr}_F) := \tilde{\xi}\varphi(\mathrm{Fr}_F)$, where $\tilde{\xi} \in \mathrm{Z}^{\circ}_{L^\vee \rtimes I_F}$ represents ξ.

Definition 6.9 We denote by $\mathfrak{s}^\vee$ the $\mathbf{G}^\vee$-conjugacy class of $({}^L L, X_{nr}({}^L L) \cdot (\varphi, \varepsilon))$, where L is a Levi subgroup of G, and (φ, ε) is a cuspidal enhanced Langlands parameter for ${}^L L$. We write

$$\mathfrak{s}^\vee := \mathfrak{s}^\vee_G = [{}^L L, (\varphi, \varepsilon)]_{G^\vee} \tag{46}$$

and we call $\mathfrak{s}^\vee$ an *inertial class* for $\Phi_e({}^L G)$. We denote by $\mathfrak{B}^\vee({}^L G)$ the set of such $\mathfrak{s}^\vee$.

6.2 *The Cuspidal Support Map for Enhanced Langlands Parameters*

Let $(\phi, \rho) \in \Phi({}^L G)$. We apply the constructions of Sect. 4.2 with $\mathcal{G}$ the group $\mathcal{G}_\phi$ defined in (18). We set

$$[\mathcal{M}_\phi, (\mathcal{O}_0, \varepsilon)]_{\mathcal{G}_\phi} := \Psi_{\mathcal{G}_\phi}(u_\phi, \rho) \in \mathfrak{B}(\mathrm{U_e}(\mathcal{G}_\phi)), \tag{47}$$

where the map $\Psi_{\mathcal{G}_\phi}$ is that of Theorem 4.19. Hence $\mathcal{M}_\phi$ is a quasi-Levi subgroup of $\mathcal{G}_\phi$ and $(\mathcal{O}_0, \varepsilon)$ is a cuspidal enhanced unipotent class in $\mathcal{M}_\phi$. For simplicity, we will often refer to the unipotent class $\mathcal{O}_0$ in $\mathcal{M}_\phi$ by a unipotent element v in it.

Let $\mathrm{Z}(\mathcal{M}_\phi)$ denote the center of $\mathcal{M}_\phi$. The torus $\mathrm{Z}(\mathcal{M}_\phi)^\circ$ commutes with $\mathcal{M}_\phi$, so $\mathrm{Z}_{\mathbf{G}^\vee}(\mathrm{Z}(\mathcal{M}_\phi)^\circ)$ is a Levi subgroup of $\mathbf{G}^\vee$ which contains the image of $\mathcal{M}_\phi$ in $\mathbf{G}^\vee$. As $\mathrm{Z}(\mathcal{M}_\phi)^\circ \subset \mathrm{Z}^1_{\mathbf{G}^\vee_{\mathrm{sc}}}(\phi|_{W_F})$, the subgroup $\mathrm{Z}_{\mathbf{G}^\vee \rtimes W_F}(\mathrm{Z}(\mathcal{M}_\phi)^\circ)$ is a Levi L-subgroup of $\mathbf{G}^\vee \rtimes W_F$. As proved in [15, Proposition 7.3], upon replacing (ϕ, ε) by a $\mathbf{G}^\vee$-conjugate, there exists a Levi subgroup L of G such that $(\phi_{|W_F}, v, \varepsilon)$ is a cuspidal enhanced Langlands parameter for L, and

$${}^L L := \mathbf{L}^\vee \rtimes W_F = \mathrm{Z}_{{}^L G}(\mathrm{Z}^\circ_{\mathcal{M}_\phi}). \tag{48}$$

Then $({}^L L, \phi|_{W_F}, v, \varepsilon)$ is a G-relevant cuspidal datum for ${}^L G$.

Let ${}^L\Psi\colon \Phi_e({}^L G) \to \mathfrak{B}^\vee({}^L G)$ be the map defined by

$${}^L\Psi(\phi, \rho) := [{}^L L, (\phi_{|W_F}, v, \varepsilon)]_{\mathbf{G}^\vee}. \tag{49}$$

Let (ϕ, ρ) be an enhanced Langlands parameter for G and write $[{}^L L, (\phi_{|W_F}, v, \varepsilon)]_{\mathbf{G}^\vee} = {}^L\Psi(\phi, \rho)$. Up to $\mathcal{G}_\phi$-conjugacy there exists a unique algebraic group homomorphism $\gamma_v\colon \mathrm{SL}_2(\mathbb{C}) \to \mathcal{M}^\circ_\phi$ such that

- $\gamma_v \begin{pmatrix} 1 & 1 \\ 0 & 1 \end{pmatrix} = v$,
- $\gamma_v(\mathrm{SL}_2(\mathbb{C}))$ commutes with $\phi\big(1, \begin{pmatrix} z & 0 \\ 0 & z^{-1} \end{pmatrix}\big)\gamma_v \begin{pmatrix} z^{-1} & 0 \\ 0 & z \end{pmatrix}$ for all $z \in \mathbb{C}^\times$,

and the cocharacter

$$\chi_{\phi,v} : z \mapsto \phi\big(1, \begin{pmatrix} z & 0 \\ 0 & z^{-1} \end{pmatrix}\big) \cdot \gamma_v \begin{pmatrix} z^{-1} & 0 \\ 0 & z \end{pmatrix}$$

has image in $\mathrm{Z}(\mathcal{M}_\phi)^\circ$, see [15, lemma 7.12].

Let $\|\cdot\| : W_F \to \mathbb{R}_{>0}$ be the group homomorphism with $\|w\| = q$ if $w(f) = f^q$ for all f in the algebraic closure of the residue field of F.

We define a Langlands parameter $\phi'_v : W'_F \to \mathrm{Z}_{\mathbf{G}^\vee \rtimes W_F}(\mathrm{Z}(\mathcal{M}_\phi)^\circ)$ by

$$\phi'_v(w, x) := \phi(w) \cdot \chi_{\phi,v}(\|w\|^{1/2}) \cdot \gamma_v(x), \quad \text{where } w \in W_F \text{ and } x \in \mathrm{SL}_2(\mathbb{C}).$$

Let ^{L} be as in (48). Then $({}^L L, (\phi'_v|_{W_F}, v, \varepsilon))$ is also G-relevant cuspidal datum for ${}^L G$.

Definition 6.10 The *cuspidal support* of (ϕ, ρ) is

$$\mathbf{Sc}(\phi, \rho) := [{}^L L, \phi'_v|_{W_F}, v, \varepsilon]_{\mathbf{G}^\vee} \in \mathfrak{B}^\vee({}^L G)$$

and the map $\mathbf{Sc} : \Phi_e({}^L G) \to \mathfrak{B}^\vee({}^L G)$ is called the *cuspidal support map* for $\Phi_e({}^L G)$.

Given $\mathfrak{s}^\vee \in \mathfrak{B}^\vee({}^L G)$, we define $\Phi_e({}^L G)^{\mathfrak{s}^\vee}$ to the fiber of $\mathfrak{s}^\vee$ under the map $\mathbf{Sc}$. We proved in [15] that $\Phi_e({}^L G)$ decomposes as

$$\Phi_e({}^L G) = \bigsqcup_{\mathfrak{s}^\vee \in \mathfrak{B}^\vee({}^L G)} \Phi_e({}^L G)^{\mathfrak{s}^\vee}. \tag{50}$$

In this sense, we may consider $\Phi_e({}^L G)^{\mathfrak{s}^\vee}$ as *Bernstein series* in the space of enhanced Langlands parameters for ${}^L G$. Moreover, for a fixed $\mathfrak{s} \in \mathfrak{B}^\vee({}^L G)$, the series $\Phi_e({}^L G)^{\mathfrak{s}^\vee}$ is in bijection with the simple modules of an extended affine Hecke algebra $\mathcal{H}(\mathfrak{s}^\vee, \mathbf{z})$ (see [16, 17]).

6.3 A Generalized Springer Correspondence for Enhanced L-Parameters

Let $\mathfrak{s}^\vee = [{}^L L, (\varphi, \varepsilon)]_{\mathbf{G}^\vee} \in \mathfrak{B}^\vee({}^L G)$. We associate to $\mathfrak{s}^\vee$ the finite group

$$W_{\mathfrak{s}^\vee} := \text{ stabilizer of } \mathfrak{s}^\vee_{\mathbf{L}^\vee} \text{ in } \mathrm{N}_{\mathbf{G}^\vee}({}^L L)/\mathbf{L}^\vee. \tag{51}$$

Let $(\phi, \rho) \in \Phi_e({}^L G)^{\mathfrak{s}^\vee}$ and write as above $\Psi_{\mathcal{G}_\phi}(u_\phi, \rho) =: [\mathcal{M}_\phi, v, \varepsilon]_{\mathcal{G}_\phi}$. Fix a G-relevant datum cuspidal datum $({}^L L, \varphi, \varepsilon)$ and set $v := u_\varphi \in {}^L L$.

We write

$$\mathfrak{t}_\phi := [\mathcal{G}_\phi \cap \mathbf{L}_c^\vee, v, \varepsilon]_{\mathcal{G}_\phi}, \quad \mathfrak{t}_\phi^\circ := [\mathcal{G}_\phi^\circ \cap \mathbf{L}_c^\vee, v, \varepsilon_0]_{\mathcal{G}_\phi^\circ}. \tag{52}$$

The next result may be viewed as a version of the generalized Springer correspondence for enhanced L-parameters instead of enhanced unipotent classes.

Proposition 6.11 *[15, Proposition 9.1]*

(a) There is a bijection

$$\begin{array}{cccc} {}^L\Sigma_{\mathfrak{t}_\phi} : {}^L\Psi^{-1}({}^LL, \phi|_{W_F}, v, \varepsilon) & \longleftrightarrow & \mathrm{Irr}(\mathbb{C}[W_{\mathfrak{t}_\phi}, \kappa_{\mathfrak{t}_\phi}]) \\ (\phi, \rho) & \mapsto & \Sigma_{\mathfrak{t}_\phi}(u_\phi, \rho) \\ (\phi|_{W_F}, \Sigma_{\mathfrak{t}_\phi}^{-1}(\tau)) & \leftarrow\!\shortmid & \tau, \end{array}$$

where $\Sigma_{\mathfrak{t}_\phi}$ is the bijection defined in Notation 4.20.

(b) The canonical bijection $\Sigma_{\mathfrak{t}_\phi^\circ}$ between $\Psi_{\mathcal{G}^\circ}^{-1}(\mathfrak{t}_\phi^\circ) \subset \mathrm{U_e}(\mathcal{G}^\circ)$ and $\mathrm{Irr}_{\mathbb{C}}(W_{\mathfrak{t}^\circ})$ relates to part (a) by

$${}^L\Sigma_{\mathfrak{t}_\phi}(\phi, \rho)|_{W_{\mathfrak{t}_\phi^\circ}} = \bigoplus_i \Sigma_{\mathfrak{t}_\phi^\circ}(u_\phi, \rho_i),$$

where $\rho = \bigoplus_i \rho_i$ is a decomposition into irreducible $A_{\mathcal{G}_\phi^\circ}(u_\phi)$-subrepresentations.

(c) The $\mathbf{G}^\vee$-conjugacy class of $(\phi|_{W_F}, u_\phi, \rho_i)$ is determined by any irreducible $\mathbb{C}[W_{\mathfrak{t}_\phi^\circ}]$-subrepresentation of ${}^L\Sigma_{\mathfrak{t}_\phi}(\phi, \rho)$.

7 A Galois Version of the ABPS Conjecture

7.1 Twisted Extended Quotients

Let Γ be a group acting on a topological space X, and let $\natural$ be a given function which assigns to each $x \in X$ a 2-cocycle

$$\natural_x : \Gamma_x \times \Gamma_x \to \mathbb{C}^\times, \text{ where } \Gamma_x = \{\gamma \in \Gamma \; : \; \gamma x = x\}.$$

We assume that $\natural_{\gamma x}$ and $\gamma_* \natural_x$ define the same class in $H^2(\Gamma_{\gamma x}, \mathbb{C}^\times)$, where $\gamma_* : \Gamma_x \to \Gamma_{\gamma x}$ sends α to $\gamma\alpha\gamma^{-1}$. Let $\mathbb{C}[\Gamma_x, \natural_x]$ be the corresponding twisted algebra as defined in Sect. 4.3. We set

$$\widetilde{X}_\natural := \{(x, E) \; : \; x \in X, E \in \mathrm{Irr}\, \mathbb{C}[\Gamma_x, \natural_x]\}.$$

and we topologize it by decreeing that a subset of $\widetilde{X}_\natural$ is open if and only if its projection to the first coordinate is open in X.

We require, for every $(\gamma, x) \in \Gamma \times X$, a definite algebra isomorphism

$$\phi_{\gamma,x} : \mathbb{C}[\Gamma_x, \natural_x] \to \mathbb{C}[\Gamma_{\gamma x}, \natural_{\gamma x}]$$

such that:

- if $\gamma x = x$, then $\phi_{\gamma,x}$ is conjugation by an element of $\mathbb{C}[\Gamma_x, \natural_x]^\times$;
- $\phi_{\gamma',\gamma x} \circ \phi_{\gamma,x} = \phi_{\gamma'\gamma,x}$ for all $\gamma', \gamma \in \Gamma, x \in X$.

Then we can define a Γ-action on $\widetilde{X}_\natural$ by

$$\gamma \cdot (x, E) := (\gamma x, E \circ \phi_{\gamma,x}^{-1}).$$

We form the *twisted extended quotient*

$$(X/\!/\Gamma)_\natural := \widetilde{X}_\natural / \Gamma.$$

Remark 7.1 Furthermore, we note that $(X/\!/\Gamma)_\natural$ reduces to the extended quotient of the second kind $(X/\!/\Gamma)_2$ from [11, § 2] if $\natural_x$ is trivial for all $x \in X$ and $\phi_{\gamma,x}$ is conjugation by γ.

The extended quotient of the second kind is an extension of the ordinary quotient in the sense that it keeps track of the duals of the isotropy groups. Namely, in $(X/\!/\Gamma)_2$ every point $x \in X/\Gamma$ has been replaced by the set $\mathrm{Irr}(\Gamma_x)$.

7.2 The Bernstein Decomposition of the Category of Smooth Representations

Let $\mathrm{Rep}(G)$ denote the category of smooth representations of G. For P a parabolic subgroup of G with Levi factor L, we write I_P^G for the functor of normalized parabolic induction from $\mathrm{Rep}(L)$ to $\mathrm{Rep}(G)$.

Notation 7.2 Let $\mathrm{Irr}_{\mathrm{cusp}}(L)$ be the set of isomorphism classes of irreducible supercuspidal representations in $\mathrm{Rep}(L)$.

Let $\sigma \in \mathrm{Irr}_{\mathrm{cusp}}(L)$. We call (L, σ) a cuspidal pair, and we consider such pairs up to *inertial equivalence*. This is the equivalence relation generated by:

- unramified twists, $(L, \sigma) \sim (L, \sigma \otimes \chi)$ for $\chi \in X_{\mathrm{nr}}(L)$, where $X_{\mathrm{nr}}(L)$ is the group of unramified (not necessarily unitary) characters $L \to \mathbb{C}^\times$;
- G-conjugation, $(L, \sigma) \sim (gLg^{-1}, g \cdot \sigma)$ for $g \in G$.

We denote a typical inertial equivalence class by $\mathfrak{s} = [L, \sigma]_G$. In particular,

$$\mathfrak{s}_L := [L, \sigma]_L = \{\sigma \otimes \chi \in \mathrm{Irr}(L) \,:\, \chi \in X_{\mathrm{nr}}(L)\}.$$

Bernstein attached to every $\mathfrak{s}$ a block in the category $\mathrm{Rep}(G)$ in the following way. First, if π is an irreducible object in $\mathrm{Rep}(G)$, then there is a cuspidal pair (L, σ) such that π is a composition factor of $\mathrm{I}_P^G(\sigma)$, for any parabolic subgroup P with Levi factor L. The conjugacy class of the pair (L, σ) is uniquely determined by π. It is called the *supercuspidal support* of π. Thus the corresponding inertial class $[L, \sigma]_G$ is also uniquely determined by π. It is the *inertial support* of π. We now define

$$\begin{aligned}\mathrm{Irr}(G)^{\mathfrak{s}} &= \{\pi \in \mathrm{Irr}(G) \ : \ \pi \text{ has inertial support } \mathfrak{s}\},\\ \mathrm{Rep}(G)^{\mathfrak{s}} &= \{\pi \in \mathrm{Rep}(G) \ : \ \text{every irreducible subquotient of } \pi \text{ belongs to } \mathrm{Irr}(G)^{\mathfrak{s}}\}.\end{aligned}$$

We denote the set of all inertial equivalence classes for G by $\mathfrak{B}(G)$.

Theorem 7.3 [18, Proposition 2.10]
The category $\mathrm{Rep}(G)$ *decomposes as*

$$\mathrm{Rep}(G) = \prod_{\mathfrak{s} \in \mathfrak{B}(G)} \mathrm{Rep}(G)^{\mathfrak{s}}.$$

In particular, the space of irreducible G-representations decomposes as the disjoint union

$$\mathrm{Irr}(G) = \bigsqcup_{\mathfrak{s} \in \mathfrak{B}(G)} \mathrm{Irr}(G)^{\mathfrak{s}}.$$

The group

$$X_{\mathrm{nr}}(L, \sigma) := \{\chi \in X_{\mathrm{nr}}(L) \ : \ \sigma \otimes \chi \cong \sigma\}$$

is finite. Thus there is a bijection

$$X_{\mathrm{nr}}(L)/X_{\mathrm{nr}}(L, \sigma) \to \mathrm{Irr}(L)^{\mathfrak{s}_L} \ : \ \chi \mapsto \sigma \otimes \chi, \tag{53}$$

which endows $\mathrm{Irr}(L)^{\mathfrak{s}_L}$ with the structure of a complex torus. Up to isomorphism, this torus depends only on $\mathfrak{s}$, and it is known as the *Bernstein torus* $T_{\mathfrak{s}}$ attached to $\mathfrak{s}$. We note that $T_{\mathfrak{s}}$ is only an algebraic variety, it is not endowed with a natural multiplication map. In fact, it does not even possess an unambigous "unit", because in general there is no preferred choice of an element $\sigma \in \mathfrak{s}_L$.

The group $W(G, L) := \mathrm{N}_G(L)/L$ acts on $\mathrm{Irr}(L)$ by

$$w \cdot \pi = [\bar{w} \cdot \pi \ : \ l \mapsto \pi(\bar{w}^{-1} l \bar{w})] \quad \text{for any lift } \bar{w} \in \mathrm{N}_G(L) \text{ of } w \in W(G, L). \tag{54}$$

Bernstein also associated with $\mathfrak{s}$ the finite group

$$W_{\mathfrak{s}} := \{w \in W(G, L) \ : \ w \cdot \mathrm{Irr}(L)^{\mathfrak{s}_L} = \mathrm{Irr}(L)^{\mathfrak{s}_L}\}. \tag{55}$$

It acts naturally on $T_{\mathfrak{s}}$, by automorphisms of algebraic varieties.

Closely related to the Bernstein decomposition is the theory of the Bernstein center. By [18, Théorème 2.13], the categorical centre of the Bernstein block $\mathrm{Rep}^{\mathfrak{s}}(G)$ is

$$\mathrm{Z}(\mathrm{Rep}(G)^{\mathfrak{s}}) \cong \mathcal{O}(T_{\mathfrak{s}})^{W_{\mathfrak{s}}} = \mathcal{O}(T_{\mathfrak{s}}/W_{\mathfrak{s}}), \tag{56}$$

where $\mathcal{O}$ denotes the ring of the regular functions on the given affine variety. Then taking supercuspidal support gives a map

$$\mathbf{sc}\colon \mathrm{Irr}(\mathbf{G})^{\mathfrak{s}} \to \mathbf{T}_{\mathfrak{s}}/\mathbf{W}_{\mathfrak{s}}, \tag{57}$$

which is surjective and has finite fibers [18, § 3]. Via the map **sc** (called the cuspidal support map), $\mathrm{Irr}^{\mathfrak{s}}(G)$ can be regarded as a non-separated algebraic variety lying over $T_{\mathfrak{s}}/W_{\mathfrak{s}}$.

7.3 The ABPS Conjecture

Let $\mathfrak{s} = [L, \sigma]_G \in \mathfrak{B}(G)$. Let $W_{\mathfrak{s},t}$ be the stabilizer in $W_{\mathfrak{s}}$ of a point $t \in T_{\mathfrak{s}}$.

The ABPS Conjecture from [7, § 15] and [12, Conjecture 2] in its roughest form asserts that there exists a family of 2-cocycles

$$\natural_t\colon W_{\mathfrak{s},t} \times W_{\mathfrak{s},t} \to \mathbb{C}^\times \qquad t \in T_{\mathfrak{s}},$$

and a bijection

$$\mathrm{Irr}(G)^{\mathfrak{s}} \longleftrightarrow (T_{\mathfrak{s}}/\!/W_{\mathfrak{s}})_\natural \tag{58}$$

such that:

- it restricts to a bijection between tempered representations and the unitary part of the extended quotient (as explained below);
- it is canonical up to permutations within L-packets, that is, for any $\phi \in \Phi(G)$, the image of $\Pi_\phi(G) \cap \mathrm{Irr}^{\mathfrak{s}}(G)$ is canonically defined (assuming a LLC for G exists).

The set $\mathrm{Irr}_{\mathrm{cusp}}(L)$ is stable under the $W(G, L)$-action (54). The definitions of $W_{\mathfrak{s}}$ and of extended quotients imply that for a fixed Levi subgroup L of G there is a natural bijection

$$\bigsqcup_{\mathfrak{s}=[L,\sigma]_G} (T_{\mathfrak{s}}/\!/W_{\mathfrak{s}})_\natural \to \left(\mathrm{Irr}_{\mathrm{cusp}}(L)/\!/W(G, L)\right)_\natural. \tag{59}$$

Let $\mathfrak{Lev}(G)$ be a set of representatives for the conjugacy classes of Levi subgroups of G. In view of Theorem 7.3, the ABPS Conjecture can also be formulated in terms of a bijection

$$\mathrm{Irr}(G) \longleftrightarrow \bigsqcup_{L\in\mathfrak{Lev}(G)} \left(\mathrm{Irr}_{\mathrm{cusp}}(L)/\!/W(G, L)\right)_\natural. \tag{60}$$

In this version, the conjecture asserts that $\mathrm{Irr}(G)$ is determined by a much smaller set of data, namely the supercuspidal representations of Levi subgroups L of G, and the actions of the Weyl groups $W(G, L)$ on those.

It is expected that the group cohomology classes $\natural_t \in H^2(W_{\mathfrak{s},t}, \mathbb{C}^\times)$ reflect the character of $Z(G^\vee_{\mathrm{sc}})^{W_F}$ which via the Kottwitz isomorphism determines how G is an inner twist of a quasi-split group. In particular $\natural$ should be trivial whenever G is quasi-split. The simplest known example of a nontrivial cocycle involves a non-split inner form of $\mathrm{SL}_{10}(F)$ [10, Example 5.5]. That example also shows that it is sometimes necessary to use *twisted* extended quotients in the ABPS Conjecture (see [13]).

7.4 A Version for Enhanced L-Parameters

The main goal of this section is to state an analog of (58) and (59) for enhanced Langlands parameters.

Let $L \in \mathfrak{Lev}(G)$. We write

$$\Phi_{\mathrm{e},\zeta_G}(G, L) := \{(\phi, \rho) \in \Phi_{\mathrm{e},\zeta_G}(G) \ : \ \mathbf{Sc}(\phi, \rho) \in \Phi_{\mathrm{cusp}}(L)\},$$

where $\Phi_{\mathrm{e},\zeta_G}(G)$ was defined in (23).

Let $\varphi\colon W_F' \to {}^L L$ be a Langlands parameter for L. Let $\mathcal{R}^L_\varphi$ the analog for L of the group defined in (12), that is, $\mathcal{R}^L_\varphi := \pi_0(Z_{\mathbf{L}^\vee}(\varphi)/Z(\mathbf{L}^\vee)^{W_F})$. There exists a natural injection $\mathcal{R}^L_\varphi \to \mathcal{R}_\varphi$. We extend the character ζ_G to a character of $Z(\mathbf{G}^\vee_{\mathrm{sc}})Z(L^\vee_{\mathrm{c}})^\circ$ which is trivial on $Z(L^\vee_{\mathrm{c}})^\circ$, and denote by ζ^L_G the restriction of the latter character to $Z(\mathbf{L}^\vee_{\mathrm{sc}})$. Then we write

$$\Phi_{\mathrm{cusp},\zeta^L_G}(L) := \Phi_{\mathrm{e},\zeta^L_G}(L) \cap \Phi_{\mathrm{cusp}}(L).$$

The following results are proved in [53, Theorem 3.3] for G a split classical group, and in [15, Theorem 9.3] for general G.

Theorem 7.4 [15, Theorem 9.3]

(a) Let $\mathfrak{s}^\vee_L = [{}^L L, \phi|_{W_F}, v, \varepsilon]_{{}^L L}$ be n G-relevant inertial equivalence class for the Levi L-subgroup ${}^L L$ of ${}^L G$. The maps ${}^L\Sigma_{\mathfrak{t}}$ from Proposition 6.11.(a) combine to a bijection

$$\begin{array}{ccc} \Phi_{\mathrm{e}}({}^L G)^{\mathfrak{s}^\vee} & \longleftrightarrow & \left(\Phi_{\mathrm{e}}({}^L L)^{\mathfrak{s}^\vee_L} /\!/ W_{\mathfrak{s}^\vee}\right)_\kappa \\ (\phi, \rho) & \mapsto & \left({}^L\Psi(\phi, \rho), \Sigma_{\mathfrak{t}}(u_\phi, \rho)\right) \\ \left(\phi_v|_{W_F}, \Sigma_{\mathfrak{t}}^{-1}(\tau)\right) & \leftarrow & ({}^L L, \phi_v, \varepsilon, \tau). \end{array}$$

(b) The bijection from part (a) has notably the following properties:

- *It preserves boundedness of (enhanced) L-parameters.*
- *The restriction of τ to $W_{\mathfrak{t}^\circ}$ canonically determines the (non-enhanced) L-parameter in ${}^L\Sigma_{q\mathfrak{t}}(\tau)$.*

(c) The bijections from part (a) give a bijection

$$\Phi_{e,\zeta_G}(G,L) \longleftrightarrow \left(\Phi_{\mathrm{cusp},\zeta_G^L}(L)//W(G,L)\right)_\kappa.$$

(d) The maps from part (c) combine to a bijection

$$\Phi_{e,\zeta_G}(G) \longleftrightarrow \bigsqcup_{L\in\mathfrak{Lev}(G)} \left(\Phi_{\mathrm{cusp},\zeta_G^L}(L)//W(G,L)\right)_\kappa.$$

(e) Assume that $\mathrm{Z}(\mathbf{L}_{\mathrm{sc}}^\vee)$ *is fixed by* W_F *for every* $L\in\mathfrak{Lev}(G)$ *(for instance, G is an inner twist of a split group). The union of part (d) for all such* $\vartheta\in H^1(F,G_{\mathrm{ad}})\cong \mathrm{Irr}_{\mathbb{C}}\left(\mathrm{Z}(\mathbf{G}_{\mathrm{sc}}^\vee)^{W_F}\right)$ *is a bijection*

$$\Phi_e({}^LG) \longleftrightarrow \bigsqcup_{\vartheta\in H^1(F,G_{\mathrm{ad}})}\ \bigsqcup_{L_\vartheta\in\mathfrak{Lev}(G_\vartheta)} \left(\Phi_{\mathrm{cusp}}(L_\vartheta)//W(G_\vartheta,L_\vartheta)\right)_\kappa.$$

7.5 A Conjectural Diagram

The above result leads to the following conjecture stated in [15]:

Conjecture 7.5 *There is a commutative diagram of bijections*

$$\begin{array}{ccc} \mathrm{Irr}(G) & \longleftrightarrow & \Phi_{e,\zeta_G}(G) \\ \updownarrow & & \updownarrow \\ \bigsqcup_{L\in\mathfrak{Lev}(G)}\left(\mathrm{Irr}_{\mathrm{cusp}}(L)//W(G,L)\right)_\kappa & \longleftrightarrow & \bigsqcup_{L\in\mathfrak{Lev}(G)}\left(\Phi_{\mathrm{cusp},\zeta_G}(L)//W(G,L)\right)_\kappa \end{array}$$

in which the arrows are given as follows:

- *the right-hand side is Theorem 7.4.(d),*
- *the upper horizontal map is a local Langlands correspondence for G,*
- *the lower horizontal map is obtained from local Langlands correspondences for* $\mathrm{Irr}_{\mathrm{cusp}}(L)$ *by applying* $(\cdot//W(G,L))_\kappa$*,*
- *the left-hand side is the conjectural bijection* (60).

With this conjecture, one can reduce the problem of finding a LLC for G to that of finding local Langlands correspondences for supercuspidal representations of its Levi subgroups. Conjecture 7.5 is currently known in the following cases:

- inner forms of $\mathrm{GL}_n(F)$ [9, Theorem 5.3],
- inner forms of $\mathrm{SL}_n(F)$ [9, Theorem 5.6],
- split classical groups [52, § 5.3], when the characteristic of F is 0,
- principal series representations of split groups [11, § 16].

References

1. P. Achar, A. Henderson, D. Juteau, S. Riche, Modular generalized Springer correspondence I: the general linear group. J. Eur. Math. Soc. (JEMS) **18**(7), 1405–1436 (2016)
2. J. Arthur, A note on L-packets. Pure Appl. Math. Q. **2**(1), 199–217 (2006)
3. J. Arthur, *The Endoscopic Classification of Representations: Orthogonal and Symplectic Groups*, Colloquium Publications, vol. 61 (American Mathematical Society, 2013)
4. M. Asgari, K. Choiy, The local Langlands conjecture for p-adic GSpin_4, GSpin_6, and their inner forms. Forum Math. **29**(6), 1261–1290 (2017)
5. A.-M. Aubert, Around the Langlands program. Jahresber. Dtsch. Math.-Ver. **120**(1), 3–40 (2018)
6. A.-M. Aubert, P.F. Baum, R.J. Plymen, M. Solleveld, On the local Langlands correspondence for non-tempered representations. Münster J. Math. **7**(1), 27–50 (2014)
7. A.-M. Aubert, P.F. Baum, R.J. Plymen, M. Solleveld, Geometric structure in smooth dual and local Langlands correspondence. Japan. J. Math. **9**, 99–136 (2014)
8. A.-M. Aubert, P.F. Baum, R.J. Plymen, M. Solleveld, Depth and the local Langlands correspondence, in Arbeitstagung Bonn 2013, in *Memory of Friedrich Hirzebruch* W. Ballmann, C. Blohmann, G. Faltings, P. Teichner, et D. Zagier, Progr. Math. vol. 319 (Birkhäuser/Springer, Berlin, 2016), pp. 17–41
9. A.-M. Aubert, P.F. Baum, R.J. Plymen, M. Solleveld, The local Langlands correspondence for inner forms of SL_n. Res. Math. Sci. **3** (2016), Paper No. 32, 34 p
10. A.-M. Aubert, P.F. Baum, R.J. Plymen, M. Solleveld, Hecke algebras for inner forms of p-adic special linear groups. J. Inst. Math. Jussieu **16**(2), 351–419 (2017)
11. A.-M. Aubert, P.F. Baum, R.J. Plymen, M. Solleveld, The principal series of p-adic groups with disconnected centre. Proc. Lond. Math. Soc. (3) **114**(5), 798–854 (2017)
12. A.-M. Aubert, P.F. Baum, R.J. Plymen, M. Solleveld, Conjectures about p-adic groups and their noncommutative geometry, in *Around Langlands Correspondences*, Contemp. Math. vol. 691 (Amer. Math. Soc., Providence, RI, 2017), pp. 15–51
13. A.-M. Aubert, P.F. Baum, R.J. Plymen, M. Solleveld, The noncommutative geometry of inner forms of p-adic special linear groups. arXiv:1505.04361 [math.RT]
14. A.-M. Aubert, S. Mendes, R. Plymen, M. Solleveld, L-packets and depth for $\mathrm{SL}_2(K)$ with K a local function field of characteristic 2. Int. J. Number Theory **13**(10), 2545–2568 (2017)
15. A.-M. Aubert, A. Moussaoui, M. Solleveld, Generalizations of the Springer correspondence and cuspidal Langlands parameters. Manuscripta Math. **1–72** (2018). https://doi.org/10.1007/s00229-018-1001-8
16. A.-M. Aubert, A. Moussaoui, M. Solleveld, Graded Hecke algebras for disconnected reductive groups, to appear in the *Proceedings of the Simons Symposium 2017, Geometric Aspects of the Trace Formula*
17. A.-M. Aubert, A. Moussaoui, M. Solleveld, Affine Hecke algebras for Langlands parameters. arXiv:1701.03593
18. J. Bernstein, P. Deligne, Le "centre" de Bernstein, in *Représentations des groupes réductifs sur un corps local* (Travaux en cours, Hermann, 1984), pp. 1–32
19. J. Bernstein, V. Lunts, *Equivariant Sheaves and Functors*. Lecture Notes in Mathematics, vol. 1578 (Springer, Berlin, 1994)
20. J. Bernstein, A. Zelevinsky, representations of the group $\mathrm{GL}(n, F)$ where F is a non-Archimedean local field. Russian Math. Surveys **31**(3), 1–68 (1976)
21. A. Borel, Automorphic L-functions. Proc. Symp. Pure Math **33**(2), 27–61 (1979)
22. C. Bushnell, G. Henniart, The essentially tame local Langlands correspondence, III: the general case. Proc. Lond. Math. Soc. (3) **101**(2), 497–553 (2010)
23. K. Choiy, The local Langlands conjecture for the p-adic inner form of Sp_4. Int. Math. Res. Not. IMRN **6**, 1830–1889 (2017)
24. C.W. Curtis, I. Reiner, *Representation Theory of Finite Groups and Associative Algebras*, Pure and Applied Mathematics, vol 11 (Wiley, New York, 1962)

25. S. DeBacker, M. Reeder, Depth-zero supercuspidal L-packets and their stability. Ann. of Math. (2) **169**(3), 795–901 (2009)
26. Y. Feng, E. Opdam, M. Solleveld, Supercuspidal unipotent representations: L-packets and formal degrees. arXiv:1805.01888
27. W.T. Gan, S. Takeda, The local Langlands conjecture for GSp(4). Ann. Math. (2) **173**(3), 1841–1882 (2011)
28. W.T. Gan, S. Takeda, The local Langlands conjecture for Sp(4). Int. Math. Res. Not. IMRN **15**, 2987–3038 (2010)
29. W.T. Gan, W. Tantono, The local Langlands conjecture for GSp(4), II: the case of inner forms. Am. J. Math. **136**(3), 761–805 (2014)
30. R. Ganapathy, The local Langlands correspondence for GSp_4 over local function fields. Amer. J. Math. **137**(6), 1441–1534 (2015)
31. R. Ganapathy, S. Varma, On the local Langlands correspondence for split classical groups over local function fields. J. Inst. Math. Jussieu **16**(5), 987–1074 (2017)
32. R. Godement, H. Jacquet, *Zeta Functions of Simple Algebras*. Lecture Notes in Mathematics, vol. 260 (Springer, Berlin, 1972)
33. T.J. Haines, The stable Bernstein center and test functions for Shimura varieties, in *Automorphic Forms and Galois Representations*, London Mathematical Society Lecture Note Series, vol. 415 (Cambridge University Press, Cambridge, 2014), pp. 118–186
34. M. Harris, R. Taylor, *The Geometry and Cohomology of Some Simple Shimura Varieties*, Annals of Math. Studies, vol. 151 (Princeton University Press, Princeton, 2001)
35. G. Henniart, Une preuve simple des conjectures de Langlands pour $GL(n)$ sur un corps p-adique. Inventiones Mathematicae **139**, 439–455 (2000)
36. K. Hiraga, H. Saito, On L-packets for inner forms of SL_n. Mem. Amer. Math. Soc. **1013**, vol. 215 (2012)
37. D. Jiang, C. Nien, On the local Langlands conjecture and related problems over p-adic local fields, in *Proceedings of the Sixth International Congress of Chinese Mathematicians* vol I, 309–325, Adv. Lect. Math. (ALM), vol. 36 (Int. Press, Somerville, MA, 2017)
38. D. Jiang, C. Nien, The local Langlands conjectures for non-quasi-split groups, in *Families of Automorphic Forms and the Trace Formula* (Springer, Berlin, 2016), pp. 217–257 (Simons Symp.)
39. T. Kaletha, Global rigid inner forms and multiplicities of discrete automorphic representations. Invent. Math. **213**(1), 271–369 (2018)
40. T. Kaletha, A. Minguez, S.W. Shin, P.-J. White, Endoscopic classification of representations: inner forms of unitary groups. arXiv:1409.3731
41. R. Kottwitz, Stable trace formula: cuspidal tempered terms. Duke Math. J. **51**(3), 611–650 (1984)
42. G. Laumon, M. Rapoport, U. Stuhler, $\mathcal{D}$-elliptic sheaves and the Langlands correspondence. Invent. Math. **113**, 217–238 (1993)
43. L. Lomeli, Langlands program and Ramanujan conjecture: a survey (Preprint, 2018)
44. G. Lusztig, Some examples of square-integrable representations of semisimple p-adic groups. Trans. Am. Math. Soc. **277**, 623–653 (1983)
45. G. Lusztig, Intersection cohomology complexes on a reductive group. Invent. Math. **75**(2), 205–272 (1984)
46. G. Lusztig, Fourier transforms on a semisimple Lie algebra over $\mathbb{F}_q$, in *Algebraic Groups*
47. G. Lusztig, Classification of unipotent representations of simple p-adic groups. Int. Math. Res. Notices **11**, 517–589 (1995)
48. G. Lusztig, Classification of unipotent representations of simple p-adic groups. II. Represent. Theory **6**, 243–289 (2002)
49. G. Lusztig, Character sheaves on disconnected groups. V. Represent. Theory **8**, 346–376 (2004)
50. M. Mishra, B. Pattanayak, A note on depth preservation. To appear in J Ramanujan Math. Soc. arXiv:1903.10771
51. C.P. Mok, Endoscopic classification of representations of quasi-split unitary groups. Mem. Amer. Math. Soc. **235** (2015), no. 1108, vi+248 p

52. A. Moussaoui, Centre de Bernstein dual pour les groupes classiques. Represent. Theory **21**, 172–246 (2017)
53. A. Moussaoui, Proof of the Aubert-Baum-Plymen-Solleveld conjecture for split classical groups, in *Around Langlands correspondences*, Contemp. Math., vol. 691 (Amer. Math. Soc., Providence, RI, 2017), pp. 257–281
54. A. Moy, G. Prasad, Jacquet functors and unrefined minimal K-types. Comment. Math. Helv. **71**(1), 98–121 (1996)
55. M. Oi, Depth preserving property of the local Langlands correspondence for quasi-split classical groups in a large residual characteristic. arXiv:1804.10901
56. P. Scholze, The local Langlands correspondence for GL_n over p-adic fields. Invent. Math. **192**, 663–715 (2013)
57. J.-P. Serre, *Corps Locaux* (Hermann, Paris, 1962)
58. A. Silberger, W. Zink, Langlands classification for L-parameters. J. Algebra
59. T.A. Springer, *Linear Algebraic Groups* 2nd ed., Progress in Mathematics, vol. 9 (Birkhäuser, 1998)
60. J. Tate, Number theoretic background. Proc. Symp. Pure Math **33**(2), 3–26 (1979)
61. D. Vogan, The local Langlands conjecture, in *Representation Theory of Groups and Algebras*, Contemp. Math., vol. 145 (American Mathematical Society, 1993), pp. 305–379
62. B. Xu, On a lifting problem of L-packets. Compos. Math. **152**(9), 1800–1850 (2016)
63. J.-K. Yu, On the local Langlands correspondence for tori, in *Ottawa Lectures on Admissible Representations of Reductice p-adic Groups*, Fields Institute Monographs (American Mathemical Society, 2009), pp. 177–183
64. A.V. Zelevinsky, Induced representations of reductive p-adic groups II. On irreducible representations of $\mathrm{GL}(n)$, Ann. Sci. École Norm. Sup. (4) **13**(2), 165–210 (1980)

Arithmetic of Cuspidal Representations

Colin J. Bushnell

Contents

1 Basic Ideas and Examples ... 42
1.1 Intertwining and Induction ... 43
1.2 Level Zero Representations ... 44
1.3 Hereditary Orders ... 45
1.4 Duality and Strata ... 48
1.5 A Fundamental Dichotomy ... 49
2 Simple Characters ... 52
2.1 Adjoint and Tame Corestriction ... 53
2.2 Critical Exponent and Simple Strata ... 54
2.3 Structure of Simple Strata ... 55
2.4 Groups Defined by Simple Strata ... 57
2.5 Simple Characters ... 58
3 Types and Classification ... 61
3.1 Maximal Types ... 62
3.2 Inertial Uniqueness ... 66
4 Encounter with the Langlands Correspondence ... 67
4.1 The Langlands Correspondence ... 67
4.2 Inertial Parameters ... 69
4.3 Local Constants ... 70
5 Transfer, Endo-equivalence and Applications ... 73
5.1 Transfer of Simple Characters ... 74
5.2 Endo-Equivalence ... 76
5.3 Conductors of Pairs ... 80
6 Tame Lifting and Functoriality ... 82
6.1 Tame Lifting ... 83
6.2 Functoriality Relations ... 85
6.3 Ramification Theorem ... 86
7 Tame Parameters ... 89
7.1 Structure of Galois Representations ... 89
7.2 Tame Parameter Fields ... 92
7.3 Comparison of Tame Parameters ... 95

C. J. Bushnell (✉)
Department of Mathematics, King's College London, Strand, London WC2R 2LS, UK
e-mail: colin.bushnell@kcl.ac.uk

A.-M. Aubert et al. (eds.), *Representations of Reductive p-adic Groups*,
Progress in Mathematics 328, https://doi.org/10.1007/978-981-13-6628-4_2

8 Description of the Langlands Correspondence . . . 96
8.1 Complements . . . 96
8.2 Algebraic Induction Maps . . . 98
8.3 Naïve Correspondence . . . 100
8.4 Main Theorems . . . 103
9 Higher Ramification and the Herbrand Function . . . 105
9.1 Ramification and Conductors . . . 105
9.2 Comparison of Ultrametrics . . . 109
9.3 Lifting and Interpolation . . . 111
9.4 Some Preliminary Estimates . . . 113
10 Carayol Representations . . . 113
10.1 Symmetry . . . 114
10.2 Conformity and the Main Theorem . . . 116
10.3 Galois Representations of Carayol Type . . . 117
10.4 Parameter Fields . . . 119
References . . . 123

Abstract Here, F is a non-Archimedean local field of residual characteristic p. We are concerned with the irreducible, cuspidal representations of the general linear groups $\mathrm{GL}(n, F)$. A complete classification of these representations has been known for a long time. It is achieved using rather complicated objects, the simple types and simple characters. The methods it requires have been useful more widely, and the general scheme is now known to apply to many more groups, including $\mathrm{GL}(m, D)$ (where D is a central F-division algebra), orthogonal groups $\mathrm{SO}(n)$, symplectic groups $\mathrm{Sp}(2n)$, (both for p not equal to 2) and even a couple of exceptional groups. In some cases, it is known that the common classification conforms to the requirements of Functoriality. The most interesting, and presently the most difficult, instance of Functoriality is the basic connection between the irreducible cuspidal representations of $\mathrm{GL}(n, F)$ and the irreducible, n-dimensional representations of the Weil group of F. These notes describe the classification of the cuspidal representations, introducing the results and techniques currently necessary for making this connection more explicit, given that it is known to exist.

Let F be a non-Archimedean local field *of residual characteristic* p. Let $\mathcal{W}_F = \mathcal{W}_{\bar{F}/F}$ be the Weil group of F, formed relative to a chosen separable algebraic closure $\bar{F}/F$, and let $\widehat{\mathcal{W}}_F$ be the set of equivalence classes of irreducible representations[1] of $\mathcal{W}_F$. For an integer $n \geqslant 1$, let $\mathcal{A}_n^0(F)$ be the set of equivalence classes of irreducible *cuspidal* representations of the general linear group $\mathrm{GL}_n(F)$ and set $\widehat{\mathrm{GL}}_F = \bigcup_{n\geqslant 1} \mathcal{A}_n^0(F)$. The Langlands correspondence thus provides a canonical bijection

$$\begin{aligned} \widehat{\mathrm{GL}}_F &\longrightarrow \widehat{\mathcal{W}}_F, \\ \pi &\longmapsto {}^L\pi. \end{aligned}$$

[1]When speaking of a *representation,* I invariably mean a *smooth complex representation.*

Following [25], the representations $\pi \in \widehat{\mathrm{GL}}_F$ are classified in terms of explicit data. In these notes, I explore the manner in which this classification translates, via the Langlands correspondence, into information concerning the structure of representations of $\mathcal{W}_F$. Ideally, one would like to be able to read off ${}^L\pi$ directly from π and the other way round. I report on the progress made towards this goal in a long-running joint project with Guy Henniart.

The notes are effectively in two halves. The first, occupying Sects. 1–3, extracts the necessary background from [25]. The important part for us is the procedure for constructing cuspidal representations, particularly the definition and properties of *simple characters*. These are complicated objects, and one has to be able to manipulate them freely. On the other hand, the "Exhaustion Theorem", asserting that the construction yields *all* cuspidal representations, can be taken as read. The amount of detail provided is therefore heavily skewed to the construction side.

However, the strategy of proof followed in [25] dictates the shape of the construction, particularly the theory of simple characters. The basic idea is traceable, in various parts, to Howe [43], Kutzko [50, 51] and Carayol [30], along with [7], but the effective starting point is the combination of [6, 54]. That leads directly to an iterative problem of which simple characters, and then simple types, are the solution. I have summarized this train of thought in Sect. 1 and illustrated it with key examples.

The main material of the notes starts with Sect. 4. In the introductory first section, I list properties of the Langlands correspondence that can be interpreted in terms of the classification theory. I examine them successively in Sects. 5–8.

To summarize the situation at the end of Sect. 8, let $\pi \in \mathcal{A}_n^0(F)$. According to [17, 25], π contains a simple character θ_π, and only one (up to conjugation in $G = \mathrm{GL}_n(F)$). The G-normalizer of θ_π is an open, compact mod. centre subgroup $\boldsymbol{J}$ of G, the natural representation Λ_π of $\boldsymbol{J}$ on the θ_π-isotypic subspace of π is irreducible, and $\pi \cong c\text{-}\mathrm{Ind}_{\boldsymbol{J}}^G \Lambda_\pi$. (The pair $(\boldsymbol{J}, \Lambda_\pi)$ is an *extended maximal simple type* in G.) The group $\boldsymbol{J}$ has a unique maximal compact subgroup J, and the restriction $\lambda_\pi = \Lambda_\pi \,|\, J$ is irreducible: the pair (J, λ_π) is a *maximal simple type* in G. Write $\sigma = {}^L\pi \in \widehat{\mathcal{W}}_F$.

(1) The conjugacy class of θ_π determines the restriction of σ to the wild inertia subgroup of $\mathcal{W}_F$, and vice versa, but we have little idea of how this process works.
(2) The conjugacy class of the maximal simple type λ_π determines the restriction of σ to the inertia subgroup of $\mathcal{W}_F$ and vice versa. The two are connected by an explicit formula.
(3) Starting from σ, one can define $\pi' \in \mathcal{A}_n^0(F)$ by an explicit formula and show that π' is almost (in a rather precise sense) the same as π.

In (3), the representation π' is a sort of tamely ramified, internal twist of π, much as in the essentially tame case of [13, 14, 16]. In the general case, we have only partial control of the twisting character or "rectifier", to use the language of those papers.

In Sects. 9 and 10, I give a framework for analysing the relation between simple characters and ramification. I treat the first step of what will have to be an iterative process, mirroring the structure of simple characters.

Throughout, I give very few proofs. Those in the first three sections are included for purely illustrative purposes. Particularly at the beginning, I concentrate on producing a coherent narrative of the concepts and results that contribute directly to the main purpose. I pay scant attention to the expansive logical structures and background necessary for their proofs. Later on, I tend to follow the literature more closely but proofs remain rare. The main results in Sect. 8 depend on comparing an explicit algebraic induction process with automorphic induction, via some quite elaborate character calculations. I decided to omit them completely, as they do not condense readily.

Since I've chosen to keep the focus very narrow, it is difficult to say much about any broader context. That context exhibits a coherence quite unexpected at the beginning. The entire scheme of [25] can be transported systematically to an inner form $\mathrm{GL}_m(D)$ of $\mathrm{GL}_n(F)$, after a first step in [3]. As reported in Vincent Sécherre's conference talk, we are (at least) very close to knowing that this transfer implements the Jacquet–Langlands correspondence [68]. Stevens [73] has shown that the cuspidal representations of a classical group can be classified in a scheme transferred from that of [25], although the extra complications are considerable. In a recent development [2], it has emerged that the explicit relation between general linear groups and symplectic groups, implemented by transfer of simple characters, is that demanded by Functoriality.

Prerequisites

I assume some familiarity with the smooth complex representation theory of locally profinite groups, particularly $\mathrm{GL}_n(F)$. The requirements here are modest and mostly gathered together, in adequate generality, in [15].

I use standard notation for objects associated with the field F. Thus $\mathfrak{o}_F$ is the discrete valuation ring in F, while $\mathfrak{p}_F$ is the maximal ideal of $\mathfrak{o}_F$. The residue field $\Bbbk_F = \mathfrak{o}_F/\mathfrak{p}_F$ is a finite field of $q = q_F$ elements and characteristic p. Also, $U_F = \mathfrak{o}_F^\times$ is the unit group and, for $k \geqslant 1$, $U_F^k = 1 + \mathfrak{p}_F^k$. Finally, $\upsilon_F : F^\times \twoheadrightarrow \mathbb{Z}$ is the normalized valuation. On occasion, μ_F will be the group of roots of unity in F of order not divisible by p.

By convention, all representations to be considered are *smooth* and *complex*.

1 Basic Ideas and Examples

This section outlines first steps towards the systematic analysis of irreducible cuspidal representations of the group $G = \mathrm{GL}_n(F)$, along with a couple of examples indicative of the scope of the problem. Recall that "representation" means "smooth complex representation".

1.1 Intertwining and Induction

For this section only, G may denote the group $\boldsymbol{G}(F)$ of F-points of a connected reductive algebraic group $\boldsymbol{G}$ defined over F: we say that "G is a connected reductive F-group". We view G as endowed with the locally profinite topology inherited from F. The basic material of this section is all covered in [15].

1.1.1 Let Z be the centre of G. Let K be an open subgroup of G that is either compact or compact modulo Z. If ρ is an irreducible representation of K and π one of G, say that π *contains* ρ if the spaces $\mathrm{Hom}_K(\rho, \pi) \cong \mathrm{Hom}_K(\pi, \rho)$ are nonzero.

For $i = 1, 2$, let K_i be such an open subgroup of G and let ρ_i be an irreducible representation of K_i. An element g of G *intertwines* ρ_1 with ρ_2 if

$$\mathrm{Hom}_{K_1^g \cap K_2}(\rho_1^g, \rho_2) \neq 0.$$

Here, ρ_1^g is the representation $x \mapsto \rho_1(gxg^{-1})$ of the group $K_1^g = g^{-1}K_1 g$. Observe that the representations $\rho_1^g \mid K_1^g \cap K_2$ and $\rho_2 \mid K_1^g \cap K_2$ are semisimple.

Write $I_G(\rho_1, \rho_2)$ for the set of $g \in G$ that intertwine ρ_1 with ρ_2. Thus $I_G(\rho_1, \rho_2)$ is either empty or a union of double cosets $K_1 x K_2$. We abbreviate $I_G(\rho_1, \rho_1) = I_G(\rho_1)$ and say that an element of $I_G(\rho_1)$ intertwines ρ_1.

The set $I_G(\rho_1)$ is certainly non-empty, and it is the support of the Hecke algebra $\mathcal{H}(G, \rho_1)$. That function algebra is canonically identified with the algebra of G-endomorphisms of the smooth, compactly induced representation $c\text{-}\mathrm{Ind}_{K_1}^G\, \rho_1$.

If there exists an irreducible smooth representation containing both ρ_1 and ρ_2, then $I_G(\rho_1, \rho_2)$ is non-empty. However, $I_G(\rho_1, \rho_2)$ being non-empty does not imply that an irreducible representation of G containing ρ_1 also contains ρ_2.

1.1.2 Our overall strategy is based on the following fact.

Proposition *Let K be an open subgroup of G that is compact modulo the centre of G. Let ρ be an irreducible representation of K. If $I_G(\rho) = K$, then*

(1) the compactly induced representation $\pi = c\text{-}\mathrm{Ind}_K^G\, \rho$ is irreducible and cuspidal;
(2) the space $\mathrm{Hom}_K(\rho, \pi)$ has dimension one.

The intertwining hypothesis implies that the $\mathbb{C}$-algebra $\mathrm{End}_G(\pi)$ has dimension one. The proof then requires the fact that an irreducible smooth representation of G is admissible. From there, it follows the case of $G = \mathrm{GL}_2(F)$ in [15] 11.4.

The proposition provides the only method available for producing families of cuspidal representations in any degree of generality. Currently, it is known to yield all the cuspidal representations of G when G is $\mathrm{GL}_n(F)$ [25], $\mathrm{SL}_n(F)$ [26, 27], an inner form $\mathrm{GL}_m(D)$ of $\mathrm{GL}_n(F)$ [67], or a classical group when the residual characteristic p of F is not 2 [73]. If G is an arbitrary connected reductive F-group G, Yu [77] uses the technique to construct a series of "tame" cuspidal representations of G. If p is sufficiently large, in a sense depending on G, Kim [48] shows that Yu's tame series comprises all the irreducible cuspidal representations of G.

In all of these cases, the argument proceeds by systematically constructing the inducing datum (K, ρ): a mere existence theorem would be much less interesting. For arbitrary G, I am unaware of any compelling reason for even believing in an existence theorem.

1.2 Level Zero Representations

We revert to the case $G = \mathrm{GL}_n(F)$ to recall an example from an extreme end of the spectrum. The main features of the general theory are discernible here, albeit in an attenuated form.

1.2.1 In the group $G = \mathrm{GL}_n(F)$, take the standard maximal compact subgroup $K = \mathrm{GL}_n(\mathfrak{o}_F)$ and its first congruence subgroup $K^1 = 1 + \mathfrak{p}_F \mathrm{M}_n(\mathfrak{o}_F)$. Identify K/K^1 with $\mathrm{GL}_n(\Bbbk_F)$. Let Z be the centre of G: thus Z is the group of scalar matrices in G and $Z \cong F^\times$. The group KZ is compact modulo Z and is the G-normalizer of K.

Let ρ be an irreducible representation of K that is trivial on K^1. Thus ρ is the inflation of an irreducible representation $\bar{\rho}$ of $\mathrm{GL}_n(\Bbbk_F)$. Say that ρ is *residually cuspidal* if $\bar{\rho}$ is cuspidal in the usual sense. Otherwise, say that ρ is *residually split.*

Proposition *(1) If ρ is residually cuspidal, then $I_G(\rho, \rho) = KZ$.*
(2) If $\bar{\rho}$ is residually cuspidal and if σ is the inflation to K of an irreducible representation $\bar{\sigma}$ of $\mathrm{GL}_n(\Bbbk_F)$, then $I_G(\rho, \sigma)$ is non-empty if and only if $\bar{\sigma} \cong \bar{\rho}$.

The proof is a straightforward exercise, for which [15] 11.5 provides an adequate hint.

1.2.2 If R is an irreducible representation of KZ, then $R \mid K$ is irreducible. Conversely, an irreducible representation of K may be extended to a representation of KZ.

Say that an irreducible representation R of KZ is of level zero if $R \mid K^1$ is trivial. If (π, V) is an irreducible representation of G, say that (π, V) is *of level zero* if the space V^{K^1} of K^1-fixed vectors in V is nonzero.

Proposition (Construction Theorem) *Let R be an irreducible representation of KZ, of level zero, such that $R \mid K$ is residually cuspidal. The representation*

$$\pi_R = c\text{-}\mathrm{Ind}_{KZ}^G R$$

is irreducible, cuspidal and of level zero. If R' is a representation of KZ with the same properties as R, then $\pi_{R'} \cong \pi_R$ if and only if $R' \cong R$.

The first assertion here follows from the propositions of 1.1.2 and 1.2.1. If $\pi_R \cong \pi_{R'}$, the representations R, R' must intertwine in G and 1.2.1 Proposition implies that $R \cong R'$. Indeed, one can recover R from π_R directly: it is the natural representation of KZ on the space of K^1-fixed vectors in π_R.

Corollary *Let π be an irreducible representation of G, of level zero. Suppose π contains an irreducible representation ρ of K that is trivial on K^1 and residually cuspidal. The representation π is then cuspidal and of the form π_R, where $R \,|\, K \cong \rho$.*

1.2.3 The Construction Theorem of 1.2.2 gives an unambiguous classification of a certain family of irreducible, cuspidal, level zero representations of G in terms of explicit data.

Exhaustion Theorem *Let (π, V) be an irreducible representation of G of level zero. If the natural representation of K on V^{K^1} has an irreducible component that is residually split, then π is not cuspidal.*

Consequently, all irreducible, level zero, cuspidal representations are given by 1.2.2. The structure of the cuspidal representations of $\mathrm{GL}_n(\mathbb{k})$, when $\mathbb{k}$ is a finite field, is thoroughly understood, as in [34] (recalled in 8.1.1 below). So, this parametrization can reasonably claim to be explicit.

The Exhaustion Theorem is much harder to prove and, in line with our policy in these notes, we say no more if it. In [25], it is treated in Sect. 8.3 as an instance of a more complex argument. The main ideas are already present in [44, 76].

Results 1.2.2 Proposition and 1.2.3 Theorem admit generalization to *any* connected reductive F-group [59, 61].

1.3 Hereditary Orders

To deal with cuspidal representations in general, we have to specify a canonical family of compact open subgroups of $\mathrm{GL}_n(F)$. To do this, it is better to work with an F-vector space V of dimension n, along with the group $G = \mathrm{Aut}_F(V)$ and the algebra $A = \mathrm{End}_F(V)$. For background on the hereditary orders that dominate this section, Chapter 9 of [63] and the early pages of [7] contain everything necessary in a moderate number of pages.

1.3.1 Let L be an $\mathfrak{o}_F$*-lattice in* V. By definition, L is a finitely generated $\mathfrak{o}_F$-submodule of V that contains an F-basis of V. Equivalently, L is the $\mathfrak{o}_F$-span of an F-basis of V. In particular, an $\mathfrak{o}_F$-lattice in V is a compact open subgroup of V.

An $\mathfrak{o}_F$*-order in* A is a subring $\mathfrak{a}$ of A that contains 1 and is an $\mathfrak{o}_F$-lattice in the vector space A.

There is a fundamental example. Let L be an $\mathfrak{o}_F$-lattice in V and set

$$\mathfrak{m}(L) = \{x \in A : xL \subset L\}.$$

This is an $\mathfrak{o}_F$-order in A, canonically identified with $\mathrm{End}_{\mathfrak{o}_F}(L)$. Identifying V with F^n via a basis of L, the order $\mathfrak{m}(L)$ becomes $\mathrm{M}_n(\mathfrak{o}_F)$. If L' is some other lattice in V, there exists $g \in G$ such that $L' = gL$, so $\mathfrak{m}(L') = g\mathfrak{m}(L)g^{-1}$. Comparing Haar measures, we see that $\mathfrak{m}(L') \subset \mathfrak{m}(L)$ if and only if $\mathfrak{m}(L') = \mathfrak{m}(L)$. An elementary argument shows that $\mathfrak{m}(L') = \mathfrak{m}(L)$ if and only if $L' = xL$, for some $x \in F^\times$.

Orders of the form $\mathfrak{m}(L)$ are called the *maximal orders* in A: there is no non-trivial containment relation between them and, if $\mathfrak{a}$ is an $\mathfrak{o}_F$-order in A, there exists an $\mathfrak{a}$-stable lattice L in V, whence $\mathfrak{a} \subset \mathfrak{m}(L)$.

1.3.2 An $\mathfrak{o}_F$-*lattice chain* in V is a non-empty set $\mathcal{L}$ of $\mathfrak{o}_F$-lattices in V that is linearly ordered under inclusion and satisfies $xL \in \mathcal{L}$ for any $L \in \mathcal{L}$ and $x \in F^\times$. One can enumerate the elements of $\mathcal{L}$, say $\mathcal{L} = \{L_i : i \in \mathbb{Z}\}$, so that $L_i \supsetneq L_{i+1}$, for all i. In particular, if $L \in \mathcal{L}$, then $\mathfrak{p}_F L \in \mathcal{L}$ and there is a unique integer $e = e_F(\mathcal{L})$ such that $\mathfrak{p}_F L_i = L_{i+e}$, for all i. This is the F-*period* of $\mathcal{L}$. Note that $1 \leqslant e \leqslant n$.

1.3.3 There is a list of rings and groups associated with the lattice chain $\mathcal{L}$. Let

$$\mathfrak{a} = \mathfrak{a}_F(\mathcal{L}) = \{x \in A : xL \subset L,\ L \in \mathcal{L}\}.$$

This is surely an $\mathfrak{o}_F$-order in A. An order of this form is called a *hereditary* order. The order $\mathfrak{a}$ and the lattice chain $\mathcal{L}$ determine each other. Indeed, $\mathcal{L}$ is the set of all $\mathfrak{a}$-lattices in V.

If we impose an enumeration $\{L_i\}_{i\in\mathbb{Z}}$ on $\mathcal{L}$ (as in 1.3.2) and take $j \in \mathbb{Z}$, we may set

$$\mathfrak{a}_F^j(\mathcal{L}) = \{x \in A : xL_i \subset L_{i+j},\ i \in \mathbb{Z}\}.$$

Thus $\mathfrak{a} = \mathfrak{a}_F^0(\mathcal{L})$. The lattice $\mathfrak{a}_F^1(\mathcal{L})$ is an ideal of $\mathfrak{a}$, in fact the *Jacobson radical* $\mathrm{rad}\,\mathfrak{a}$ of $\mathfrak{a}$. Abbreviating $\mathfrak{a}_F^1(\mathcal{L}) = \mathfrak{p}$, one has $\mathfrak{a}_F^j(\mathcal{L}) = \mathfrak{p}^j$, $j \in \mathbb{Z}$: this makes sense for negative j, since $\mathfrak{p}$ is an invertible fractional ideal of $\mathfrak{a}$ (one of the definitions of "hereditary"). The period $e = e(\mathfrak{a}) = e_F(\mathcal{L})$ appears in the relation

$$\mathfrak{p}_F\,\mathfrak{a} = \mathfrak{p}^e.$$

The unit group $U_\mathfrak{a} = \mathfrak{a}^\times$ is a compact open subgroup of $G = \mathrm{Aut}_F(V)$. It has a canonical series of open normal subgroups

$$U_\mathfrak{a}^k = 1 + \mathfrak{p}^k, \quad k \geqslant 1.$$

One also sets

$$\begin{aligned}\mathcal{K}_\mathfrak{a} = \mathrm{Aut}\,\mathcal{L} &= \{g \in G : g\mathcal{L} = \mathcal{L}\}\\ &= \{g \in G : g\mathfrak{a}g^{-1} = \mathfrak{a}\}.\end{aligned}$$

Thus $\mathcal{K}_\mathfrak{a}$ is open in G and compact modulo the centre Z of G. It normalizes each of the groups $U_\mathfrak{a}$ and $U_\mathfrak{a}^k$, $k \geqslant 1$. It is indeed the G-normalizer of any one of them, *cf.* [7] 1.3.10. In the same vein, one can recover the order $\mathfrak{a}$ from any one of the groups $U_\mathfrak{a}$, $U_\mathfrak{a}^k$.

Remark Fix an $\mathfrak{o}_F$-lattice L in V. The set of lattice chains $\mathcal{L}$ in V, such that $L \in \mathcal{L}$, is in canonical bijection with the set of flags of subspaces of the n-dimensional $\Bbbk_F$-space $L/\mathfrak{p}_F L$. This leads to a useful block matrix description of hereditary orders and their associated groups: see [25] (2.5.1) for some pictures.

1.3.4 There is a more general context. In the language of [5, 75], these groups $U_{\mathfrak{a}}$ are the *parahoric subgroups* of G and $\{U_{\mathfrak{a}}^k\}_{k\geqslant 1}$ is the *standard filtration* of $U_{\mathfrak{a}}$. It is often necessary to use other filtrations given by *lattice sequences* in V. A lattice sequence is a map $\boldsymbol{L} : i \mapsto L_i$, from $\mathbb{Z}$ to the set of $\mathfrak{o}_F$-lattices in V, such that $L_i \supseteq L_{i+1}$, and with the periodicity property $\mathfrak{p}_F L_i = L_{i+e}$, for an integer $e = e(\boldsymbol{L}) \geqslant 1$. (Note the non-strict containment here.) One defines

$$\mathfrak{a}^j(\boldsymbol{L}) = \{x \in A : xL_i \subset L_{i+j}\}$$

as before. The image of $\boldsymbol{L}$ is a lattice chain $\mathcal{L}$, and clearly $\mathfrak{a}^0(\boldsymbol{L}) = \mathfrak{a}(\mathcal{L})$. However, the groups $1 + \mathfrak{a}^j(\boldsymbol{L}), j \geqslant 1$, provide another filtration of $U_{\mathfrak{a}(\mathcal{L})}$.

These non-standard filtrations are instances of the very general filtrations of [60]. The filtrations derived from lattice sequences are, in many cases, known to be the Moy–Prasad filtrations with rational jumps [4, 55].

Lattice sequences are necessary for the systematic analysis of general representations of $\mathrm{GL}_n(F)$ [29]. They become more important in inner forms of $\mathrm{GL}_n(F)$ [64–66] and are central to the analysis of classical groups [72].

1.3.5 We return to the main theme. There are some useful incidence relations. For example, a lattice chain $\mathcal{L}$ contains exactly $e_F(\mathcal{L})$ lattice chains $\mathcal{L}'$ with $e_F(\mathcal{L}') = 1$. It is contained in a finite number of lattice chains $\mathcal{L}''$ satisfying $e_F(\mathcal{L}'') = n$ (*cf.* 1.3.3 Remark). Using the obvious notation for the associated orders and their radicals, we have

$$\mathfrak{a}' \supset \mathfrak{a} \supset \mathfrak{a}'' \quad \text{and} \quad \mathfrak{p}' \subset \mathfrak{p} \subset \mathfrak{p}''.$$

1.3.6 There is a special case we need to mention. Say that the lattice chain $\mathcal{L} = \{L_i\}$ is *uniform* if the index $(L_i{:}L_{i+1})$ is independent of i. Equivalently, there exists $g \in G$ such that $gL_i = L_{i+1}$ for all i. Any such element g satisfies

$$g\mathfrak{a} = \mathfrak{a}g = \mathfrak{p},$$

so one says that the associated hereditary order $\mathfrak{a} = \mathfrak{a}_F(\mathcal{L})$ is *principal.* In this case, the group $\mathcal{K}_{\mathfrak{a}}$ is generated by g and $U_{\mathfrak{a}}$. The groups $\mathcal{K}_{\mathfrak{a}}$, for principal orders $\mathfrak{a}$, are the *maximal compact modulo centre subgroups of* G, [7] (1.3.2) Theorem.

1.3.7 A certain relative structure arises repeatedly. Let E/F be a finite field extension, and suppose that V is given as an E-vector space. Set $B = \mathrm{End}_E(V)$. If $A = \mathrm{End}_F(V)$ as before, then E is an F-subalgebra of A while B is the centralizer of E in A.

Let $\mathcal{L}$ be an $\mathfrak{o}_E$-lattice chain in V and form the hereditary $\mathfrak{o}_E$-order $\mathfrak{b} = \mathfrak{a}_E(\mathcal{L})$ in B. The set $\mathcal{L}$ is also an $\mathfrak{o}_F$-lattice chain in V and, if $\mathfrak{a} = \mathfrak{a}_F(\mathcal{L})$, then $\mathfrak{b} = \mathfrak{a} \cap B$. Indeed, if $\mathfrak{p} = \mathrm{rad}\,\mathfrak{a}$ and $\mathfrak{q} = \mathrm{rad}\,\mathfrak{b}$, then $\mathfrak{q}^j = \mathfrak{p}^j \cap B, j \in \mathbb{Z}$. We have the relation

$$e_F(\mathfrak{a}) = e_E(\mathfrak{b})\, e(E|F).$$

As $E^\times \subset \mathfrak{K}_\mathfrak{a}$, so $E^\times$ normalizes $\mathfrak{a}$ and also the groups $U_\mathfrak{a}$, $U_\mathfrak{a}^k$.

These properties are reversible. Suppose we are given an $\mathfrak{o}_F$-lattice chain $\mathfrak{L}$ in V and a subfield E/F of A. Say that $\mathfrak{a} = \mathfrak{a}_F(\mathfrak{L})$ is *E-pure* if $E^\times \subset \mathfrak{K}_\mathfrak{a}$. This is equivalent to $\mathfrak{L}$ being an $\mathfrak{o}_E$-lattice chain and we are in the previous situation.

Example Let E/F be a finite field extension. View E as an F-subalgebra of $A_{E/F} = \mathrm{End}_F(E)$. There is then a unique E-pure, hereditary $\mathfrak{o}_F$-order $\mathfrak{a}_{E/F}$ in $A_{E/F}$: it is the one given by the lattice chain $\{\mathfrak{p}_E^j : j \in \mathbb{Z}\}$.

1.4 Duality and Strata

We keep the notation of Sect. 1.3. We examine a class of *characters* of the groups $U_\mathfrak{a}^{1+k}$, $k \geqslant 0$, where $\mathfrak{a}$ is a hereditary $\mathfrak{o}_F$-order in A. To do this, we choose a smooth character ψ_F of F that is *of level one*, meaning that ψ_F is non-trivial on $\mathfrak{o}_F$ but trivial on $\mathfrak{p}_F$. This condition serves mainly to keep the bookkeeping simple: changing it has only trivial consequences. Details of everything in this section can be found in [6, 7, 25].

1.4.1 Let $\mathrm{tr}_A : A \to F$ be the matrix trace map and set $\psi_A = \psi_F \circ \mathrm{tr}_A$. Thus ψ_A is a non-trivial smooth character of A. Write $\widehat{A}$ for the Pontrjagin dual of A.

Lemma *If $a, x \in A$, write $\psi_A \circ a(x) = \psi_A(ax)$. The map*

$$\begin{aligned} A &\longrightarrow \widehat{A}, \\ a &\longmapsto \psi_A \circ a, \end{aligned}$$

is an isomorphism of topological groups.

If L is an $\mathfrak{o}_F$-lattice in A, the set

$$L^\perp = \{a \in A : \psi_A(aL) = 1\}$$

is an $\mathfrak{o}_F$-lattice in A, and ψ_A induces a topological isomorphism of $A/L^\perp$ with the Pontrjagin dual $\widehat{L}$ of L.

Proposition *If $\mathfrak{a}$ is a hereditary $\mathfrak{o}_F$-order in A with* rad $\mathfrak{a} = \mathfrak{p}$, *then*

$$\left(\mathfrak{p}^{1+m}\right)^\perp = \mathfrak{p}^{-m}, \quad m \in \mathbb{Z}.$$

1.4.2 For $a \in A$, define a function $\psi_A * a$ on A by

$$\psi_A * a(1+x) = \psi_A(ax),$$

and take $\mathfrak{a}$, $\mathfrak{p}$ as in 1.4.1 Proposition.

Proposition *Suppose* $0 \leqslant j < k \leqslant 2j+1$. *The isomorphism*

$$\begin{aligned} \mathfrak{p}^{1+j}/\mathfrak{p}^{1+k} &\longrightarrow U_{\mathfrak{a}}^{1+j}/U_{\mathfrak{a}}^{1+k}, \\ x &\longmapsto 1+x, \end{aligned}$$

induces an isomorphism

$$\begin{aligned} \mathfrak{p}^{-k}/\mathfrak{p}^{-j} &\longrightarrow \big(U_{\mathfrak{a}}^{1+j}/U_{\mathfrak{a}}^{1+k}\big)\widehat{\ }, \\ a &\longmapsto \psi_A * a \,\big|\, U_{\mathfrak{a}}^{1+j}. \end{aligned}$$

This follows directly from 1.4.1 Proposition.

1.4.3 There is a convenient administrative device for handling 1.4.2 Proposition.

Definition A *stratum* in A is a quadruple $\mathfrak{s} = [\mathfrak{a}, m, l, a]$ in which $\mathfrak{a}$ is a hereditary $\mathfrak{o}_F$-order in A, with $\mathfrak{p} = \operatorname{rad}\mathfrak{a}$, while $m > l$ are integers and $a \in \mathfrak{p}^{-m}$.

A stratum $[\mathfrak{a}, m, l, a']$ is deemed *equivalent to* $\mathfrak{s}$ if $a \equiv a' \pmod{\mathfrak{p}^{-l}}$. If $0 \leqslant l < m \leqslant 2l+1$, the equivalence class of the stratum $\mathfrak{s}$ and the character

$$\chi_{\mathfrak{s}} = \psi_A * a \,\big|\, U_{\mathfrak{a}}^{1+l}$$

determine each other uniquely (by 1.4.2 Proposition). For $i = 1, 2$, let $\mathfrak{s}_i = [\mathfrak{a}_i, m_i, l_i, a_i]$ be a stratum in A and write $\mathfrak{p}_i = \operatorname{rad}\mathfrak{a}_i$. An element $g \in G$ is said to *formally intertwine* $\mathfrak{s}_1$ with $\mathfrak{s}_2$ if

$$g^{-1}(a_1 + \mathfrak{p}_1^{-l_1})g \cap (a_2 + \mathfrak{p}_2^{-l_2}) \neq \emptyset$$

Proposition *Suppose that* $0 \leqslant l_i < m_i \leqslant 2l_i + 1$, $i = 1, 2$. *An element* g *of* G *formally intertwines* $\mathfrak{s}_1$ *with* $\mathfrak{s}_2$ *if and only if it intertwines* $\chi_{\mathfrak{s}_1}$ *with* $\chi_{\mathfrak{s}_2}$.

Formal intertwining of strata often provides a practical method for computing intertwining of the associated characters.

1.5 A Fundamental Dichotomy

We use the machinery of Sects. 1.3 and 1.4 in a preliminary analysis of the irreducible representations of $G = \operatorname{Aut}_F(V)$.

1.5.1 Here, we consider only strata of the form $\mathfrak{s} = [\mathfrak{a}, m, m-1, a]$, $m \geqslant 1$. Let $\operatorname{rad}\mathfrak{a} = \mathfrak{p}$ and set $\ell(\mathfrak{s}) = m/e_F(\mathfrak{a})$.

Let π be an irreducible representation of G. Define the *level* $\ell(\pi)$ *of* π as the minimum value of $m/e_{\mathfrak{a}}$, where $\mathfrak{a}$ ranges over the hereditary $\mathfrak{o}_F$-orders in $A = \operatorname{End}_F(V)$

and m over the non-negative integers such that π contains the trivial character of $U_{\mathfrak{a}}^{1+m}$. Thus $\ell(\pi) = 0$ in this sense if and only if π is of level zero as in 1.2.2.

Say that π *contains* the stratum $\mathfrak{s} = [\mathfrak{a}, m, m-1, a]$ if $m \geqslant 1$ and π contains the character $\chi_{\mathfrak{s}} = \psi_A * a \,|\, U_{\mathfrak{a}}^m$.

Definition The stratum $\mathfrak{s} = [\mathfrak{a}, m, m-1, a]$ is *fundamental* if the coset $a + \mathfrak{p}^{1-m}$ contains no nilpotent element of A.

Equivalently, $\mathfrak{s}$ is fundamental if and only if $a^t \notin \mathfrak{p}^{1-tm}$ for any $t \geqslant 1$ [6].

Theorem *Let π be an irreducible representation of G such that $\ell(\pi) > 0$. Let $\mathfrak{s} = [\mathfrak{a}, m, m-1, a]$ be a stratum contained in π. The following conditions are equivalent.*

(1) $\ell(\mathfrak{s}) = \ell(\pi)$.
(2) $\mathfrak{s}$ is fundamental.

If $\ell(\pi) = 0$, the π contains no fundamental stratum $\mathfrak{a}$ with $\ell(\mathfrak{s}) > 0$.

The result nominates a special family of characters of certain compact open subgroups of G, at least one of which is guaranteed to occur in a given representation π of positive level.

Comments The theorem, in this form, for $\mathrm{GL}_n(F)$ is proved in [6]. In a far-reaching generalization, Moy and Prasad [60] prove an analogue for representations of any connected reductive F-group G.

1.5.2 Let $\mathfrak{s}$ be a stratum $[\mathfrak{a}, m, m-1, a]$, $m \geqslant 1$. Abbreviate $e = e_F(\mathfrak{a})$, set $g = \gcd(m, e)$ and choose a prime element ϖ of F. The element $a_0 = a^{e/g}\varpi^{m/g}$ then lies in $\mathfrak{a}$, and the equivalence class of $\mathfrak{s}$ determines the coset $a_0 + \mathfrak{p}$. If $\mathfrak{a}$ is given by the lattice chain $\{L_i\}_{i\in\mathbb{Z}}$, then

$$\mathfrak{a}/\mathfrak{p} \cong \prod_{i=0}^{e-1} \mathrm{End}_{\Bbbk_F}(L_i/L_{i+1}).$$

As an endomorphism of the vector space $\prod_{i=0}^{e-1} L_i/L_{i+1}$, the coset $a_0 + \mathfrak{p}$ has a characteristic polynomial, which is the reduction, modulo $\mathfrak{p}_F$, of the characteristic polynomial of the F-endomorphism $a_0 \in A = \mathrm{End}_F(V)$. This polynomial, call it $\mathrm{ch}_{\mathfrak{s}}(t) \in \Bbbk_F[t]$, is determined by the equivalence class of $\mathfrak{s}$. Any monic polynomial of degree $n = \dim_F V$ can arise here, but $\mathfrak{s}$ is fundamental if and only if $\mathrm{ch}_{\mathfrak{s}}(t) \neq t^n$ (1.5.1 Theorem).

If $\mathfrak{s}$, $\mathfrak{s}'$ are fundamental strata occurring in the same irreducible representation of G, they must intertwine formally. We deduce:

Proposition *Let π be an irreducible representation of G with $\ell(\pi) > 0$. If $\mathfrak{s}$, $\mathfrak{s}'$ are fundamental strata occurring in π, then $\ell(\mathfrak{s}) = \ell(\mathfrak{s}') = \ell(\pi)$ and* $\mathrm{ch}_{\mathfrak{s}}(t) = \mathrm{ch}_{\mathfrak{s}'}(t)$.

Definition A fundamental stratum $\mathfrak{s}$ is called *split* if $\mathrm{ch}_{\mathfrak{s}}(t)$ has at least two distinct irreducible factors. Otherwise, it is *non-split*.

Theorem *Let π be an irreducible representation of G with $\ell(\pi) > 0$ containing a fundamental stratum $\mathfrak{s}$. If $\mathfrak{s}$ is split, then π is not cuspidal.*

So, if π is cuspidal, it contains a non-split fundamental stratum. The proof of the theorem is an instance of the arguments of [25] 8.2 and is quite elaborate. Observe a parallel with 1.2.3 Theorem.

1.5.3 Since we are only concerned with cuspidal irreducible representations π, we restrict to the case in which π contains a *non-split* fundamental stratum. There is a generic example of such strata.

Example Let E/F be a subfield of A. Let $\alpha \in E$ satisfy $E = F[\alpha]$ and $\nu = -\upsilon_E(\alpha) > 0$. Let $\mathfrak{a}$ be an E-pure hereditary $\mathfrak{o}_F$-order in A with rad $\mathfrak{a} = \mathfrak{p}$. In particular, $\alpha\mathfrak{a} = \mathfrak{p}^{-m}$, where $m = \nu e(\mathfrak{a})/e(E|F)$. The quadruple $[\mathfrak{a}, m, m-1, \alpha]$ is a non-split fundamental stratum in A.

A non-split fundamental stratum need not be of this form. On the other hand, the example itself can be refined.

Definition Let $\alpha \in G$. Say that α is *minimal over F* if

(1) the algebra $E = F[\alpha]$ is a field,
(2) $\upsilon = \upsilon_E(\alpha)$ is relatively prime to $e = e(E|F)$ and
(3) if ϖ is a prime element of F, the coset $\alpha^e\varpi^{-\upsilon} + \mathfrak{p}_E$ generates the residue field extension $\Bbbk_E/\Bbbk_F$.

A stratum of the form $[\mathfrak{a}, m, m-1, \alpha]$ will be called *simple* if α is minimal over F, the hereditary order $\mathfrak{a}$ is $F[\alpha]$-pure and $\alpha\mathfrak{a} = \mathfrak{p}^{-m}$.

Theorem 1 *Let π be an irreducible representation of G such that $\ell(\pi) > 0$. If π contains a non-split fundamental stratum, in particular if π is cuspidal, then it contains a simple stratum $\mathfrak{s} = [\mathfrak{a}, m, m-1, \alpha]$.*

Theorem 1 relies on matrix calculations carried through in section 2.5 of [25]. The proof has an important feature. If π contains the non-split fundamental stratum $[\mathfrak{a}, m, m-1, a]$, a simple stratum contained in π will take the form $[\mathfrak{a}', m', m'-1, a']$ *for a possibly different hereditary order* $\mathfrak{a}'$. Indeed, the proof starts by taking $\mathfrak{a}$ as small as possible.

To analyse the irreducible representations π containing a fixed simple stratum $\mathfrak{s} = [\mathfrak{a}, m, m-1, \alpha]$, we will need to know the formal intertwining of $\mathfrak{s}$.

Theorem 2 *Let $\mathfrak{s} = [\mathfrak{a}, m, m-1, \alpha]$ be a simple stratum in A. Let B be the A-centralizer of α. An element $g \in G$ intertwines $\mathfrak{s}$ formally if and only if $g \in U^1_{\mathfrak{a}} B^\times U^1_{\mathfrak{a}}$.*

At this point, we give an example of how one can deduce properties of strata from information about their intertwining.

Corollary *Let $[\mathfrak{a}, m, m-1, \alpha]$ be a simple stratum in A, and suppose that the field extension $E = F[\alpha]/F$ has degree $n = \dim_F V$. If $\beta \in \alpha + \mathfrak{p}^{1-m}$, then $[\mathfrak{a}, m, m-1, \beta]$ is simple, the field extension $F[\beta]/F$ has degree $n = \dim_F V$ and $e(F[\beta]|F) = e(F[\alpha]|F)$.*

Proof The A-centralizer of α is E itself, so the formal intertwining set of $[\mathfrak{a}, m, m-1, \alpha]$ is $E^\times U^1_\mathfrak{a}$ and this set is compact modulo centre. On the other hand, the formal intertwining set of the equivalent stratum $[\mathfrak{a}, m, m-1, \beta]$ contains $U^1_\mathfrak{a} C^\times U^1_\mathfrak{a}$, where C is the A-centralizer of β. We conclude that $C^\times$ is compact modulo $F^\times$, whence $[F[\beta] : F] = n$ and $C = F[\beta]$. Moreover, $C^\times \subset E^\times U^1_\mathfrak{a}$, so $e(C|F) \leqslant e(E|F)$ and $f(C|F) \leqslant f(E|F)$. Also, $\upsilon_F(\det \beta) = \upsilon_F(\det \alpha)$. The remaining assertions follow readily. □

1.5.4 The strata of 1.5.3 Corollary provide a family of examples complementary to those in Sect. 1.2. This case was first observed by Carayol [30].

Theorem *Let* $\mathfrak{s} = [\mathfrak{a}, m, m-1, \alpha]$ *be a simple stratum in* A, *and suppose that* $E = F[\alpha]/F$ *is of degree* n. *An irreducible representation* π *of* G *that contains* $\mathfrak{s}$ *is then cuspidal and of the form*

$$\pi = c\text{-}\mathrm{Ind}_{\boldsymbol{K}}^G \Lambda,$$

where $\boldsymbol{K} = E^\times U^1_\mathfrak{a}$ *and* Λ *is an irreducible representation of* $\boldsymbol{K}$ *that contains* $\psi_A * \alpha \,\big|\, U^m_\mathfrak{a}$.

This follows directly from 1.5.3 Theorem 2. Observe that any irreducible representation of $\boldsymbol{K}$ containing $\psi_A * \alpha \,\big|\, U^m_\mathfrak{a}$ induces irreducibly to a cuspidal representation of G.

Comment We begin to see the emergence of a pattern in the general case, echoing that of the level zero case in Sect. 1.2. We have two families of representations of certain compact open subgroups of G, namely the split fundamental strata and the simple strata. An irreducible representation π of G contains a member of one family, but not both. If it contains a split fundamental stratum, then it is not cuspidal. So, to isolate and analyse the cuspidal representations of G, we need to refine the concept of simple stratum and the associated character. This occupies the following two sections.

2 Simple Characters

This section is an account of the *simple characters* in $G = \mathrm{GL}_n(F)$. These constitute a canonical family of characters of compact open subgroups of G, so named because their intertwining features simple algebras.

The concept of simple character is not amenable to direct definition. One starts from a *simple stratum* in $A = \mathrm{M}_n(F)$, in a sense more general than that of Sect. 1.5. A simple stratum $\mathfrak{s}$ is a comparatively straightforward object attached to a hereditary order $\mathfrak{a}$ in A. It defines an open subgroup of the unit group $U^1_\mathfrak{a}$ and then, after choosing a character ψ_F as in Sect. 1.4, a finite family $\mathcal{C}(\mathfrak{s}, \psi_F)$ of characters of that group: these are the *simple characters* defined by $\mathfrak{s}$. Simple characters are defined inductively, via explicit formulæ following the structure of $\mathfrak{s}$. This process involves many choices, but one shows them to be ultimately insignificant. The complexity of the process derives from the fact that one cannot recover $\mathfrak{s}$ from the set $\mathfrak{C}(\mathfrak{s})$. In consequence, the evolution in Chapters 2 and 3 of [25] is indirect and convoluted, with several different ideas being developed in an interwoven pattern. Here, we pay no regard to this logical structure. We aim for a coherent narrative of the important definitions and basic results, occasionally informed by later papers. We have chosen results that give insight into the structure or, as is often the same, are subsequently re-used frequently. The aim is to minimize the need for a reader to return to the detailed arguments of [25].

Throughout this section, V will denote an F-vector space, of finite dimension n, with $A = \mathrm{End}_F(V)$ and $G = \mathrm{Aut}_F(V)$.

2.1 Adjoint and Tame Corestriction

We introduce a pair of maps $A \to A$. To do this, we need a subfield E/F of A and an element β of G such that $E = F[\beta]$. Let $B = \mathrm{End}_E(V)$ be the A-centralizer of E.

2.1.1 The standard *adjoint map*

$$\begin{aligned} a_\beta : A &\longrightarrow A, \\ x &\longmapsto \beta x - x\beta, \end{aligned}$$

provides a B, B-bimodule map with kernel B.

2.1.2 There is a canonical family of B, B-bimodule maps $A \to B$ that are, in a certain sense, complementary to the adjoint. The space $\mathrm{Hom}_{B,B}(A, B)$ of bimodule homomorphisms $A \to B$ is an E-vector space of dimension one. Any map $f \in \mathrm{Hom}_{B,B}(A, B)$ satisfies $f(E) \subset E$ and $f(A) = B$, provided $f \neq 0$.

Definition A *tame corestriction,* on A relative to E/F, is a B, B-bimodule homomorphism $s : A \to B$ such that, if $\mathfrak{a}$ is an E-pure hereditary $\mathfrak{o}_F$-order in A, then $s(\mathfrak{a}) = \mathfrak{a} \cap B$.

A tame corestriction exists and is unique up to multiplication by a unit of E. The proof proceeds via a duality property. As in Sect. 1.4, let ψ_F be a character of F of level one, and set $\psi_A = \psi_F \circ \mathrm{tr}_A$. Likewise, let ψ_E be a character of E of level one and put $\psi_B = \psi_E \circ \mathrm{tr}_B$. For $a \in A$, the map $b \mapsto \psi_A(ab)$ is a character of B, so there is a unique element $s(a) \in B$ such that

$$\psi_A(ab) = \psi_B(s(a)b), \quad b \in B.$$

This map s has all required properties.

The kernel of s is $a_\beta(A)$, while its image is $B = \mathrm{Ker}\, a_\beta$. Tame corestrictions have a finer property relative to hereditary orders.

Proposition *Let $\mathfrak{a}$ be an E-pure hereditary order in A. Set $\mathfrak{b} = \mathfrak{a} \cap B$, $\mathfrak{p} = \text{rad}\,\mathfrak{a}$ and $\mathfrak{q} = \mathfrak{p} \cap B = \text{rad}\,\mathfrak{b}$. If s is a tame corestriction on A, relative to E/F, then $s(\mathfrak{p}^r) = \mathfrak{q}^r$, $r \in \mathbb{Z}$.*

Example Suppose that E/F is separable. The non-degenerate, symmetric F-bilinear form $(x, y) \mapsto \text{tr}_A(xy)$ on A then restricts to a non-degenerate form on B. The orthogonal projection $\pi : A \to B$ is a B, B-homomorphism $A \to B$, but *it is a tame corestriction if and only if E/F is tamely ramified.* Indeed, let $d(E|F)$ be the exponent of the different of E/F and let $w \in E$ have valuation $d(E|F) - e(E|F) + 1$. The map $w\pi$ is then a tame corestriction.

Of course, if E/F is not separable, then $B \subset a_\beta(A)$ and $\text{tr}_A \,\big|\, B = 0$. There is no projection with which to compare, but the tame corestriction exists nonetheless.

2.2 Critical Exponent and Simple Strata

We have already mentioned examples of simple strata in Sect. 1.5. We now give the general definition.

2.2.1 We need a preliminary definition.

Definition A stratum $[\mathfrak{a}, m, r, a]$ in A is *pure* if

(1) the algebra $F[a]$ is a field;
(2) the order $\mathfrak{a}$ is $F[a]$-pure;
(3) $a\mathfrak{a} = \mathfrak{p}^{-m}$, where $\mathfrak{p} = \text{rad}\,\mathfrak{a}$.

The property of purity is not invariably preserved under equivalence of strata.

2.2.2 Let $[\mathfrak{a}, m, r, \beta]$ be a pure stratum in A and write $E = F[\beta]$. Let B denote the A-centralizer of E and set $\mathfrak{b} = \mathfrak{a} \cap B$, $\mathfrak{p} = \text{rad}\,\mathfrak{a}$. Intertwining properties depend on the behaviour of a_β relative to the lattices $\mathfrak{p}^k$, $k \in \mathbb{Z}$. This can be quite unstable as β varies over its $\mathfrak{p}^{-r}$-coset. It is controlled via a *critical exponent* $k_0(\beta, \mathfrak{a})$ as follows.

For $k \in \mathbb{Z}$, set

$$\mathfrak{N}_k(\beta, \mathfrak{a}) = \{x \in \mathfrak{a} : a_\beta(x) \in \mathfrak{p}^k\}.$$

As $a_\beta(\mathfrak{a}) \subset \mathfrak{p}^{-m}$, so $\mathfrak{N}_k(\beta, \mathfrak{a}) = \mathfrak{a}$ for $k \leqslant -m$. On the other hand, a short argument shows that $\mathfrak{N}_k(\beta, \mathfrak{a}) \subset \mathfrak{b} + \mathfrak{p}$, for k sufficiently large. We accordingly define

$$k_0(\beta, \mathfrak{a}) = \max\{k : \mathfrak{N}_k(\beta, \mathfrak{a}) \not\subset \mathfrak{b} + \mathfrak{p}\}.$$

Remark In the case $\beta \in F^\times$, we have $\{0\} = a_\beta(\mathfrak{a}) \subset \mathfrak{p}^k$ for all k, while $\mathfrak{b} + \mathfrak{p} = \mathfrak{a}$. One therefore sets $k_0(\beta, \mathfrak{a}) = -\infty$ in this case.

The exponent $k_0(\beta, \mathfrak{a})$ depends on $\mathfrak{a}$ in a straightforward way. Let $E = F[\beta]$, let V_1 be a finite-dimensional E-vector space and let $\mathfrak{a}_1$ be an E-pure hereditary order in

$\mathrm{End}_F(V_1)$. If $\mathfrak{p}_1 = \mathrm{rad}\,\mathfrak{a}_1$, then $\beta\mathfrak{a}_1 = \mathfrak{p}_1^{m_1}$, where $m_1/e(\mathfrak{a}_1) = m/e(\mathfrak{a})$. The critical exponent has a similar property:

$$\frac{k_0(\beta, \mathfrak{a}_1)}{e(\mathfrak{a}_1)} = \frac{k_0(\beta, \mathfrak{a})}{e(\mathfrak{a})}, \tag{2.2.1}$$

[25] (1.4.13).

Definition A *simple stratum* is a pure stratum $[\mathfrak{a}, m, r, \beta]$ such that $r < -k_0(\beta, \mathfrak{a})$.

The following result illustrates the rôle played by the lattices $\mathcal{N}_k(\beta, \mathfrak{a})$. Recall that $g \in G$ is said to formally intertwine the stratum $\mathfrak{s} = [\mathfrak{a}, m, r, \beta]$ if

$$g^{-1}(\beta + \mathfrak{p}^{-r})g \cap (\beta + \mathfrak{p}^{-r}) \neq \emptyset.$$

We let $I_G(\mathfrak{s})$ denote the set of such elements g.

Proposition *Let* $\mathfrak{s} = [\mathfrak{a}, m, r, \beta]$ *be a simple stratum in* A. *Let* B *be the* A*-centralizer of* β, *write* $\mathfrak{p} = \mathrm{rad}\,\mathfrak{a}$, $\mathfrak{b} = \mathfrak{a} \cap B$ *and* $\mathfrak{q} = \mathrm{rad}\,\mathfrak{b} = \mathfrak{p} \cap B$. *If* $k = k_0(\beta, \mathfrak{a})$ *and* $\mathcal{N} = \mathcal{N}_k(\beta, \mathfrak{a})$, *then*

$$I_G(\mathfrak{s}) = (1 + \mathfrak{q}^{-(r+k)}\mathcal{N})\, B^\times (1 + \mathfrak{q}^{-(r+k)}\mathcal{N}).$$

Remark The formal intertwining set $I_G(\mathfrak{s})$ depends only on the equivalence class, call it $\mathfrak{S}$, of the stratum $\mathfrak{s}$. The last formula describes this set in terms of one simple element $\mathfrak{s}$ of $\mathfrak{S}$. So, the formula gives a loose relation between the simple elements of the equivalence class $\mathfrak{S}$. This looseness is intrinsic to the situation. Two simple strata $[\mathfrak{a}, m, r, \beta_i]$ may be equivalent while the field extensions $F[\beta_i]/F$ are not isomorphic: see 2.3.1 below for an extreme example. At this stage, the intertwining formula is the only available method for comparing them.

The proof of the proposition requires a certain apparatus but is basically a formal result, not requiring any real knowledge of simple strata. However, it has a consequence from which the structure theory is developed. Using the notation of the proposition, let $s : A \to B$ be a tame corestriction with respect to $F[\beta]/F$.

Corollary *For* $i = 1, 2$, *let* $a_i \in \mathfrak{p}^{-r}$. *The strata* $[\mathfrak{b}, r, r-1, s(a_i)]$ *in* B *are equivalent if and only if there exists* $x \in 1 + \mathfrak{q}^{-(r+k)}\mathcal{N}$ *such that the strata* $[\mathfrak{a}, m, r-1, x^{-1}(\beta + a_1)x]$, $[\mathfrak{a}, m, r-1, \beta + a_2]$ *are equivalent.*

This follows from the formalism of the critical exponent. It indicates the utility of passing to the centralizer via a tame corestriction.

2.3 Structure of Simple Strata

We give a procedure for constructing simple strata in A and indicate how, up to equivalence, it yields them all.

2.3.1 Let $[\mathfrak{a}, m, r, \beta]$ be a pure stratum in A. We know that $k_0(\beta, \mathfrak{a})$ is either $\geqslant -m$ or $= -\infty$, the latter case corresponding to $\beta \in F$. We characterize this extreme case.

Proposition *If* $[\mathfrak{a}, m, r, \beta]$ *is a pure stratum in* A, *then* $k_0(\beta, \mathfrak{a}) \leqslant -m$ *if and only if* β *is minimal over* F. *In particular, a pure stratum* $[\mathfrak{a}, m, m-1, \beta]$ *is simple if and only if* β *is minimal over* F.

Example Recall that p is the residual characteristic of F. Let E/F be totally ramified of degree p^r, for some $r \geqslant 1$. Let $\mathfrak{a}$ be the unique E-pure hereditary $\mathfrak{o}_F$-order in $A = \mathrm{End}_F(E)$ (as in 1.3.7 Example). If $\alpha \in E$ has valuation -1, then α is minimal over F and $\mathfrak{s} = [\mathfrak{a}, 1, 0, \alpha]$ is a simple stratum in A. Let E'/F be totally ramified of degree p^r. We may embed E' in A so that $\mathfrak{a}$ is E'-pure. Choose $\alpha' \in E'$, of valuation -1, such that $\det \alpha' \equiv \det \alpha \pmod{U_F^1}$. The stratum $[\mathfrak{a}, 1, 0, \alpha']$ is again simple and it is conjugate, under $\mathcal{K}_\mathfrak{a}$, to a simple stratum equivalent to $\mathfrak{s}$. (See [18] section 2 for a full discussion of this example.)

2.3.2 The proposition of 2.3.1 serves as the first step in an iterative procedure for constructing more complex simple strata. Starting from a simple stratum $[\mathfrak{a}, m, r, \gamma]$ in $A = \mathrm{End}_F(V)$, set $E = F[\gamma]$ and let B denote the A-centralizer of γ. Set $\mathfrak{b} = \mathfrak{a} \cap B$, $\mathfrak{p} = \mathrm{rad}\, \mathfrak{a}$, $\mathfrak{q} = \mathrm{rad}\, \mathfrak{b}$. Let $s : A \to B$ be a tame corestriction relative to E/F.

Theorem *Let* $[\mathfrak{b}, r, r-1, \alpha]$ *be a simple stratum in* B *and let* $a \in \mathfrak{p}^{-r}$ *satisfy* $s(a) \equiv \alpha \pmod{\mathfrak{q}^{1-r}}$. *The stratum* $[\mathfrak{a}, m, r-1, \gamma + a]$ *is equivalent to a simple stratum* $[\mathfrak{a}, m, r-1, \beta]$. *Moreover,*

(1) $k_0(\beta, \mathfrak{a}) = \max\{k_0(\alpha, \mathfrak{b}), k_0(\gamma, \mathfrak{a})\}$;
(2) $[F[\beta] : F] = [E : F]\,[E[\alpha] : E]$ *and* $e(F[\beta]|F) = e(E|F)e(E[\alpha]|E)$.

That is,

$$k_0(\beta, \mathfrak{a}) = \begin{cases} k_0(\gamma, \mathfrak{a}) & \text{if } \alpha \in E, \\ -r > k_0(\gamma, \mathfrak{a}) & \text{if } \alpha \notin E. \end{cases}$$

Remarks There are some special cases worthy of comment.

(1) If $E[\alpha]$ is a maximal subfield of B, then $[\mathfrak{a}, m, r-1, \gamma + a]$ *is* a simple stratum in A, not just equivalent to one. Also, $F[\gamma + a]$ is a maximal subfield of A.
(2) If E/F is *tamely ramified*, one may take s to be the orthogonal projection $A \to B$ and $a = \alpha$. One sees that $F[\beta] = E[\alpha]$, so $F[\beta]$ contains E. When E/F is not tamely ramified and $[F[\beta] : F] > [E : F]$, the element a can *never* commute with γ since that would imply $s(a) \in \mathfrak{q}^{1-r}$ (*cf.* 2.1.2 Example).

2.3.3 We now quote a sequence of results from [25] 2.4. Together, these give a complete account of the structure of simple strata and show that they are all obtained from the case of 2.3.1 Proposition via the refinement process of 2.3.2.

Theorem 1 *Let* $\mathfrak{s} = [\mathfrak{a}, m, r, \beta]$ *be a pure stratum in* A. *There exists a simple stratum* $\mathfrak{s}_0 = [\mathfrak{a}, m, r, \gamma]$ *equivalent to* $\mathfrak{s}$. *For any such* $\mathfrak{s}_0$, $e(F[\gamma]|F)$ *divides* $e(F[\beta]|F)$ *and* $f(F[\gamma]|F)$ *divides* $f(F[\beta]|F)$.

We remark that in Theorem 1, if $\mathfrak{s}$ is not simple, then $-k_0(\beta, \mathfrak{a}) \leqslant r < -k_0(\gamma, \mathfrak{a})$.

Corollary *Let* $\mathfrak{s} = [\mathfrak{a}, m, r, \beta]$ *be a pure stratum in A. Among the pure strata* $[\mathfrak{a}, m, r, \gamma]$ *in A that are equivalent to* $\mathfrak{s}$*, the simple ones are those for which the degree* $[F[\gamma] : F]$ *is minimal.*

Next, we compare the basic invariants of two equivalent simple strata.

Theorem 2 *For* $i = 1, 2$*, let* $\mathfrak{s}_i = [\mathfrak{a}, m, r, \gamma_i]$ *be a simple stratum in A. If* $\mathfrak{s}_1$ *is equivalent to* $\mathfrak{s}_2$*, then*

(1) $k_0(\gamma_1, \mathfrak{a}) = k_0(\gamma_2, \mathfrak{a})$,
(2) $e(F[\gamma_1]|F) = e(F[\gamma_2]|F)$ *and* $f(F[\gamma_1]|F) = f(F[\gamma_2]|F)$.
(3) Let s_1 *be a tame corestriction on A relative to* $F[\gamma_1]/F$*. There exists* $\delta \in F[\gamma_1]$ *such that*

$$s_1(\gamma_1 - \gamma_2) \equiv \delta \pmod{\mathfrak{p}^{1-r}},$$

where $\mathfrak{p} = \mathrm{rad}\,\mathfrak{a}$.

The final result in the sequence shows that the construction in 2.3.2 yields all simple strata, up to equivalence.

Theorem 3 *Let* $\mathfrak{s} = [\mathfrak{a}, m, r, \beta]$ *be a pure stratum in A with* $r = -k_0(\beta, \mathfrak{a})$*. Let* $[\mathfrak{a}, m, r, \gamma]$ *be a simple stratum equivalent to* $\mathfrak{s}$*. Let B denote the A-centralizer of* γ*, set* $\mathfrak{b} = \mathfrak{a} \cap B$*, and let* $s : A \to B$ *be a tame corestriction relative to* $F[\gamma]/F$*. The stratum* $[\mathfrak{b}, r, r-1, s(\beta - \gamma)]$ *in B is equivalent to a simple stratum* $[\mathfrak{b}, r, r-1, \delta]$ *such that* $\delta \notin F[\gamma]$.

2.3.4 We conclude this survey with a fundamental, "intertwining implies conjugacy", property. This is the first of a sequence of such results, essential to any classification scheme.

Theorem *For* $i = 1, 2$*, let* $\mathfrak{s}_i = [\mathfrak{a}, m, r, \beta_i]$ *be a simple stratum in A. Suppose there exists* $g \in G$ *that formally intertwines* $\mathfrak{s}_1$ *with* $\mathfrak{s}_2$*. There then exists* $x \in U_{\mathfrak{a}}$ *such that* $[\mathfrak{a}, m, r, x^{-1}\beta_1 x]$ *is equivalent to* $[\mathfrak{a}, m, r, \beta_2]$.

2.4 Groups Defined by Simple Strata

We take a simple stratum in $A = \mathrm{End}_F(V)$, of the form $[\mathfrak{a}, m, 0, \beta]$, and attach to it a pair $H^1(\beta, \mathfrak{a}) \subset J^1(\beta, \mathfrak{a})$ of open subgroups of $U^1_{\mathfrak{a}}$. In all cases, we put

$$\left.\begin{aligned} H^k(\beta, \mathfrak{a}) &= H^1(\beta, \mathfrak{a}) \cap U^k_{\mathfrak{a}}, \\ J^k(\beta, \mathfrak{a}) &= J^1(\beta, \mathfrak{a}) \cap U^k_{\mathfrak{a}}, \end{aligned}\right\} \quad k \geqslant 1.$$

2.4.1 The definition of these groups is *inductive along* β. For the first step, let $[\mathfrak{a}, m, 0, \alpha]$ be a simple stratum in which α is *minimal over* F. Thus $k_0(\alpha, \mathfrak{a})$ is either $-m$ or $-\infty$. Let B denote the A-centralizer of α and $\mathfrak{b} = \mathfrak{a} \cap B$. Define

$$H^1(\alpha, \mathfrak{a}) = U_{\mathfrak{b}}^1\, U_{\mathfrak{a}}^{1+[m/2]},$$
$$J^1(\alpha, \mathfrak{a}) = U_{\mathfrak{b}}^1\, U_{\mathfrak{a}}^{[m+1/2]},$$

where $x \mapsto [x]$ is the greatest integer function. In particular, if $\alpha \in F$ then $H^1(\alpha, \mathfrak{a}) = J^1(\alpha, \mathfrak{a}) = U_{\mathfrak{a}}^1$.

2.4.2 For the general step, let $[\mathfrak{a}, m, 0, \beta]$ be a simple stratum in A such that $r = -k_0(\beta, \mathfrak{a}) < m$. Let B denote the A-centralizer of β and put $\mathfrak{b} = \mathfrak{a} \cap B$. Choose a simple stratum $[\mathfrak{a}, m, r, \gamma]$ equivalent to $[\mathfrak{a}, m, r, \beta]$ (as in 2.3.3 Theorem 1), and define

$$H^1(\beta, \mathfrak{a}) = U_{\mathfrak{b}}^1\, H^{1+[r/2]}(\gamma, \mathfrak{a}),$$
$$J^1(\beta, \mathfrak{a}) = U_{\mathfrak{b}}^1\, J^{[r+1/2]}(\gamma, \mathfrak{a}).$$

The definition is independent of the choice of γ. Each of these groups is normalized by $\mathfrak{K}_{\mathfrak{b}}$. We have $H^{1+k}(\beta, \mathfrak{a}) = H^{1+k}(\gamma, \mathfrak{a})$ when $k \geqslant [r/2]$, and similarly for the groups J.

The groups $H^1(\beta, \mathfrak{a}), J^1(\beta, \mathfrak{a})$ depend only on the equivalence class of the simple stratum $[\mathfrak{a}, m, 0, \beta]$. Stronger statements, following from the intertwining properties of simple strata, are given in [25]. They are necessary for the inductive construction but we may omit them.

2.5 *Simple Characters*

Let ψ_F be a character of F of level one, and use the notation ψ_A, $\psi_A * a$ of 1.4.1, 1.4.2. Following another inductive procedure, we define the set of *simple characters* attached to ψ_F and a simple stratum in A of the form $[\mathfrak{a}, m, 0, \beta]$.

2.5.1 Let $[\mathfrak{a}, m, 0, \beta]$ be a simple stratum in A. Let B be the A-centralizer of β and $\mathfrak{b} = \mathfrak{a} \cap B$.

Definition 1 Suppose that β is minimal over F. Define $\mathcal{C}(\mathfrak{a}, \beta, \psi_F)$ to be the set of characters θ of $H^1(\beta, \mathfrak{a}) = U_{\mathfrak{b}}^1 U_{\mathfrak{a}}^{1+[m/2]}$ such that

(1) $\theta(u) = \psi_A * \beta(u)$, $u \in U_{\mathfrak{a}}^{1+[m/2]}$, and
(2) $\theta \,\big|\, U_{\mathfrak{b}}^1$ factors through the determinant map $\det_B : B^\times \to F[\beta]^\times$.

Definition 2 Suppose that β is not minimal over F and set $r = -k_0(\beta, \mathfrak{a})$. Let $[\mathfrak{a}, m, r, \gamma]$ be a simple stratum equivalent to $[\mathfrak{a}, m, r, \beta]$. Define $\mathcal{C}(\mathfrak{a}, \beta, \psi_F)$ to be the set of characters θ of the group $H^1(\beta, \mathfrak{a}) = U_{\mathfrak{b}}^1\, H^{1+[r/2]}(\gamma, \mathfrak{a})$ such that

(1) $\theta \,|\, U^1_{\mathfrak{b}}$ factors through the determinant map $\det_B$ and
(2) there exists $\vartheta \in \mathcal{C}(\mathfrak{a}, \gamma, \psi_F)$ such that

$$\theta(x) = \psi_A * (\beta - \gamma)(x)\, \vartheta(x), \quad x \in H^{1+[r/2]}(\gamma, \mathfrak{a}).$$

This definition is equivalent to that in [25], but that equivalence can only be verified at a later stage. In all cases, the set $\mathcal{C}(\mathfrak{a}, \beta, \psi_F)$ is finite and non-empty.

Remark The dependence of the definition on the character ψ_F is straightforward. For, if $\psi'_F \in \widehat{F}$ has level one, there exists $u \in U_F$ so that $\psi'_F(x) = \psi_F(ux)$, $x \in F$. The quadruple $[\mathfrak{a}, m, 0, u^{-1}\beta]$ is again a simple stratum and

$$\mathcal{C}(\mathfrak{a}, \beta, \psi_F) = \mathcal{C}(\mathfrak{a}, u^{-1}\beta, \psi'_F).$$

There is no harm, therefore, in fixing ψ_F once for all and omitting it from the notation.

In more formal contexts, the following reformulation of the definition is to be preferred.

Definition 3 A *simple character in G* is a smooth character θ of a compact open subgroup K of G with the following property: there exists a simple stratum $[\mathfrak{a}, m, 0, \beta]$ in A such that $K = H^1(\beta, \mathfrak{a})$ and $\theta \in \mathcal{C}(\mathfrak{a}, \beta)$.

2.5.2 A lengthy computation yields a concise formula for the intertwining set $I_G(\theta)$ of a simple character θ. It implies, among other things, that the set $\mathcal{C}(\mathfrak{a}, \beta)$ depends only on the equivalence class of $[\mathfrak{a}, m, 0, \beta]$.

Intertwining Theorem *Let* $[\mathfrak{a}, m, 0, \beta]$ *be a simple stratum in* A *and let* $\theta \in \mathcal{C}(\mathfrak{a}, \beta)$. *If* B *denotes the* A*-centralizer of* β, *then* $I_G(\theta) = J^1 B^\times J^1$, *where* $J^1 = J^1(\beta, \mathfrak{a})$. *The* G*-normalizer of* θ *is* $N_G(\theta) = \mathcal{K}_{\mathfrak{b}} J^1$, *where* $\mathfrak{b} = \mathfrak{a} \cap B$.

It follows that $\mathfrak{a}$ is the unique hereditary $\mathfrak{o}_F$-order in A such that $N_G(\theta) \subset \mathcal{K}_{\mathfrak{a}}$. In other words, the simple character θ determines the order $\mathfrak{a}$ and therefore the parameter m: the character θ is trivial on $U^{1+m}_{\mathfrak{a}}$, and m is the least integer with this property.

2.5.3 There is a more general notion of "truncated" simple character. If $[\mathfrak{a}, m, 0, \beta]$ is a simple stratum in A and l is an integer, $0 \leqslant l < -k_0(\beta, \mathfrak{a})$, we may define $\mathcal{C}(\mathfrak{a}, l, \beta)$ as the set of characters of $H^{1+l}(\beta, \mathfrak{a})$ of the form $\theta \,|\, H^{1+l}(\beta, \mathfrak{a})$, $\theta \in \mathcal{C}(\mathfrak{a}, \beta)$.

If $-k_0(\beta, \mathfrak{a}) \leqslant l < m$, we may equally define $\mathcal{C}(\mathfrak{a}, l, \beta)$ as the set of restrictions $\theta \,|\, H^{1+l}(\beta, \mathfrak{a})$, $\theta \in \mathcal{C}(\mathfrak{a}, \beta)$. We have the following connection.

Proposition *If* $[\mathfrak{a}, m, l, \gamma]$ *is a simple stratum equivalent to* $[\mathfrak{a}, m, l, \beta]$, *then* $H^{1+l}(\beta, \mathfrak{a}) = H^{1+l}(\gamma, \mathfrak{a})$ *and* $\mathcal{C}(\mathfrak{a}, l, \beta) = \mathcal{C}(\mathfrak{a}, l, \gamma)$.

Comment The development in [25] works throughout in terms of truncated simple characters, as a matter of necessity. Once the results are complete, one can use the simpler version here.

2.5.4 A feature of the definition is that a simple stratum gives rise to a finite *set* of simple characters. This relation has a certain coherence.

Proposition 1 *For $i = 1, 2$, let $[\mathfrak{a}, m, 0, \beta_i]$ be a simple stratum in A and suppose that $H^1(\beta_1, \mathfrak{a}) = H^1(\beta_2, \mathfrak{a})$. If $\mathcal{C}(\mathfrak{a}, \beta_1) \cap \mathcal{C}(\mathfrak{a}, \beta_2) \neq \emptyset$, then $\mathcal{C}(\mathfrak{a}, \beta_1) = \mathcal{C}(\mathfrak{a}, \beta_2)$.*

The proposition applies equally to sets of truncated simple characters.

One can recover some properties of the simple stratum $[\mathfrak{a}, m, 0, \beta]$ from the set $\mathcal{C}(\mathfrak{a}, \beta)$ or, therefore, any one of its elements.

Proposition 2 *For $i = 1, 2$, let $[\mathfrak{a}, m, 0, \beta_i]$ be a simple stratum in A and let l be an integer, $0 \leqslant l < m$. Suppose that $H^{1+l}(\beta_1, \mathfrak{a}) = H^{1+l}(\beta_2, \mathfrak{a})$ and that $\mathcal{C}(\mathfrak{a}, l, \beta_1) \cap \mathcal{C}(\mathfrak{a}, l, \beta_2) \neq \emptyset$. If $[\mathfrak{a}, m, l, \gamma_i]$ is a simple stratum equivalent to $[\mathfrak{a}, m, l, \beta_i]$, then*

$$\begin{aligned} k_0(\gamma_1, \mathfrak{a}) &= k_0(\gamma_2, \mathfrak{a}), \\ e(F[\gamma_1] \,|\, F) &= e(F[\gamma_2] \,|\, F), \\ f(F[\gamma_1] \,|\, F) &= f(F[\gamma_2] \,|\, F). \end{aligned}$$

2.5.5 Simple characters have a remarkable "intertwining implies conjugacy" property analogous to that for simple strata (2.3.4).

Theorem *For $i = 1, 2$, let $[\mathfrak{a}, m_i, 0, \beta_i]$ be a simple stratum in A and let $\theta_i \in \mathcal{C}(\mathfrak{a}, \beta_i)$. If θ_1 intertwines in G with θ_2, then*

(1) $m_1 = m_2$ and
(2) there exists $x \in U_{\mathfrak{a}}$ such that $\mathcal{C}(\mathfrak{a}, \beta_2) = \mathcal{C}(\mathfrak{a}, x^{-1}\beta_1 x)$ and $\theta_2 = \theta_1^x$.

Observe here that $\mathcal{C}(\mathfrak{a}, x^{-1}\beta_1 x) = \mathcal{C}(\mathfrak{a}, \beta_1)^x$.

Remark The theorem implies that if an irreducible representation of G contains two simple characters, *attached to isomorphic hereditary orders,* then those two characters are conjugate in G. If the two underlying hereditary orders are not isomorphic, one can only conclude that the two characters intertwine in G. As it turns out, intertwining of simple characters is a relation every bit as strong as conjugation: see 5.2.2 Complement 3.

2.5.6 Let $[\mathfrak{a}, m, 0, \beta]$ be a simple stratum in A and let $\theta \in \mathcal{C}(\mathfrak{a}, \beta)$. An irreducible representation π of G that contains θ necessarily contains an irreducible representation $\tilde{\theta}$ of $J^1 = J^1(\beta, \mathfrak{a})$ that contains θ. The intertwining formula of 2.5.2 implies that $\tilde{\theta} \,|\, H^1(\beta, \mathfrak{a})$ is a multiple of θ. We now account for all such representations $\tilde{\theta}$.

From the definitions, $H^1(\beta, \mathfrak{a})$ is a normal subgroup of $J^1(\beta, \mathfrak{a})$ and the quotient is a finite, elementary abelian p-group. Indeed, it is a finite-dimensional vector space over the residue field $\Bbbk_F$.

Proposition 1 *Let $\theta \in \mathcal{C}(\mathfrak{a}, \beta)$. There is a unique irreducible representation $\eta = \eta_\theta$ of $J^1(\beta, \mathfrak{a})$ such that $\eta \,|\, H^1(\beta, \mathfrak{a})$ contains θ. Further, $\eta \,|\, H^1(\beta, \mathfrak{a})$ is a multiple of θ and $I_G(\eta) = J^1 B^\times J^1 = I_G(\theta)$, where $J^1 = J^1(\beta, \mathfrak{a})$.*

In particular, θ induces to a multiple of η on $J^1(\beta, \mathfrak{a})$. The reasons for the form of the proposition are elementary and familiar. Write $\mathfrak{J}$ for the $\mathbb{F}_p$-vector space $J^1(\beta, \mathfrak{a})/H^1(\beta, \mathfrak{a})$. The pairing

$$(x, y) \longmapsto \theta[x, y] = \theta(xyx^{-1}y^{-1})$$

on $J^1(\beta, \mathfrak{a})$ takes its values in the group of p-th roots of unity in $\mathbb{C}$, which we identify with $\mathbb{F}_p$. The pairing induces an alternating bilinear form on the $\mathbb{F}_p$-space $\mathfrak{J}$ that is *non-degenerate.* It follows that the finite p-group $J^1(\beta, \mathfrak{a})/\mathrm{Ker}(\theta)$ is extra special of class 2 with centre $\mathrm{Im}\,\theta$. The proposition thus comes down to a familiar variation on Heisenberg representations. We sometimes say that η_θ is the 1-*Heisenberg representation* defined by θ.

The intertwining of the representation η has a singular feature of technical importance.

Proposition 2 *For $b \in B^\times$, the space* $\mathrm{Hom}_{(J^1)^b\cap J^1}(\eta^b, \eta)$ *has dimension* 1.

In another direction, the pairing on $J^1(\beta, \mathfrak{a})$ actually only depends on the underlying stratum. If we take elements $1+x, 1+y \in J^1(\beta, \mathfrak{a})$, the value of θ on the commutator $[1+x, 1+y]$ is

$$\theta[1+x, 1+y] = \psi_A\big(\beta(xy - yx)\big), \qquad (2.5.1)$$

[11] 6.1 Proposition.

2.5.7 In order to accommodate representations of level zero, it is convenient to introduce the notion of a *trivial simple character.*

Definition A *trivial simple character* in A is the trivial character of $U^1_\mathfrak{a}$, where $\mathfrak{a}$ is a hereditary $\mathfrak{o}_F$-order in A.

We can treat these on the same basis as before, agreeing that $F[\beta]$ means F, $m = 0$ and $H^1(\beta, \mathfrak{a}) = J^1(\beta, \mathfrak{a}) = U^1_\mathfrak{a}$.

3 Types and Classification

We complete our narrative account of the construction of the cuspidal representations of $\mathrm{GL}_n(F)$, accepting from [25] that the process gives them all. We then see that the method we have used is effectively the only one.

Throughout, $G = \mathrm{GL}_n(F)$ and $A = \mathrm{M}_n(F)$.

3.1 Maximal Types

Let $\mathfrak{s} = [\mathfrak{a}, m, 0, \beta]$ be a simple stratum in A. Let B denote the A-centralizer of β and set $\mathfrak{b} = B \cap \mathfrak{a}$. Write $E = F[\beta]$.

3.1.1 Let $\theta \in \mathcal{C}(\mathfrak{a}, \beta)$ and let $\eta = \eta_\theta$ be the 1-Heisenberg representation over θ, as in 2.5.6. Write

$$J = J(\beta, \mathfrak{a}) = U_{\mathfrak{b}} J^1, \quad J^1 = J^1(\beta, \mathfrak{a}).$$

By the intertwining theorem of 2.5.2, the group J is the $U_{\mathfrak{a}}$-normalizer of the simple character θ. We now have one of the major results of [25].

Theorem *There exists a representation κ of $J(\beta, \mathfrak{a})$ such that $\kappa \mid J^1(\beta, \mathfrak{a}) \cong \eta$ and*

$$I_G(\kappa) = I_G(\eta) = I_G(\theta) = J^1 B^\times J^1.$$

For any such κ and $g \in G$, the space $\mathrm{Hom}_{J^g \cap J}(\kappa^g, \kappa)$ *has dimension at most one.*

The proof of the main assertion, in [25] 5.2, is quite elaborate. The assertion concerning intertwining spaces follows from 2.5.6 Proposition 2.

Definition Let $\mathbb{H}^0(\theta)$ be the set of equivalence classes of irreducible representations of $J(\beta, \mathfrak{a})$ satisfying the conditions of the theorem.

Remarks (1) There are cases in which the group J admits a representation ξ such that $\xi \mid J^1 \cong \eta$ but $\xi \notin \mathbb{H}^0(\theta)$.

(2) The defining condition on κ is independent of the choice of simple stratum giving rise to θ, since it is expressed in terms of intertwining.

(3) In [25], a representation κ satisfying the conditions of the theorem is called a "β-extension" of η. No replacement for this unsatisfactory term has ever been agreed so we avoid the issue by using the notation $\mathbb{H}^0(\theta)$.

The intertwining property of the theorem has a useful consequence.

Corollary *Let $\kappa \in \mathbb{H}^0(\theta)$. For $i = 1, 2$, let σ_i be an irreducible representation of J that is trivial on J^1. The representations $\kappa \otimes \sigma_i$ are irreducible and*

$$I_G(\kappa \otimes \sigma_1, \kappa \otimes \sigma_2) = J^1 I_{B^\times}(\sigma_1, \sigma_2) J^1.$$

The corresponding intertwining spaces have the same dimensions.

The representations $\kappa \in \mathbb{H}^0(\theta)$ have uniqueness properties derived from the intertwining properties. The inclusion $U_{\mathfrak{b}} \subset J$ induces an isomorphism $U_{\mathfrak{b}}/U^1_{\mathfrak{b}} \cong J/J^1$. If ϕ is a character of U_E trivial on U^1_E, we form $\phi \circ \det_B \mid U_{\mathfrak{b}}$ and transfer it to a character $\tilde{\phi}$ of J, trivial on J^1. Clearly, if $\kappa \in \mathbb{H}^0(\theta)$, then $\tilde{\phi} \otimes \kappa \in \mathbb{H}^0(\theta)$.

Proposition *(1) If $\kappa, \kappa' \in \mathbb{H}^0(\theta)$, then κ intertwines with κ' in G if and only if $\kappa = \kappa'$.*

(2) *The pairing* $(\phi, \kappa) \mapsto \tilde{\phi} \otimes \kappa$ *endows* $\mathbb{H}^0(\theta)$ *with the structure of a principal homogeneous space over the group* $(U_E/U_E^1)\widehat{}$.
(3) *There is a unique* $\kappa_0 \in \mathbb{H}^0(\theta)$ *such that the character* $\det \kappa_0$ *has order a power of* p.

The action of $(U_E/U_E^1)\widehat{}$ on $\mathbb{H}^0(\theta)$ cannot be regarded as canonical, owing to the ill-defined nature of the relation between E and θ. We will remedy this defect in Sect. 7.2 below. The representation κ_0 is canonically determined, but is not always the most natural choice.

3.1.2 Since we focus entirely on cuspidal representations, we restrict to a special case. Let $\mathfrak{s} = [\mathfrak{a}, m, 0, \beta]$ be a simple stratum in A. Write $E = F[\beta]$ and let B denote the A-centralizer of E. Set $\mathfrak{b} = \mathfrak{a} \cap B$. Say that $\mathfrak{s}$ is *max-simple* if $\mathfrak{b}$ is a *maximal* $\mathfrak{o}_E$-order in B or, equivalently, if $e(\mathfrak{a}) = e(E|F)$. In this case,

$$U_\mathfrak{b}/U_\mathfrak{b}^1 = J^0(\beta, \mathfrak{a})/J^1(\beta, \mathfrak{a}) \cong \mathrm{GL}_t(\Bbbk_E),$$

where $[E : F] = n/t$. A *max-simple character* in G is a simple character in G defined by a max-simple stratum in A.

We remark that, if $[\mathfrak{a}, m, 0, \beta]$ is max-simple, the hereditary order $\mathfrak{a}$ is *principal:* if $\mathfrak{p} = \mathrm{rad}\, \mathfrak{a}$, then $\mathfrak{p} = \varpi_E \mathfrak{a}$, for any prime element ϖ_E of E.

Definition A *maximal simple type* in G is a pair (J, λ), where J is an open compact subgroup of G and λ is a representation of J, satisfying the following conditions.

(1) There is a max-simple stratum $[\mathfrak{a}, m, 0, \beta]$ in A such that $J = J(\beta, \mathfrak{a})$.
(2) There is a simple character $\theta \in \mathcal{C}(\mathfrak{a}, \beta)$ such that $\lambda = \kappa \otimes \sigma$, where
    (a) $\kappa \in \mathbb{H}^0(\theta)$, and
    (b) σ is inflated from an irreducible cuspidal representation of the finite general linear group $J^0(\beta, \mathfrak{a})/J^1(\beta, \mathfrak{a})$.

Note that, by 3.1.1 Corollary, the representation λ is irreducible. Taken together with 1.2.1 Proposition, the same result implies:

Proposition *Let* (J, λ) *be a maximal simple type in* G *and use the notation of the definition. An element* g *of* G *intertwines* λ *if and only if* $g \in \boldsymbol{J}(\beta, \mathfrak{a}) = E^\times J$.

In particular, the set $\boldsymbol{J}(\beta, \mathfrak{a})$ is the G-normalizer of λ and an open, compact mod. centre subgroup of G. Indeed, $\boldsymbol{J}(\beta, \mathfrak{a})$ is the G-normalizer of the simple character θ contained in λ (2.5.2). The group $J = J(\beta, \mathfrak{a})$ is the unique maximal compact subgroup of $\boldsymbol{J} = \boldsymbol{J}(\beta, \mathfrak{a})$. The quotient $\boldsymbol{J}/J$ is infinite cyclic and $\boldsymbol{J} = F[\beta]^\times J$ contains $F^\times J$ with index $e(\mathfrak{a})$. One can equally recognize $J^1(\beta, \mathfrak{a})$ in canonical terms: it is the unique maximal, normal, pro-p subgroup of J.

3.1.3 Continue in the same situation. Since $\boldsymbol{J}/J$ is infinite cyclic and contains $F^\times J$ with finite index, the representation λ admits extension to a representation Λ of $\boldsymbol{J}$. This extension is uniquely determined up to tensoring with a character $\chi \circ \det_G$, where χ is an *unramified* character of $F^\times$ (*cf.* 3.1.1 Proposition).

Lemma *Let Λ be a representation of $\boldsymbol{J}$ that extends λ. Let χ be an unramified character of $F^\times$ and write $\chi\Lambda$ for the representation $x \mapsto \chi(\det_G x)\Lambda(x)$, $x \in \boldsymbol{J}$. The following conditions are equivalent.*

(1) The representations Λ, $\chi\Lambda$ intertwine in G.
(2) The representations Λ, $\chi\Lambda$ are equivalent.
(3) The character χ satisfies $\chi^{n/e(F[\beta]|F)} = 1$.

This follows readily from the intertwining properties of κ. A pair of this form $(\boldsymbol{J}, \Lambda)$ will be called an *extended maximal simple type* in G. We abbreviate this term to *EMST*.

3.1.4 We summarize the implications of the discussion in 3.1.2, 3.1.3.

Construction Theorem *If $(\boldsymbol{J}, \Lambda)$ is an EMST in G, the representation $\pi_\Lambda = c\text{-}\mathrm{Ind}_{\boldsymbol{J}}^G\,\Lambda$ is irreducible and cuspidal.*

There is a complementary uniqueness property.

Uniqueness Theorem *For $i = 1, 2$, let $(\boldsymbol{J}_i, \Lambda_i)$ be an EMST in G. The following conditions are equivalent.*

(1) The representations π_{Λ_1}, π_{Λ_2} are equivalent.
(2) The representations Λ_1, Λ_2 intertwine in G.
(3) The representations Λ_1, Λ_2 are conjugate in G.

Proof The equivalence of (1) and (2) is a formal property of intertwining and compact induction, given that the induced representations c-Ind Λ_i are both irreducible. Surely (3) implies (1), so we show that (1) implies (3). The representation Λ_i is given in terms of a max-simple stratum $[\mathfrak{a}_i, m_i, 0, \beta_i]$ in A, in which $\mathfrak{a}_i$ is a principal order of period n/d_i, where d_i is the number of unramified characters χ such that $\chi\pi_{\Lambda_i} = \pi_{\Lambda_i}$. Statement (1) implies that $d_1 = d_2$ whence $\mathfrak{a}_1 \cong \mathfrak{a}_2$. The simple characters θ_1, θ_2 contained in Λ_1, Λ_2 intertwine in G, and so are conjugate in G (2.5.5). We may as well assume, therefore, that $\mathfrak{a}_1 = \mathfrak{a}_2 = \mathfrak{a}$ say, $\theta_1 = \theta_2 = \theta$ and $\beta_1 = \beta_2 = \beta$. The representations $\lambda_i = \Lambda_i \,|\, J(\beta, \mathfrak{a})$ intertwine in G. The intertwining element g, say, intertwines θ so we may take $g \in B^\times$, where B is the A-centralizer of β. Write $\lambda_i = \kappa \otimes \sigma_i$, $\kappa \in \mathbb{H}^0(\theta)$ as in 3.1.1. The Corollary of 3.1.1 implies that g intertwines σ_1 with σ_2 and 1.2.1 Proposition implies $\sigma_1 \cong \sigma_2$. We may as well take $\sigma_1 = \sigma_2$ whence $\lambda_1 = \lambda_2$ and $\Lambda_2 = \chi\Lambda_1$, for some unramified character χ of $F^\times$. The result now follows from 3.1.3 Lemma. □

3.1.5 To complete the picture, we need the following result that combines Theorems (6.2.1) and (8.4.1) of [25].

Exhaustion Theorem *If π is an irreducible cuspidal representation of $G = \mathrm{GL}_n(F)$, then π contains some EMST $(\boldsymbol{J}, \Lambda)$ in G. In particular,*

$$\pi \cong \pi_\Lambda = c\text{-}\mathrm{Ind}_{\boldsymbol{J}}^G\,\Lambda.$$

We give an outline of the strategy of the proof of the Exhaustion Theorem in 5.1.4, after we have recalled some more machinery.

With the Exhaustion Theorem in hand, we can extract a consequence of earlier arguments.

Corollary *Let π_1, π_2 be irreducible cuspidal representations of $G = \mathrm{GL}_n(F)$. The following conditions are equivalent.*

(1) There is an unramified character χ of $F^\times$ such that $\pi_2 = \chi\pi_1$.
(2) The representations π_1, π_2 contain isomorphic maximal simple types.

3.1.6 If π is an irreducible cuspidal representation of G, containing the EMST $(\boldsymbol{J}, \Lambda)$, it surely contains the max-simple character θ on which $\boldsymbol{J}$ and Λ are constructed. Any two simple characters obtained in this way are conjugate. However, it is not clear whether π could contain an essentially different simple character θ'. Back-engineering of the arguments of [25] Chapter 8 yields the following clarification [17].

Theorem *Let π be an irreducible cuspidal representation of $G = \mathrm{GL}_n(F)$.*

(1) A simple character contained in π is max-simple.
(2) Any two simple characters contained in π are conjugate in G.

So, let π contain a simple character θ. Let $\boldsymbol{J}$ be the G-normalizer of θ. The natural representation of $\boldsymbol{J}$ on the θ-isotypic subspace of π is then the EMST Λ for which $\pi = \pi_\Lambda$.

Comment Let θ be a simple character in $G = \mathrm{GL}_n(F)$. If θ is max-simple, we know that θ occurs in an irreducible cuspidal representation of G and also that any irreducible cuspidal representation contains some max-simple character. However, except in the trivial case of $n = 1$, a max-simple character will occur in a non-cuspidal irreducible representation. More generally, for arbitrary θ, one can describe exactly the irreducible representations of G that contain θ [17].

3.1.7 To conclude the section, we record some technical developments that will be needed later in the notes.

If E/F is a finite field extension, write $X_1(E)$ (resp. $X_0(E)$) for the group of tamely ramified (resp. unramified) characters of $E^\times$. If s is a positive integer, let $X_0(E)_s$ be the group of $\chi \in X_0(E)$ such that $\chi^s = 1$.

Let θ be a max-simple character in $G = \mathrm{GL}_n(F)$. Let $\boldsymbol{J}_\theta$ denote the G-normalizer of θ. Let J_θ be the maximal compact subgroup of $\boldsymbol{J}_\theta$ and J^1_θ the pro-p radical of J_θ.

Definition 1 Let $X_1(\theta)$ be the group of characters χ of $\boldsymbol{J}_\theta$ such that

(1) χ is trivial on J^1_θ and
(2) χ is intertwined every $g \in I_G(\theta)$.

To describe the group $X_1(\theta)$, choose a simple stratum $[\mathfrak{a}, m, 0, \beta]$ such that $\theta \in \mathcal{C}(\mathfrak{a}, \beta)$. Write $E = F[\beta]$, let G_E be the G-centralizer of $E^\times$ and let $\det_E : G_E \to E^\times$ be the determinant map. Thus

$$\boldsymbol{J}_\theta = \boldsymbol{J}(\beta, \mathfrak{a}) = G_E \cap \boldsymbol{J}_\theta \cdot J^1_\theta.$$

Let $s = n/[E : F]$. From [19] 3.1, we have:

Lemma *Let $\phi \in X_1(E)$. There exists a unique character $\phi_\theta \in X_1(\theta)$ such that*

$$\phi_\theta(x) = \phi(\det_E x), \quad x \in G_E \cap \boldsymbol{J}_\theta.$$

The map $\phi \mapsto \phi_\theta$ is a surjective homomorphism $X_1(E) \to X_1(\theta)$ with kernel $X_0(E)_s$.

Let $\eta = \eta_\theta$ be the unique irreducible representation of J^1_θ that contains θ.

Definition 2 Let $\mathbb{H}(\theta)$ be the set of equivalence classes of representations $\boldsymbol{\kappa}$ of $\boldsymbol{J}_\theta$ such that

(1) $\boldsymbol{\kappa} \,\big|\, J^1_\theta = \eta$ and
(2) $\boldsymbol{\kappa}$ is intertwined by every element of $I_G(\theta)$.

Thus we have a restriction map $\mathbb{H}(\theta) \to \mathbb{H}^0(\theta)$ (notation of 3.1.1) and the following stronger version of 3.1.1 Proposition, adapted from [19] 3.2.

Proposition *The group $X_1(\theta)$ acts on $\mathbb{H}(\theta)$ by $(\chi, \boldsymbol{\kappa}) \mapsto \chi \otimes \boldsymbol{\kappa}$. Relative to this action, $\mathbb{H}(\theta)$ is a principal homogeneous space over $X_1(\theta)$. The restriction map $\mathbb{H}(\theta) \to \mathbb{H}^0(\theta)$ is surjective.*

3.2 *Inertial Uniqueness*

At this stage, it is natural to enquire whether the classification we have just given is the best available. We recall a result of Paskunas indicating that it is, in effect, the only one.

3.2.1 We quote from [62] Theorem 1.3.

Theorem *Let K be a maximal compact subgroup of $G = \mathrm{GL}_n(F)$ and let π be an irreducible cuspidal representation of G. There exists at most one irreducible representation ϱ of K such that*

(a) π contains ϱ and,
(b) an irreducible representation π' of G contains ϱ if and only if $\pi' \cong \chi\pi$, for some unramified character χ of $F^\times$.

The assertion of the theorem is independent of the maximal compact subgroup K, since any two such subgroups are conjugate in G.

A given irreducible cuspidal representation π of G contains a maximal simple type (J, λ). The group J is contained in a finite number of maximal compact subgroups K. Choosing one at random, the representation $\varrho = c\text{-Ind}_J^K \lambda$ is irreducible and satisfies the requirements of the theorem.

4 Encounter with the Langlands Correspondence

This section introduces the second part of the notes, concerning connections between the Langlands correspondence and the classification theory of the first three sections. Here, I give a preliminary list of the available techniques, looking quickly at the Galois side for orientation. Before they can be used effectively, most of these techniques require further development of the theory of simple characters. I treat them one at a time in sections to follow. In this one, I give just one application, namely a formula for the Godement–Jacquet local constant of an irreducible cuspidal representation of $\mathrm{GL}_n(F)$ in terms of the extended maximal simple type it contains.

4.1 The Langlands Correspondence

Recall that $\mathcal{A}_n^0(F)$, $n \geqslant 1$, is the set of equivalence classes of irreducible cuspidal representations of $\mathrm{GL}_n(F)$ and $\widehat{\mathrm{GL}}_F = \bigcup_{n \geqslant 1} \mathcal{A}_n^0(F)$. Write $\mathrm{gr}(\pi) = n$ to indicate that $\pi \in \mathcal{A}_n^0(F)$. Let $\widehat{\mathcal{W}}_F$ be the set of equivalence classes of irreducible representations of $\mathcal{W}_F$ and let

$$\begin{aligned} \widehat{\mathrm{GL}}_F &\longrightarrow \widehat{\mathcal{W}}_F, \\ \pi &\longmapsto {}^L\pi, \end{aligned}$$

denote the Langlands correspondence.

4.1.1 We first recall the standard characterization of the Langlands correspondence. Let $\pi_1, \pi_2 \in \widehat{\mathrm{GL}}_F$. Let ψ be a non-trivial character of F and s a complex variable. To the pair (π_1, π_2) are attached the L-function $L(\pi_1 \times \pi_2, s)$ and the local constant $\varepsilon(\pi_1 \times \pi_2, s, \psi)$, as in [47, 71]. On the other side, let $\sigma_1, \sigma_2 \in \widehat{\mathcal{W}}_F$. To the semisimple representation $\sigma_1 \otimes \sigma_2$ of $\mathcal{W}_F$ are attached the Artin L-function $L(\sigma_1 \otimes \sigma_2, s)$ and the Langlands–Deligne local constant $\varepsilon(\sigma_1 \otimes \sigma_2, s, \psi)$ [74] or [15] 29.4.

Theorem *The Langlands correspondence is the unique map $\widehat{\mathrm{GL}}_F \to \widehat{\mathcal{W}}_F$ with the following properties.*

(1) If $\pi \in \widehat{\mathrm{GL}}_F$, then $\dim {}^L\pi = \mathrm{gr}(\pi)$.
(2) If χ is a character of $F^\times = \mathrm{GL}_1(F)$, then ${}^L\chi$ is the character of $\mathcal{W}_F$ attached to χ by local class field theory.

(3) If $\pi_1, \pi_2 \in \widehat{\mathrm{GL}}_F$ *and* $1 \leqslant \mathrm{gr}(\pi_1) < \mathrm{gr}(\pi_2)$*, then*

$$\begin{aligned} L({}^L\pi_1 \otimes {}^L\pi_2, s) &= L(\pi_1 \times \pi_2, s), \\ \varepsilon({}^L\pi_1 \otimes {}^L\pi_2, s, \psi) &= \varepsilon(\pi_1 \times \pi_2, s, \psi). \end{aligned} \tag{4.1.1}$$

The correspondence is bijective, and the relations (4.1.1) hold for all $\pi_1, \pi_2 \in \widehat{\mathrm{GL}}_F$*.*

For this version, see [38]. Recent work of Jacquet and Liu [46] shows that the bound in (3) can be replaced by $1 \leqslant \mathrm{gr}(\pi_1) \leqslant \frac{1}{2}\mathrm{gr}(\pi_2)$. In (2), we often fail to distinguish between χ and ${}^L\chi$. Observe the following consequences.

Corollary *If* $\pi \in \widehat{\mathrm{GL}}_F$ *and* $\sigma = {}^L\pi$*, then:*

(1) ${}^L\check{\pi} = \check{\sigma}$;
(2) ${}^L(\chi\pi) = {}^L\chi \otimes \sigma$*, for any character* χ *of* $F^\times$*;*
(3) $\det\sigma = {}^L\omega_\pi$*, where* ω_π *is the central character of* π*.*

Recall also that $L(\pi_1 \times \pi_2, s) = 1$ except when π_2 is an unramified twist $\chi\check{\pi}_1$ of the contragredient $\check{\pi}_1$ of π_1.

4.1.2 A major obstruction is that the local constant $\varepsilon(\pi \times \pi', s, \psi)$ remains resistant to effective calculation: given the amount of information it contains, this is perhaps unsurprising. However, there are a number of other techniques that can be used to investigate the correspondence.

(1) Take $\pi \in \widehat{\mathrm{GL}}_F$ and let 1 denote the trivial character of $\mathrm{GL}_1(F) \cong F^\times$. The local constant

$$\varepsilon(\pi, s, \psi) = \varepsilon(\pi \times 1, s, \psi)$$

is that of Godement and Jacquet [33]. If π contains the extended maximal simple type $(\boldsymbol{J}, \Lambda)$, there is an explicit formula for $\varepsilon(\pi, s, \psi)$ in terms of Λ (Sect. 4.3).

(2) Let $\pi_i \in \mathcal{A}^0_{n_i}(F)$, $i = 1, 2$, and let $q = q_F$ be the cardinality of the residue field $\Bbbk_F$. The local constant has an expression

$$\varepsilon(\pi_1 \times \pi_2, s, \psi) = \varepsilon(\pi_1 \times \pi_2, 0, \psi)\, q^{-(a(\pi_1 \times \pi_2) + n_1 n_2 c(\psi))s}$$

in which $c(\psi)$ is an integer depending only on ψ. The term $a(\pi_1 \times \pi_2)$ is an integer, called the *Rankin–Selberg exponent* of the pair (π_1, π_2). It is equal to the Artin exponent $a({}^L\pi_1 \otimes {}^L\pi_2)$. There is an explicit formula for $a(\pi_1 \times \pi_2)$ in terms of the simple characters occurring in the factors π_i (Sect. 5.3).

Continuing with our list, let $\mathcal{A}^{\mathrm{tm}}_n(F)$ be the set of equivalence classes of irreducible *tempered* representations of $\mathrm{GL}_n(F)$, and set $\widehat{\mathrm{GL}}{}^{\mathrm{tm}}_F = \bigcup_{n \geqslant 1} \mathcal{A}^{\mathrm{tm}}_n(F)$. On the other side, let $\widehat{\mathcal{W}}{}^{\mathrm{ss}}_F$ be the set of equivalence classes of finite-dimensional *semisimple* representations of $\mathcal{W}_F$. Thus $\widehat{\mathrm{GL}}_F \subset \widehat{\mathrm{GL}}{}^{\mathrm{tm}}_F$ and $\widehat{\mathcal{W}}_F \subset \widehat{\mathcal{W}}{}^{\mathrm{ss}}_F$. The Langlands correspondence extends canonically to a bijection $\widehat{\mathrm{GL}}{}^{\mathrm{tm}}_F \to \widehat{\mathcal{W}}{}^{\mathrm{ss}}_F$ that we continue to denote by $\pi \mapsto {}^L\pi$.

If K/F is a finite *cyclic* field extension, there are canonical maps

$$\begin{aligned} \boldsymbol{b}_{K/F} &: \widehat{\mathrm{GL}}_F^{\mathrm{tm}} \longrightarrow \widehat{\mathrm{GL}}_K^{\mathrm{tm}}, \\ \boldsymbol{a}_{K/F} &: \widehat{\mathrm{GL}}_K^{\mathrm{tm}} \longrightarrow \widehat{\mathrm{GL}}_F^{\mathrm{tm}}, \end{aligned}$$

known as *base change* and *automorphic induction.* They correspond respectively to the restriction and induction maps

$$\begin{aligned} \mathrm{Res}_{K/F} &: \widehat{\mathcal{W}}_F^{\mathrm{ss}} \longrightarrow \widehat{\mathcal{W}}_K^{\mathrm{ss}}, \\ \mathrm{Ind}_{K/F} &: \widehat{\mathcal{W}}_K^{\mathrm{ss}} \longrightarrow \widehat{\mathcal{W}}_F^{\mathrm{ss}}, \end{aligned}$$

on representations of the Weil groups. That is,

$$\begin{aligned} {}^L\boldsymbol{b}_{K/F}(\pi) &= {}^L\pi \,\big|\, \mathcal{W}_K, \quad \pi \in \widehat{\mathrm{GL}}_F^{\mathrm{tm}}, \\ {}^L\boldsymbol{a}_{K/F}(\rho) &= \mathrm{Ind}_{K/F}\, {}^L\rho, \quad \rho \in \widehat{\mathrm{GL}}_K^{\mathrm{tm}}. \end{aligned}$$

In the case of F having characteristic zero, see [1] for base change and [39] for automorphic induction. The extension of both theories to characteristic p is achieved in [40, 41]. The final item in our list of available techniques is:

(3) If K/F is finite and tamely ramified (but not necessarily Galois), there are analogues of base change and automorphic induction for simple characters (Sect. 6.2).

In some circumstances, one can even identify the effect of base change or automorphic induction on *extended maximal simple types,* as in [9].

4.2 *Inertial Parameters*

We make a first, straightforward connection.

4.2.1 Let $\mathcal{I}_F$ be the inertia subgroup of $\mathcal{W}_F$ and $\mathcal{P}_F$ the wild inertia subgroup. Let $\widehat{\mathcal{I}}_F$ be the set of equivalence classes of irreducible representations of the profinite group $\mathcal{I}_F$. Likewise define $\widehat{\mathcal{P}}_F$. The group $\mathcal{W}_F$ acts on each of these sets by conjugation. There are surjective restriction maps

$$\widehat{\mathcal{W}}_F \xrightarrow{\ r_F\ } \mathcal{W}_F\backslash\widehat{\mathcal{I}}_F$$

and

$$\widehat{\mathcal{W}}_F \xrightarrow{\ r_F^+\ } \mathcal{W}_F\backslash\widehat{\mathcal{P}}_F.$$

For example, if $\sigma \in \widehat{\mathcal{W}}_F$, then $r_F(\sigma)$ is the $\mathcal{W}_F$-orbit of an irreducible component of $\sigma \,\big|\, \mathcal{I}_F$.

On the other side, let $n \geqslant 1$ and define $\mathcal{T}_n(F)$ to be the set of conjugacy classes of extended maximal simple types in $\mathrm{GL}_n(F)$. Set $\mathcal{T}(F) = \bigcup_{n\geqslant 1} \mathcal{T}_n(F)$. Define $\mathcal{T}^0_n(F)$ to be the set of conjugacy classes of maximal simple types in $\mathrm{GL}_n(F)$. Set $\mathcal{T}^0(F) = \bigcup_{n\geqslant 1} \mathcal{T}^0_n(F)$. We have a diagram

$$\begin{array}{ccc} \mathcal{T}(F) & \xrightarrow{\approx} & \widehat{\mathcal{W}}_F \\ \big\downarrow & & \big\downarrow{\scriptstyle r_F} \\ \mathcal{T}^0(F) & & \mathcal{W}_F\backslash\widehat{\mathcal{I}}_F \end{array}$$

in which the horizontal arrow combines the Langlands correspondence and Classification Theorem, while the vertical arrows are restriction maps.

Proposition *The Langlands correspondence induces a canonical bijection*

$$\mathcal{T}^0(F) \xrightarrow{\approx} \mathcal{W}_F\backslash\widehat{\mathcal{I}}_F$$

so that the diagram

$$\begin{array}{ccc} \mathcal{T}(F) & \xrightarrow{\approx} & \widehat{\mathcal{W}}_F \\ \big\downarrow & & \big\downarrow \\ \mathcal{T}^0(F) & \xrightarrow{\approx} & \mathcal{W}_F\backslash\widehat{\mathcal{I}}_F \end{array}$$

commutes.

Proof Let $X_0(F)$ denote the group of unramified characters of $F^\times$ (or of $\mathcal{W}_F$). The group $X_0(F)$ acts on $\widehat{\mathcal{W}}_F$ by $(\chi, \sigma) \mapsto \chi \otimes \sigma$, and the orbit space $X_0(F)\backslash\widehat{\mathcal{W}}_F$ is canonically identified with $\mathcal{W}_F\backslash\widehat{\mathcal{I}}_F$. On the other hand, $X_0(F)$ acts on $\widehat{\mathrm{GL}}_F$ by $(\chi, \pi) \mapsto \chi\pi$. We transfer this to an action of $X_0(F)$ on $\mathcal{T}(F)$, as in 3.1.3. The orbit space $X_0(F)\backslash\mathcal{T}(F)$ is then identified with $\mathcal{T}^0(F)$ (3.1.5 Corollary). Since the Langlands correspondence respects these actions of $X_0(F)$, we have the result. □

We will dedicate several sections to identifying, finally in Sect. 6.3, the automorphic object that corresponds to the space $\mathcal{W}_F\backslash\widehat{\mathcal{P}}_F$. Once that is done, we give an explicit account of the map $\mathcal{T}^0(F) \to \mathcal{W}_F\backslash\widehat{\mathcal{I}}_F$ (8.4.3 Theorem).

4.3 Local Constants

We deal with item (1) on the list in 4.1.2.

Let $\pi \in \mathcal{A}_n^0(F)$, $n \geqslant 2$, and let ψ_F be a character of F of level one: in terms of 4.1.2 (2), this is equivalent to $c(\psi_F) = -1$. We give a formula for the Godement–Jacquet local constant $\varepsilon(\pi, s, \psi_F)$, leading to a Galois-theoretic interpretation of some properties of the simple character contained in π.

We shall exclude the case $n = 1$ for administrative convenience, although everything is valid there provided π is not unramified: see, for example, [15] section 23 for the exceptional case. The effect of changing the additive character is minor and well-understood, so we stick with the one we have been using.

4.3.1 Let $\pi \in \mathcal{A}_n^0(F), n \geqslant 2$, and let $(\boldsymbol{J}, \Lambda)$ be an EMST contained in π. In particular, Λ is attached to a principal $\mathfrak{o}_F$-order $\mathfrak{a}$ in $A = \mathrm{M}_n(F)$. If π has level zero, we set $m = 0$. Otherwise, Λ contains a simple character $\theta \in \mathcal{C}(\mathfrak{a}, \beta, \psi_F)$, for some simple stratum $[\mathfrak{a}, m, 0, \beta]$ in A.

Let ϱ be the representation of $\mathcal{K}_\mathfrak{a}$ induced by Λ. Thus ϱ is irreducible, it is trivial on $U_\mathfrak{a}^{1+m}$, and m is the least integer with this property.

Lemma *Let ϱ act on the vector space X, and let $c \in \mathcal{K}_\mathfrak{a}$ satisfy $c\mathfrak{a} = \mathfrak{p}^{-m}$. The operator*

$$T(\varrho, \psi_F) = \sum_{x \in U_\mathfrak{a}/U_\mathfrak{a}^{1+m}} \check{\varrho}(cx)\, \psi_A(cx)$$

on $X^\vee$ is scalar. Its unique eigenvalue $\tau(\varrho, \psi_F)$ is nonzero and independent of the choice of c.

Allowing for some minor changes of convention, this is (2.3.4) Lemma of [7]. In the language of [7], the representation ϱ is non-degenerate, so (3.3.8) Theorem of that paper asserts the following.

Theorem *Using the preceding notation,*

$$\varepsilon(\pi, s, \psi_F) = (\mathfrak{p}^{-m} : \mathfrak{a})^{(\frac{1}{2}-s)/n}\, \frac{\tau(\varrho, \psi_F)}{(\mathfrak{a} : \mathfrak{p}^{1+m})^{\frac{1}{2}}}\,.$$

4.3.2 We recast 4.3.1 Theorem in a more usable form. In the notation of 4.1.2, the character ψ_F satisfies $c(\psi_F) = -1$ so

$$\varepsilon(\pi, s, \psi_F) = \varepsilon(\pi, \tfrac{1}{2}, \psi_F)\, q^{(\frac{1}{2}-s)(a(\pi)-n)}, \quad \pi \in \mathcal{A}_n^0(F),\ n \geqslant 2,$$

where $a(\pi)$ is the *Artin exponent* of π. For $n \geqslant 2$, one writes $\mathrm{sw}(\pi) = a(\pi) - n$, and calls $\mathrm{sw}(\pi)$ the *Swan exponent* of π. For $n = 1$, the same applies except when π is an *unramified* character of $F^\times = \mathrm{GL}_1(F)$. In that case, $\mathrm{sw}(\pi) = a(\pi) = 0$.

Corollary *Let $\pi \in \mathcal{A}_n^0(F)$, $n \geqslant 2$.*

(1) If π has level zero, then $a(\pi) = n$ and $\mathrm{sw}(\pi) = 0$.

(2) *Let π have positive level and contain a simple character $\theta \in \mathcal{C}(\mathfrak{a}, \beta)$, for some simple stratum $[\mathfrak{a}, m, 0, \beta]$ in $A = \mathrm{M}_n(F)$. If $d(\pi)$ is the number of unramified characters χ of $F^\times$ for which $\chi\pi \cong \pi$, then*

$$\mathrm{sw}(\pi) = md(\pi). \tag{4.3.1}$$

Proof The condition $\ell(\pi) = 0$ is equivalent to $m = 0$, whence follows (1). In part (2), note that $\mathfrak{a}/\mathfrak{p}$ is a $\Bbbk_F$-vector space of dimension $e(\mathfrak{a})d(\pi)^2 = nd(\pi)$. The order $\mathfrak{a}$ is principal, so $(\mathfrak{p}^{-m} : \mathfrak{a}) = q^{md(\pi)}$, $q = |\Bbbk_F|$, as required. □

For $\sigma \in \widehat{\mathcal{W}}_F$, $\dim\sigma \geqslant 2$, the Langlands–Deligne local constant has the form

$$\varepsilon(\sigma, s, \psi_F) = q^{(\frac{1}{2}-s)\mathrm{sw}(\sigma)}\,\varepsilon(\sigma, \tfrac{1}{2}, \psi_F).$$

So, if $\sigma = {}^L\pi$, with π as in the corollary, the Swan exponent $\mathrm{sw}(\sigma)$ of σ is

$$\mathrm{sw}(\sigma) = md(\sigma) = md(\pi), \tag{4.3.2}$$

in the obvious notation.

4.3.3 Continue with the notation of 4.3.1.

Suppose first that π has level zero. Thus $\mathcal{K}_\mathfrak{a} = \boldsymbol{J}$, $\varrho = \Lambda$ and we may as well take $\mathfrak{a} = \mathrm{M}_n(\mathfrak{o}_F)$. The integer m is zero and, in 4.3.1 Lemma, we may take $c = 1$. The Gauss sum $\tau(\varrho, \psi_F)$ is the unique eigenvalue of the operator

$$\sum_{x \in U_\mathfrak{a}/U^1_\mathfrak{a}} \Lambda^\vee(x)\,\psi_A(x).$$

Taking traces,

$$\varepsilon(\pi, \tfrac{1}{2}, \psi_F) = \frac{1}{q^n \dim\Lambda} \sum_{x \in U_\mathfrak{a}/U^1_\mathfrak{a}} \mathrm{tr}\,\lambda(x^{-1})\,\psi_A(x), \tag{4.3.3}$$

where $\lambda = \Lambda \,\big|\, U_\mathfrak{a}$. The representation λ is inflated from an irreducible cuspidal representation $\bar\lambda$ of $\mathrm{GL}_n(\Bbbk_F)$, and (4.3.3) only depends on the representation $\bar\lambda$. The sum in (4.3.3) reduces to a Kondo Gauss sum [49, 57].

Suppose therefore that π has positive level, and use the notation of 4.3.2 Corollary 2. The natural representation of $\boldsymbol{J} = \boldsymbol{J}(\beta, \mathfrak{a})$ on the θ-isotypic subspace of ϱ is equivalent to Λ. Consequently,

$$\varepsilon(\pi, \tfrac{1}{2}, \psi_F) = \frac{1}{(\mathfrak{a} : \mathfrak{p}^{m+1})^{\frac{1}{2}} \dim\Lambda} \sum_{x \in J/U^{1+m}_\mathfrak{a}} \mathrm{tr}\,\Lambda^\vee(\beta x)\,\psi_A(\beta x),$$

where $J = J(\beta, \mathfrak{a})$. Equivalently:

Proposition *The constant* $\varepsilon(\pi, \frac{1}{2}, \psi_F)$ *is the unique eigenvalue* $\tau(\Lambda, \beta, \psi_F)$ *of the operator*

$$T(\Lambda, \beta, \psi_F) = (\mathfrak{a} : \mathfrak{p}^{1+m})^{-\frac{1}{2}} \sum_{x \in J/U_{\mathfrak{a}}^{1+m}} \Lambda^\vee(\beta x)\, \psi_A(\beta x). \tag{4.3.4}$$

The formula for the Gauss sum $\tau(\Lambda, \beta, \psi_F)$ can be simplified to various degrees. To give the easiest example, write $x = y(1+z)$, where $y \in J/U_{\mathfrak{a}}^m$ and $1+z \in U_{\mathfrak{a}}^m/U_{\mathfrak{a}}^{1+m}$. For a fixed y, consider the sum

$$\sum_z \Lambda^\vee(\beta y(1+z))\, \psi_A(\beta y(1+z)) = \Lambda^\vee(\beta y)\psi_A(\beta y) \sum_z \Lambda^\vee(1+z)\, \psi_A(\beta yz).$$

For $z \in \mathfrak{p}^m$, the operator $\Lambda^\vee(1+z)$ is scalar with eigenvalue $\psi_A(-\beta z)$. So, the inner sum vanishes if $y \notin J^1 = J^1(\beta, \mathfrak{a})$. Otherwise, it takes the value $(U_{\mathfrak{a}}^m : U_{\mathfrak{a}}^{1+m}) = q^{nd(\pi)}$. Therefore $\tau(\Lambda, \beta, \psi_F)$ is the unique eigenvalue of the operator

$$q^{nd(\pi)}\, \Lambda^\vee(\beta y)\, \psi_A(\beta y) \sum_{1+z \in J^1/U_{\mathfrak{a}}^m} \Lambda^\vee(1+z)\, \psi_A(\beta z). \tag{4.3.5}$$

Corollary *Let* $\pi \in \mathcal{A}_n^0(F)$, $n \geqslant 2$, *contain a simple character* $\theta \in \mathcal{C}(\mathfrak{a}, \beta, \psi_F)$, *for a simple stratum* $[\mathfrak{a}, m, 0, \beta]$. *If* χ *is a tamely ramified character of* $F^\times$, *then*

$$\varepsilon(\chi\pi, s, \psi_F) = \chi(\det \beta)^{-1}\, \varepsilon(\pi, s, \psi_F).$$

One can use the same device to further restrict the range of summation in (4.3.5) and so generalize the corollary. See [9] section 6 for some examples.

5 Transfer, Endo-equivalence and Applications

This section introduces a fundamental property of "transfer" of simple characters between different hereditary orders. A first version (here in Sect. 5.1) is used in [25] to analyse a single representation and its Jacquet modules. It is the foundation on which the proof of the Exhaustion Theorem is erected: we give a brief outline of its rôle in 5.1.4. In Sect. 5.2, we follow [8] to organize the transfer process into an equivalence relation on the entire corpus of simple characters in all groups $\mathrm{GL}_n(F)$, $n \geqslant 1$. This version provides a framework for comparing representations of different groups and also, in some circumstances, different base fields. In this section, we use it in an explicit formula for the Rankin–Selberg exponent $a(\pi_1 \times \pi_2)$, $\pi_1, \pi_2 \in \widehat{\mathrm{GL}}_F$.

5.1 Transfer of Simple Characters

5.1.1 Let $[\mathfrak{a}, m, 0, \beta]$ be a simple stratum in $A = \mathrm{End}_F(V)$, where V is some finite-dimensional F-vector space. Write $E = F[\beta]$ and let V' be a finite-dimensional E-vector space. Let $\mathfrak{a}'$ be an E-pure hereditary $\mathfrak{o}_F$-order in $A' = \mathrm{End}_F(V')$. If $m' = me(\mathfrak{a}')/e(\mathfrak{a})$, then m' is an integer and $[\mathfrak{a}', m', 0, \beta]$ is a simple stratum in A' (2.2.1). We follow [25] 3.6 to construct a bijection

$$\mathrm{t}^{\beta}_{\mathfrak{a},\mathfrak{a}'} : \mathcal{C}(\mathfrak{a}, \beta, \psi_F) \xrightarrow{\approx} \mathcal{C}(\mathfrak{a}', \beta, \psi_F). \tag{5.1.1}$$

This bijection, called a "transfer", is to be transitive with regard to the hereditary orders $\mathfrak{a}$, $\mathfrak{a}'$ and canonical for a fixed choice of β.

Example Suppose that β is *minimal* over F. Let $E = F[\beta]$, let B be the A-centralizer of E and $\mathfrak{b} = \mathfrak{a} \cap B$. Thus $H^1(\beta, \mathfrak{a}) = U^1_{\mathfrak{b}} U^{1+[m/2]}_{\mathfrak{a}}$. If $\theta \in \mathcal{C}(\mathfrak{a}, \beta)$, there is a unique character χ of U^1_E such that $\theta \mid U^1_{\mathfrak{b}} = \chi \circ \det_B$. We write $\theta = \theta_\chi$. There is a unique $\theta'_\chi \in \mathcal{C}(\mathfrak{a}', \beta)$ such that $\theta' \mid U^1_{\mathfrak{b}'} = \chi \circ \det_{B'}$, in the obvious notation. Here we may set $\mathrm{t}^{\beta}_{\mathfrak{a},\mathfrak{a}'}(\theta_\chi) = \theta'_\chi$ (but this *ad hoc* approach is inefficient in general).

5.1.2 Revert to a general simple stratum $[\mathfrak{a}, m, 0, \beta]$ in $A = \mathrm{End}_F(V)$, and set $E = F[\beta]$. We describe the transfer in two special cases from which all others can be deduced.

Theorem *Let $\mathfrak{a}'$ be an E-pure hereditary $\mathfrak{o}_F$-order in the algebra A. If $\theta \in \mathcal{C}(\mathfrak{a}, \beta)$, there exists a unique $\theta' \in \mathcal{C}(\mathfrak{a}', \beta)$ such that*

$$\theta'(x) = \theta(x), \quad x \in H^1(\beta, \mathfrak{a}) \cap H^1(\beta, \mathfrak{a}').$$

The map

$$\begin{aligned} \mathcal{C}(\mathfrak{a}, \beta) &\longrightarrow \mathcal{C}(\mathfrak{a}', \beta), \\ \theta &\longmapsto \theta', \end{aligned} \tag{5.1.2}$$

is a bijection.

This result is [25] (3.6.1) Theorem. In this case, (5.1.2) provides the transfer map $\mathrm{t}^{\beta}_{\mathfrak{a},\mathfrak{a}'}$.

Suppose we have a simple stratum $[\mathfrak{a}, m, 0, \beta']$ such that $\mathcal{C}(\mathfrak{a}, \beta') = \mathcal{C}(\mathfrak{a}, \beta)$. Suppose that $\mathfrak{a}'$ is also $F[\beta']$-pure and that $\mathcal{C}(\mathfrak{a}', \beta') = \mathcal{C}(\mathfrak{a}', \beta)$. The theorem implies that $\mathrm{t}^{\beta'}_{\mathfrak{a},\mathfrak{a}'} = \mathrm{t}^{\beta}_{\mathfrak{a},\mathfrak{a}'}$. (Compare 5.1.1 Example here.)

5.1.3 Let $\mathfrak{a}_{E/F}$ be the unique E-pure, hereditary $\mathfrak{o}_F$-order in $A_{E/F} = \mathrm{End}_F(E)$. With $[\mathfrak{a}, m, 0, \beta]$ as in 5.1.2, we exhibit the map $\mathrm{t}^{\beta}_{\mathfrak{a},\mathfrak{a}_{E/F}}$.

Let $\mathcal{L}$ be the chain of $\mathfrak{a}$-lattices in V. Choose an $\mathfrak{o}_E$-*basis of* $\mathcal{L}$, in the following sense. Each $L \in \mathcal{L}$ is an $\mathfrak{o}_E$-lattice. Fixing an L at random, it has an $\mathfrak{o}_E$-basis $\{v_1, v_2, \dots, v_d\}$ with the following property: any $L' \in \mathcal{L}$ has an $\mathfrak{o}_E$-basis of the form $\{x_1v_1, x_2v_2, \dots, x_dv_d\}$ for some $x_j \in E^\times$. (This is a variation on the standard block matrix picture of a hereditary order.) Write $W_j = Ev_j$, so that $V = W_1 \oplus W_2 \oplus \cdots \oplus W_d$. Let M be the Levi subgroup of G that stabilizes each W_j and choose parabolic subgroups P, P' of G so that $P \cap P' = M$. Let N and N' be the unipotent radicals of P and P' respectively. By construction, the set $\{X \cap W_j : X \in \mathcal{L}\}$ is the unique $\mathfrak{o}_E$-lattice chain in the one-dimensional E-space W_j.

Proposition *Let $\theta \in \mathcal{C}(\mathfrak{a}, \beta)$.*

(1) The group $H^1 = H^1(\beta, \mathfrak{a})$ satisfies

$$H^1 = H^1 \cap N' \cdot H^1 \cap M \cdot H^1 \cap N,$$

the factors in the product being disjoint.

(2) The character θ is trivial on the subgroups $H^1 \cap N'$ and $H^1 \cap N$.

(3) The group $H^1 \cap M$ takes the form

$$H^1 \cap M = \prod_{j=1}^{d} H^1 \cap \mathrm{Aut}_F(W_j).$$

Further,

$$H^1 \cap \mathrm{Aut}_F(W_j) = H^1(\beta, \mathfrak{a}_j),$$

where $\mathfrak{a}_j$ is the unique E-pure hereditary $\mathfrak{o}_F$-order in $\mathrm{End}_{F(W_j)}$.

(4) Identify each $\mathfrak{a}_j$ with $\mathfrak{a}_{E/F}$ via the obvious isomorphism $W_j \to E$. There is then a unique $\theta_0 \in \mathcal{C}(\mathfrak{a}_{E/F}, \beta)$ such that

$$\theta \,\big|\, H^1 \cap M = \theta_0 \otimes \theta_0 \otimes \cdots \otimes \theta_0.$$

In this situation, set

$$t^{\beta}_{\mathfrak{a},\mathfrak{a}_{E/F}}\,\theta = \theta_0 \quad\text{and}\quad t^{\beta}_{\mathfrak{a}_{E/F},\mathfrak{a}}\,\theta_0 = \theta.$$

If $\mathfrak{a}'$ is an E-pure hereditary $\mathfrak{o}_F$-order in $\mathrm{End}_F(V')$, say, we then have

$$t^{\beta}_{\mathfrak{a},\mathfrak{a}'} = t^{\beta}_{\mathfrak{a}_{E/F},\mathfrak{a}'} \circ t^{\beta}_{\mathfrak{a},\mathfrak{a}_{E/F}}.$$

5.1.4 By way of an aside, we sketch the manner in which transfer of simple characters intervenes in the proof of the Exhaustion Theorem.

Let π be an irreducible smooth representation of $G = \mathrm{GL}_n(F)$, of positive level. Thus π contains a fundamental stratum (1.5.1 Theorem). If that stratum is split, then π is not cuspidal and we discard it. Otherwise, it contains a simple stratum $[\mathfrak{a}_1, m_1, m_1 - 1, a]$ (by 1.5.3 Theorem 1). So, there exists a simple stratum $[\mathfrak{a}_2, m_2, 0, \beta]$ in $A = \mathrm{M}_n(F)$, a character $\theta \in \mathcal{C}(\mathfrak{a}_2, \beta)$ and an integer t, $0 \leqslant t < m$, such that π contains

$\theta \mid H^{1+t}(\beta, \mathfrak{a}_2)$. Choose this data to minimize $t/e(\mathfrak{a}_2)$. Suppose that $t > 0$, and choose a simple stratum $[\mathfrak{a}_2, m_2, t, \gamma]$ equivalent to $[\mathfrak{a}_2, m_2, t, \beta]$. Thus $\theta \mid H^{1+t}(\beta, \mathfrak{a}_2)$ lies in $\mathcal{C}(\mathfrak{a}_2, \gamma)$. The representation π contains a character of the form $\theta\, \psi_A * a \mid H^t(\beta, \mathfrak{a}_2)$, for some $a \in \mathfrak{p}^{-t}$. Let B be the A-centralizer of γ and set $\mathfrak{b} = \mathfrak{a}_2 \cap B$. Consider the stratum $[\mathfrak{b}, t, t-1, s_\gamma(a)]$ in B, where s_γ is a tame corestriction. If this stratum is not fundamental, one can reduce the relative level $t/e(\mathfrak{a}_2)$. Assume therefore that $[\mathfrak{b}, t, t-1, s_\gamma(a)]$ is fundamental. If it is split, one shows that π is not cuspidal. If it is non-split, one may adjust the original choice of $[\mathfrak{a}_2, m_2, 0, \beta]$ and θ to reduce $t/e(\mathfrak{a}_2)$. Overall, π contains some simple character so, at the beginning, we could have taken $t = 0$.

Each of those steps requires a change to the hereditary order $\mathfrak{a}_2$ and relies on the ability to change θ with it. That done, π contains a simple character $\theta \in \mathcal{C}(\mathfrak{a}, \beta)$, for some choice of simple stratum $[\mathfrak{a}, m, 0, \beta]$. The final step in the argument imitates the proof of 1.2.3 Theorem, but has to be conducted "over θ", again requiring the facility of changing the order $\mathfrak{a}$.

5.2 *Endo-Equivalence*

The transfer structure of Sect. 5.1 leads to an equivalence relation on the class of all simple characters. We follow the original treatment in [8], but use a more condensed terminology.

5.2.1 We make an elementary observation.

Lemma *Let E/F be a finite field extension. If $\mathfrak{a}$ is a hereditary order in $A = \mathrm{End}_F(V)$, defined by a lattice chain $\{L_i\}_{i\in\mathbb{Z}}$ in V, the following are equivalent.*

(1) There exists an F-embedding $\iota : E \to A$ such that $\mathfrak{a}$ is $\iota(E)$-pure.
(2) The integer $e(\mathfrak{a})$ is divisible by $e(E|F)$ and $\dim_{\Bbbk_F} L_i/L_{i+1}$ is divisible by $f(E|F)$, for all i.

Suppose these conditions are satisfied, and let $\iota' : E \to A$ be an F-embedding. The order $\mathfrak{a}$ is then $\iota'(E)$-pure if and only if $\iota' = \mathrm{Ad}\, x \circ \iota$, for some $x \in \mathcal{K}_{\mathfrak{a}}$.

Say that $\mathfrak{a}$ is *E/F-conformal* if the conditions of the lemma are satisfied.

5.2.2 For $i = 1, 2$, we are given a finite-dimensional F-vector space V_i and a simple character θ_i in $G_i = \mathrm{Aut}_F(V_i)$. This θ_i is invariantly attached to a hereditary $\mathfrak{o}_F$-order $\mathfrak{a}_i$ in $A_i = \mathrm{End}_F(V_i)$. Relative to our fixed character $\psi_F \in \widehat{F}$, we choose the following data.

5.2.2.1 Data

(1) For $i = 1, 2$, let $[\mathfrak{a}_i, m_i, 0, \beta_i]$ be a simple stratum in A_i such that $\theta_i \in \mathcal{C}(\mathfrak{a}_i, \beta_i, \psi_F)$.

(2) Let V be a finite-dimensional F-vector space and let $\mathfrak{a}$ be a hereditary $\mathfrak{o}_F$-order in $A = \mathrm{End}_F(V)$ that is $F[\beta_i]/F$-conformal for $i = 1, 2$.
(3) Let $\iota_i : F[\beta_i] \to A$ be an F-embedding such that $\mathfrak{a}$ is $\iota_i(F[\beta_i])$-pure.

Set

$$\theta_i^V = \mathrm{t}_{\mathfrak{a}_i,\mathfrak{a}}^{\beta_i}\,\theta_i. \tag{5.2.1}$$

We remark that, in (5.2.2.1), the choice of embeddings ι_i is immaterial since, for example, any two embeddings ι_1 are conjugate under $\mathcal{K}_{\mathfrak{a}_1}$.

Theorem *Suppose there exists a choice of data (5.2.2.1) for which the simple characters θ_1^V, θ_2^V are conjugate in $G = \mathrm{Aut}_F(V)$. The same then holds for all choices of such data.*

When these conditions holds, we say that θ_1 is *endo-equivalent to* θ_2. The transitivity of the constructions ensures that endo-equivalence is indeed an equivalence relation on the class of all simple characters in all groups $\mathrm{Aut}_F(V)$, as V ranges over the class of finite-dimensional F-vector spaces. We write $\mathcal{E}(F)$ for the set of equivalence classes relative to the relation of endo-equivalence (and tend to use the word "endo-classes" for the elements of the set $\mathcal{E}(F)$).

Complements (1) The relation of endo-equivalence does not depend on the choice of character ψ_F (*cf.* 2.5.1 Remark).
(2) Two simple characters in $G = \mathrm{Aut}_F(V)$, attached to the same hereditary order in $\mathrm{End}_F(V)$, are endo-equivalent if and only if they are conjugate in G.
(3) Following [17], one knows that two simple characters in $G = \mathrm{Aut}_F(V)$, attached to arbitrary hereditary orders, intertwine in G if and only if they are endo-equivalent.
(4) It is convenient to adjoin to $\mathcal{E}(F)$ a *trivial element* $\mathbf{0}$, which is the endo-class of *trivial* simple characters. Thus $\mathbf{0}$ is the class of trivial characters of groups $U_{\mathfrak{a}}^1$, as $\mathfrak{a}$ ranges over all hereditary $\mathfrak{o}_F$-orders in all matrix algebras $\mathrm{End}_F(V)$.

Terminology Let $[\mathfrak{a}, m, 0, \beta]$ be a simple stratum and let $\theta \in \mathcal{C}(\mathfrak{a}, \beta)$. If $\Theta \in \mathcal{E}(F)$ is the endo-class of θ, say that θ is a *realization* of Θ on $\mathfrak{a}$ (or $[\mathfrak{a}, m, 0, \beta]$).

The set of simple characters contained in a representation $\pi \in \mathcal{A}_n^0(F)$ constitutes a single conjugacy class in $\mathrm{GL}_n(F)$. In particular, all such characters are endo-equivalent. We therefore have a canonical map

$$\begin{aligned} \widehat{\mathrm{GL}}_F &\longrightarrow \mathcal{E}(F), \\ \pi &\longmapsto \Theta(\pi). \end{aligned} \tag{5.2.2}$$

Proposition *The map (5.2.2) is surjective.*

Proof The trivial element $\mathbf{0}$ is the endo-class simple characters contained in any $\pi \in \widehat{\mathrm{GL}}_F$ of level zero, so let $\Theta \in \mathcal{E}(F)$ be non-trivial. It is the endo-class of a simple character $\theta \in \mathcal{C}(\mathfrak{a}, \beta)$, for a simple stratum $[\mathfrak{a}, m, 0, \beta]$ in some $\mathrm{End}_F(V)$. Transferring θ, we may assume $\dim_F V = [F[\beta] : F]$. Thus θ is max-simple and, if Λ is an irreducible representation of $\boldsymbol{J}(\beta, \mathfrak{a})$ containing θ, then Λ is an EMST. The representation π_Λ of $\mathrm{Aut}_F(V)$ induced by Λ is then irreducible and cuspidal. It satisfies $\Theta(\pi_\Lambda) = \Theta$. □

5.2.3 An endo-class $\Theta \in \mathcal{E}(F)$ carries a panoply of invariants.

Proposition *Let $\Theta \in \mathcal{E}(F)$, $\Theta \neq \mathbf{0}$. Let $[\mathfrak{a}, m, 0, \beta]$ be a simple stratum in $A = \mathrm{End}_F(V)$ carrying a realization $\theta \in \mathcal{C}(\mathfrak{a}, \beta)$ of Θ. Write $E = F[\beta]$. The following quantities depend only on Θ and not on the choice of θ:*

$$\begin{aligned} \varsigma_\Theta &= m/e(\mathfrak{a}); \\ e_\Theta &= e(E|F); \\ f_\Theta &= f(E|F); \\ \deg \Theta &= [E : F]; \\ k_\Theta &= k_0(\beta, \mathfrak{a})/e(\mathfrak{a}). \end{aligned} \tag{5.2.3}$$

These assertions follow readily from properties of simple characters (2.5.4 Proposition 2) and the definition of endo-equivalence. To complete this part of the picture, we set

$$\varsigma_\mathbf{0} = 0, \;\; e_\mathbf{0} = f_\mathbf{0} = 1, \;\; k_\mathbf{0} = -\infty. \tag{5.2.4}$$

In Sect. 9, we will need some finer invariants of a non-trivial endo-class Θ. We first return to simple strata for an inductive definition.

Definition Let $\mathfrak{s} = [\mathfrak{a}, m, 0, \beta]$ be a simple stratum in an algebra $\mathrm{End}_F(V)$. The set $j(\mathfrak{s})$ of (normalized) *jumps* of $\mathfrak{s}$ is given as follows.

(1) If $\beta \in F$, then $j(\mathfrak{s})$ is empty.
(2) If β is minimal over F and $\beta \notin F$, then $j(\mathfrak{s}) = \{m/e(\mathfrak{a})\}$.
(3) Otherwise, let $r = -k_0(\beta, \mathfrak{a})$, choose a simple stratum $[\mathfrak{a}, m, r, \gamma]$ equivalent to $[\mathfrak{a}, m, r, \beta]$ and set
$$j(\mathfrak{s}) = \{r/e(\mathfrak{a})\} \cup j(\mathfrak{s}'),$$
where $\mathfrak{s}' = [\mathfrak{a}, m, 0, \gamma]$.

Proposition *Let $\Theta \in \mathcal{E}(F)$, $\Theta \neq \mathbf{0}$. Let $\theta \in \mathcal{C}(\mathfrak{a}, \beta)$ be a realization of Θ, for a simple stratum $\mathfrak{s} = [\mathfrak{a}, m, 0, \beta]$ in an algebra $\mathrm{End}_F(V)$. The set of rational numbers*

$$j(\Theta) = j(\mathfrak{s}) \tag{5.2.5}$$

depends only on Θ and not on the choices of θ and $\mathfrak{s}$.

This follows readily (5.2.3) and the definitions.

The set $\boldsymbol{j}(\Theta)$ is the set of *jumps of* Θ. We could define $\boldsymbol{j}(\mathbf{0}) = \emptyset$, but the need never arises.

5.2.4 The set $\mathcal{E}(F)$ carries a canonical pairing $\mathbf{A}$ with non-negative rational values. To define it, let $\Theta_1, \Theta_2 \in \mathcal{E}(F)$. There is a hereditary order $\mathfrak{a}$ in some $A = \mathrm{End}_F(V)$, occurring in simple strata $[\mathfrak{a}, m_i, 0, \beta_i]$, such that $\mathcal{C}(\mathfrak{a}, \beta_i)$ contains a simple character θ_i of endo-class Θ_i. (If, for example, $\Theta_1 = \mathbf{0}$, we may take θ_1 to be the trivial character of $U^1_{\mathfrak{a}}$.) Let $t \geqslant 0$ be the least integer such that the restrictions $\theta_i \,\big|\, H^{1+t}(\beta_i, \mathfrak{a})$ intertwine (and are therefore conjugate) in $\mathrm{Aut}_F(V)$. Set

$$\mathbf{A}(\Theta_1, \Theta_2) = t/e(\mathfrak{a}). \tag{5.2.6}$$

Proposition *Let* $\Theta_1, \Theta_2 \in \mathcal{E}(F)$.

(1) $\mathbf{A}(\Theta_1, \Theta_2) \leqslant \max\{\varsigma_{\Theta_1}, \varsigma_{\Theta_2}\}$, *with equality if* $\varsigma_{\Theta_1} \neq \varsigma_{\Theta_2}$.
(2) $\mathbf{A}(\Theta_1, \Theta_2) = 0$ *if and only if* $\Theta_1 = \Theta_2$.
(3) *If* $\Theta \in \mathcal{E}(F)$, *then*

$$\mathbf{A}(\Theta_1, \Theta_2) \leqslant \max\big\{\mathbf{A}(\Theta_1, \Theta), \mathbf{A}(\Theta, \Theta_2)\big\}.$$

A proof is given in [21] 5.1 Proposition. Combined with the obvious property $\mathbf{A}(\Theta_1, \Theta_2) = \mathbf{A}(\Theta_2, \Theta_1)$, parts (2) and (3) say that $\mathbf{A}$ provides an *ultrametric* on the set $\mathcal{E}(F)$.

5.2.5 It is sometimes helpful to use a slightly different language in connection with the ultrametric $\mathbf{A}$. Let $\epsilon > 0$. The binary relation

$$\Theta_1 \sim_\epsilon \Theta_2 \quad\iff\quad \mathbf{A}(\Theta_1, \Theta_2) < \epsilon \tag{5.2.7}$$

is an equivalence relation on $\mathcal{E}(F)$. Denote by $\mathrm{tc}_\epsilon(\Theta)$ the equivalence class of Θ. We refer to $\mathrm{tc}_\epsilon(\Theta)$ as the *ϵ-truncation* of Θ.

Likewise,

$$\Theta_1 \simeq_\epsilon \Theta_2 \quad\iff\quad \mathbf{A}(\Theta_1, \Theta_2) \leqslant \epsilon \tag{5.2.8}$$

is an equivalence relation, the class of Θ being denoted $\mathrm{tc}^+_\epsilon(\Theta)$.

Proposition *Let* $\Theta, \Upsilon \in \mathcal{E}(F)$ *and* $\epsilon > 0$. *The following are equivalent.*

(1) $\mathrm{tc}_\epsilon(\Theta) = \mathrm{tc}_\epsilon(\Upsilon)$.
(2) *Let* $\mathfrak{a}$ *be a hereditary* $\mathfrak{o}_F$*-order in a matrix algebra* $\mathrm{End}_F(V)$, *admitting realizations* $\theta \in \mathcal{C}(\mathfrak{a}, \beta)$, $\upsilon \in \mathcal{C}(\mathfrak{a}, \gamma)$ *of* Θ, Υ *respectively. If* t *is the largest integer such that* $t/e(\mathfrak{a}) < \epsilon$, *the characters* $\theta \,\big|\, H^{1+t}(\beta, \mathfrak{a})$, $\upsilon \,\big|\, H^{1+t}(\gamma, \mathfrak{a})$ *are conjugate in* $\mathrm{Aut}_F(V)$.

The proposition is a re-statement of the definition of $\mathbf{A}$. The exact analogue holds with tc^+_ϵ replacing tc_ϵ and $t/e(\mathfrak{a}) \leqslant \epsilon$ instead of $t/e(\mathfrak{a}) \leqslant \epsilon$.

5.3 Conductors of Pairs

As a first application of the machinery of endo-classes of simple characters, we give an explicit formula for the Rankin–Selberg exponent $a(\pi_1 \times \pi_2)$ of a pair of representations $\pi_1, \pi_2 \in \widehat{\mathrm{GL}}_F$. It is based on Shahidi's account [70, 71] of $L(\pi_1 \times \pi_2, s)$ and $\varepsilon(\pi_1 \times \pi_2, s, \psi)$ rather than [47]. The proof [23] uses [71] along with representation-theoretic structures [28, 29] not discussed here.

5.3.1 Before stating the result, we need to introduce a certain constant.

Let E/F be a finite field extension and let $A = \mathrm{End}_F(E)$. Let $\mathfrak{a}$ be the unique E-pure hereditary $\mathfrak{o}_F$-order in A. Let $\beta \in E^\times$ satisfy $E = F[\beta]$ and $m = -\upsilon_E(\beta) > 0$. Assume moreover that the quadruple $[\mathfrak{a}, m, 0, \beta]$ is a simple stratum in A. Let a_β denote the adjoint map $A \to A$, $x \mapsto \beta x - x\beta$, and let $s : A \to E$ be a tame corestriction relative to E/F (2.1.2). The sequence

$$0 \to E \longrightarrow A \xrightarrow{\ a_\beta\ } A \xrightarrow{\ s\ } E \to 0$$

is then exact. There exist $\mathfrak{o}_F$-lattices $\mathfrak{l}$, $\mathfrak{l}'$ in E and $\mathfrak{m}$, $\mathfrak{m}'$ in A such that the sequence

$$0 \to \mathfrak{l} \longrightarrow \mathfrak{m} \xrightarrow{\ a_\beta\ } \mathfrak{m}' \xrightarrow{\ s_{E/F}\ } \mathfrak{l}' \to 0$$

is exact. For Haar measures μ_E and μ_A on E and A respectively, the quantity

$$C(\beta) = \frac{\mu_E(\mathfrak{l})\,\mu_A(\mathfrak{m}')}{\mu_E(\mathfrak{l}')\,\mu_A(\mathfrak{m})}$$

is independent of all these choices. If $q = |\Bbbk_F|$, there is an integer $\mathfrak{c}(\beta)$ such that

$$C(\beta) = q^{\mathfrak{c}(\beta)}. \tag{5.3.1}$$

We give a method for calculating $\mathfrak{c}(\beta)$ in 5.3.4 below.

5.3.2 Let $\pi \in \mathcal{A}_n^0(F) \subset \widehat{\mathrm{GL}}_F$. Now set

$$\varsigma(\pi) = \varsigma_\Theta,$$

where $\Theta = \Theta(\pi)$ is the endo-class of a simple character θ_π contained in π. So, if $\theta_\pi \in \mathcal{C}(\mathfrak{a}, \beta)$, for a simple stratum $[\mathfrak{a}, m, 0, \beta]$ in $\mathrm{M}_n(F)$, then

$$\varsigma(\pi) = m/e(\mathfrak{a}) = md(\pi)/n = \mathrm{sw}(\pi)/n, \tag{5.3.2}$$

by 4.3.2 Corollary. In the notation of 1.5.1, we have

$$\ell(\pi) = \varsigma(\pi) = \mathbf{A}(\Theta_\pi, \mathbf{0}). \tag{5.3.3}$$

5.3.3 Let $\pi_i \in \mathcal{A}^0_{n_i}(F)$, $i = 1, 2$. Let $d(\pi_1, \pi_2)$ be the number of unramified characters χ of $F^\times$ such that $\pi_2 \cong \chi\pi_1$. Let Θ_i be the endo-class of a simple character contained in π_i.

To give the formula for the Rankin–Selberg exponent, we divide into cases.

Theorem 1 *Let* $\pi_i \in \mathcal{A}^0_{n_i}(F)$, $i = 1, 2$. *If*

$$\mathbf{A} = \mathbf{A}(\Theta_1, \Theta_2) = \max\{\varsigma(\pi_1), \varsigma(\pi_2)\}, \tag{5.3.4}$$

then

$$a(\check{\pi}_1 \times \pi_2) = n_1 n_2 (1 + \mathbf{A}) - d(\pi_1, \pi_2). \tag{5.3.5}$$

We remark here that if $\mathbf{A} > 0$ then $d(\pi_1, \pi_2) = 0$. To deal with the case where (5.3.4) fails, that is, where

$$0 \leqslant \mathbf{A} = \mathbf{A}(\Theta_1, \Theta_2) < \max\{\varsigma(\pi_1), \varsigma(\pi_2)\}, \tag{5.3.6}$$

we need some further notation. Note that (5.3.6) implies $\varsigma(\pi_1) = \varsigma(\pi_2)$.

Notation Let $\mathfrak{a}$ be a hereditary $\mathfrak{o}_F$-order in an algebra $A = \mathrm{End}_F(V)$ admitting realizations of both Θ_1 and Θ_2.

(1) For $i = 1, 2$, let $[\mathfrak{a}, m_i, 0, \beta_i]$ be a simple stratum in A such that $\mathcal{C}(\mathfrak{a}, \beta_i)$ contains a simple character θ_i of endo-class Θ_i: here, $m_1 = m_2$ by (5.3.6).
(2) Let $t \geqslant 0$ be the least integer for which the characters $\theta_i \,\big|\, H^{1+t}(\beta_i, \mathfrak{a})$ intertwine in $G = \mathrm{Aut}_F(V)$.
(3) Let $[\mathfrak{a}, m_1, t, \gamma]$ be a simple stratum equivalent to $[\mathfrak{a}, m_1, t, \beta_1]$.
(4) Set $d_\gamma = [F[\gamma] : F]$ and $e = e(\mathfrak{a})$.

Theorem 2 *Under the hypothesis (5.3.6) and using the preceding notation, we have*

$$a(\check{\pi}_1 \times \pi_2) = n_1 n_2 \left(1 + \frac{\mathfrak{c}(\gamma)}{d_\gamma^2} + \frac{t}{e d_\gamma}\right) - d(\pi_1, \pi_2).$$

Remarks (1) We have $d(\pi_1, \pi_2) = 0$ *except* in certain cases where $n_1 = n_2$ and $\mathbf{A} = 0$.
(2) The formula becomes a little tidier when it is expressed in terms of the Swan exponent,

$$\mathrm{sw}(\check{\pi}_1 \times \pi_2) = a(\check{\pi}_1 \times \pi_2) - n_1 n_2 + d(\pi_1, \pi_2).$$

(3) Suppose that $\pi \in \mathcal{A}^0_n(F)$ contains a simple character $\theta \in \mathcal{C}(\mathfrak{a}, \beta)$. We then have

$$\frac{\mathrm{sw}(\check{\pi} \times \pi)}{n^2} = \frac{\mathfrak{c}(\beta)}{d_\beta^2},$$

where $d_\beta = [F[\beta] : F]$.

One can re-phrase the theorems in terms of the Swan conductor and the ultrametric $\mathbf{A}$. This will be used in Sect. 9.

Theorem 3 *Let $\Theta \in \mathcal{E}(F)$. There is a unique continuous function $\Phi_\Theta(x)$, $x \geqslant 0$, with the following property. If $\pi \in \mathcal{A}_n^0(F)$ satisfies $\Theta(\pi) = \Theta$ and $\rho \in \mathcal{A}_m^0(F)$, then*

$$\frac{\mathrm{sw}(\check{\pi} \times \rho)}{mn} = \Phi_\Theta(\mathbf{A}(\Theta, \Theta(\rho))). \tag{5.3.7}$$

The function Φ_Θ is strictly increasing, piece-wise linear and convex.

See [21] 5.4 Proposition. An explicit formula for the function Φ_Θ is given in [21] (4.4.1). In particular, the jumps of Φ_Θ, that is, the discontinuities of the derivative $\Phi'_\Theta(x)$, $x > 0$, are the jumps of Θ in the sense of 5.2.3.

5.3.4 As promised in 5.3.1, we shed some light on the integer $\mathfrak{c}(\beta)$.

Proposition 1 *Let $E = F[\alpha]/F$ be a finite field extension and suppose that α is minimal over F. If $m = -\upsilon_E(\alpha)$, $e = e(E|F)$, $f = f(E|F)$, then $\mathfrak{c}(\alpha) = mf(ef - 1)$.*

Proof Write $\mathfrak{a} = \mathfrak{a}_{E/F}$ (1.3.7 Example) and let $\mathfrak{p} = \mathrm{rad}\,\mathfrak{a}$. There is an exact sequence

$$0 \to \mathfrak{o}_E \longrightarrow \mathfrak{a} \xrightarrow{a_\alpha} \mathfrak{p}^{-m} \xrightarrow{s} \mathfrak{p}_E^{-m} \to 0$$

in which $s : A_{E/F} \to E$ is a tame corestriction. The result then follows from a short calculation. □

To compute $\mathfrak{c}(\beta)$ in general, one uses a reduction argument.

Proposition 2 *Let $[\mathfrak{a}, m, 0, \beta]$ be a simple stratum in $\mathrm{M}_d(F)$ for which the field extension $E = F[\beta]/F$ has degree d. Set $e = e(E|F)$. Let $r = -k_0(\beta, \mathfrak{a})$ and assume $r < m$. Let $[\mathfrak{a}, m, r, \gamma]$ be a simple stratum equivalent to $[\mathfrak{a}, m, r, \beta]$. If $d(\gamma) = [F[\gamma] : F]$, then*

$$\frac{\mathfrak{c}(\beta)}{d^2} + \frac{r}{ed} = \frac{\mathfrak{c}(\gamma)}{d(\gamma)^2} + \frac{r}{ed(\gamma)}.$$

Proof See [12] 3.1 Proposition. □

6 Tame Lifting and Functoriality

Let K/F be a finite, *tamely ramified* field extension. We construct a canonical relation between the sets $\mathcal{E}(K)$, $\mathcal{E}(F)$ of endo-classes of simple characters over the fields K, F. This *tame lifting* relation is closely connected with the operations of base change and automorphic induction recalled in 4.1.2. In the first instance, it yields a Galois-theoretic interpretation of the set $\mathcal{E}(F)$ of endo-classes of simple characters over F.

6.1 Tame Lifting

We start with a transparent, albeit *ad hoc,* construction. The results of this section are taken from [8], mostly section 7.

6.1.1 Let $[\mathfrak{a}, m, 0, \beta]$ be a simple stratum in $A = \mathrm{End}_F(V)$ and write $E = F[\beta]$.

Data We are given a subfield K/F of A such that

(1) K commutes with β,
(2) the algebra $EK = K[\beta]$ is a field, and
(3) the hereditary $\mathfrak{o}_F$-order $\mathfrak{a}$ is EK-pure.

Let A_K be the centralizer of K in A and set $\mathfrak{a}_K = \mathfrak{a} \cap A_K$.

Proposition *The quadruple* $[\mathfrak{a}_K, m, 0, \beta]$ *is a simple stratum in* A_K *satisfying*

$$k_0(\beta, \mathfrak{a}_K) \leqslant k_0(\beta, \mathfrak{a}). \tag{6.1.1}$$

Proof See [8], 2.4 Theorem. □

There are cases in which the inequality (6.1.1) is strict.

6.1.2 Following 6.1.1 Proposition, we form the subgroups $H^1(\beta, \mathfrak{a}_K)$, $J^1(\beta, \mathfrak{a}_K)$ of $G_K = A_K^\times$.

Proposition *There are relations*

$$H^1(\beta, \mathfrak{a}_K) = H^1(\beta, \mathfrak{a}) \cap G_K,$$
$$J^1(\beta, \mathfrak{a}_K) = J^1(\beta, \mathfrak{a}) \cap G_K.$$

Proof See [8], 2.2.1 Proposition. □

Fix a character ψ_F of F of level one, as before. Set

$$\psi_K = \psi_F \circ \mathrm{Tr}_{K/F}. \tag{6.1.2}$$

Thus ψ_K is a character of K of level one.

Theorem *If* $\theta \in \mathcal{C}(\mathfrak{a}, \beta, \psi_F)$, *the restriction*

$$\theta_K = \theta \,\big|\, H^1(\beta, \mathfrak{a}_K) \tag{6.1.3}$$

lies in $\mathcal{C}(\mathfrak{a}_K, \beta, \psi_K)$.

This is [8] 7.7 Theorem. The restriction map $\mathcal{C}(\mathfrak{a}, \beta, \psi_F) \to \mathcal{C}(\mathfrak{a}_K, \beta, \psi_K)$ need not be surjective.

6.1.3 From the foundation of 6.1.2, we outline a more systematic approach following [8] section 9. Let $\Theta \in \mathcal{E}(F)$, $\Theta \neq \mathbf{0}$. Let $\theta \in \mathcal{C}(\mathfrak{a}, \beta, \psi_F)$ be a realization of Θ on a simple stratum $[\mathfrak{a}, m, 0, \beta]$ in some algebra $A = \mathrm{End}_F(V)$. Write $F[\beta] = E$.

Let K/F be a finite, tamely ramified field extension: *we do not assume that K/F is Galois.* The K-algebra $E \otimes_F K$ is semisimple and so the direct sum of its minimal ideals,

$$E \otimes_F K = E^1 \oplus E^2 \oplus \cdots \oplus E^r,$$

for an integer r, $1 \leqslant r \leqslant [K : F]$. Each ideal E^j is a field extension of K. If β^j denotes the projection of β in E^j, then $E^j = K[\beta^j]$.

For each j, let V^j be a finite-dimensional E^j-vector space and let $\mathfrak{a}^j$ be an E^j-pure hereditary $\mathfrak{o}_F$-order in $A^j = \mathrm{End}_F(V^j)$. There is an integer m^j so that $[\mathfrak{a}^j, m^j, 0, \beta^j]$ is a simple stratum in A_j. Using the F-embedding $E \to A^j$ induced by $\beta \mapsto \beta^j$, we transfer θ to a simple character $\theta^j \in \mathcal{C}(\mathfrak{a}^j, \beta^j, \psi_F)$. Let A^j_K be the centralizer of K in A^j and set $\mathfrak{a}^j_K = \mathfrak{a}^j \cap A^j_K$. Following 6.1.2 Theorem, we get a character $\theta^j_K \in \mathcal{C}(\mathfrak{a}^j_K, \beta^j, \psi_K)$. Define $\Theta^j_K \in \mathcal{E}(K)$ to be the endo-class of θ^j_K.

Theorem 1 *The endo-classes Θ^j_K, $1 \leqslant j \leqslant r$, are distinct. The set $\{\Theta^j_K : 1 \leqslant j \leqslant r\}$ depends only on Θ, not on any of the choices made in the definition.*

Further, the set of Θ^j_K does not depend on the choice of character ψ_F (*cf.* 2.5.1 Remark). The endo-classes Θ^j_K, $1 \leqslant j \leqslant r$, are the *$K/F$-lifts of* Θ.

To complete the picture, let $\mathbf{0}_F$ be the trivial element of $\mathcal{E}(F)$. The class $\mathbf{0}_F$ has a unique K/F-lift, namely the trivial element $\mathbf{0}_K$ of $\mathcal{E}(K)$.

We summarize the main properties of the lifting operation.

Theorem 2 *Let K/F be a finite, tamely ramified field extension.*

(1) If $\Xi \in \mathcal{E}(K)$, there is a unique $\Xi^F \in \mathcal{E}(F)$ of which Ξ is a K/F-lift. The map

$$\begin{aligned} \boldsymbol{i}_{K/F} : \mathcal{E}(K) &\longrightarrow \mathcal{E}(F), \\ \Xi &\longmapsto \Xi^F, \end{aligned}$$

is surjective and, for $\Theta \in \mathcal{E}(F)$, the fibre $\boldsymbol{i}^{-1}_{K/F}(\Theta)$ is the set of K/F-lifts of Θ.

(2) If L/K is a finite tame extension, then $\boldsymbol{i}_{L/F} = \boldsymbol{i}_{K/F} \circ \boldsymbol{i}_{L/K}$.

(3) If the tame extension K/F is Galois, then any two K/F-lifts of Θ are conjugate under Gal(K/F).

(4) The map $\boldsymbol{i}_{K/F}$ does not depend on the choice of character ψ_F.

6.1.4 Example A character χ of U^1_F is a simple character in $\mathrm{GL}_1(F)$. If not trivial, it is attached to a simple stratum $[\mathfrak{o}_F, m, 0, c]$ in $\mathrm{M}_1(F) = F$. Either way, its endo-class has degree 1. If K/F is a finite, tame extension, then $\chi \circ \mathrm{N}_{K/F}$ is a simple character over K, and its endo-class is the unique K/F-lift of the endo-class of χ.

The other direction is more interesting. Let K/F be a finite tame extension and χ a character of U_K^1, of endo-class X say. Suppose that χ does not factor through the norm $\mathrm{N}_{K/L}$, for any intermediate field $F \subset L \subsetneq K$. The endo-class $\boldsymbol{i}_{K/F}\mathrm{X}$ then has degree $[K : F]$. As in [13], one can write down an explicit representative of $\boldsymbol{i}_{K/F}\mathrm{X}$. It is the simple character attached to $(K/F, \chi)$ via the classical construction of [42].

6.2 Functoriality Relations

We establish a fundamental relation between tame lifting and base change, as promised in item (3) of the list in 4.1.2.

6.2.1 Let K/F be a finite cyclic extension. As in 4.1.2, let

$$\begin{aligned} \boldsymbol{b}_{K/F} &: \widehat{\mathrm{GL}}_F^{\mathrm{tm}} \longrightarrow \widehat{\mathrm{GL}}_K^{\mathrm{tm}}, \\ \boldsymbol{a}_{K/F} &: \widehat{\mathrm{GL}}_K^{\mathrm{tm}} \longrightarrow \widehat{\mathrm{GL}}_F^{\mathrm{tm}} \end{aligned}$$

be respectively the maps of *base change* and *automorphic induction.*

If $\pi \in \widehat{\mathrm{GL}}_F^{\mathrm{tm}}$, we may realize π as a Langlands quotient in the standard way,

$$\pi = \pi_1 \boxplus \pi_2 \boxplus \dots \boxplus \pi_r,$$

where $\pi_1, \pi_2, \dots, \pi_r \in \widehat{\mathrm{GL}}_F$. Thus

$${}^L\pi = {}^L\pi_1 \oplus {}^L\pi_2 \oplus \dots \oplus {}^L\pi_r.$$

6.2.2 The relation between the pair of maps $\boldsymbol{b}_{K/F}, \boldsymbol{a}_{K/F}$ and the operation of tame lifting is contained in two parallel results. For $\pi \in \widehat{\mathrm{GL}}_F$, let $\Theta(\pi)$ denote the endo-class of simple characters contained in π.

Theorem 1 *Let K/F be a finite, tamely ramified, cyclic extension and let $\pi \in \widehat{\mathrm{GL}}_F$. If*

$$\boldsymbol{b}_{K/F}\,\pi = \rho_1 \boxplus \rho_2 \boxplus \dots \boxplus \rho_s,$$

where $\rho_j \in \widehat{\mathrm{GL}}_K$, then

$$\Theta(\pi) = \boldsymbol{i}_{K/F}\,\Theta(\rho_j), \quad 1 \leqslant j \leqslant s.$$

The K/F-lifts of $\Theta(\pi)$ occur among the $\Theta(\rho_j)$ with equal multiplicities.

Theorem 2 *Let K/F be a finite, tamely ramified, cyclic extension and let $\rho \in \widehat{\mathrm{GL}}_K$. If*

$$\boldsymbol{a}_{K/F}\,\rho = \pi_1 \boxplus \pi_2 \boxplus \ldots \boxplus \pi_r,$$

where $\pi_i \in \widehat{\mathrm{GL}}_F$, then

$$\Theta(\pi_i) = \boldsymbol{i}_{K/F}\,\Theta(\rho), \quad 1 \leqslant i \leqslant r.$$

Remark Suppose for a moment that K/F is a finite, Galois, tame extension and let K_0/K be the maximal unramified sub-extension of K/F. We may formally write $\boldsymbol{b}_{K/F} = \boldsymbol{b}_{K/K_0} \circ \boldsymbol{b}_{K_0/F}$. For $\pi \in \widehat{\mathrm{GL}}_F$, we again have an expansion

$$\boldsymbol{b}_{K/F}\,\pi = \rho_1 \boxplus \rho_2 \boxplus \ldots \boxplus \rho_s,$$

for various $\rho_j \in \widehat{\mathrm{GL}}_K$, the representations ${}^L\rho_j$ being the irreducible components of ${}^L\pi \,\big|\, \mathcal{W}_K$. A similar comment holds for automorphic induction. The transitivity property of 6.1.3 Theorem 2 shows that the theorems hold unchanged for K/F. See [24] section 2 for a discussion of these matters.

6.2.3 The proofs of the theorems of 6.2.2 are closely entwined and spread over the two papers [8, 12]. The first paper deals with some special cases in which $s = 1 = r$. One works directly with the character relations that define base change [1] and automorphic induction [39]. The general case is then deduced in [12], without appeal to the character relations. It relies instead on more formal aspects of the Langlands correspondence and the conductor formula of 5.3.3 above.

Both [1, 39] assume that the base field F has characteristic zero. This assumption is removed in [41]. The general machinery of simple characters, at the level of detail required for these results, is completely insensitive to characteristic, so everything in [8, 12] (and their derivatives) holds unchanged in positive characteristic.

6.3 Ramification Theorem

We explore some implications of 6.2.2, following [12].

6.3.1 We return to the canonical surjection $\widehat{\mathrm{GL}}_F \to \mathcal{E}(F)$, $\pi \mapsto \Theta(\pi)$, of (5.2.2). We compare it with the surjective map

$$r_F^+ : \widehat{\mathcal{W}}_F \longrightarrow \mathcal{W}_F \backslash \widehat{\mathcal{P}}_F$$

that takes $\sigma \in \widehat{\mathcal{W}}_F$ to the $\mathcal{W}_F$-conjugacy of an irreducible component of $\sigma \,\big|\, \mathcal{P}_F$.

Ramification Theorem *Let $\Theta \in \mathcal{E}(F)$ and let $\pi \in \widehat{\mathrm{GL}}_F$ satisfy $\Theta(\pi) = \Theta$. If $\sigma = {}^L\pi \in \widehat{\mathcal{W}}_F$, the conjugacy class ${}^L\Theta = r_F^+(\sigma)$ depends only on Θ and not on the choice of π. The map*

$$\begin{aligned} \mathcal{E}(F) &\longrightarrow \mathcal{W}_F\backslash\widehat{\mathcal{P}}_F, \\ \Theta &\longmapsto {}^L\Theta, \end{aligned}$$

is a bijection and the diagram

$$\begin{array}{ccc} \widehat{\mathrm{GL}}_F & \xrightarrow{\pi \mapsto {}^L\pi} & \widehat{\mathcal{W}}_F \\ {\scriptstyle\Theta}\big\downarrow & & \big\downarrow{\scriptstyle r_F^+} \\ \mathcal{E}(F) & \xrightarrow[\Theta \mapsto {}^L\Theta]{} & \mathcal{W}_F\backslash\widehat{\mathcal{P}}_F \end{array}$$

commutes.

Proof The core assertion of the theorem amounts to the following.

Proposition *For $i = 1, 2$, let $\pi_i \in \widehat{\mathrm{GL}}_F$ and set $\sigma_i = {}^L\pi_i$. The following conditions are equivalent.*

(1) $\Theta(\pi_1) = \Theta(\pi_2)$.
(2) $\mathrm{Hom}_{\mathcal{P}_F}(\sigma_1, \sigma_2) \neq 0$.

The proposition implies straightaway that the map $\Theta \mapsto {}^L\Theta$ is well-defined and injective. The diagram commutes, whence the map is surjective.

To prove the proposition, we need a lemma. Say that $\sigma \in \widehat{\mathcal{W}}_F$ is *totally wild* if $\sigma \mid \mathcal{P}_F$ is irreducible. Equivalently, $\dim \sigma$ is a power of p (the residual characteristic of F) and $\chi \otimes \sigma \ncong \sigma$ for any unramified character $\chi \neq 1$ of $\mathcal{W}_F$. Let $\widehat{\mathcal{W}}_F^{\mathrm{wr}}$ be the set of totally wild $\sigma \in \widehat{\mathcal{W}}_F$.

Lemma *For $i = 1, 2$, let $\pi_i \in \widehat{\mathrm{GL}}_F$ and suppose that $\sigma_i = {}^L\pi_i$ is totally wild. The following are equivalent.*

(1) $\Theta(\pi_1) = \Theta(\pi_2)$.
(2) $\sigma_1 \mid \mathcal{P}_F \cong \sigma_2 \mid \mathcal{P}_F$.
(3) There is a tamely ramified character ϕ of $\mathcal{W}_F$ such that $\sigma_2 \cong \phi \otimes \sigma_1$.

Proof The equivalence of (2) and (3) is immediate while (3) surely implies (1). Assume (1) holds. Since σ_i is totally wild, π_i is a representation of $\mathrm{GL}_{m_i}(F)$, where $m_i = p^{r_i}$, for some $r_i \geqslant 0$. Also $\chi\pi_i \ncong \pi_i$ for any unramified character $\chi \neq 1$ of $F^\times$. We conclude that $\deg \Theta(\pi_i) = e(\Theta(\pi_i)) = p^{r_i}$, whence (1) implies $r_1 = r_2 = r$, say. It follows that $\pi_2 \cong \phi\pi_1$, for some tamely ramified character ϕ of $F^\times$, so (3) holds. □

To prove the proposition, write $\sigma_i = \mathrm{Ind}_{E_i/F}\,\tau_i$, where E_i/F is a finite tame extension and $\tau_i \in \widehat{\mathcal{W}}^{\mathrm{wr}}_{K_i}$. Let E/F be a finite, tame Galois extension containing E_1 and E_2. Thus

$$\sigma_i^E = \sigma_i \,\big|\, \mathcal{W}_E = \bigoplus_{1\leqslant j\leqslant r_i} \tau_j^i,$$

for integers r_1, r_2 and representations $\tau_j^i \in \widehat{\mathcal{W}}^{\mathrm{wr}}_E$. As E/F is tamely ramified, so $\mathcal{P}_E = \mathcal{P}_F$ and

$$\mathrm{Hom}_{\mathcal{P}_F}(\sigma_1, \sigma_2) = \mathrm{Hom}_{\mathcal{P}_E}(\sigma_1^E, \sigma_2^E) = \bigoplus_{j,k} \mathrm{Hom}_{\mathcal{P}_E}(\tau_j^1, \tau_k^2).$$

As in 6.2.2 Remark, if K/F is the maximal unramified sub-extension of E/F, then

$$\boldsymbol{b}_{E/F}(\pi_i) = \boxplus_j \rho_j^i,$$

where $\rho_j^i \in \widehat{\mathrm{GL}}_E$ and ${}^L\rho_j^i = \tau_j^i$. If $\mathrm{Hom}_{\mathcal{P}_F}(\sigma_1, \sigma_2) \neq 0$, there exist indices j, k such that $\mathrm{Hom}_{\mathcal{P}_E}(\tau_j^1, \tau_k^2) \neq 0$. Applying the lemma, we get $\Theta(\tau_j^1) = \Theta(\tau_k^2)$, whence by 6.2.2 Theorem 1,

$$\Theta(\pi_1) = \boldsymbol{i}_{E/F}\,\Theta(\tau_j^1) = \boldsymbol{i}_{E/F}\,\Theta(\tau_k^2) = \Theta(\pi_2).$$

The argument reverses, and the proof is complete. □

6.3.2 If K/F is a finite tame extension, then $\mathcal{P}_K = \mathcal{P}_F$ as subgroups of $\mathcal{W}_F$, so we have a canonical map

$$\mathcal{W}_K\backslash\widehat{\mathcal{P}}_K \longrightarrow \mathcal{W}_F\backslash\widehat{\mathcal{P}}_F$$

taking the $\mathcal{W}_K$-conjugacy class of an irreducible representation of $\mathcal{P}_F = \mathcal{P}_K$ to its $\mathcal{W}_F$-conjugacy class.

Theorem *If K/F is a finite tame extension, the diagram*

$$\begin{array}{ccc} \mathcal{E}(K) & \xrightarrow{\approx} & \mathcal{W}_K\backslash\widehat{\mathcal{P}}_K \\ {\scriptstyle \boldsymbol{i}_{K/F}}\downarrow & & \downarrow \\ \mathcal{E}(F) & \xrightarrow[\approx]{} & \mathcal{W}_F\backslash\widehat{\mathcal{P}}_F \end{array}$$

commutes.

Proof The assertion is transitive with respect to the field extension K/F. It is enough, therefore, to treat the case where K/F is *of prime degree.*

Let $\xi \in \widehat{\mathcal{P}}_K$ and write $O_K(\xi)$ for the $\mathcal{W}_K$-conjugacy class of ξ. Take $\tau \in \widehat{\mathcal{W}}_K$ such that $\tau \,\big|\, \mathcal{P}_K$ contains ξ and write $\tau = {}^L\rho$, for $\rho \in \widehat{\mathrm{GL}}_K$. The Ramification Theorem gives ${}^L\Theta(\rho) = O_K(\xi)$.

Suppose, in the first instance, that K/F is cyclic and set $\mathrm{Gal}(K/F) = \Gamma$. Consider the representation $\sigma = \mathrm{Ind}_{K/F}\,\tau$. If σ is not irreducible, it is the direct sum of representations $\chi \otimes \sigma_1$, where $\sigma_1 \,|\, \mathcal{W}_K = \tau$ and χ ranges over the characters of $\mathcal{W}_F/\mathcal{W}_K = \Gamma$. Likewise

$$\boldsymbol{a}_{K/F}\,\rho = \boxplus_{\chi}\, \chi\pi_1,$$

where ${}^L\pi_1 = \sigma_1$. By 6.2.1 Theorem 2, $\Theta(\pi_1) = \boldsymbol{i}_{K/F}\,\Theta(\rho)$. However, $r_F^+(\sigma_1) = O_F(\xi)$, in the obvious notation, and the result follows in this case. The case where σ is irreducible follows the same course. We conclude that the result holds whenever K/F is a Galois extension of any finite degree.

There remains the case where K/F is not Galois and hence totally tamely ramified. Let E/F be the normal closure of K/F and let E_0/F be the maximal unramified subextension of E/F. Our last remark shows that the result holds for the extensions E/F and E/K. Chasing diagrams, it holds for K/F. □

7 Tame Parameters

In the first part of this section, we review a construction of the irreducible representations of $\mathcal{W}_F$ in terms of their restriction to the wild inertia group $\mathcal{P}_F$. Attached to $\alpha \in \widehat{\mathcal{P}}_F$ is a finite tame extension E/F and a spectrum of actions of the group of tame characters of $\mathcal{W}_E$ on the set of representations containing α. In the second part, we seek parallel structures on the GL-side. In the final part, we connect the two sides. Most of the material derives from [19].

7.1 Structure of Galois Representations

We give a systematic description of the representations $\sigma \in \widehat{\mathcal{W}}_F$.

7.1.1 Let $\alpha \in \widehat{\mathcal{P}}_F$. Define $\mathcal{Z}_F(\alpha)$ to be the group of $x \in \mathcal{W}_F$ such that $\alpha^x = \alpha$.

Proposition *There is a finite tame extension E/F such that $\mathcal{W}_E = \mathcal{Z}_F(\alpha)$.*

Proof The kernel of α is an open normal subgroup of $\mathcal{P}_F$, and so contains a group $\mathcal{P}_F \cap \mathcal{W}_K$, where K/F is a finite Galois extension. Thus α extends uniquely to a smooth representation $\bar{\alpha}$ of the group $\mathcal{P}_F\mathcal{W}_K$ such that $\bar{\alpha} \,|\, \mathcal{W}_K$ is trivial. The representation $\bar{\alpha}$ is then stable under the group $\mathcal{P}_F\mathcal{W}_K$, which therefore stabilizes $\bar{\alpha} \,|\, \mathcal{P}_F = \alpha$. The group $\mathcal{P}_F\mathcal{W}_K$ is of the form $\mathcal{W}_L$, for a finite, Galois, tamely ramified extension L/F, so the result follows. □

We refer to E/F as the *F-centralizer field* of α and use the notation $E = Z_F(\alpha)$. Note that $\mathcal{P}_E = \mathcal{P}_F$, since E/F is tame.

7.1.2 Let $\alpha \in \widehat{\mathcal{P}}_F$. We use the F-centralizer field of α to analyse the representations of $\mathcal{W}_F$ that contain α.

Proposition *Let $\alpha \in \widehat{\mathcal{P}}_F$ and write $E = Z_F(\alpha)$.*

(1) There exists a unique representation $\alpha_{\mathcal{I}} \in \widehat{\mathcal{I}}_E$ such that $\alpha_{\mathcal{I}} \,\big|\, \mathcal{P}_F \cong \alpha$ and $(\det \alpha_{\mathcal{I}})^{p^r} = 1$ for some $r \geqslant 0$.

(2) Let ρ be an irreducible representation of $\mathcal{I}_E$ such that $\rho \,\big|\, \mathcal{P}_F$ contains α. There is a unique character ϕ of $\mathcal{I}_E$, trivial on $\mathcal{P}_E = \mathcal{P}_F$, such that $\rho \cong \phi \otimes \alpha_{\mathcal{I}}$.

Proof The group $\mathcal{I}_E/\mathcal{P}_F$ is pro-cyclic, whence α admits extension to a representation ρ of $\mathcal{I}_E$. For the same reason, any extension of α to $\mathcal{I}_E$ is of the form $\phi \otimes \rho$, for a character ϕ of $\mathcal{I}_E$ trivial on $\mathcal{P}_E = \mathcal{P}_F$. Taking determinants, $\det(\phi \otimes \rho) = \phi^{p^r} \det \rho$, where $p^r = \dim \alpha$. The character ϕ has finite order not divisible by p, so $\phi \otimes \rho \cong \rho$ if and only if $\phi = 1$. Moreover, there is a unique choice of ϕ so that $\phi^{p^r} \det \rho = 1$, as required for (1). This also implies, in (2), that ρ is of the form $\phi \otimes \alpha_{\mathcal{I}}$, for a character ϕ of $\mathcal{I}_E/\mathcal{P}_E$. We can recover ϕ from the character $\det(\phi \otimes \alpha_{\mathcal{I}})$, so it is uniquely determined. □

Corollary *(1) There exists a representation τ of $\mathcal{W}_E$ such that $\tau \,\big|\, \mathcal{I}_E = \alpha_{\mathcal{I}}$.*

(2) If $\tau' \in \widehat{\mathcal{W}}_E$ also satisfies $\tau' \,\big|\, \mathcal{I}_E = \alpha_{\mathcal{I}}$, there is a unique unramified character χ of $\mathcal{W}_E$ such that $\tau' = \chi \otimes \tau$.

(3) The induced representation $\mathrm{Ind}_{E/F}\, \tau$ is irreducible and contains α with multiplicity one.

Proof Its uniqueness property implies that $\alpha_{\mathcal{I}}$ is stable under conjugation by $\mathcal{W}_E$. That $\mathcal{W}_E/\mathcal{I}_E$ is cyclic implies the first assertion and the others follow. □

In the corollary, one may recover τ from $\sigma = \mathrm{Ind}_{E/F}\, \tau$ as the natural representation of $\mathcal{W}_E$ on the α-isotypic subspace of $\sigma \,\big|\, \mathcal{P}_F$.

7.1.3 As a minor digression, 7.1.2 Proposition implies a convenient description of the space $\mathcal{W}_F \backslash \widehat{\mathcal{I}}_F$.

Definition Let $\mathrm{IP}(F)$ be the set of pairs (α, χ), where $\alpha \in \widehat{\mathcal{P}}_F$ and χ is a character of $\mathcal{I}_{Z_F(\alpha)}$ trivial on $\mathcal{P}_{Z_F(\alpha)} = \mathcal{P}_F$.

The group $\mathcal{W}_F$ acts on $\mathrm{IP}(F)$ by conjugation.

Proposition *Let $(\alpha, \chi) \in \mathrm{IP}(F)$ and write $E = Z_F(\alpha)$. The representation*

$$\lambda_{(\alpha,\chi)} = \mathrm{Ind}_{\mathcal{I}_E}^{\mathcal{I}_F}\, \chi \otimes \alpha_{\mathcal{I}}$$

of $\mathcal{I}_F$ is irreducible. The map $(\alpha, \chi) \mapsto \lambda_{(\alpha,\chi)}$ induces a canonical bijection

$$\mathcal{W}_F \backslash \mathrm{IP}(F) \xrightarrow{\ \approx\ } \mathcal{W}_F \backslash \widehat{\mathcal{I}}_F.$$

Proof An element x of $\mathcal{J}_F$ that intertwines $\chi \otimes \alpha_{\mathcal{J}}$ must intertwine α and so lie in $\mathcal{W}_E \cap \mathcal{J}_F = \mathcal{J}_E$. The induced representation is therefore irreducible and $(\alpha, \chi) \mapsto \lambda_{(\alpha,\chi)}$ gives a well-defined map $\mathcal{W}_F\backslash \mathrm{IP}(F) \to \mathcal{W}_F\backslash\widehat{\mathcal{J}}_F$. To show it is surjective, take $\sigma \in \widehat{\mathcal{W}}_F$, choose an irreducible component α of $\sigma \,|\, \mathcal{P}_F$ and set $E = Z_F(\alpha)$. It follows that $\sigma \,|\, \mathcal{J}_E$ has an irreducible component containing α. That component is of the form $\chi \otimes \alpha_{\mathcal{J}}$, for a character χ of $\mathcal{J}_E/\mathcal{P}_F$. Thus $\mathrm{IP}(F)$ maps onto $\mathcal{W}_F\backslash\widehat{\mathcal{J}}_F$.

To show the map is injective, take $(\alpha, \chi), (\beta, \psi) \in \mathrm{IP}(F)$ and suppose that $\lambda_{(\alpha,\chi)}$ is $\mathcal{W}_F$-conjugate to $\lambda_{(\beta,\psi)}$. In particular, α is conjugate to β so we may take $\beta = \alpha$. The conjugation between $\lambda_{(\alpha,\chi)}$ and $\lambda_{(\alpha,\psi)}$ is then implemented by an element of $\mathcal{W}_E$. Since α occurs in $\lambda_{(\alpha,\chi)}$ with multiplicity one, it follows that ψ is $\mathcal{W}_E$-conjugate to χ, as required. □

7.1.4 For $\alpha \in \widehat{\mathcal{P}}_F$, denote by $O_F(\alpha) \in \mathcal{W}_F\backslash\widehat{\mathcal{P}}_F$ the $\mathcal{W}_F$-conjugacy class of α. If $m \geqslant 1$ is an integer, define $\widehat{\mathcal{W}}_F(\alpha; m)$ to be the set of $\sigma \in \widehat{\mathcal{W}}_F$ such that $\mathrm{Hom}_{\mathcal{P}_F}(\alpha, \sigma)$ has dimension m. The set $\widehat{\mathcal{W}}_F(\alpha; m)$ depends on the conjugacy class $O_F(\alpha)$ rather than α.

With $\alpha_{\mathcal{J}}$ as in 7.1.2, consider a representation $\phi \otimes \alpha_{\mathcal{J}}$, where ϕ is a character of $\mathcal{J}_E/\mathcal{P}_F$, $E = Z_F(\alpha)$. The $\mathcal{W}_E$-centralizer of ϕ (or of $\phi \otimes \alpha_{\mathcal{J}}$) is of the form $\mathcal{W}_K$, for a finite unramified extension K/E.

Let $E = Z_F(\alpha)$ and let E_m/E be the unramified extension of E, inside $\bar{F}$, of degree m. Set $\Delta_m = \mathrm{Gal}(E_m/E)$. The group Δ_m acts on the set $\widehat{\mathcal{W}}_{E_m}(\alpha; 1)$. Let $\widehat{\mathcal{W}}_{E_m}(\alpha; 1)^{\Delta_m\text{-reg}}$ be the set of Δ_m-regular elements of $\widehat{\mathcal{W}}_{E_m}(\alpha; 1)$, that is, the set of τ for which the conjugates τ^δ, $\delta \in \Delta_m$, are distinct.

Proposition *If $\xi \in \widehat{\mathcal{W}}_{E_m}(\alpha; 1)^{\Delta_m\text{-reg}}$, the representation $\mathrm{Ind}_{E_m/F}\,\xi$ is irreducible and lies in $\widehat{\mathcal{W}}_F(\alpha; m)$. The map*

$$\begin{aligned} \Delta_m\backslash\widehat{\mathcal{W}}_{E_m}(\alpha; 1)^{\Delta_m\text{-reg}} &\longrightarrow \widehat{\mathcal{W}}_F(\alpha; m), \\ \xi &\longmapsto \mathrm{Ind}_{E_m/F}\,\xi, \end{aligned} \tag{7.1.1}$$

is a bijection.

Proof Let $\theta \in \widehat{\mathcal{W}}_{E_m}(\alpha; 1)^{\Delta_m\text{-reg}}$. The representation $\mathrm{Ind}_{E_m/E}\,\theta$ is then irreducible and contains α with multiplicity m. The definition of E ensures that $\sigma = \mathrm{Ind}_{E/F}\,\theta$ is irreducible. It lies in $\widehat{\mathcal{W}}_F(\alpha; m)$ and depends only on the Δ_m-conjugacy class of θ. On the other hand, starting from $\sigma \in \widehat{\mathcal{W}}_F(\alpha; m)$, the natural representation θ_α of $\mathcal{W}_E$ on the α-isotypic subspace of σ is irreducible. It induces σ and contains α with multiplicity m. By 7.1.2 Corollary, there exists $\tau \in \widehat{\mathcal{W}}_E(\alpha; 1)$, so $\theta_\alpha = \xi \otimes \tau$, where ξ is an irreducible, m-dimensional representation of $\mathcal{W}_E$ trivial on $\mathcal{P}_F$. The factor ξ is uniquely determined by θ_α. The representation ξ is induced from an Δ_m-regular character ϕ of $\mathcal{W}_{E_m}/\mathcal{J}_E$, so σ is induced from an element of $\widehat{\mathcal{W}}_{E_m}(\alpha; 1)$. The map (7.1.1) is therefore surjective. The uniqueness of ξ implies it is injective. □

Remark Let $\sigma \in \widehat{\mathcal{W}}_F(\alpha; m)$, and write $\sigma = \mathrm{Ind}_{E_m/F}\,\tau$, $\tau \in \widehat{\mathcal{W}}_{E_m}(\alpha; 1)$. The restriction of τ to $\mathcal{J}_{E_m} = \mathcal{J}_E$ takes the form $\phi \otimes \alpha_{\mathcal{J}}$, for a character ϕ of $\mathcal{J}_E/\mathcal{P}_F$. The image of σ in $\mathcal{W}_F\backslash\widehat{\mathcal{J}}_F$ is then parametrized by the pair $(\alpha, \phi) \in \mathrm{IP}(F)$.

7.1.5 If F'/F is a finite extension inside $\bar{F}$, let $X_1(F')$ (resp. $X_0(F')$) be the group of tamely ramified (resp. unramified) characters of $\mathcal{W}_{F'}$. If s is a positive integer, let $X_0(F')_s$ be the group of $\chi \in X_0(F')$ such that $\chi^s = 1$.

Let $\alpha \in \widehat{\mathcal{P}}_F$ and write $Z_F(\alpha) = E$. Define a pairing

$$\begin{aligned} X_1(E) \times \widehat{\mathcal{W}}_F(\alpha; m) &\longrightarrow \widehat{\mathcal{W}}_F(\alpha; m), \\ (\xi, \sigma) &\longmapsto \xi \odot_\alpha \sigma, \end{aligned} \tag{7.1.2}$$

as follows. Let $\sigma \in \widehat{\mathcal{W}}_F(\alpha; m)$. Following 7.1.4 Proposition, there is an irreducible representation $\mu \in \widehat{\mathcal{W}}_{E_m}(\alpha; 1)^{\Delta_m\text{-reg}}$ such that $\sigma = \mathrm{Ind}_{E_m/F}\, \mu$. We define

$$\chi \odot_\alpha \sigma = \mathrm{Ind}_{E_m/F}\, \mu \otimes \chi \,\big|\, \mathcal{W}_{E_m}, \quad \chi \in X_1(E). \tag{7.1.3}$$

This action *does* depend on α, not just on the orbit $O_F(\alpha)$. Note that $\chi \odot_\alpha \sigma = \sigma$ if and only if $\xi \in X_0(E)_m$. There is a useful special case.

Proposition *Under the action (7.1.3), the set $\widehat{\mathcal{W}}_F(\alpha; 1)$ is a principal homogeneous space over $X_1(E)$.*

In the general case, take E_m/E and $\Delta_m = \mathrm{Gal}(E_m/E)$ as before. Take $\tau \in \widehat{\mathcal{W}}_E(\alpha; 1)$ (*cf.* 7.1.2 Corollary). The set $\widehat{\mathcal{W}}_{E_m}(\alpha; 1)^{\Delta_m\text{-reg}}$ comprises the representations $\phi \otimes \tau \,\big|\, \mathcal{W}_{E_m}$, where ϕ ranges over $X_1(E_m)^{\Delta_m\text{-reg}}$. Restriction of characters induces an isomorphism $X_1(E)/X_0(E)_m \cong X_1(E_m)^{\Delta_m}$. The action of $X_1(E)$ on $\widehat{\mathcal{W}}_F(\alpha; m)$ follows the natural action of $X_1(E)$ on $\Delta_m \backslash X_1(E_m)^{\Delta_m\text{-reg}}$.

7.2 *Tame Parameter Fields*

In this section, we attach to an endo-class $\Theta \in \mathcal{E}(F)$ an isomorphism class of finite tame extensions E/F. We construct a family of actions of the group $X_1(E)$ on the set of representations $\pi \in \widehat{\mathrm{GL}}_F$ such that $\Theta(\pi) = \Theta$. This is intended to mirror Sect. 7.1, but the technicalities are more elaborate.

7.2.1 Let $\Theta \in \mathcal{E}(F)$, $\Theta \neq \mathbf{0}$. Let $\theta \in \mathcal{C}(\mathfrak{a}, \beta)$ be a realization of Θ on a simple stratum $[\mathfrak{a}, m, 0, \beta]$ in a matrix algebra $A = \mathrm{End}_F(V)$. Any field extension $F[\beta]/F$, obtained in this way, is a *parameter field* for Θ. (If $\Theta = \mathbf{0}$, its only parameter field is F, and there is nothing to say.)

Definitions (1) Let $\Theta \in \mathcal{E}(F)$. Say that Θ is *totally wild* if $\deg \Theta = e_\Theta = p^r$, for an integer $r \geqslant 0$.

(2) Let $\Theta \in \mathcal{E}(F)$ and let K/F be a finite, tamely ramified field extension. Say that K/F is a *tame parameter field* for Θ if

(a) Θ has a totally wild K/F-lift and
(b) if $F \subset K' \subsetneq K$, then Θ has no totally wild K'/F-lift.

Let $\Theta \in \mathcal{E}(F)$ be the endo-class of $\theta \in \mathcal{C}(\mathfrak{a}, \beta)$, say. We remark that Θ is totally wild if and only if the field extension $F[\beta]/F$ is totally wildly ramified. Otherwise, let T/F be the maximal tame sub-extension of $F[\beta]/F$ and define the simple character θ_T by restriction, as in (6.1.3). The endo-class Θ_T of θ_T is then a totally wild T/F-lift of Θ.

Proposition *Let $\Theta \in \mathcal{E}(F)$ be the endo-class of $\theta \in \mathcal{C}(\mathfrak{a}, \beta)$. Let T/F be the maximal tamely ramified sub-extension of $F[\beta]/F$ and let K/F be a finite, tamely ramified field extension. The following are equivalent:*

(1) Θ has a totally wild K/F-lift;
(2) there exists an F-embedding $T \to K$.

In particular, T/F is a tame parameter field for Θ and, up to isomorphism, it is the only one.

The proof is a straightforward consequence of the definitions. Let $\mathrm{Aut}(T|F)$ be the group of F-automorphisms of the field T. The group $\mathrm{Aut}(T|F)$ acts on the set $\mathcal{E}(T)$ in a natural way. The definitions imply:

Corollary *The set of totally wild T/F-lifts of Θ is a principal homogeneous space over the group* $\mathrm{Aut}(T|F)$.

7.2.2 An endo-class Θ determines its tame parameter field up to isomorphism over F, but a specific realization of Θ gives a specific tame extension of F. Reconciliation of these divergent uniqueness properties has a degree of subtlety. We follow [19] in examining this phenomenon only for max-simple characters, that being all we need.

We can remove some potential ambiguity at the start.

Lemma *For $i = 1, 2$, let $[\mathfrak{a}, m, 0, \beta_i]$ be a max-simple stratum in $\mathrm{M}_n(F)$ and suppose that $\mathcal{C}(\mathfrak{a}, \beta_1) = \mathcal{C}(\mathfrak{a}, \beta_2)$. Let T_i/F be the maximal tame sub-extension of $F[\beta_i]/F$. There is a unique F-isomorphism $T_1 \to T_2$ of the form $x \mapsto j^{-1}xj$, where $j \in J^1(\beta_1, \mathfrak{a}) = J^1(\beta_2, \mathfrak{a})$.*

The proof is in [19] 2.6. In the notation of the lemma, say that T_i/F is an *internal* tame parameter field for $\theta \in \mathcal{C}(\mathfrak{a}, \beta_i)$. Using the restriction construction of 6.1.2, we form the simple character θ_{T_i} over T_i. The isomorphism $T_1 \to T_2$ of the lemma fixes θ and carries θ_{T_1} to θ_{T_2}.

If $\iota : E \to E'$ is an isomorphism of local fields, denote by ι_* the induced bijection $\mathcal{E}(E) \to \mathcal{E}(E')$.

Proposition *Let $\Theta \in \mathcal{E}(F)$ have tame parameter field E/F. Let θ be a max-simple realization of Θ in some $G = \mathrm{GL}_n(F)$ with internal tame parameter field T/F, and define θ_T as in (6.1.3). Let $\Xi \in \mathcal{E}(E)$ be a totally wild E/F-lift of Θ. There is a unique F-isomorphism $\iota : E \to T$ such that θ_T has endo-class $\iota_*\Xi$.*

Proof This combines the lemma with 7.2.1 Corollary. □

7.2.3 We use these considerations to normalize certain character actions. We first give the internal version of the construction. Let θ be a max-simple character in some group $G = \mathrm{GL}_n(F)$. Let $\boldsymbol{J}_\theta$ denote the G-normalizer of θ. Let J_θ be the maximal compact subgroup of $\boldsymbol{J}_\theta$ and J^1_θ the pro-p radical of J_θ. Let $X_1(\theta)$ be the character group defined in 3.1.7.

Proposition *Let T/F be an internal tame parameter field for θ and let G_T be the G-centralizer of $T^\times$. Define $s = n/\deg\Theta$, where Θ is the endo-class of θ.*

(1) Let $\phi \in X_1(T)$. There exists a unique character $\phi_\theta \in X_1(\theta)$ such that

$$\phi_\theta(x) = \phi(\det_T x), \quad x \in G_T \cap \boldsymbol{J}_\theta.$$

(2) The map $\phi \mapsto \phi_\theta$ is a surjective homomorphism $X_1(E) \to X_1(\theta)$ with kernel $X_0(T)_s$.

Proof Suppose θ is attached to the max-simple stratum $[\mathfrak{a}, m, 0, \beta]$ in which T/F is the maximal tame sub-extension of $F[\beta]/F$. Write $P = F[\beta]$. Composition with the norm $\mathrm{N}_{P/T}$ is an isomorphism $X_1(T) \to X_1(P)$, carrying $X_0(T)_s$ to $X_0(P)_s$, so the proposition follows from 3.1.7 Lemma. □

By 7.2.2 Lemma, this construction is effectively independent of the choice of T. We remark that, if $\theta, \theta' \in \mathcal{C}(\mathfrak{a}, \beta)$, then $\phi_{\theta'} = \phi_\theta$.

7.2.4 We externalize the construction. To recap, $\theta \in \mathcal{C}(\mathfrak{a}, \beta)$ is a max-simple character, $P = F[\beta]$ and T/F is the maximal tame sub-extension of P/F. Let G_T be the G-centralizer of $T^\times$. Define $\mathbb{H}(\theta)$ as in 3.1.7. The endo-class of θ is $\Theta \in \mathcal{E}(F)$. The character θ_T is the restriction of θ to $H^1(\beta, \mathfrak{a}) \cap G_T$. Let $\Theta_T \in \mathcal{E}(T)$ be the endo-class of θ_T.

Let E/F be a tame parameter field for the endo-class Θ, and let $\Xi \in \mathcal{E}(E)$ be a totally wild E/F-lift of Θ. From 7.2.2 Proposition, there is an F-isomorphism $\iota : E \to T$ such that $\Theta_T = \iota_*\Xi$. The map ι induces an isomorphism $\iota_* : X_1(E) \to X_1(T)$. Write

$$\phi \odot_\Xi \boldsymbol{\kappa} = \iota_*\phi \otimes \boldsymbol{\kappa}, \quad \phi \in X_1(E),\ \boldsymbol{\kappa} \in \mathbb{H}(\theta). \tag{7.2.1}$$

This defines an action of $X_1(E)$, or of $X_1(E)/X_0(E)_s$, on $\mathbb{H}(\theta)$. It *does* depend on the choice of Ξ.

7.2.5 We re-formulate 7.2.4 in terms of representations of general linear groups.

Let $\Theta \in \mathcal{E}(F)$, let $s \geqslant 1$ be an integer and set $n = s\deg\Theta$. Define

$$\widehat{\mathrm{GL}}_F(\Theta; s) = \{\pi \in \mathcal{A}^0_n(F) : \Theta(\pi) = \Theta\}.$$

Let $\pi \in \widehat{\mathrm{GL}}_F(\Theta; s)$, let E/F be a tame parameter field for Θ and let Ξ be a totally wild E/F-lift of Θ. Choose a max-simple realization θ of Θ in $G = \mathrm{GL}_n(F)$. Let $(\boldsymbol{J}_\theta, \Lambda)$ be the EMST in G that contains θ. In particular, $\pi = c\text{-}\mathrm{Ind}_{\boldsymbol{J}_\theta}^G \Lambda$.

Take $\boldsymbol{\kappa} \in \mathbb{H}(\theta)$, and let λ be the natural representation of $\boldsymbol{J}_\theta$ on the space of J^1_θ-fixed vectors in $\Lambda \otimes \check{\boldsymbol{\kappa}}$. Thus $\Lambda = \boldsymbol{\kappa} \otimes \lambda$. If $\phi \in X_1(E)$, we may form the representation $\phi \odot_\Xi \boldsymbol{\kappa} \in \mathbb{H}(\theta)$ and then

$$\phi \odot_\Xi \pi = c\text{-Ind}\big((\phi \odot_\Xi \boldsymbol{\kappa}) \otimes \lambda\big).$$

This gives a canonical action

$$\begin{aligned} X_1(E) \times \widehat{\mathrm{GL}}_F(\Theta; s) &\longrightarrow \widehat{\mathrm{GL}}_F(\Theta; s), \\ (\phi, \pi) &\longmapsto \phi \odot_\Xi \pi, \end{aligned} \tag{7.2.2}$$

of $X_1(E)$ on $\widehat{\mathrm{GL}}_F(\Theta; s)$. In particular, $\widehat{\mathrm{GL}}_F(\Theta; 1)$ is a principal homogeneous space over $X_1(E)/X_0(E)_s$.

We note that when $\Theta = \mathbf{0}$, we have $E = F$ and (7.2.2) reduces to the standard twisting action

$$(\phi, \pi) \longmapsto \phi\pi, \quad \phi \in X_1(F), \ \pi \in \widehat{\mathrm{GL}}_F(\mathbf{0}; s).$$

In this case, $\widehat{\mathrm{GL}}_F(\mathbf{0}; s)$ is the set of $\pi \in \mathcal{A}^0_s(F)$ of level zero.

7.3 Comparison of Tame Parameters

We relate the two preceding lines of thought.

7.3.1 Recall that, if $\alpha \in \widehat{\mathcal{P}}_F$, then $O_F(\alpha)$ is the $\mathcal{W}_F$-conjugacy class of α.

Tame Parameter Theorem *Let $\alpha \in \widehat{\mathcal{P}}_F$ and set $Z_F(\alpha) = E$. If $\Theta \in \mathcal{E}(F)$ satisfies ${}^L\Theta = O_F(\alpha)$, then*

(1) $\deg \Theta = [E : F] \dim \alpha$ *and*
(2) E/F is a tame parameter field for Θ.

Proof Start with the case $E = F$. In this case, α extends to a representation σ of $\mathcal{W}_F$. The extension has dimension $p^r = \dim \alpha$, $r \geqslant 0$, and is totally ramified: if $\chi \neq 1$ is an unramified character of $\mathcal{W}_F$, then $\chi \otimes \sigma \not\cong \sigma$. If $\sigma = {}^L\pi$, where $\pi \in \mathcal{A}^0_{p^r}(F)$, then π has the same property: $\chi\pi \not\cong \pi$ for any unramified character $\chi \neq 1$ of $F^\times$. We deduce that the endo-class $\Theta = \Theta(\pi)$ is totally wild, in that $\deg \Theta = e_\Theta = p^r$. It follows that F provides a tame parameter field for $\Theta(\pi)$. Since $O_F(\alpha) = {}^L\Theta(\pi)$, we are done in this case.

We turn to the general case $E \neq F$.

Lemma *Let $O_F(\alpha) = {}^L\Theta$, $\Theta \in \mathcal{E}(F)$, and let T/F be a tame parameter field for Θ. There exists an F-embedding $T \to E$ and* $\deg \Theta = [T : F] \dim \alpha$.

Proof By 6.3.2 Theorem, we have $\Theta = \boldsymbol{i}_{E/F}\, \Theta_E$, where $\Theta_E \in \mathcal{E}(E)$ satisfies ${}^L\Theta_E = O_E(\alpha)$. By the first case above, the endo-class Θ_E is totally wild of degree $p^r = \dim \alpha$. □

Set $d = \deg \Theta = [T : F] \dim \alpha$ and take $\pi \in \mathcal{A}_d^0(F)$ with $\Theta(\pi) = \Theta$. The representation $\sigma = {}^L\pi$ thus has dimension d. However, $\dim \sigma$ is divisible by $[E : F] \dim \alpha$, so $[E : F] = [T : F]$ and the lemma implies $E \cong T$, as required. □

Corollary *Let m be a positive integer, let $\Theta \in \mathcal{E}(F)$ and let $\alpha \in {}^L\Theta$. The Langlands correspondence induces a bijection*

$$\widehat{\mathrm{GL}}_F(\Theta; m) \xrightarrow{\approx} \widehat{\mathcal{W}}_F(\alpha; m).$$

This is an immediate consequence of the theorem. If $E = Z_F(\alpha)$, each of the sets $\widehat{\mathrm{GL}}_F(\Theta; m)$, $\widehat{\mathcal{W}}_F(\alpha; m)$ carries a collection of actions of the group $X_1(E)$. We show in the next section that the Langlands correspondence also matches these actions.

8 Description of the Langlands Correspondence

Starting from our "black box" correspondence $\mathcal{E}(F) \to \mathcal{W}_F\backslash\widehat{\mathcal{P}}_F$, we use an explicit method to define a canonical bijection

$$\begin{aligned} \widehat{\mathrm{GL}}_F &\longrightarrow \widehat{\mathcal{W}}_F, \\ \pi &\longmapsto {}^N\pi, \end{aligned}$$

that we call the *naïve correspondence*. We estimate the difference between the naïve correspondence and the Langlands correspondence. The material of this section is almost entirely taken from [19].

8.1 Complements

By way of preparation, we re-examine the concept of (extended) maximal simple type in light of the ideas of Sect. 7.2.

8.1.1 Recall the *Green parametrization* [34, 56] of the irreducible cuspidal representations of $\mathrm{GL}_n(k)$, where k is a finite field.

Let l/k be a field extension of degree n and set $\Gamma = \mathrm{Gal}(l/k)$. Embed l in $\mathrm{M}_n(k)$ as a k-subalgebra: any two such embeddings are conjugate, so the choice will be immaterial. Let χ be a Γ-regular character of $l^\times$. There is then a unique irreducible cuspidal representation λ_χ of $\mathrm{GL}_n(k)$ such that

$$\mathrm{tr}\,\lambda_\chi(\zeta) = (-1)^{n-1} \sum_{\gamma\in\Gamma} \chi(\zeta^\gamma), \tag{8.1.1}$$

for all Γ-regular elements ζ of $l^\times$. The map $\chi \mapsto \lambda_\chi$ is a bijection between the set of Γ-orbits of Γ-regular characters χ of $l^\times$ and the set of equivalence classes of irreducible cuspidal representations of $\mathrm{GL}_n(k)$. The representation λ_χ has central character $\chi \mid k^\times$.

8.1.2 Returning to our standard situation, let E/F be unramified of degree n and set $\Gamma = \mathrm{Gal}(E/F)$. Let $\boldsymbol{\mu}_E$ be the group of roots of unity in E of order not divisible by p. Reduction modulo $\mathfrak{p}_E$ induces an isomorphism $\boldsymbol{\mu}_E \to \Bbbk_E^\times$.

Let $X_1(E)^{\Gamma\text{-reg}}$ be the set of Γ-regular characters $\chi \in X_1(E)$. Let $\mathbf{1}$ denote the trivial character of $\mathcal{P}_F$. There are then canonical bijections

$$\Gamma\backslash X_1(E)^{\Gamma\text{-reg}} \longrightarrow \widehat{\mathcal{W}}_F(\mathbf{1}; n), \qquad \Gamma\backslash X_1(E)^{\Gamma\text{-reg}} \longrightarrow \widehat{\mathrm{GL}}_F(\mathbf{0}; n),$$
$$\chi \longmapsto \sigma_\chi, \qquad\qquad \chi \longmapsto \pi_\chi.$$

The first map is given by $\sigma_\chi = \mathrm{Ind}_{E/F}\,\chi$. For the second, take a maximal order $\mathfrak{m}$ in $A = \mathrm{M}_n(F)$ and embed E in A so that $E^\times \subset \mathcal{K}_\mathfrak{m} = F^\times U_\mathfrak{m}$. (Any two such configurations are conjugate in $G = \mathrm{GL}_n(F)$.) Such a map induces an embedding of $\boldsymbol{\mu}_E$ in $U_\mathfrak{m}/U_\mathfrak{m}^1 \cong \mathrm{GL}_n(\Bbbk_F)$. Define an EMST $(\mathcal{K}_\mathfrak{m}, \Lambda_\chi^F)$ in G by deeming that $\Lambda_\chi^F \mid U_\mathfrak{m}$ be the inflation of the representation $\lambda_{\chi|\boldsymbol{\mu}_E}$ of $U_\mathfrak{m}/U_\mathfrak{m}^1$ and that $\Lambda_\chi^F \mid F^\times$ be a multiple of $\chi \mid F^\times$. The representation π_χ is then

$$\pi_\chi = c\text{-}\mathrm{Ind}_{\mathcal{K}_\mathfrak{m}}^G\, \Lambda_\chi^F.$$

Proposition *If χ is a Γ-regular character of $E^\times$, then ${}^L\pi_\chi = \sigma_{\chi'}$, where*

$$\chi'(x) = \chi(x)\,(-1)^{(n-1)v_E(x)}, \quad x \in E^\times.$$

This description of the Langlands correspondence $\widehat{\mathrm{GL}}_F(\mathbf{0}; n) \to \widehat{\mathcal{W}}_F(\mathbf{1}; n)$ is classical. It is a first exercise in the application of automorphic induction, as in [3, 41].

Remark Let P/F be totally wildly ramified. With the notation otherwise as before, define $\chi^P = \chi \circ \mathrm{N}_{PE/E}$. The map $\Lambda_\chi \mapsto \Lambda_\chi^P = \Lambda_{\chi^P}$ is then a bijection $\widehat{\mathrm{GL}}_F(\mathbf{0}; n) \to \widehat{\mathrm{GL}}_P(\mathbf{0}; n)$, corresponding to restriction of representations from $\mathcal{W}_F$ to $\mathcal{W}_P$. Observe also that, if ζ is a Γ-regular element of $\boldsymbol{\mu}_E$, it is also a Γ-regular element of $\boldsymbol{\mu}_{PE}$ and $\mathrm{tr}\,\Lambda_{\chi^P}(\zeta) = \mathrm{tr}\,\Lambda_\chi(\zeta^{[P:F]})$.

8.1.3 It is necessary to re-phrase the discussion of extended maximal simple types (3.1.2, 3.1.3) in terms of tame parameter fields.

Let θ be a max-simple character in $G = \mathrm{GL}_n(F)$, say $\theta \in \mathcal{C}(\mathfrak{a}, \beta)$. Write $P = F[\beta]$ and let T/F be the maximal tame sub-extension of P/F. Let G_P be the G-centralizer of $P^\times$ and $\mathfrak{a}_P$ the $\mathfrak{a}$-centralizer of P. Recall that $\mathcal{T}(\theta)$ is the set of equivalence classes of extended maximal simple types in G containing θ.

Put $d = [P : F]$ and $n = ds$. We use the bijection

$$\widehat{\mathrm{GL}}_T(\mathbf{0};s) \longrightarrow \widehat{\mathrm{GL}}_P(\mathbf{0};s)$$

of 8.1.2 Remark, induced by the inclusion $T \to P$. Let $\mathcal{T}(\mathbf{0};\,\mathfrak{a}_P)$ be the set of equivalence classes of EMST's in G_P of level zero, based on the maximal $\mathfrak{o}_P$-order $\mathfrak{a}_P$. Likewise define $\mathcal{T}(\mathbf{0};\,\mathfrak{m}_T)$, for a maximal $\mathfrak{o}_T$-order $\mathfrak{m}_T$ in $\mathrm{M}_s(T)$. Re-phrasing 8.1.2 Remark, we get a canonical bijection

$$\begin{aligned} \mathcal{T}(\mathbf{0};\,\mathfrak{m}_T) &\longrightarrow \mathcal{T}(\mathbf{0};\,\mathfrak{a}_P), \\ \Lambda &\longmapsto \Lambda^P. \end{aligned} \tag{8.1.2}$$

The definition of EMST amounts to a canonical pairing

$$\begin{aligned} \mathcal{T}(\mathbf{0};\,\mathfrak{a}_P) \times \mathbb{H}(\theta) &\longrightarrow \mathcal{T}(\theta), \\ (\Lambda, \boldsymbol{\kappa}) &\longmapsto \Lambda \otimes \boldsymbol{\kappa}. \end{aligned}$$

(To form the tensor product here, we extend Λ, by triviality on $J^1(\theta)$, to a representation of $\boldsymbol{J}(\theta)$.) Combining with (8.1.2), we get a pairing

$$\begin{aligned} \mathcal{T}(\mathbf{0};\,\mathfrak{m}_T) \times \mathbb{H}(\theta) &\longrightarrow \mathcal{T}(\theta), \\ (\Lambda, \boldsymbol{\kappa}) &\longmapsto \Lambda^P \otimes \boldsymbol{\kappa} = \Lambda \ltimes_{\Theta_T} \boldsymbol{\kappa}. \end{aligned} \tag{8.1.3}$$

Here, Θ_T is the T/F-lift of the endo-class Θ of θ defined by the inclusion of T in P. The pairing is compatible with the $\odot_{\Theta_T}$-actions of $X_1(T)$ on $\mathcal{T}(\mathbf{0};\,\mathfrak{m}_T)$ and $\mathbb{H}(\theta)$.

8.2 Algebraic Induction Maps

Let E/F be a finite tame extension. We construct a family of canonical induction-like maps from certain EMST's over E to EMST's over F, refining the map $\boldsymbol{i}_{E/F} : \mathcal{E}(E) \to \mathcal{E}(F)$. The material of this section is taken from section 5 of [19], to which we refer for the copious detail.

8.2.1 Fix a max-simple character θ in $G = \mathrm{GL}_n(F)$. Thus $\theta \in \mathcal{C}(\mathfrak{a}, \beta)$, for a max-simple stratum $[\mathfrak{a}, m, 0, \beta]$ in $\mathrm{M}_n(F)$. Let $P = F[\beta]$ and let E/F be the maximal tame sub-extension of P/F. As usual, θ_E is the restriction of θ to $H^1(\beta, \mathfrak{a}) \cap G_E$, where G_E is the G-centralizer of $E^\times$. Let Θ, Θ_E be the endo-classes of θ, θ_E respectively. The class Θ_E is totally wild.

In this subsection, we concentrate on the case where E/F is *totally ramified and* $\deg \Theta = [P : F] = n$. Write $\boldsymbol{J} = \boldsymbol{J}(\theta) = \boldsymbol{J}(\beta, \mathfrak{a})$ and $\boldsymbol{J}_E = \boldsymbol{J}(\theta_E)$. Again, $\mathcal{T}(\theta)$ is the set of equivalence classes of EMST's in G that contain θ, and similarly over E.

Theorem *Let $\Lambda \in \mathcal{T}(\theta)$. There exists a unique $\Lambda_E \in \mathcal{T}(\theta_E)$ such that*

$$\operatorname{tr} \Lambda(x) = \epsilon_{E/F} \operatorname{tr} \Lambda_E(x),$$

for all $x \in \boldsymbol{J}_E$ *such that* $\upsilon_F(\det_G x)$ *is relatively prime to* n *and a constant* $\epsilon_{E/F} = \pm 1$. *The map*

$$\begin{aligned} \mathcal{T}(\theta) &\longrightarrow \mathcal{T}(\theta_E), \\ \Lambda &\longmapsto \Lambda_E, \end{aligned} \tag{8.2.1}$$

is a bijection and an isomorphism of $X_1(E)$*-spaces.*

This result is 5.3 Proposition of [19]. It is based on a construction from the representation theory of finite groups, known as the "Glauberman correspondence": see Glauberman's original paper [32] or the development in Isaacs' book [45].

We translate in terms of representations of general linear groups.

Corollary *Let* $\Theta \in \mathcal{E}(F)$ *satisfy* $\deg \Theta = e_\Theta$ *and let* E/F *be a tame parameter field for* Θ. *Let* Θ_E *be a totally wild* E/F*-lift of* Θ. *The inverse of the map (8.2.1) induces a canonical bijection*

$$ind_{E/F} : \widehat{\mathrm{GL}}_E(\Theta_E; 1) \xrightarrow{\ \approx\ } \widehat{\mathrm{GL}}_F(\Theta; 1) \tag{8.2.2}$$

satisfying

$$ind_{E/F}(\phi\rho) = \phi \odot_{\Theta_E} ind_{E/F}(\rho),$$

for $\phi \in X_1(E)$ *and* $\rho \in \widehat{\mathrm{GL}}_E(\Theta_E; 1)$.

Recall that $\widehat{\mathrm{GL}}_E(\Theta_E; 1)$ and $\widehat{\mathrm{GL}}_F(\Theta; 1)$ are principal homogeneous spaces over $X_1(E)$ and observe the parallel with the definition of the operation $\odot_\alpha$ in 7.1.5.

Remark Let $F \subset E' \subset E$. The map $ind_{E/F}$ factors through $ind_{E/E'}$. In the context of the theorem, we have

$$\mathrm{tr}\, \Lambda(x) = \epsilon_{E/F}\, \mathrm{tr}\, \Lambda_E(x) = \epsilon_{E/F}\, \epsilon_{E/E'}^{-1}\, \mathrm{tr}\, \Lambda_{E'}(x),$$

where $\Lambda_{E'} = ind_{E/E'}\, \Lambda_E$. We may write $\Lambda = ind_{E'/F}\, \Lambda_{E'}$, giving a relation

$$ind_{E/F} = ind_{E'/F} \circ ind_{E/E'}.$$

8.2.2 There is also an unramified induction map, for which we need a more elaborate notational scheme. We follow that of [19] 5.4 for convenience of reference.

Notation (1) $\Theta \in \mathcal{E}(F)$ has degree d and $\theta \in \mathcal{C}(\mathfrak{a}, \beta)$ is a max-simple realization of Θ in $G = \mathrm{GL}_n(F)$, where $n = ds$, $s \geqslant 1$.
(2) $P_0 = F[\beta]$ and E_0/F is the maximal tame sub-extension of P_0/F.
(3) P/P_0 is an unramified extension of degree s, such that $P^\times \subset \boldsymbol{J}(\beta, \mathfrak{a})$.
(4) E/F is the maximal tame sub-extension of P/F and K/F is the maximal unramified sub-extension of E/F.
(5) $\Delta = \mathrm{Gal}(P/P_0) = \mathrm{Gal}(E/E_0) = \mathrm{Gal}(K/K \cap E_0)$.

We remark that any two choices of P/P_0 are conjugate by an element of $U_{\mathfrak{a}}$ that commutes with P_0. Define the simple character θ_K over K as in (6.1.3), and likewise θ_E. The set $\mathbb{H}(\theta_K)$ is a principal homogeneous space over $X_1(E)$ and hence, by restriction, a space over $X_1(E_0)$. The heart of the matter is the following lemma.

Lemma 1 *There is a canonical $X_1(E_0)$-map*

$$\ell_{K/F} : \mathbb{H}(\theta) \longrightarrow \mathbb{H}(\theta_K). \tag{8.2.3}$$

It is injective and its image is the space $\mathbb{H}(\theta_K)^{\Delta}$ of Δ-fixed points in $\mathbb{H}(\theta_K)$.

The construction of $\ell_{K/F}$ is an elaborate exercise in the use of the Glauberman correspondence. If $\det \boldsymbol{\kappa} \,\big|\, J(\beta, \mathfrak{a})$ has p-power order, the same is true for $\ell_{K/F}(\boldsymbol{\kappa})$. In other words, $\ell_{K/F}$ preserves the "standard" elements of the spaces $\mathbb{H}^0$, as in 3.1.1 Proposition (3).

It is also necessary to consider the set $\mathbb{H}(\theta_K)^{\Delta\text{-reg}}$ of Δ-regular elements of $\mathbb{H}(\theta_K)$.

Lemma 2 *A representation $\boldsymbol{\kappa} \in \mathbb{H}(\theta_K)$ is Δ-regular if and only if there exists $\boldsymbol{\kappa}_0 \in \mathbb{H}(\theta)$ and $\xi \in X_1(E)^{\Delta\text{-reg}}$ such that $\boldsymbol{\kappa} = \xi \odot_{\Theta_E} \ell_{K/F}(\boldsymbol{\kappa}_0)$.*

The $\odot$-product here is taken relative to the endo-class Θ_E of the simple character θ_E.

The simple character θ_K lies in $\mathcal{C}(\mathfrak{a}_K, \beta)$, where $\mathfrak{a}_K$ is the $\mathfrak{a}$-centralizer of K. The extension $K[\beta] = P/K$ is totally ramified so $\mathcal{T}(\theta_K) = \mathbb{H}(\theta_K)$. Let $\Phi = \varphi \odot_{\Theta_E} \ell_{K/F}(\boldsymbol{\kappa}) \in \mathcal{T}(\theta_K)^{\Delta\text{-reg}}$. The character $\varphi \in X_1(E)$ defines an EMST Λ_φ, of level zero, over the field $E_0 = E^{\Delta}$. Set

$$ind_{K/F}\, \Phi = \Lambda_\varphi \ltimes_{\Theta_E} \boldsymbol{\kappa}. \tag{8.2.4}$$

Theorem *The map (8.2.4) induces a canonical bijection*

$$ind_{K/F} : \Delta\backslash\mathcal{T}(\theta_K)^{\Delta\text{-reg}} \xrightarrow{\;\approx\;} \mathcal{T}(\theta), \tag{8.2.5}$$

and hence a canonical bijection

$$\Delta\backslash\widehat{\mathrm{GL}}_K(\Theta_K; 1)^{\Delta\text{-reg}} \xrightarrow{\;\approx\;} \widehat{\mathrm{GL}}_F(\Theta; s). \tag{8.2.6}$$

8.3 Naïve Correspondence

The algebraic induction maps (8.2.2), (8.2.6) enable the construction of an approximate version of the Langlands correspondence.

8.3.1 Let $\pi \in \widehat{\mathrm{GL}}_F$ be *totally wild,* that is, $\deg \Theta = e_\Theta$ is a power of p. Thus $\pi \in \widehat{\mathrm{GL}}_F(\Theta; 1)$, where the endo-class $\Theta \in \mathcal{E}(F)$ is totally wild. It follows that the conjugacy class ${}^L\Theta \in \mathcal{W}_F\backslash\widehat{\mathcal{P}}_F$ has exactly one element, α say. Defining ${}^N\pi = {}^L\pi$, we get a canonical bijection

$$\begin{aligned} \widehat{\mathrm{GL}}_F(\Theta; 1) &\longrightarrow \widehat{\mathcal{W}}_F(\alpha; 1), \\ \pi &\longmapsto {}^N\pi = {}^L\pi. \end{aligned} \tag{8.3.1}$$

This map commutes with isomorphisms of the base field F and respects twisting with characters χ of $F^\times$,

$$ {}^N(\chi\pi) = {}^L\chi \otimes {}^N\pi. $$

Since the tame parameter field of Θ is F, this says that the map also preserves the $\odot$-actions of $X_1(F)$.

Remark The map (8.3.1) is almost completely determined by the relation

$$ {}^L\Theta = O_F(\alpha). \tag{8.3.2}$$

First, if $\pi \in \widehat{\mathrm{GL}}_F(\Theta; 1)$ and $\sigma = {}^L\Theta$, then

$$ {}^L\omega_\pi = \det\sigma. \tag{8.3.3}$$

Let $\dim\sigma = \deg\Theta = p^r$, for some $r \geqslant 1$. The relations (8.3.2), (8.3.3) determine π up to twisting with an unramified character of $F^\times$ of order dividing p^r. Following [10], there exists a totally ramified representation $\pi' \in \mathcal{A}^0_{p^s}(F)$, with $0 \leqslant s < r$, such that $\mathrm{sw}(\check{\pi}' \times \pi)$ is not divisible by p. This distinguishes the various π satisfying the two relations. Further, if $\mathrm{sw}(\pi)$ is not divisible by p, one may take π' to be the trivial character of $\mathrm{GL}_1(F)$: this follows from 4.3.3 Corollary.

8.3.2 Consider next the case where $\pi \in \widehat{\mathrm{GL}}_F$ is *totally ramified.* That is, $\pi \in \widehat{\mathrm{GL}}_F(\Theta; 1)$, where $e_\Theta = \deg\Theta$. Let E/F be a tame parameter field for Θ, and choose a totally wild E/F-lift Θ_E of Θ. Embed E in $\bar{F}$ in some way, and define $\alpha \in \widehat{\mathcal{P}}_F = \widehat{\mathcal{P}}_E$ by the condition ${}^L\Theta_E = \{\alpha\}$. Following (8.2.2), there is a unique $\rho \in \widehat{\mathrm{GL}}_E(\Theta_E; 1)$ such that $\pi = ind_{E/F}\,\rho$. The representation ρ is totally wild, so we may set

$$ {}^N\pi = {}^N(ind_{E/F}\,\rho) = \mathrm{Ind}_{E/F}\,{}^N\rho. $$

This gives a bijection

$$\begin{aligned} \widehat{\mathrm{GL}}_F(\Theta; 1) &\longrightarrow \widehat{\mathcal{W}}_F(\alpha; 1), \\ \pi &\longmapsto {}^N\pi. \end{aligned} \tag{8.3.4}$$

It satisfies

$$ {}^N(\chi \odot_{\Theta_E} \pi) = \chi \odot_\alpha {}^N\pi, \quad \chi \in X_1(E), \tag{8.3.5}$$

and commutes with isomorphisms of the base field.

8.3.3 Finally, let $\pi \in \widehat{\mathrm{GL}}_F(\Theta; s)$, for an arbitrary $\Theta \in \mathcal{E}(F)$ and an integer $s \geqslant 1$. Let E/F be a tame parameter field for Θ, viewed as a subfield of $\bar{F}$. Let Θ_E be a totally wild E/F-lift of Θ and define $\alpha \in \widehat{\mathcal{P}}_F$ by the condition ${}^L\Theta_E = \{\alpha\}$. Let E_s/E be unramified of degree s, and put $\Delta = \mathrm{Gal}(E_s/E)$. Let K/F be the maximal unramified sub-extension of E_s/F and identify Δ with $\mathrm{Gal}(K/K \cap E)$.

As in (8.2.6), there exists $\rho \in \widehat{\mathrm{GL}}_K(\Theta_K; 1)^{\Delta\text{-reg}}$ such that $\pi = ind_{K/F}\, \rho$. The representation ρ is uniquely determined up to conjugation by Δ. The endo-class Θ_K is $\boldsymbol{i}_{E_s/K}\, \Theta_{E_s}$, where Θ_{E_s} is the unique E_s/E-lift of Θ_E.

The representation ${}^N\rho$ lies in $\widehat{\mathcal{W}}_K(\alpha; 1)^{\Delta\text{-reg}}$, so we may define

$$ {}^N\pi = \mathrm{Ind}_{K/F}\, {}^N\rho. $$

We therefore have a bijection

$$ \begin{aligned} \widehat{\mathrm{GL}}_F(\Theta; s) &\longrightarrow \widehat{\mathcal{W}}_F(\alpha; s), \\ \pi &\longmapsto {}^N\pi \end{aligned} \tag{8.3.6} $$

with the property

$$ {}^N(\phi \odot_{\Theta_E} \pi) = \phi \odot_\alpha {}^N\pi. \tag{8.3.7} $$

8.3.4 Altogether, we have produced a canonical bijection

$$ \begin{aligned} \widehat{\mathrm{GL}}_F &\longrightarrow \widehat{\mathcal{W}}_F, \\ \pi &\longmapsto {}^N\pi. \end{aligned} $$

It commutes with isomorphisms of the base field and twisting by characters of $F^\times$. However, the determinant character det ${}^N\pi$ is not in general the same as the central character ω_π. The naïve correspondence and the Langlands correspondence behave differently with respect to automorphisms of the coefficient field $\mathbb{C}$.

On the other hand, the naïve correspondence has the "higher twisting" property:

Theorem *Let $\Theta \in \mathcal{E}(F)$ and let E/F be a tame parameter field for Θ. Let $\Xi \in \mathcal{E}(E)$ be a totally wild E/F-lift of Θ. If $\alpha = {}^L\Xi \in \mathcal{W}_E\backslash\widehat{\mathcal{P}}_E$, then*

$$ {}^N(\phi \odot_\Xi \pi) = {}^L\phi \odot_\alpha {}^N\pi, \quad \pi \in \widehat{\mathrm{GL}}_F(\Theta; s),\ s \geqslant 1. $$

8.4 Main Theorems

We compare the naïve correspondence $\pi \mapsto {}^N\pi$ with the Langlands correspondence $\pi \mapsto {}^L\pi$ and then deduce some consequences for the Langlands correspondence itself.

8.4.1 The correspondences are related as follows.

Comparison Theorem *Let $\Theta \in \mathcal{E}(F)$ and let $s \geqslant 1$ be an integer. Let E/F be a tame parameter field for Θ, identified with a subfield of $\bar{F}$, and let Θ_E be a totally wild E/F-lift of Θ. Define $\alpha \in \widehat{\mathcal{P}}_F$ by the condition ${}^L\Theta_E = \{\alpha\}$. There is a character $\mu = \mu^F_{\alpha,s} \in X_1(E)$, uniquely determined modulo $X_0(E)_s$, such that*

$$ {}^L\pi = \mu \odot_\alpha {}^N\pi, \quad \pi \in \widehat{\mathrm{GL}}_F(\Theta; s). $$

In the essentially tame case $\dim \alpha = 1$, the character $\mu^F_{\alpha,s}$ is worked out completely in [13, 14, 16]. The answer is somewhat complex, although μ (modulo $X_0(E)_s$) has order at most 4. The general case is similar in many respects, although more complex in others. We discuss it briefly in 8.4.4 below.

Remark Let ℓ be a prime number, $\ell \neq p$. The Comparison Theorem has a direct application to the theory of ℓ-modular representations. In brief, it implies that two representations $\pi_1, \pi_2 \in \mathcal{A}^0_n(F)$, with central characters of finite order, are congruent modulo ℓ if and only if the representations ${}^L\pi_i$ are congruent mod. ℓ. See [20] for a full and self-contained account.

8.4.2 Comparing the theorem with the higher twisting property of the naïve correspondence in 8.3.4 Theorem, we get the following corollary. We retain the notation of the Comparison Theorem.

Homogeneity Theorem *The Langlands correspondence satisfies*

$$ {}^L(\phi \odot_{\Theta_E} \pi) = \phi \odot_\alpha {}^L\pi, $$

for $\pi \in \widehat{\mathrm{GL}}_F(\Theta; s)$ and $\phi \in X_1(E)$.

8.4.3 We determine exactly how the Langlands correspondence behaves relative to maximal simple types (or inertial Galois parameters).

We use the same notation as in the Comparison Theorem. In particular, $\pi \in \widehat{\mathrm{GL}}_F(\Theta; s) \cap \mathcal{A}^0_n(F)$, where $n = s \deg \Theta$. Thus π contains a simple character θ of endo-class Θ. We may assume E is a tame parameter field for θ in the internal sense and let Θ_E be the endo-class of θ_E.

Write $\sigma = {}^L\pi$. Let $\alpha_{\mathcal{I}}$ be the unique representation of the inertia group $\mathcal{I}_E$ such that $\alpha_{\mathcal{I}} \,|\, \mathcal{P}_F = \alpha$ and $\det \alpha_{\mathcal{I}}$ has p-power order (7.1.2 Proposition). The representation $\sigma \,|\, \mathcal{I}_E$ thus has an irreducible component $\chi \otimes \alpha_{\mathcal{I}}$, where χ is a character of $\mathcal{I}_E/\mathcal{P}_F$. The $\mathcal{W}_E$-stabilizer of χ is $\mathcal{W}_{E_s}$, where E_s/E is unramified of degree s. The character

χ determines a maximal simple type λ_χ in $\mathrm{GL}_s(E)$, as in 8.1.2. Using the obvious, restricted, analogue of the notation (8.1.3), we get a maximal simple type $\lambda_\chi \ltimes_{\Theta_E} \kappa$ in $\mathrm{GL}_n(F)$, for any $\kappa \in \mathbb{H}^0(\theta)$.

Types Theorem *Let* $\Delta_s = Gal(E_s/E)$.

(1) *There is a unique representation* $\kappa_1 \in \mathbb{H}^0(\theta)$ *such that* $\mathrm{tr}\,\kappa_1$ *is constant on the set of* Δ_s*-regular elements of* $\boldsymbol{\mu}_{E_s}$.
(2) *Let* δ *be the character of* $J(\theta)/J^1(\theta)$ *that has order* 2 *when* $e(E|F)$ *is even, and is trivial otherwise.*
(3) *The representation* π *contains the maximal simple type* $\lambda_\chi \ltimes_{\Theta_E} \delta\kappa_1$.

We remark that the factor δ is trivial when $p = 2$, because E/F is tamely ramified. The inertial parameter (7.1.3) of σ is the class of (α, χ) in $\mathcal{W}_F\backslash \mathrm{IP}(F)$.

8.4.4 The character $\mu^F_{\alpha,s}$ is quite complex in structure. The restriction of $\mu = \mu^F_{\alpha,s}$ to units is readily computable, as encapsulated here in the Types Theorem. Using the notation of 8.3.3, the contribution from the unramified step K/F is relatively straightforward. It is of order 2 (mod. $X_0(E_0)_s$), and its value at a suitable prime element is discussed in [19] 10.7 Corollary. The totally ramified step seems quite difficult. Using the notation of 8.3.2, one has to work inductively, along the group Γ of F-automorphisms of E. If Γ is trivial, the existence of a character $\mu = \mu^F_{\alpha,1}$, with the required properties, is relatively straightforward [19] 8.3. However, it is not clear how to compute it: certainly the base change method of the essentially tame case is not completely effective. If Γ is not trivial, let l be the largest prime divisor of $|\Gamma|$ and let L/F be the unique sub-extension of E/F of degree l. Inductively, one can assume that $\mu^L_{\alpha,1}$ is known. The contribution from L/F is then given by an automorphic induction calculation leading to a formula discussed in [19] 8.9.

8.4.5 This totally ramified case has an interesting feature. Again in the notation of 8.3.2, let $\pi \in \widehat{\mathrm{GL}}_F(\Theta; 1)$ and write $\pi = ind_{E/F}\,\rho$, where $\rho \in \widehat{\mathrm{GL}}_E(\Theta_E; 1)$. We view π as a representation of a group $G \cong \mathrm{GL}_n(F)$ containing $E^\times$, and ρ as a representation of the G-centralizer G_E of $E^\times$. With Γ as before, one has [19] 8.1

$$\mathrm{tr}\,\pi(h) = \epsilon_{E/F} \sum_{\gamma\in\Gamma} \mathrm{tr}\,\rho^\gamma(h), \tag{8.4.1}$$

valid for all elements h of G_E that are elliptic regular in G and such that $\upsilon_F(\det_G h)$ is relatively prime to n. Here, $\epsilon_{E/F} = \pm 1$ is the constant sign coming from the Glauberman correspondence, as in 8.2.1. The characters appearing here are uniquely determined by their values on the specified set of elements h.

Put another way, let $\tau = {}^L\rho \in \widehat{\mathcal{W}}_F(\alpha; 1)$. Write $\sigma = \mathrm{Ind}_{E/F}\,\tau$. Thus $\sigma \in \widehat{\mathcal{W}}_F(\alpha; 1)$ and $\sigma = {}^N\pi = {}^L\pi'$, for some $\pi' \in \widehat{\mathrm{GL}}_F(\Theta; 1)$. Indeed, $\pi' = \mu \odot \pi$, $\mu = \mu^F_{\alpha,1}$, so

$$\mathrm{tr}\,\mu^{-1} \odot \pi'(h) = \epsilon_{E/F} \sum_{\gamma\in\Gamma} \mathrm{tr}\,\rho^\gamma(h), \tag{8.4.2}$$

for elements h as before. Equally,

$$\operatorname{tr} \pi'(h) = \epsilon_{E/F} \sum_{\gamma \in \Gamma} \operatorname{tr} (\mu\rho)^{\gamma}(h). \tag{8.4.3}$$

In the case where E/F is cyclic, we have $\pi' = \boldsymbol{a}_{E/F}\, \rho$ and we have retrieved the automorphic induction equation in a novel version. In addition, (8.4.3) is valid when E/F is not cyclic, or even Galois.

9 Higher Ramification and the Herbrand Function

Following the Types Theorem of 8.4.3, we know how the structure of a cuspidal representation $\pi \in \widehat{\mathrm{GL}}_F$ is reflected in the behaviour of $\sigma = {}^L\pi \in \widehat{\mathcal{W}}_F$ on the standard subgroups $\mathcal{P}_F \subset \mathcal{I}_F \subset \mathcal{W}_F$. We look to next steps.

The group $\mathcal{P}_F$ carries a canonical filtration by *ramification subgroups,* here denoted $\mathcal{R}_F(x)$, $x \in \mathbb{R}$, $x \geqslant 0$. The behaviour of the restriction $\sigma \,|\, \mathcal{R}_F(x)$ provides penetrating insight into the structure of σ. In this section, we follow [21] to show how, in principle, it is predicted by the endo-class $\Theta(\pi)$ of simple characters contained in π.

9.1 Ramification and Conductors

This section is a review of ramification theory and its connections with conductors. We start from the standard ramification theory, for finite Galois extensions, as in [69].

9.1.1 Let E/F be a finite Galois extension, and set $\mathrm{Gal}(E/F) = \Gamma$. If $x \geqslant -1$ is a real number, let Γ_x be the associated ramification subgroup of Γ, in the lower numbering, and Γ^x that in the upper numbering. The two filtrations are connected by

$$\Gamma^x = \Gamma_{\psi_{E/F}(x)}, \qquad \Gamma_y = \Gamma^{\varphi_{E/F}(y)},$$

where $\varphi_{E/F}$, $\psi_{E/F}$ are the Herbrand functions for E/F. Both functions are piecewise linear with $\varphi_{E/F}$ concave and $\psi_{E/F}$ convex. The Herbrand functions are defined for $x \geqslant -1$, but we focus entirely on the range $x \geqslant 0$.

If K/F is a Galois sub-extension of E/F, then $\psi_{E/F} = \psi_{E/K} \circ \psi_{K/F}$. If, on the other hand, E/F is a finite separable extension, not necessarily Galois, there is a finite Galois extension E'/F containing E. The expression

$$\psi_{E/F} = \varphi_{E'/E} \circ \psi_{E'/F} \tag{9.1.1}$$

defines a continuous, strictly increasing function $\psi_{E/F}$.

Proposition *Let E/F be a finite, separable extension.*

(1) The definition (9.1.1) of $\psi_{E/F}$ does not depend on the choice of E'.
(2) The function $\psi_{E/F}(x)$ is convex in the region $x \geqslant 0$, and its derivative $\psi'_{E/F}(x)$ has only finitely many discontinuities.
(3) If $F \subset K \subset E$, then

$$\psi_{E/F} = \psi_{E/K} \circ \psi_{K/F}. \tag{9.1.2}$$

For the proof and some more detail, see [22] 1.2. The definition of $\psi_{E/F}$ via (9.1.1) is the same as that of [31].

The discontinuities of $\psi'_{E/F}(x)$, in the region $x > 0$, are called the *jumps* of the extension E/F. We note a particularly simple case.

Example If K/F is tamely ramified, then $\psi_{K/F}(x) = e(K|F)x$, $x \geqslant 0$.

9.1.2 For $x \in \mathbb{R}$, $x \geqslant 0$, let

$$\mathcal{R}_F(x) = \varprojlim_{E/F} \Gamma^x_{E/F},$$

where E/F ranges over the finite Galois extensions in $\bar{F}$ and $\Gamma_{E/F} = \mathrm{Gal}(E/F)$. Thus $\mathcal{R}_F(x)$ is a compact subgroup of $\mathcal{R}_F(0) = \mathcal{I}_F$, normalized by $\mathcal{W}_F$. In parallel, let $\mathcal{R}^+_F(x)$ be the closure in $\mathcal{R}_F(x)$ of the group $\bigcup_{y>x} \mathcal{R}_F(y)$. This is a compact normal subgroup of $\mathcal{W}_F$. In particular, $\mathcal{R}^+_F(0)$ is the wild inertia group $\mathcal{P}_F$. For $x > 0$, $\mathcal{R}^+_F(x) \neq \mathcal{R}_F(x)$ if and only if x is *rational:* see [21] 2.4 Corollary. The quotient $\mathcal{R}_F(x)/\mathcal{R}^+_F(x)$ is a compact abelian group of exponent p.

The groups $\mathcal{R}_F(x)$ are the *ramification subgroups* of $\mathcal{W}_F$, often denoted $\mathcal{W}^x_F$. This traditional notation is unusable in the context of representation theory.

9.1.3 We insert a couple of technical results taken from [22] 1.6. We do not refer to them until late in the next section, but they influence the course of the intervening arguments.

If E/F is a finite separable extension, define the *wild exponent* $w_{E/F}$ of E/F by

$$w_{E/F} = d_{E/F} + 1 - e(E|F), \tag{9.1.3}$$

where $d_{E/F}$ is the differential exponent of E/F. That is, the different of E/F is $\mathfrak{p}_E^{d_{E/F}}$. If E/F is not separable, one may take $w_{E/F} = \infty$.

Proposition 1 *Let E/F be a separable, totally wildly ramified extension of degree p^r. If $j_\infty = j_\infty(E|F)$ is the largest jump of E/F, then*

$$\psi_{E/F}(x) = p^r x - w_{E/F}, \quad x \geqslant j_\infty. \tag{9.1.4}$$

The proposition is equivalent to the standard result [69] IV Proposition 4 connecting the Artin conductor with ramification groups.

Corollary *If E/F is totally wildly ramified of degree p^r, and $j_\infty = j_\infty(E|F)$, then*

$$(p^r - 1)j_\infty \geqslant w_{E/F} \geqslant p^{r-1}(p-1)j_\infty.$$

One has $(p^r - 1)j_\infty = w_{E/F}$ if and only if j_∞ is the unique jump of E/F.

Inseparable extensions. If E/F is purely inseparable of degree n, one may define $\psi_{E/F}(x) = nx$. The definition can then be extended to arbitrary finite extensions via the transitivity relation (9.1.2). If E/F is not separable, the proposition and corollary remain true with $j_\infty = w_{E/F} = \infty$. Inseparable extensions play no rôle on the Galois side. On the GL-side, a parameter field of a simple character need not be separable over F but we will need to use the classical Herbrand functions in that context. The problem can be avoided via an approximation argument, but this *ad hoc* approach is simpler.

The largest jump appears in another rôle.

Proposition 2 *If E/F is a finite separable extension, then*

$$\mathcal{R}_F(x) \cap \mathcal{W}_E = \mathcal{R}_E(\psi_{E/F}(x)), \quad x \geqslant 0,$$

and similarly for the groups $\mathcal{R}^+$. Moreover, $\mathcal{R}^+_F(x) \subset \mathcal{W}_E$ if and only if $x \geqslant j_\infty(E|F)$.

9.1.4 Let σ be an irreducible representation of $\mathcal{W}_F$, $\sigma \in \widehat{\mathcal{W}}_F$. For $x \geqslant 0$, the group $\mathcal{R}_F(x)$ is compact, so $\sigma \,|\, \mathcal{R}_F(x)$ is semisimple. Any two irreducible components of the restriction are $\mathcal{W}_F$-conjugate and occur with the same multiplicity. The same applies with $\mathcal{R}^+_F(x)$ replacing $\mathcal{R}_F(x)$.

The definition of ramification groups implies the existence of $x > 0$ such that $\mathcal{R}_F(x) \subset \operatorname{Ker}\sigma$. We may accordingly define the *slope* $\varsigma(\sigma)$ of σ by

$$\varsigma(\sigma) = \inf\{x > 0 : \mathcal{R}_F(x) \subset \operatorname{Ker}\sigma\}. \tag{9.1.5}$$

In particular, $\varsigma(\sigma) \geqslant 0$. One has

$$\varsigma(\sigma) = \operatorname{sw}(\sigma)/\dim\sigma, \tag{9.1.6}$$

by [36] Théorème 3.5.

Proposition *Let $\sigma \in \widehat{\mathcal{W}}_F$. If $\varsigma = \varsigma(\sigma)$, then σ is trivial on $\mathcal{R}^+_F(\varsigma)$. If $\varsigma \neq 0$, then $\sigma \,|\, \mathcal{R}_F(\varsigma)$ is a direct sum of non-trivial characters.*

The point here is that, for $x \geqslant 0$, the commutator group $[\mathcal{R}_F(x), \mathcal{R}_F(x)]$ is contained in $\mathcal{R}^+_F(x)$. If $\varsigma(\sigma) = 0$, then $\sigma \,|\, \mathcal{R}_F(0)$ is a direct sum of $\mathcal{W}_F$-conjugate characters, but they may be trivial.

Remark Let $\pi \in \mathcal{A}_n^0(F)$ and let Θ be the endo-class of a simple character contained in π. If $\sigma = {}^L\pi$, then

$$\varsigma(\sigma) = \mathrm{sw}(\pi)/n = \varsigma_\Theta, \tag{9.1.7}$$

in the notation of (5.2.3).

9.1.5 For $\sigma, \tau \in \widehat{\mathcal{W}}_F$, define

$$\Delta(\sigma, \tau) = \inf\{x > 0 : \mathrm{Hom}_{\mathcal{R}_F(x)}(\sigma, \tau) \neq 0\}.$$

The pairing Δ is symmetric. If $\delta = \Delta(\sigma, \tau) > 0$, then $\sigma \,|\, \mathcal{R}_F^+(\delta)$ and $\tau \,|\, \mathcal{R}_F^+(\delta)$ have an irreducible component in common, while $\mathrm{Hom}_{\mathcal{R}_F(\delta)}(\sigma, \tau) = 0$. Consequently, Δ takes non-negative rational values. It satisfies an ultrametric inequality,

$$\Delta(\sigma_1, \sigma_3) \leqslant \max\{\Delta(\sigma_1, \sigma_2), \Delta(\sigma_2, \sigma_3)\},$$

for all $\sigma_i \in \widehat{\mathcal{W}}_F$.

The value $\Delta(\sigma, \tau)$, $\sigma, \tau \in \widehat{\mathcal{W}}_F$, depends only on the restrictions of σ and τ to $\mathcal{P}_F = \mathcal{R}_F^+(0)$. It is therefore often more convenient to regard Δ as a pairing on the orbit space $\mathcal{W}_F\backslash\widehat{\mathcal{P}}_F$. If r_F^+ denotes the canonical map $\widehat{\mathcal{W}}_F \to \mathcal{W}_F\backslash\widehat{\mathcal{P}}_F$ as in 6.3.1, the two versions are related by

$$\Delta(\sigma, \tau) = \Delta(r_F^+(\sigma), r_F^+(\tau)).$$

The second version has the property

$$\Delta(\zeta, \xi) = 0 \qquad \Longleftrightarrow \qquad \zeta = \xi,$$

for $\zeta, \xi \in \mathcal{W}_F\backslash\widehat{\mathcal{P}}_F$. That is, Δ *defines an ultrametric on* $\mathcal{W}_F\backslash\widehat{\mathcal{P}}_F$.

9.1.6 Following [35] and [21] section 3, the ultrametric pairing Δ leads to a formula for the conductor of a tensor product.

Proposition 1 *Let* $\sigma \in \widehat{\mathcal{W}}_F$. *There is a unique continuous function* $\Sigma_\sigma(x)$, $x \geqslant 0$, *such that*

$$\frac{\mathrm{sw}(\check{\sigma} \otimes \tau)}{\dim \sigma \, \dim \tau} = \Sigma_\sigma\big(\Delta(\sigma, \tau)\big), \quad \tau \in \widehat{\mathcal{W}}_F. \tag{9.1.8}$$

The *decomposition function* Σ_σ is given as follows. Let V be the representation space of σ. Thus $\check{\sigma}$ acts on the space $\check{V} = \mathrm{Hom}_{\mathbb{C}}(V, \mathbb{C})$, and the space $X = \check{V} \otimes V$ carries the representation $\check{\sigma} \otimes \sigma$ of $\mathcal{W}_F$. For $\delta \geqslant 0$, let $X(\delta)$ be the space of $\mathcal{R}_F^+(\delta)$-fixed points in X. Let $X'(\delta)$ be the unique $\mathcal{R}_F^+(\delta)$-complement of $X(\delta)$ in X. Each of the spaces $X(\delta)$, $X'(\delta)$ is $\mathcal{W}_F$-stable so, in particular, the conductor $\mathrm{sw}\, X'(\delta)$ is defined. The function Σ_σ is then

$$\Sigma_\sigma(\delta) = \big(\delta \dim X(\delta) + \operatorname{sw} X'(\delta)\big)\big/(\dim\sigma)^2, \quad \delta \geqslant 0. \tag{9.1.9}$$

Therefore:

Proposition 2 *The continuous function Σ_σ is strictly increasing, piecewise linear and convex. The derivative $\Sigma'_\sigma(x)$ has only finitely many discontinuities in the region $x > 0$.*

The discontinuities of Σ'_σ are the *jumps* of σ: they all lie in the interval $0 < x < \varsigma(\sigma)$. We remark that if x is not a jump of σ, then

$$\Sigma'_\sigma(x) = \dim \operatorname{End}_{\mathcal{R}_F(x)}(\sigma)/(\dim\sigma)^2. \tag{9.1.10}$$

This formula (9.1.9) shows that Σ_σ depends only on the orbit $r_F^+(\sigma) \in \mathcal{W}_F\backslash\widehat{\mathcal{P}}_F$. It is sometimes more convenient to use the notation $\Sigma_\sigma = \Sigma_\xi$, where $\xi = r_F^+(\sigma)$.

9.1.7 One may characterize the jumps of $\sigma \in \widehat{\mathcal{W}}_F$ in representation-theoretic terms.

Proposition *Let $\sigma \in \widehat{\mathcal{W}}_F$, let $\epsilon > 0$ and let σ_ϵ be an irreducible component of $\sigma \,\big|\, \mathcal{R}_F(\epsilon)$. Let Γ_ϵ be the group of characters of $\mathcal{R}_F(\epsilon)/\mathcal{R}_F^+(\epsilon)$. The following are equivalent.*

(1) The function $\Sigma'_\sigma(x)$ is continuous at $x = \epsilon$.
(2) The representation σ_ϵ is not $\mathcal{W}_F$-conjugate to $\chi \otimes \sigma_\epsilon$, for any $\chi \in \Gamma_\epsilon$, $\chi \neq 1$.

In particular, if $\sigma_\epsilon \,\big|\, \mathcal{R}_F^+(\epsilon)$ is not irreducible then ϵ is a jump of σ.

This result is 8.1 Proposition of [21]. Note that σ_ϵ can be irreducible on $\mathcal{R}_F^+(\epsilon)$ even when ϵ is a jump.

9.2 Comparison of Ultrametrics

The Ramification Theorem of 6.3.1 asserts that the Langlands correspondence specializes to a canonical bijection $\mathcal{E}(F) \to \mathcal{W}_F\backslash\widehat{\mathcal{P}}_F$. These sets carry respectively the ultrametrics $\mathbf{A}$, Δ, but the Langlands correspondence does *not give an isometry* $(\mathcal{E}(F), \mathbf{A}) \to (\mathcal{W}_F\backslash\widehat{\mathcal{P}}_F, \Delta)$. If $\Theta \in \mathcal{E}(F)$ and $\delta > 0$, the δ-neighbourhood of Θ (in the obvious sense) is mapped to the ϵ-neighbourhood of ${}^L\Theta$, for a certain $\epsilon > 0$. However, ϵ depends on both δ and Θ. The relation takes the form $\delta = \Psi_\Theta(\epsilon)$, for a certain "Herbrand function" Ψ_Θ attached to Θ. In the rest of this section, we review the definition and main properties of this function.

9.2.1 To $\Theta \in \mathcal{E}(F)$ we have attached two functions, namely the structure function Φ_Θ (5.3.3 Theorem 3) and the decomposition function $\Sigma_{{}^L\Theta}$ (9.1.6). Define the *Herbrand function* $\Psi_\Theta(x)$, $x \geqslant 0$, by

$$\Psi_\Theta = \Phi_\Theta^{-1} \circ \Sigma_{{}^L\Theta}. \tag{9.2.1}$$

Proposition *(1) The function Ψ_Θ is continuous, strictly increasing and piecewise linear.*
(2) It satisfies $\Psi_\Theta(0) = 0$, while $\Psi_\Theta(x) = x$ for $x \geqslant \varsigma_\Theta$.
(3) The derivative Ψ'_Θ is continuous except at a finite number of points y, all of which satisfy $0 < y < \varsigma_\Theta$.

Proof Part (1) is clear. In (2), one has $\Phi_\Theta(x) = x$ for $x \geqslant \varsigma_\Theta$, either from the definition in [21] or as a consequence of 5.3.3 Theorem 1. If $\pi \in \mathcal{A}^0_n(F)$ satisfies $\Theta(\pi) = \Theta$ and $\sigma = {}^L\sigma$, (9.1.9) implies $\Sigma_\sigma(x) = x$ for $x \geqslant \varsigma_\sigma = \varsigma_\Theta$. This proves the first assertion. With this same notation,

$$\Phi_\Theta(0) = \mathrm{sw}(\check{\pi} \times \pi)/n^2 = \mathrm{sw}(\check{\sigma} \otimes \sigma)/n^2 = \Sigma_\sigma(0),$$

as the result follows. Part (3) follows from the corresponding properties of Φ_Θ and Σ_σ. □

Over the range $0 \leqslant x < \infty$, the Herbrand function Ψ_Θ is convex if and only if it is linear (see 9.4.2 below). In the interesting range $0 \leqslant x \leqslant \varsigma_\Theta$, it is convex only in special cases (*cf.* 10.1.2 below).

Going forward, we tend to view the structure function Φ_Θ as known, on the basis that we usually start from a simple character. We use the Herbrand function as a means of calculating the decomposition function Σ_σ.

9.2.2 We make a fundamental connection.

Higher Ramification Theorem *If $\Theta, \Upsilon \in \mathcal{E}(F)$, then*

$$\mathbf{A}(\Theta, \Upsilon) = \Psi_\Theta(\Delta({}^L\Theta, {}^L\Upsilon)).$$

In particular, if $\epsilon > 0$ and $\delta = \Psi_\Theta(\epsilon)$, then

$$\begin{aligned} \Delta({}^L\Theta, {}^L\Upsilon) < \epsilon \quad &\Longleftrightarrow \quad \mathbf{A}(\Theta, \Upsilon) < \delta, \\ \Delta({}^L\Theta, {}^L\Upsilon) \leqslant \epsilon \quad &\Longleftrightarrow \quad \mathbf{A}(\Theta, \Upsilon) \leqslant \delta, \end{aligned} \tag{9.2.2}$$

for any $\Upsilon \in \mathcal{E}(F)$.

Proof Choose $\pi \in \mathcal{A}^0_m(F)$ and $\rho \in \mathcal{A}^0_n(F)$ so that $\Theta(\pi) = \Theta$ and $\Theta(\rho) = \Upsilon$. According to 5.3.3 Theorem 3,

$$\frac{\mathrm{sw}(\check{\pi} \times \rho)}{mn} = \Phi_\Theta(\mathbf{A}(\Theta, \Upsilon)).$$

Setting $\sigma = {}^L\pi$ and $\tau = {}^L\rho$, (9.1.8) says that

$$\frac{\mathrm{sw}(\check{\sigma} \otimes \tau)}{mn} = \Sigma_\sigma(\Delta(\sigma, \tau)).$$

In the alternate notation, $\Sigma_\sigma(\Delta(\sigma,\tau)) = \Sigma_{^L\Theta}(\Delta(^L\Theta,\, ^L\Upsilon))$, while the Langlands correspondence has the property

$$\frac{\mathrm{sw}(\check{\pi}\times\rho)}{mn} = \frac{\mathrm{sw}(\check{\sigma}\otimes\tau)}{mn}.$$

The first assertion follows. The function Ψ_Θ is strictly increasing, whence follows the second. □

To interpret (9.2.2) in more concrete terms, use the notation of the proof and choose an irreducible component σ_ϵ of $\sigma \,\big|\, \mathcal{R}_F(\epsilon)$. The condition $\Delta(^L\Theta,\, ^L\Upsilon) = \Delta(\sigma,\tau) < \epsilon$ is equivalent to σ_ϵ being a component of $\tau \,\big|\, \mathcal{R}_F(\epsilon)$. On the other hand, in the notation of 5.2.5, $\mathbf{A}(\Theta,\Upsilon) < \delta$ is equivalent to $\mathrm{tc}_\delta(\Theta) = \mathrm{tc}_\delta(\Xi)$. In other words, it is equivalent to π, ρ containing the same δ-truncated endo-class of simple characters. A similar comment applies to the non-strict inequality.

9.2.3 There is a useful technical consequence of the theorem.

Corollary *Let $\Theta,\Upsilon \in \mathcal{E}(F)$. If $x \geqslant \mathbf{A}(\Theta,\Upsilon)$, then $\Psi_\Theta^{-1}(x) = \Psi_\Upsilon^{-1}(x)$. Equivalently, $\Psi_\Theta(y) = \Psi_\Upsilon(y)$ for $y \geqslant \Delta(^L\Theta, {^L\Upsilon})$.*

Proof Let $\delta > \mathbf{A}(\Theta,\Upsilon)$ and set $\epsilon = \Psi_\Theta^{-1}(\delta)$. The ultrametric inequality then implies that an endo-class $\Xi \in \mathcal{E}(F)$ satisfies $\mathbf{A}(\Xi,\Upsilon) < \delta$ if and only if $\mathbf{A}(\Xi,\Theta) < \delta$. The second condition is equivalent to $\Delta(^L\Xi, {^L\Theta}) < \epsilon$ by the theorem, while the first is equivalent to $\Delta(^L\Upsilon, {^L\Xi}) < \Psi_\Upsilon^{-1}(\delta)$. On the other hand,

$$\Delta(^L\Upsilon, {^L\Xi}) \leqslant \max\big\{\Delta(^L\Upsilon, {^L\Theta}), \Delta(^L\Theta, {^L\Xi})\big\} < \epsilon.$$

It follows that $\Psi_\Upsilon^{-1}(\delta) \leqslant \epsilon = \Psi_\Theta^{-1}(\delta)$ for $\delta > \mathbf{A}(\Theta,\Upsilon)$. By symmetry,

$$\Psi_\Upsilon^{-1}(\delta) = \Psi_\Theta^{-1}(\delta), \quad \delta > \mathbf{A}(\Theta,\Upsilon).$$

By continuity, the relation holds for $\delta \geqslant \mathbf{A}(\Theta,\Upsilon)$. □

9.3 *Lifting and Interpolation*

The Herbrand function Ψ_Θ, $\Theta \in \mathcal{E}(F)$, has a tidy property relative to tame lifting that is not shared by its factors Φ_Θ, $\Sigma_{^L\Theta}$. This leads to a method of calculating Ψ_Θ directly from Θ, without reference to the Galois side.

9.3.1 The basic lifting property is the following (7.1 Proposition of [21]).

Proposition *Let $\Theta \in \mathcal{E}(F)$. Let K/F be a finite tame extension and set $e = e(K|F)$. If Θ^K is a K/F-lift of Θ, then*

$$\Psi_{\Theta^K}(x) = e\Psi_\Theta(x/e), \quad x \geqslant 0.$$

The proposition is a detailed refinement of the Tame Parameter Theorem of 7.3.1.

9.3.2 We recall an elementary structure. Let $\Theta \in \mathcal{E}(F)$ and let ϕ be a character of $F^\times$. If θ is a realization of Θ, on a simple stratum $[\mathfrak{a}, m, 0, \beta]$ in a matrix algebra A, the character

$$\phi\theta : x \longmapsto \phi(\det_A x)\, \theta(x), \quad x \in H^1(\beta, \mathfrak{a}),$$

is a simple character that is either trivial or attached to a simple stratum $[\mathfrak{a}, m', 0, \beta + c]$, for an integer m' and some $c \in F$: see the appendix to [27]. The endo-class of $\phi\theta$ depends only on $\phi \mid U_F^1$ and Θ and so is denoted $\phi\Theta$. We note:

Lemma *If ϕ is a character of $F^\times$ and $\Theta \in \mathcal{E}(F)$, then $\Psi_{\phi\Theta} = \Psi_\Theta$.*

9.3.3 Recall that $\Theta \in \mathcal{E}(F)$ is called totally wild if $\deg \Theta = e_\Theta = p^r$, for some $r \geqslant 0$. If $\theta \in \mathcal{C}(\mathfrak{a}, \beta)$ is a realization of Θ on a simple stratum $[\mathfrak{a}, m, 0, \beta]$, then Θ is totally wild if and only if the field extension $F[\beta]/F$ is totally wildly ramified.

For a general $\Theta \in \mathcal{E}(F)$, it follows from the Definition in 6.1.3 that there exists a finite tame extension K/F such that Θ has a totally wild K/F-lift. Indeed, we may choose K so that every K/F-lift of Θ is totally wild. On the other hand, if Θ is totally wild and K/F is a finite tame extension, then Θ has a unique K/F-lift and that lift is totally wild.

The effect of 9.3.1. Proposition is that we need only compute Ψ_Θ for totally wild endo-classes Θ. If K/F is a finite tame extension, let $\mathbf{A}_K$ be the canonical ultrametric on $\mathcal{E}(K)$.

Interpolation Theorem *Let $\Theta \in \mathcal{E}(F)$ be totally wild. The function Ψ_Θ has the following properties.*

(1) It is continuous, strictly increasing and piecewise linear.
(2) The derivative Ψ'_Θ is continuous except at a finite number of points.
(3) There is a finite set D of positive real numbers such that

(a) if K/F is a finite tame extension, with $e = e(K|F)$, and
(b) if χ is a character of $K^\times$ such that $e^{-1}\mathrm{sw}(\chi) \notin D$, then

$$\mathbf{A}_K(\Theta^K, \chi\Theta^K) = e\Psi_\Theta(e^{-1}\mathrm{sw}(\chi)) \tag{9.3.1}$$

where Θ^K is the unique K/F-lift of Θ.

These properties determine the function Ψ_Θ uniquely.

The Interpolation Theorem is proved in section 7 of [21].

The theorem is quite difficult to use as a tool for computing the Herbrand function over its entire range. However, it provides essential information on part of that range, notably in [22] 2.6 and 5.3.

9.4 Some Preliminary Estimates

We record a couple of properties of the Herbrand function that are accessible without detailed analysis.

9.4.1 Let $\Theta \in \mathcal{E}(F)$. Following part (2) of 9.2.1 Proposition, the function $\Psi_\Theta(x)$ is only interesting in the region $0 < x < \varsigma_\Theta$. We describe its derivatives towards the ends of this interval.

By 9.3.1 Proposition, it is enough to treat the case where Θ is *totally wild.* In the case $\deg \Theta = 1$, there is nothing to do as $\Psi_\Theta(x) = x$, for all x. So, let $\Theta \in \mathcal{E}(F)$ be totally wild, of degree p^r, $r \geqslant 1$. Thus $\varsigma_\Theta = mp^{-r}$, for an integer m which we write as $m = ap^t$, where $t \geqslant 0$ and a is not divisible by p. Replacing Θ by $\chi\Theta$, for an appropriate character χ of $F^\times$, we may assume that $t < r$: since $\Psi_{\chi\Theta} = \Psi_\Theta$ (9.3.2 Lemma), this changes nothing. Using this notation, 7.6 Proposition of [21] asserts:

Proposition *There exist* $\epsilon, \epsilon' > 0$ *such that*

$$\begin{aligned} \Psi'_\Theta(x) &= p^{-r}, \quad 0 < x < \epsilon, \\ \Psi'_\Theta(x) &= p^{r-t}, \quad m - \epsilon' < x < m. \end{aligned}$$

9.4.2 At the other extreme, let $\Theta \in \mathcal{E}(F)$ be *essentially tame:* in other words, e_Θ is not divisible by p. In particular, there is a finite tame extension K/F such that Θ has a K/F-lift of degree 1.

Proposition *Let* $\Theta \in \mathcal{E}(F)$. *The following are equivalent.*

(1) Θ *is essentially tame.*
(2) $\Psi_\Theta(x) = x$, *for all* $x \geqslant 0$.

This follows from the Propositions of 9.3.1 and 9.4.1.

10 Carayol Representations

In this section, we look into the problem of computing the Herbrand function Ψ_Θ of an endo-class $\Theta \in \mathcal{E}(F)$, and uncovering what it tells us about the Langlands correspondence. From 9.3.2 Proposition, we know it is enough to work only with endo-classes Θ that are *totally wild,* in that $\deg \Theta = e_\Theta = p^r$. In view of the hierarchical nature of simple characters, one has to start with the case of endo-classes given by a minimal element. In the notation of (5.2.3), this amounts to $k_\Theta = -\varsigma_\Theta$. An endo-class Θ satisfying all of these conditions, and of degree > 1, will be called *totally wild of Carayol type.* We denote the set of all such endo-classes by $\mathcal{E}^{\mathrm{C}}(F)$. Here, we summarize the results of [22] leading to a complete description of Ψ_Θ, $\Theta \in \mathcal{E}^{\mathrm{C}}(F)$. The corresponding Galois representations have a particularly simple and transparent form, and come in explicit "families" almost, but not completely, parallel to the families $\mathcal{C}$ into which simple characters are organized. This discrepancy is already visible in the case $\deg\Theta = p = 2$ of Kutzko's classic works [52], [53] summarized in [15]. The general case is, at least conceptually, no worse.

10.1 Symmetry

We present a fundamental property of Ψ_Θ, where Θ is totally wild of Carayol type.

10.1.1 We assemble notation and terminology. Let $\mathcal{E}^{\mathrm{C}}(F)$ be the set of endo-classes $\Theta \in \mathcal{E}(F)$ that are totally wild, of Carayol type and degree p^r, for some $r \geqslant 1$. Equivalently, Θ is the endo-class of a simple character $\theta \in \mathcal{C}(\mathfrak{a}, \alpha)$, for a simple stratum $[\mathfrak{a}, m, 0, \alpha]$ in $\mathrm{M}_{p^r}(F)$ such that $F[\alpha]/F$ is totally ramified of degree p^r, and $m = -v_{F[\alpha]}(\alpha)$ is not divisible by p.

Lemma *Let $\Theta \in \mathcal{E}^{\mathrm{C}}(F)$. If $\deg\Theta = p^r$ and $\varsigma_\Theta = m/p^r$, then*

(1) $\Phi_\Theta(0) = p^{-2r}m(p^r-1)$;
(2) $\Phi_\Theta(x) = x$, for $x \geqslant \varsigma_\Theta$;
(3) $\Phi'_\Theta(x) = p^{-r}$, for $0 < x < \varsigma_\Theta$;
(4) Ψ_Θ is convex in the region $0 < x < \varsigma_\Theta$.

Proof Parts (1–3) are the definition [21] (4.4.1) of Φ_Θ in this case, and (4) follows. □

Let $\sigma \in \widehat{\mathcal{W}}_F$. Say that σ is *totally wild* if $\sigma \mid \mathcal{P}_F$ is irreducible. Equivalently, $\dim\sigma$ is a power of p and $\chi\otimes\sigma \not\cong \sigma$ for any non-trivial unramified character χ of $\mathcal{W}_F$. Write $\widehat{\mathcal{W}}{}^{\mathrm{wr}}_F$ for the set of totally wild $\sigma \in \widehat{\mathcal{W}}_F$.

A representation $\sigma \in \widehat{\mathcal{W}}_F$ is totally wild if and only if $r_F^+(\sigma) = {}^L\Theta$, for a totally wild endo-class Θ such that $\deg\Theta = \dim\sigma$. Moreover, $\Theta \in \mathcal{E}^{\mathrm{C}}(F)$ if and only if $\mathrm{sw}(\sigma)$ is not divisible by p (and $\dim\sigma > 1$). We say that such representations σ are *of Carayol type.*

10.1.2 Endo-classes $\Theta \in \mathcal{E}^{\mathrm{C}}(F)$ have a striking and characteristic property.

Theorem *Let $\Theta \in \mathcal{E}^{\mathrm{C}}(F)$. The Herbrand function Ψ_Θ satisfies*

$$\varsigma_\Theta - x = \Psi_\Theta(\varsigma_\Theta - \Psi_\Theta(x)), \quad 0 \leqslant x \leqslant \varsigma_\Theta. \tag{10.1.1}$$

The functional equation (10.1.1) has a more intuitive interpretation.

Corollary *In the region $0 \leqslant x \leqslant \varsigma_\Theta$, the graph $y = \Psi_\Theta(x)$ is symmetric with respect to the line $x + y = \varsigma_\Theta$.*

We give a flavour of the proof of the theorem in 10.1.3 below. The corollary simply re-states the theorem. Before passing on, we note that the theorem has a converse.

Complement *Let $\Upsilon \in \mathcal{E}(F)$ be totally wild of degree p^r, $r \geqslant 1$, and suppose that $\varsigma_\Upsilon \leqslant \varsigma_{\chi\Upsilon}$ for all characters χ of $F^\times$. The function Ψ_Υ has the property (10.1.1), namely,*

$$\varsigma_\Upsilon - x = \Psi_\Upsilon(\varsigma_\Upsilon - \Psi_\Upsilon(x)), \quad 0 \leqslant x \leqslant \varsigma_\Upsilon,$$

if and only if $\Upsilon \in \mathcal{E}^C(F)$.

Proof As in [22] 3.1, the result follows from 9.4.1 Proposition. □

10.1.3 We sketch the key steps in the proof of 10.1.2 Theorem.

Let $\sigma \in \widehat{\mathcal{W}}_F^{\mathrm{wr}}$ be of dimension p^r, for some $r \geqslant 1$, and let j be a jump of Σ_σ. Define the *height* $h = h_\sigma(j)$ of j by

$$p^h = \lim_{\epsilon \to 0} \Sigma'_\sigma(j+\epsilon)/\Sigma'_\sigma(j-\epsilon).$$

Thus h is a positive integer and $\sum_j h_\sigma(j) = 2r$, the sum being taken over all jumps of Σ_σ. It is often convenient to think of a point at which Σ_σ is smooth as being a jump of height 0.

Say that $\sigma \in \widehat{\mathcal{W}}_F^{\mathrm{wr}}$ is *absolutely wild* if it factors through a finite Galois group $\mathrm{Gal}(K/F)$, where K/F is totally wildly ramified. To describe Σ_σ in general, it is enough to treat the case where σ is absolutely wild.

For the first step of the proof, we assume only that σ is absolutely wild. Let a be the least jump of Σ_σ. There is then a character χ of $\mathcal{W}_F$, with $\mathrm{sw}(\chi) = a$, such that $\chi \otimes \sigma \cong \sigma$ [22] 3.3 Lemma 1. If $\mathrm{Ker}\,\chi = \mathcal{W}_K$, there is an absolutely wild representation $\tau \in \widehat{\mathcal{W}}_K^{\mathrm{wr}}$ such that $\sigma = \mathrm{Ind}_{K/F}\,\tau$. If k is a jump of τ, there is a jump j of σ such that $k = \psi_{K/F}(j)$. Moreover, $h_\tau(k) \leqslant h_\sigma(j)$. However, $a = \psi_{K/F}(a)$ and $h_\tau(a) < h_\sigma(a)$. Since $\sum_k h_\tau(k) = 2(r-1)$, there are only two possibilities, namely

(1) $h_\tau(a) = h_\sigma(a) - 1$ and there exists a unique jump $c > a$ of Σ_σ such that $h_\tau(\psi_{K/F}(c)) < h_\sigma(c)$, or
(2) $h_\tau(a) = h_\sigma(a) - 2$ while $h_\tau(\psi_{K/F}(j)) = h_\sigma(j)$ for all jumps $j \neq a$ of Σ_σ.

In case (1), we have $h_\tau(\psi_{K/F}(c)) = h_\sigma(c) - 1$.

Assume now that σ is also of Carayol type. The functions Ψ_Θ and Σ_σ then have the same jumps. The representation τ is either of Carayol type or of dimension 1. One proves that c is the *largest* jump of Σ_σ and that

$$c = (\mathrm{sw}(\sigma) - a)/p^r. \tag{10.1.2}$$

In case (1), this is the first step of an inductive argument: if $\dim\sigma = p$, the relation between a and c tells us Ψ_Θ completely and it satisfies the requirements of the theorem. In case (2), we see that a is the *only* jump of Σ_σ and has the value $\mathrm{sw}(\sigma)/(1+p^r)$. The graph of Ψ_Θ is made of two line segments and is easy to identify.

The proof of (10.1.2) has an interesting feature: it relies on the case of the conductor formula given by 5.3.3 Remark (3) and 5.3.4 Proposition 1.

10.1.4 In preparation for the next section, we introduce a family of special functions, as in [22] 4.2.

Definition Let E/F be totally ramified of degree p^r, $r \geqslant 1$, and let $\varsigma = m/p^r$, where m is a positive integer not divisible by p. Define

$$\begin{aligned} \Psi^{\times}_{(E/F,\varsigma)}(x) &= p^{-r}\psi_{E/F}(x), \\ \Psi^{+}_{(E/F,\varsigma)}(x) &= \varsigma - \varphi_{E/F}(p^r(\varsigma - x)), \qquad\qquad 0 \leqslant x \leqslant \varsigma. \\ {}^2\Psi_{(E/F,\varsigma)}(x) &= \max\{\Psi^{\times}_{(E/F,\varsigma)}(x), \Psi^{+}_{(E/F,\varsigma)}(x)\}, \end{aligned}$$

The function ${}^2\Psi_{(E/F,\varsigma)}$ is the *bi-Herbrand function* of the datum $(E/F, \varsigma)$.

Proposition *The function* ${}^2\Psi_{(E/F,\varsigma)}$ *is continuous, strictly increasing, piecewise linear and convex. The graph* $y = {}^2\Psi_{(E/F,\varsigma)}(x)$ *is symmetric with respect to the line* $x + y = \varsigma$.

Indeed, the graph $y = {}^2\Psi_{(E/F,\varsigma)}(x)$ is $y = \Psi^{\times}_{(E/F,\varsigma)}(x)$ in the triangular region $x + y \leqslant \varsigma$. If $x + y \geqslant \varsigma$, it is $y = \Psi^{+}_{(E/F,\varsigma)}(x)$.

10.2 Conformity and the Main Theorem

Fix a simple stratum $[\mathfrak{a}, m, 0, \alpha]$ in $\mathrm{M}_{p^r}(F)$ as in 10.1.1, and set $E = F[\alpha]$. Here, the canonical map $\mathcal{C}(\mathfrak{a}, \alpha) \to \mathcal{E}(F)$ is injective: this follows on comparing the intertwining theorems of 2.2.2 and 2.5.2. Let $\|\mathcal{C}(\mathfrak{a}, \alpha)\|$ denote its image. Thus $\|\mathcal{C}(\mathfrak{a}, \alpha)\| \subset \mathcal{E}^{\mathrm{C}}(F)$, and the map $\mathcal{C}(\mathfrak{a}, \alpha) \to \|\mathcal{C}(\mathfrak{a}, \alpha)\|$ is a bijection. The function $\Theta \mapsto \Psi_\Theta$ is not necessarily constant on $\|\mathcal{C}(\mathfrak{a}, \alpha)\|$. We deal with this by introducing a dissection of the set $\mathcal{C}(\mathfrak{a}, \alpha)$.

10.2.1 Let w_α be the wild exponent of E/F, as in 9.1.3. Write

$$\begin{aligned} l_\alpha &= \max\{0, m - w_\alpha\}, \\ \lambda_\alpha &= [l_\alpha/2]. \end{aligned} \tag{10.2.1}$$

The function $\psi_A * \alpha$ of 1.4.2 restricts to a character of the group $U_E^{1+\lambda_\alpha}$. Say that $\theta \in \mathcal{C}(\mathfrak{a}, \alpha)$ *conforms to* α if θ agrees with $\psi_A * \alpha$ on $U_E^{1+\lambda_\alpha}$. Let $\mathcal{C}^*(\mathfrak{a}, \alpha)$ be the set of $\theta \in \mathcal{C}(\mathfrak{a}, \alpha)$ that conform to α. Write $\|\mathcal{C}^*(\mathfrak{a}, \alpha)\|$ for the set of endo-classes of elements of $\mathcal{C}^*(\mathfrak{a}, \alpha)$.

Proposition *(1) The set* $\mathcal{C}^*(\mathfrak{a}, \alpha)$ *is not empty.*
(2) Let $\theta \in \mathcal{C}(\mathfrak{a}, \alpha)$*. There exists a simple stratum* $[\mathfrak{a}, m, 0, \alpha']$ *such that* $\mathcal{C}(\mathfrak{a}, \alpha') = \mathcal{C}(\mathfrak{a}, \alpha)$ *and* $\theta \in \mathcal{C}^*(\mathfrak{a}, \alpha')$.

Part (1) here is obvious. Part (2) is proved in [22] 7.1.

Remark When $m \leqslant 2w_{E/F}$, there is a wide choice of fields $E' = F[\alpha']$ for which $\mathcal{C}(\mathfrak{a}, \alpha') = \mathcal{C}(\mathfrak{a}, \alpha)$. The various choices of E' give rise to different values of $w_{E'/F}$ and different Herbrand functions $\psi_{E'/F}$: this accounts for the need to work with $\mathcal{C}^*(\mathfrak{a}, \alpha)$. When $m > 2w_{E/F}$, all choices of α' give rise to the same Herbrand function. These phenomena are analysed in [22] section 6.

10.2.2 We describe the Herbrand functions Ψ_Θ, $\Theta \in \|\mathcal{C}^*(\mathfrak{a}, \alpha)\|$, in terms of the bi-Herbrand functions of 10.1.4.

Theorem *Let $\Theta \in \mathcal{E}^C(F)$. Let $[\mathfrak{a}, m, 0, \alpha]$ be a simple stratum in $\mathrm{M}_{p^r}(F)$ such that $\Theta \in \|\mathcal{C}^*(\mathfrak{a}, \alpha)\|$, and write $E = F[\alpha]$. The Herbrand function Ψ_Θ is given by*

$$\Psi_\Theta(x) = {}^2\Psi_{(E/F,\varsigma_\Theta)}(x), \quad 0 \leqslant x \leqslant \varsigma_\Theta.$$

The proof of the theorem is intricate and occupies sections 4–7 of [22]. It starts from the almost Galois-theoretic 10.1.2 Theorem, but then works entirely on the GL-side. We give no details. Other results in [22] section 7 indicate how Ψ_Θ varies as Θ ranges over $\|\mathcal{C}(\mathfrak{a}, \alpha)\|$, although the picture is not complete.

There is a direct consequence, useful for constructing examples.

Corollary *Let E/F be a totally ramified field extension of degree p^r, $r \geqslant 1$, and let m be a positive integer not divisible by p. There exists $\Theta \in \mathcal{E}^C(F)$, with degree p^r and $\varsigma_\Theta = m/p^r$, such that*

$$\Psi_\Theta(x) = {}^2\Psi_{(E/F,m/p^r)}(x), \quad 0 \leqslant x \leqslant m/p^r.$$

Proof Let $\alpha \in E$ satisfy $v_E(\alpha) = -m$. In particular, $E = E[\alpha]$. Let $\mathfrak{a}$ be the unique E-pure hereditary $\mathfrak{o}_F$-order in $\mathrm{End}_F(E)$. Thus $[\mathfrak{a}, m, 0, \alpha]$ is a simple stratum. There exists $\theta \in \mathcal{C}(\mathfrak{a}, \alpha)$ that conforms to α. The theorem implies that the endo-class Θ of θ has the required property. □

10.2.3 Suppose for the moment that $p \neq 2$. One can then refine the notion of conformity. The function

$$ {}^2\psi_A * \alpha : 1 + x \longmapsto \psi_A\big(\alpha(x - x^2/2)\big) \qquad (10.2.2)$$

defines a character of the group $U_E^{[(1+l_\alpha)/2]}$. Let $\mathcal{C}^{**}(\mathfrak{a}, \alpha)$ be the set of $\theta \in \mathcal{C}(\mathfrak{a}, \alpha)$ agreeing with ${}^2\psi_A * \alpha$ on that group. The exact analogue of 10.2.1 Proposition holds for the set $\mathcal{C}^{**}(\mathfrak{a}, \alpha)$: this approach is used by Mœglin in [58]. The extra precision can be useful here: see 10.4.5.3 below.

10.3 Galois Representations of Carayol Type

Let $\sigma \in \widehat{\mathcal{W}}_F^{\mathrm{wr}}$ be of Carayol type and dimension p^r, $r \geqslant 1$. Thus $r_F^+(\sigma) = {}^L\Theta$, for some $\Theta \in \mathcal{E}^C(F)$. Developing ideas from the proof of 10.1.2 Theorem, we write down the structure of $\sigma \,|\, \mathcal{R}_F(x)$, $\sigma \,|\, \mathcal{R}_F^+(x)$, for all $x > 0$. For a complete account of this material, see section 9 of [22].

10.3.1 Let $\sigma \in \widehat{\mathcal{W}}_F^{\mathrm{wr}}$ be of Carayol type and dimension p^r, $r \geqslant 1$. Thus $r_F^+(\sigma) = {}^L\Theta$, for some $\Theta \in \mathcal{E}^{\mathrm{C}}(F)$: to save notation, let us write $\Psi_\sigma = \Psi_\Theta$. Define c_σ by the equation

$$c_\sigma + \Psi_\sigma(c_\sigma) = \varsigma(\sigma).$$

Thus c_σ is the x-coordinate of the intersection of the graph $y = \Psi_\sigma(x)$ with the line of symmetry $x + y = \varsigma(\sigma)$. The symmetry property of 10.1.2 Corollary induces an involution, $j \mapsto \bar{j}$ say, on the set of jumps j of Ψ_σ. We may therefore number them

$$j_1 < j_2 < \cdots < j_n < (c_\sigma) < \bar{j}_n < \ldots \bar{j}_1,$$

for some $n \geqslant 0$, where the term c_σ is included when c_σ is a jump.

Proposition *Suppose $\Theta \in \|\mathcal{C}^*(\mathfrak{a}, \alpha)\|$. The point c_σ is a jump of $\Psi_\Theta = \Psi_\sigma$ if and only if $j_\infty(F[\alpha]|F) \geqslant c_\sigma$.*

Proof If c_σ is not a jump, the symmetry property (10.1.1) implies that $\Psi_\Theta'(c_\sigma) = 1$. Together, 9.1.3 Proposition 1 and 10.2.2 Theorem imply $c_\sigma > j_\infty(F[\alpha]|F)$. The argument reverses to give the result. □

10.3.2 Using the same notation, we write down the ramification structure of σ.

Theorem 1 *(1) The restriction $\sigma \,|\, \mathcal{R}_F^+(c_\sigma)$ is a direct sum of characters.*

(2) Let ξ be a character of $\mathcal{R}_F^+(c_\sigma)$ occurring in σ, let $\mathcal{W}_{L_\xi}$ be the $\mathcal{W}_F$-normalizer of ξ and let $\rho_\xi = \rho_\xi(\sigma)$ be the natural representation of $\mathcal{W}_{L_\xi}$ on the ξ-isotypic subspace of $\sigma \,|\, \mathcal{R}_F^+(c_\sigma)$.

   (a) The field extension L_ξ/F is totally wildly ramified.
   (b) The representation ρ_ξ of $\mathcal{W}_{L_\xi}$ is irreducible and totally wild.
   (c) $\sigma = \mathrm{Ind}_{L_\xi/F}\, \rho_\xi(\sigma)$.

(3) If c_σ is not a jump, then ρ_ξ is a character.

(4) If c_σ is a jump, then dim $\rho_\xi = p^s$, *where $s = h_\sigma(c_\sigma)/2$. The representation $\rho_\xi \in \widehat{\mathcal{W}}_{L_\xi}^{wr}$ is of Carayol type and has a unique jump, at $\psi_{L_\xi/F}(c_\sigma)$.*

(5) In both cases,

$$\Psi_\sigma(x) = p^{-r}\psi_{L_\xi/F}(x), \quad 0 \leqslant x \leqslant c_\sigma. \tag{10.3.1}$$

Note that the conjugacy class of L_ξ/F is determined by σ. In all cases, $\sigma \,|\, \mathcal{R}_F(\bar{j}_k)$ is a direct sum of characters, $1 \leqslant k \leqslant n$. When c_σ is a jump, the representation $\rho_\xi \in \widehat{\mathcal{W}}_{L_\xi}^{\mathrm{wr}}$ has interesting structure, on which we comment in 10.3.3. We remark that (10.3.1) determines Ψ_σ completely, because of the symmetry property of 10.1.2.

Theorem 2 *For $1 \leqslant k \leqslant n$, the restriction $\sigma \,|\, \mathcal{R}_F^+(j_k)$ is a multiplicity-free representation. Its irreducible components restrict to the isotypic components of $\sigma \,|\, \mathcal{R}_F(\bar{j}_k)$. If c_σ is a jump, then $\sigma \,|\, \mathcal{R}_F(c_\sigma)$ is multiplicity-free.*

10.3.3 The representation ρ_ξ of 10.3.2 has a very special form. We summarize the main points from [22]. Starting in generality, let $\tau \in \widehat{\mathcal{W}}_F^{\mathrm{wr}}$, say $\tau : \mathcal{W}_F \to \mathrm{GL}_n(\mathbb{C})$, and let $\bar\tau : \mathcal{W}_F \to \mathrm{PGL}_n(\mathbb{C})$ be the associated projective representation. Let $\mathrm{Ker}\,\bar\tau = \mathcal{W}_K$. The finite Galois extension K/F is the *centric field* of τ. The *tame* centric field T/F is the maximal tame sub-extension of K/F. The representation τ is called *H-cyclic* if the finite p-group $\bar\tau(\mathcal{P}_F)$ is elementary abelian. It follows that the finite p-group $\tau(\mathcal{P}_F)$ is extra special of class 2.

One says that τ is *H-singular* if $\psi_{K/F}$ has a single jump.

Proposition *Let $\tau \in \widehat{\mathcal{W}}_F^{wr}$ be of Carayol type and have a single jump. The representation τ is then H-singular.*

The representation ρ_ξ of 10.3.2 is therefore H-singular.

10.4 Parameter Fields

We compare the theorems of 10.3.2 with the main theorem of 10.2.2. The material here is extracted from section 10 of [22].

10.4.1 We re-set the notation to more closely match the focus of the section. Fix a simple stratum $[\mathfrak{a}, m, 0, \alpha]$ in $\mathrm{M}_{p^r}(F)$ such that $E = F[\alpha]/F$ is totally ramified of degree p^r and m is not divisible by p. Abbreviate $w_\alpha = w_{E/F}$ and set

$$l_\alpha = \max\{0, m - w_\alpha\},$$
$$\lambda_\alpha = [l_\alpha/2].$$

Write

$$\widehat{\mathrm{GL}}_F^*(\alpha) = \{\pi \in \mathcal{A}_{p^r}^0(F) : \Theta(\pi) \in \|\mathcal{C}^*(\mathfrak{a}, \alpha)\|\},$$

and

$$\widehat{\mathcal{W}}_F^*(\alpha) = \{\sigma \in \widehat{\mathcal{W}}_F : \dim\sigma = p^r \text{ and } r_F^+(\sigma) \in {}^L\|\mathcal{C}^*(\mathfrak{a}, \alpha)\|\}.$$

Thus

$$\widehat{\mathcal{W}}_F^*(\alpha) = {}^L\big(\widehat{\mathrm{GL}}_F^*(\alpha)\big).$$

All endo-classes $\Theta \in \|\mathcal{C}^*(\mathfrak{a}, \alpha)\|$ have the same Herbrand function, namely $\Psi_\Theta = {}^2\Psi_{(E/F, m/p^r)}$. We now denote this by Ψ_α. In the convention of 10.3.1, we thus have $\Psi_\alpha = \Psi_\sigma$, $\sigma \in \widehat{\mathcal{W}}_F^*(\alpha)$. Likewise, define c_α by the equation

$$c_\alpha + \Psi_\alpha(c_\alpha) = m/p^r.$$

We view the ultrametric $\mathbf{A}$ as a pairing on $\widehat{\mathrm{GL}}_F$, that is,

$$\mathbf{A}(\pi_1, \pi_2) = \mathbf{A}(\Theta(\pi_1), \Theta(\pi_2)), \quad \pi_i \in \widehat{\mathrm{GL}}_F.$$

Lemma *(1) If* $\pi_0 \in \widehat{\mathrm{GL}}_F^*(\alpha)$, *then*

$$\widehat{\mathrm{GL}}_F^*(\alpha) = \{\pi \in \mathcal{A}^0_{p^r}(F) : \mathbf{A}(\pi, \pi_0) \leqslant \lambda_\alpha/p^r\}.$$

(2) If $\sigma_0 \in \widehat{\mathcal{W}}_F^*(\alpha)$, *then*

$$\widehat{\mathcal{W}}_F^*(\alpha) = \{\sigma \in \widehat{\mathcal{W}}_F : \Delta(\sigma, \sigma_0) \leqslant \epsilon_\alpha, \dim \sigma = p^r\},$$

where $\Psi_\alpha(\epsilon_\alpha) = \lambda_\alpha/p^r$.

Proof Part (1) is the definition of $\mathcal{C}^*(\mathfrak{a}, \alpha)$ whence part (2) follows from the Higher Ramification Theorem of 9.2.2. □

10.4.2 There is a simple result to bear in mind as we go forward. If K/F is a finite extension and c a non-negative integer, let $\mathrm{X}_c(K)$ be the group of characters χ of $K^\times$ such that $\mathrm{sw}(\chi) \leqslant c$.

Let $\pi \in \widehat{\mathrm{GL}}_F^*(\alpha)$ and set $G = \mathrm{GL}_{p^r}(F)$. Thus $\pi = c\text{-}\mathrm{Ind}_{\boldsymbol{J}}^G \Lambda$, where Λ is an irreducible representation of $\boldsymbol{J} = E^\times U_{\mathfrak{a}}^{[(1+m)/2]}$ containing some $\theta \in \mathcal{C}^*(\mathfrak{a}, \alpha)$. Surely $[(1+m)/2] > \lambda_\alpha$ so, if $\chi \in \mathrm{X}_{\lambda_\alpha}(E)$, we may view χ as a character of $\boldsymbol{J}$ trivial on $U_{\mathfrak{a}}^{[(1+m)/2]}$. The representation $\chi \otimes \Lambda$ of $\boldsymbol{J}$ is an extended maximal simple type in G, containing some $\theta' \in \mathcal{C}^*(\mathfrak{a}, \alpha)$. Define

$$\chi \odot_\alpha \pi = c\text{-}\mathrm{Ind}_{\boldsymbol{J}}^G \chi \otimes \Lambda. \tag{10.4.1}$$

Working through the definitions, we find:

Proposition *Relative to the pairing*

$$\begin{aligned} \mathrm{X}_{\lambda_\alpha}(E) \times \widehat{\mathrm{GL}}_F^*(\alpha) &\longrightarrow \widehat{\mathrm{GL}}_F^*(\alpha), \\ (\chi, \pi) &\longmapsto \chi \odot_\alpha \pi, \end{aligned}$$

the set $\widehat{\mathrm{GL}}_F^*(\alpha)$ *is a principal homogeneous space over* $\mathrm{X}_{\lambda_\alpha}(E)$.

10.4.3 We make a preliminary analysis of the set $\widehat{\mathcal{W}}_F^*(\alpha)$. Let $j_\infty(\alpha) = j_\infty(E|F)$ be the largest jump of the field extension E/F. (Remember that $j_\infty(\alpha) = \infty$ if E/F happens to be inseparable.)

Theorem *(1) There exists a character* ξ *of* $\mathcal{R}_F^+(c_\alpha)$, *unique up to conjugation by* $\mathcal{W}_F$, *that occurs in every representation* $\sigma \in \widehat{\mathcal{W}}_F^*(\alpha)$.

(2) Fix a character ξ *as in (1). Assume that* $j_\infty(\alpha) > c_\alpha$ *or that* l_α *is odd. There is a unique irreducible representation* ρ_ξ *of* $\mathcal{R}_F(c_\alpha)$ *that contains* ξ *and occurs in every* $\sigma \in \widehat{\mathcal{W}}_F^*(\alpha)$.

Proof We first clarify the relation between some of the numerical parameters.

Lemma 1 *(1) If* $j_\infty(\alpha) \leqslant c_\alpha$, *then* $c_\alpha = (m + w_\alpha)/2p^r$.
(2) If $j_\infty(\alpha) > c_\alpha$, *then* $c_\alpha < (m + w_\alpha)/2p^r$.
(3) If $l_\alpha = 0$, *then* $j_\infty(\alpha) > c_\alpha$.

Proof Suppose first that $j_\infty(\alpha) \leqslant c_\alpha$. The definition 10.1.4 of the bi-Herbrand function and 10.2.2 Theorem imply that $\Psi_\alpha(c_\alpha) = p^{-r}\psi_{E/F}(c_\alpha)$ and, by 9.1.3 Proposition 1, $\psi_{E/F}(c_\alpha) = p^r c_\alpha - w_\alpha$. However, the definition of c_α is that $c_\alpha + \Psi_\alpha(c_\alpha) = m/p^r$ whence assertion (1) follows.

In (2), the line $y = x - p^{-r}w_\alpha$ lies strictly below the graph $y = \Psi_\alpha(x)$ and the assertion follows on drawing a picture.

In (3), assume for a contradiction that $j_\alpha \leqslant c_\alpha$, whence $c_\alpha = (m + w_\alpha)/2p^r$. The hypothesis $l_\alpha = 0$ implies $m \leqslant w_\alpha$, whence

$$c_\alpha = (m + w_\alpha)/2p^r \leqslant w_\alpha/p^r < j_\infty(\alpha)$$

by 9.1.3 Corollary. This gives the desired contradiction. □

We translate Lemma 1 in terms of ultrametrics.

Lemma 2 *For* $i = 1, 2$, *let* $\sigma_i \in \widehat{\mathcal{W}}^*_F(\alpha)$ *and let* $r_F^+(\sigma_i) = {}^L\Theta_i$, $\Theta_i \in \|\mathcal{C}^*(\mathfrak{a}, \alpha)\|$. *We then have*

$$\Delta(\sigma_1, \sigma_2) \leqslant c_\alpha. \tag{10.4.2}$$

Moreover, $\Delta(\sigma_1, \sigma_2) = c_\alpha$ *if and only if all of the following conditions are satisfied:*

(1) $\mathbf{A}(\Theta_1, \Theta_2) = l_\alpha/2p^r$,
(2) l_α *is even, and*
(3) $j_\infty(\alpha) \leqslant c_\alpha$.

Proof Define ϵ_α by $\Psi_\alpha(\epsilon_\alpha) = \lambda_\alpha/p^r$. By 10.4.1 Lemma, we have $\Delta(\sigma_1, \sigma_2)/le\epsilon_\alpha$, with equality if and only if $\mathbf{A}(\Theta_1, \Theta_2) = \lambda_\alpha/p^r$.

Suppose that $j_\infty(\alpha) \leqslant c_\alpha$. So, by Lemma 1, $c_\alpha = (m + w_\alpha)/2p^r$ and therefore $\Psi_\alpha(c_\alpha) = l_\alpha/2p^r$. That is, $\epsilon_\alpha \leqslant c_\alpha$, with equality if and only if l_α is even. All assertions are proved in this case.

Suppose now that $j_\infty(\alpha) > c_\alpha$. The line $y = x - p^{-r}w_\alpha$ lies strictly below the graph $y = \Psi_\alpha(x)$. The three lines $y = l_\alpha/2p^r$, $y = x - p^{-r}w_\alpha$ and $x + y = \varsigma_\alpha$ all meet at $x = (m + w_\alpha)/2p^r$. By Lemma 1, $(m + w_\alpha)/2p^r > c_\alpha$, so $\Psi_\alpha(c_\alpha) > l_\alpha/2p^r$. It follows that $c_\alpha > \epsilon_\alpha \geqslant \Delta(\sigma_1, \sigma_2)$, as required. □

Remark Condition (2) in Lemma 2 is actually redundant, since $p^r\mathbf{A}(\Theta_1, \Theta_2)$ is an integer, for any $\Theta_i \in \|\mathcal{C}(\mathfrak{a}, \alpha)\|$.

We prove the theorem. Part (1) follows directly from (10.4.2). The assertion of (2) holds if and only if $\Delta(\sigma_1, \sigma_2) < c_\alpha$ for all $\sigma_i \in \widehat{\mathcal{W}}^*_F(\alpha)$. Part (2) of the theorem therefore follows from Lemma 2. □

Comments We comment further on the conclusions of the theorem, in connection with 10.3.2 Theorem 1.

(1) If $j_\infty(\alpha) < c_\alpha$, the representation ρ_ξ is a character.
(2) In the case $j_\infty(\alpha) > c_\alpha$, the representation ρ_ξ is the restriction to $\mathcal{R}_F(c_\sigma)$ of the ρ_ξ occurring in 10.3.2. The same comment applies if l_α is odd and $j_\infty(\alpha) = c_\alpha$.
(3) Suppose that $j_\infty(\alpha) = c_\alpha$ and that l_α is even (and so positive, by Lemma 2(3)). For $\sigma \in \widehat{\mathcal{W}}^*_F(\alpha)$, the natural representation $\rho_\xi(\sigma)$ of $\mathcal{R}_F(c_\alpha)$ on the ξ-isotypic subspace of σ is again irreducible of dimension p^s, $s \geqslant 1$. In this case, it does vary with σ.

10.4.4 We seek an analogue of 10.4.2 Proposition for Galois representations. With no restriction on the parameters at first, fix a character ξ of $\mathcal{R}^+_F(c_\alpha)$ that occurs in $\sigma \in \widehat{\mathcal{W}}^*_F(\alpha)$. The $\mathcal{W}_F$-stabilizer of ξ is of the form $\mathcal{W}_L$, for a totally wildly ramified extension L/F. Let σ_ξ be the natural representation of $\mathcal{W}_L$ on the ξ-isotypic subspace of σ, and let Δ_L be the canonical ultrametric pairing on $\widehat{\mathcal{W}}_L$.

Lemma 1

$$\max\{\Delta_L(\sigma_\xi, \sigma'_\xi) : \sigma, \sigma' \in \widehat{\mathcal{W}}^*_F(\alpha)\} = \lambda_\alpha. \tag{10.4.3}$$

Proof By 10.4.1 Lemma, $\max\{\mathbf{A}(\Theta, \Theta') : \Theta, \Theta' \in \|\mathcal{C}^\star(\mathfrak{a}, \alpha)\|\} = \lambda_\alpha/p^r$. So,

$$\max\left\{\Delta(\sigma, \sigma') : \sigma, \sigma' \in \mathcal{G}^\star(\alpha)\right\} = \Psi_\alpha^{-1}(\lambda_\alpha/p^r) = \varphi_{L/F}(\lambda_\alpha),$$

by (10.3.1). The relation (10.4.3) now follows from (9.1.4). □

At the same level of generality, let $\sigma \in \widehat{\mathcal{W}}^*_F(\alpha)$ and let $\chi \in \mathrm{X}_{\lambda_\alpha}(L)$. Form

$$\chi \odot_\alpha \sigma = \mathrm{Ind}_{L/F}\, \chi \otimes \sigma_\xi. \tag{10.4.4}$$

Lemma 2 *The representation $\chi \odot_\alpha \sigma$ of (10.4.4) is irreducible and lies in $\widehat{\mathcal{W}}^*_F(\alpha)$.*

Proof The restriction $\chi \otimes \sigma_\xi \,\big|\, \mathcal{R}^+_F(c_\alpha)$ is a multiple of ξ, so the irreducibility assertion follows from the definition of L. The second assertion follows from Lemma 1 and 10.4.1 Lemma. □

Theorem *Suppose that either $j_\infty(\alpha) \neq c_\alpha$ or that l_α is odd. Relative to the pairing*

$$\begin{aligned} \mathrm{X}_{\lambda_\alpha}(L) \times \widehat{\mathcal{W}}^*_F(\alpha) &\longrightarrow \widehat{\mathcal{W}}^*_F(\alpha), \\ (\chi, \sigma) &\longmapsto \chi \odot_\alpha \sigma, \end{aligned}$$

the set $\widehat{\mathcal{W}}^*_F(\alpha)$ is a principal homogeneous space.

Proof See 10.4 Proposition of [22]. □

10.4.5 Comments

10.4.5.1 The case of $j_\infty(\alpha) = c_\alpha$ and l_α even positive does arise in practice. Let $p = p^r = 2$. There is a ramified quadratic extension E/F with $w_{E/F}$ odd. Take $\alpha \in E$ of valuation $-m$, where $m = 3w_{E/F}$. This gives $j_\infty(\alpha) = w_{E/F} = c_\alpha$, while $l_\alpha = 2w_{E/F} > 0$.

10.4.5.2 Staying in the same case $j_\infty(\alpha) = c_\alpha$ with l_α even positive, let $\sigma \in \widehat{\mathcal{W}}^*_F(\alpha)$ and let $\rho_\xi(\sigma)$ be the natural representation of $\mathcal{W}_L$ on the ξ-isotypic subspace of σ. Thus $\rho_\xi(\sigma)$ is H-singular and may vary with σ. However, the representations $\rho_\xi(\sigma)$, $\sigma \in \widehat{\mathcal{W}}^*_F(\alpha)$, all have the same *tame* centric field T/L. This generalizes Kutzko's celebrated "cubic polynomial theorem" that lies at the heart of his proof of the Langlands Conjecture for $GL(2)$. See [14] section 45 for an exposition in the present language. The phenomenon is also visible in [18]. On the other hand, the *centric* field may vary with σ: see [22] 10.5 Theorem.

10.4.5.3 Suppose now that p is odd. One can carry out the same analysis using $\mathcal{C}^{**}(\mathfrak{a}, \alpha)$ in place of $\mathcal{C}^*(\mathfrak{a}, \alpha)$. Improvising notation in the obvious way, (10.4.2) is replaced by

$$\Delta(\sigma_1, \sigma_2) < c_\alpha, \quad \sigma_i \in \widehat{\mathcal{W}}^{**}_F(\alpha),$$

and the exceptional case of 10.4.3 Lemma 2 has been removed. In 10.4.3 Theorem, part (2) holds without restriction one either $j_\infty(\alpha)$ or l_α. Likewise 10.4.4 Theorem, except that λ_α is replaced by

$$\lambda'_\alpha = \begin{cases} [l_\alpha/2], & \text{if } l_\alpha \text{ is odd}, \\ (l_\alpha/2) - 1, & \text{if } l_\alpha \text{ is even}. \end{cases}$$

10.4.5.4 Suppose that $p \geqslant 5$ and work in dimension p. Mœglin [58] shows that $L = E$ *provided* the extensions E/F, L/F are *cyclic*. When $p = 2$, one knows that they need not be the same.

References

1. J. Arthur, L. Clozel, Simple algebras, base change, and the advanced theory of the trace formula. Ann. Math. Stud. **120** (1989) Princeton University Press
2. C. Blondel, G. Henniart, S.A.R. Stevens, Jordan blocks of cuspidal representations of symplectic groups. arXiv:1704.03545v1
3. P. Broussous, Extension du formalisme de Bushnell-Kutzko au cas d'une algèbre à division. Proc. Lond. Math. Soc. **77**(3), 292–326 (1998)
4. P. Broussous, B. Lemaire, Building of $GL(m, D)$ and centralizers. Transform. Groups **7**(1), 15–50 (2002)
5. F. Bruhat, J. Tits, Groupes réductifs sur un corps local. I, Données radicielles valuées. Publ. Math. Inst. Hautes Etudes Scientifiques **41**, 5–251 (1972)

6. C.J. Bushnell, Hereditary orders, Gauss sums and supercuspidal representations of GL_N. J. reine angew. Math. **375**(376), 184–210 (1987)
7. C.J. Bushnell, A. Fröhlich, Non-abelian congruence Gauss sums and p-adic simple algebras. Proc. Lond. Math. Soc. **50**(3), 207–264 (1985)
8. C.J. Bushnell, G. Henniart, Local tame lifting for GL(n) I: simple characters. Publ. Math. Inst. Hautes Etudes Scientifiques **83**, 105–233 (1996)
9. C.J. Bushnell, G. Henniart, Local tame lifting for GL(n) II: wildly ramified supercuspidals. Astérisque **254** (1999)
10. C.J. Bushnell, G. Henniart, Local Rankin-Selberg convolution for $GL(n)$: divisibility of the conductor. Math. Ann. **321**, 455–461 (2001)
11. C.J. Bushnell, G. Henniart, Local tame lifting for GL(n) III: explicit base change and Jacquet-Langlands correspondence. J. reine angew. Math. **508**, 39–100 (2005)
12. C.J. Bushnell, G. Henniart, Local tame lifting for GL(n) IV: simple characters and base change. Proc. Lond. Math. Soc. (3) **87**, 337–362 (2003)
13. C.J. Bushnell, G. Henniart, The essentially tame local Langlands correspondence, I. J. Am. Math. Soc. **18**, 685–710 (2005)
14. C.J. Bushnell, G. Henniart, The essentially tame local Langlands correspondence, II: totally ramified representations. Compos. Math. **141**, 979–1011 (2005)
15. C.J. Bushnell, G. Henniart, *The Local Langlands Conjecture for* GL(2), *Grundlehren der mathematischen Wissenschaften*, vol. 335 (Springer, 2006)
16. C.J. Bushnell, G. Henniart, The essentially tame local Langlands correspondence, III: the general case. Proc. Lond. Math. Soc. **101**(3), 497-553 (2010)
17. C.J. Bushnell, G. Henniart, Intertwining of simple characters in GL(n). Int. Math. Res. Not. (2012). https://doi.org/10.1093/imrn/rns162
18. C.J. Bushnell, G. Henniart, Langlands parameters for epipelagic representations of GL_n. Math. Ann. **358**, 433–463 (2014)
19. C.J. Bushnell, G. Henniart, To an effective local Langlands correspondence. Memoirs Am. Math. Soc. **231**(1087), v+88 (2014)
20. C.J. Bushnell, G. Henniart, Modular local Langlands correspondence for GL_n. Int. Math. Res. Not. (2014). https://doi.org/10.1093/imrn/rnt063
21. C.J. Bushnell, G. Henniart, Higher ramification and the local Langlands correspondence. Ann. Math. **185**(2)(3), 919–955 (2017)
22. C.J. Bushnell, G. Henniart, Local Langlands correspondence and ramification for Carayol representations. arXiv: 1611.09258v2
23. C.J. Bushnell, G. Henniart, P.C. Kutzko, Local Rankin-Selberg convolutions for GL_n: explicit conductor formula. J. Am. Math. Soc. **11**, 703–730 (1998)
24. C.J. Bushnell, G. Henniart, P.C. Kutzko, Correspondance de Langlands locale pour GL(n) et conducteurs de paires. Ann. Scient. École Norm. Sup. **31**(4), 537–560 (1998)
25. C.J. Bushnell, P.C. Kutzko, The admissible dual of $GL(N)$ via compact open subgroups. Ann. Math. Stud. **129** (1993) Princeton University Press
26. C.J. Bushnell, P.C. Kutzko, The admissible dual of $SL(N)$ I. Ann. Scient. École Normale Sup. **26**(4), 261–279 (1993)
27. C.J. Bushnell, P.C. Kutzko, The admissible dual of $SL(N)$ II. Proc. Lond. Math. Soc. **68**(3), 317–379 (1994)
28. C.J. Bushnell, P.C. Kutzko, Smooth representations of p-adic reductive groups; structure theory via types. Proc. Lond. Math. Soc. **77**(3), 582–634 (1998)
29. C.J. Bushnell, P.C. Kutzko, Semisimple types for $GL(N)$. Compos. Math. **119**, 53–97 (1999)
30. H. Carayol, Représentations cuspidales du groupe linéaire. Ann. Scient. École Norm. Sup. **17**(4), 191–225 (1984)
31. P. Deligne, *Les corps locaux de caractéristique p, limite de corps locaux de caractéristique* 0. *Appendice: théorie de la ramification, et fonctions de Herbrand, pour des extensions non galoisiennes* (Représentations des groupes réductifs sur un corps local, Hermann, Paris, 1984), pp. 150–157

32. G. Glauberman, Correspondences of characters for relatively prime operator groups. Can. J. Math. **20**, 1465–1488 (1968)
33. R. Godement, H. Jacquet, *Zeta Functions of Simple Algebras*. Lecture Notes in Mathematics, vol. 260 (Springer, 1972)
34. J.A. Green, The characters of the finite general linear groups. Trans. Am. Math. Soc. **80**, 402–447 (1955)
35. V. Heiermann, Sur l'espace des représentations irréductibles du groupe de Galois d'un corps local. C. R. Acad. Sci. Paris Sér. I Math. **323**(6), 571–576 (1996)
36. G. Henniart, Représentations du groupe de Weil d'un corps local. L'Ens. Math. Sér **II**(26), 155–172 (1980)
37. G. Henniart, Correspondance de Langlands-Kazhdan explicite dans le cas non ramifié. Math. Nachr. **158**, 7–26 (1992)
38. G. Henniart, Caractérisation de la correspondance de Langlands par les facteurs ε de paires. Invent. Math. **113**, 339–350 (1993)
39. G. Henniart, R. Herb, Automorphic induction for $GL(n)$ (over local non archimedean fields). Duke Math. J. **78**, 131–192 (1995)
40. G. Henniart, B. Lemaire, Formules de caractères pour l'induction automorphe. J. reine angew. Math. **645**, 41–84 (2010)
41. G. Henniart, B. Lemaire, Changement de base et induction automorphe pour GL_n en caractéristique non nulle. Mém. Soc. Math. France **108** (2010)
42. R.E. Howe, Tamely ramified supercuspidal representations of GL_n. Pac. J. Math. **73**, 437–460 (1977)
43. R.E. Howe, Some qualitative results on the representation theory of GL_n over a p -adic field. Pac. J. Math. **73**, 479–538 (1977)
44. R.E. Howe, A. Moy, Hecke algebra isomorphisms for $GL(n)$ over a p -adic field. J. Alg. **131**, 388–424 (1990)
45. I.M. Isaacs, *Character Theory of Finite Groups* (A.M.S. Chelsea Publishing, 2006)
46. H. Jacquet, B. Liu, On the local converse theorem for p-adic $GL(n)$. Am. J. Math. arXiv:1601.03656
47. H. Jacquet, I. Piatetski-Shapiro, J. Shalika, Rankin-Selberg convolutions. Am. J. Math. **105**, 367–483 (1983)
48. J.-L. Kim, Supercuspidal representations: an exhaustion theorem. J. Am. Math. Soc. **20**, 273–320 (2007)
49. T. Kondo, On Gaussian sums attached to the general linear groups over finite fields. J. Math. Soc. Jap. **15**, 244–255 (1963)
50. P.C. Kutzko, On the supercuspidal representations of GL_2, I. Am. J. Math. **100**, 43–60 (1978)
51. P.C. Kutzko, On the supercuspidal representations of GL_2, II. Am. J. Math. **100**, 705–716 (1978)
52. P.C. Kutzko, The Langlands conjecture for GL_2 of a local field. Ann. Math. **112**, 381–412 (1980)
53. P.C. Kutzko, The exceptional representations of GL_2. Compos. Math. **51**, 3–14 (1984)
54. P.C. Kutzko, Towards a classification of the supercuspidal representations of GL_N. J. Lond. Math. Soc. **37**(2), 265–274 (1988)
55. B. Lemaire, Comparison of lattice filtrations and Moy-Prasad filtrations. J. Lie Theory **19**, 29–54 (2009)
56. I.G. Macdonald, *Symmetric Functions and Hall Polynomials* (Oxford University Press, 1979)
57. I.G. Macdonald, Zeta functions attached to finite general linear groups. Math. Ann. **249**, 1–15 (1980)
58. C. Mœglin, Sur la correspondance de Langlands-Kazhdan. J. Math. Pures et Appl. **69**(9), 175–226 (1990)
59. L.E. Morris, Tamely ramified intertwining algebras. Invent. Math. **114**, 1–54 (1993)
60. A. Moy, G. Prasad, Unrefined minimal K-types for p-adic groups. Invent. Math. **116**, 393–408 (1994)

61. A. Moy, G. Prasad, Jacquet functors and unrefined minimal K-types. Comment. Math. Helv. **71**, 98–121 (1996)
62. V. Paskunas, Unicity of types for supercuspidal representations of GL_N. Proc. Lond. Math. Soc. (3) **91**, 623–654 (2005)
63. I. Reiner, *Maximal Orders, London Mathematical Society Monographs (New Series)*, vol. **28** (Oxford University Press, 2003)
64. V. Sécherre, Représentations lisses de $\mathrm{GL}(m, D)$ I. Caractères simples. Bull. Soc. Math. Fr. **132**(3), 327–396 (2004)
65. V. Sécherre, Représentations lisses de $\mathrm{GL}(m, D)$ II. β-extensions. Compos. Math. **141**(6), 1531–1550 (2005)
66. V. Sécherre, Représentations lisses de $\mathrm{GL}(m, D)$ III : types simples. Ann. Scient. Éc. Norm. Sup. **38**(4), 951–977 (2005)
67. V. Sécherre, S. Stevens, Représentations lisses de $\mathrm{GL}_m(D)$. IV. Représentations supercuspidales. J. Inst. Math. Jussieu **7**(3), 527–574 (2008)
68. V. Sécherre, S. Stevens, Towards an explicit local Jacquet-Langlands correspondence beyond the cuspidal case. arXiv: 1611.04317
69. J-P. Serre, *Corps Locaux* (Hermann, Paris, 1968)
70. F. Shahidi, On certain L-functions. Am. J. Math. **103**, 297–355 (1981)
71. F. Shahidi, Fourier transforms of intertwining operators and Plancherel measures for $GL(n)$. Am. J. Math. **106**, 67–111 (1984)
72. S. Stevens, Semisimple characters for p-adic classical groups. Duke Math. J. **127**(1), 123–173 (2005)
73. S. Stevens, The supercuspidal representations of p-adic classical groups. Invent. Math. **172**(2), 289–352 (2008)
74. J. Tate, Number theoretic background, in *Automorphic forms, representations andL-functions* ed. by A. Borel, W. Casselman. Proceedings of Symposia in Pure Mathematics, vol. 33, No. 2 (American Mathematical Society, Providence RI, 1979), pp. 3–26
75. J. Tits, Reductive groups over local fields, in *Automorphic forms, representations and L-functions*, ed. by A. Borel, W. Casselman. Proceedings of Symposia in Pure Mathematics, vol. 33, No. 1 (American Mathematical Society, Providence RI, 1979), pp. 29–69
76. J.-L. Waldspurger, Algèbres de Hecke et induites de représentations cuspidales, pour $GL(N)$. J. reine angew. Math. **370**, 127–191 (1986)
77. J.-K. Yu, Construction of tame supercuspidal representations. J. Am. Math. Soc. **14**, 579–622 (2001)

Root Data with Group Actions

Jeffrey D. Adler and Joshua M. Lansky

Contents

1 Introduction 128
2 Restrictions of Root Data 130
3 Automorphisms of Root Data and Automorphisms of Reductive Groups 134
4 Proofs of Theorems 136
5 Remarks on Cohomological Parametrizations 141
References 143

Abstract Suppose k is a field, G is a connected reductive algebraic k-group, T is a maximal k-torus in G, and Γ is a finite group that acts on (G, T). From the above, one obtains a root datum Ψ on which $\mathrm{Gal}(k) \times \Gamma$ acts. Provided that Γ preserves a positive system in Ψ, not necessarily invariant under $\mathrm{Gal}(k)$, we construct an inverse to this process. That is, given a root datum on which $\mathrm{Gal}(k) \times \Gamma$ acts appropriately, we show how to construct a pair (G, T), on which Γ acts as above. Although the pair (G, T) and the action of Γ are canonical only up to an equivalence relation, we construct a particular pair for which G is k-quasisplit and Γ fixes a $\mathrm{Gal}(k)$-stable pinning of G. Using these choices, we can define a notion of taking "Γ-fixed points" at the level of equivalence classes, and this process is compatible with a general "restriction" process for root data with Γ-action.

Keywords Reductive algebraic group · Root datum · Quasi-semisimple automorphisms

2010 Mathematics Subject Classification Primary 20G15 · 20G40 · Secondary 20C33.

J. D. Adler (✉) · J. M. Lansky
Department of Mathematics and Statistics, American University,
Washington, DC 20016-8050, USA
e-mail: jadler@american.edu

J. M. Lansky
e-mail: lansky@american.edu

A.-M. Aubert et al. (eds.), *Representations of Reductive p-adic Groups*,
Progress in Mathematics 328, https://doi.org/10.1007/978-981-13-6628-4_3

1 Introduction

Let k be a field with separable closure k^{sep}. Let Γ be a finite group.

Suppose Ψ is a (reduced) based root datum on which the absolute Galois group $\text{Gal}(k)$ acts. Then it is well known ([10, Theorem 6.2.7]) that there exists a connected, reductive, k-quasisplit k-group G, uniquely determined up to k-isomorphism, such that the root datum of G (with respect to a maximal k-torus contained in a Borel k-subgroup) is isomorphic to Ψ and carries the same action of $\text{Gal}(k)$. We generalize this result in two directions.

(A) Suppose G is a connected reductive k-group, and T is an arbitrary maximal torus. Then, the *non-based* root datum $\Psi(G, T)$ carries an action of $\text{Gal}(k)$. We show that one can reverse this process. That is, given a root datum Ψ with an action of $\text{Gal}(k)$, one can obtain a pair (G, T) that gives rise to Ψ. In general, the pair (G, T) need not be uniquely determined up to k-isomorphism. However, we can always choose G to be k-quasisplit, and all possibilities for G must be k-inner forms of each other.

(B) Now suppose that Γ acts on a pair (G, T) as above via k-automorphisms. Then Γ acts on the root datum $\Psi(G, T)$, and the actions of Γ and $\text{Gal}(k)$ commute. We show that one can reverse this process under mild conditions. That is, suppose that Ψ is a root datum with an action of $\text{Gal}(k) \times \Gamma$. Assume that Γ (but not necessarily $\text{Gal}(k)$) preserves a base. Then, one can obtain a pair (G, T) as above, carrying an action of Γ via k-automorphisms. That is, under appropriate conditions, we can lift an action of Γ from a root datum to a reductive group. Moreover, one can choose G to be k-quasisplit and can choose the action of Γ to preserve a pinning.

The above are all contained in our main result, Theorem 1. In order to state it more precisely, let us consider the collection of abstract root data Ψ that carry an action of $\text{Gal}(k) \times \Gamma$ such that the action of Γ stabilizes a base for Ψ. We consider two data Ψ and Ψ' with such actions to be *equivalent* if there is a $\text{Gal}(k) \times \Gamma$-equivariant isomorphism $\Psi \longrightarrow \Psi'$. Let $\mathscr{R}_\Gamma$ denote the set of equivalence classes of reduced data with such actions.

Let G be a connected reductive k-group and $T \subseteq G$ a maximal k-torus. Suppose there exists some Borel subgroup $B \subseteq G$ (not necessarily defined over k) containing T, and a homomorphism φ from Γ to the group $\text{Aut}_k(G, B, T)$ of k-automorphisms of G stabilizing T and B. Suppose G', T', and φ' are defined similarly. We say that the triples (G, T, φ) and (G', T', φ') are *equivalent* if there exists an isomorphism $\nu\colon G \longrightarrow G'$ whose restriction gives a Γ-equivariant k-isomorphism $T \longrightarrow T'$. (In this situation, ν must be an inner twisting by [8, §3.2].) Let $\mathscr{T}_\Gamma$ be the set of equivalence classes of such triples (G, T, φ).

A triple (G, T, φ) as above naturally determines a root datum with appropriate actions of $\text{Gal}(k)$ and Γ, hence an element of $\mathscr{R}_\Gamma$. It is easily seen that if (G', T', φ') and (G, T, φ) are equivalent, then they determine the same class in $\mathscr{R}_\Gamma$. Hence, we have a natural map $\mathrm{r}_\Gamma : \mathscr{T}_\Gamma \longrightarrow \mathscr{R}_\Gamma$.

Our main result is the following:

Theorem 1 *The map* $\mathrm{r}_\Gamma : \mathscr{T}_\Gamma \longrightarrow \mathscr{R}_\Gamma$ *is a bijection.*

In Sect. 5, we give cohomological descriptions of $\mathscr{R}_\Gamma$ and the inverse of the map r_Γ, and thus obtain a cohomological description of $\mathscr{T}_\Gamma$.

In [1], we introduce the notion of a *parascopic group* $\bar{G}$ for a pair (G, Γ), where Γ is a finite group acting via k-automorphisms (in a certain specified way) on a maximal k-torus of the reductive k-group G. This notion axiomatizes and generalizes the relationship between G and the connected part $(G^\Gamma)^\circ$ of the group of Γ-fixed points in the case where the action of Γ extends to one on G. When $\bar{G}$ is parascopic for (G, Γ), we construct a canonical lifting of semisimple stable conjugacy classes from $\bar{G}^\wedge$ to $G^\wedge$, k-groups that are in duality with $\bar{G}$ and G, respectively. Theorem 1 is a key tool in our further study of this lifting in [2]. In particular, in many settings, this result allows one to factor such a lifting into a composition of simpler ones which are much more readily understood.

We also prove a generalization of a second well-known result. Suppose Γ is a finite group acting algebraically on a connected reductive group G, fixing a Borel-torus pair (B, T) in G and a pinning for (G, B, T). Then, the root system of the connected part $\bar{G} := (G^\Gamma)^\circ$ of the group of fixed points is obtained as follows. The set of restrictions of roots of G from T to $\bar{T} := (T^\Gamma)^\circ$ is a root system, not necessarily reduced, but there is a preferred way to choose a maximal reduced subsystem. The above is well known, but in Theorem 7 and Lemma 28 we describe the root datum, not just the root system, of $\bar{G}$ with respect to $\bar{T}$.

Theorem 2 says that the process of passing from (G, T) to $(\bar{G}, \bar{T})$ is compatible with the bijection of Theorem 1. To state this result more precisely, suppose that the triple (G, T, φ) represents an element of $\mathscr{T}_\Gamma$. Then we know [1, Proposition 3.5] that $\bar{G}$ is a connected reductive k-group, and $\bar{T}$ is a maximal k-torus in $\bar{G}$. Thus, if we let "1" represent the map from the trivial group $\mathbf{1}$ to $\mathrm{Aut}(G)$, then the triple $(\bar{G}, \bar{T}, 1)$ represents an element of $\mathscr{T}_1$. The equivalence class of (G, T, φ) does not determine that of $(\bar{G}, \bar{T}, 1)$, or even the k^{sep}-isomorphism class of $\bar{G}$. Nonetheless, we can obtain a well-defined map $\mathscr{T}_\Gamma \longrightarrow \mathscr{T}_1$ as follows: From Remark 25, we will see that every class in $\mathscr{T}_\Gamma$ contains a triple (G, T, φ) such G is k-quasisplit and φ fixes a $\mathrm{Gal}(k)$-invariant pinning. Use this choice of triple to define $\bar{G}$ and $\bar{T}$, and it is straightforward to show that our choices determine $\bar{G}$ and $\bar{T}$ up to k-isomorphism.

Suppose that the root datum Ψ represents an element of $\mathscr{R}_\Gamma$. We will see in Sect. 2 that the action of Γ on Ψ allows us to construct a "restricted" root datum $\bar{\Psi}$ that has a preferred choice of reduced subdatum $\bar{\Psi}'$. We thus obtain a map $\mathscr{R}_\Gamma \longrightarrow \mathscr{R}_1$.

Theorem 2 *Our maps* $\mathscr{T}_\Gamma \longrightarrow \mathscr{T}_1$ *and* $\mathscr{R}_\Gamma \longrightarrow \mathscr{R}_1$ *above are compatible with the maps of Theorem 1, in the sense that the following diagram commutes:*

$$\begin{array}{ccc} \mathscr{T}_\Gamma & \xrightarrow{\mathrm{r}_\Gamma} & \mathscr{R}_\Gamma \\ \downarrow & & \downarrow \\ \mathscr{T}_1 & \xrightarrow{\mathrm{r}_1} & \mathscr{R}_1 \end{array}$$

We prove both theorems in Sect. 4.

Some of the results in this paper have recently appeared elsewhere, and others are generalizations of existing results. For example, as indicated above, [10] presents Statement (A) in the case where G is quasisplit over k and T is contained in a Borel k-subgroup of G. Corollary 15 has appeared as [4, Lemma 4.2(2)]. Lemma 28 overlaps to some extent with [4, Proposition 4.1], but our proof is quite different. Theorem 7 is similar to [4, Lemma 4.2(1)] and [5, Theorem A], except that we describe the full restricted root datum, and our proof does not rely on Steinberg [11].

2 Restrictions of Root Data

Let $\Psi = (X^*, \Phi, X_*, \Phi^\vee)$ be a root datum. (We do not assume that Φ is reduced.) Let Γ denote a finite group of automorphisms of Ψ. We assume that there exists a Γ-stable set Δ of simple roots in Φ. Let $V^* = X^* \otimes \mathbb{Q}$ and $V_* = X_* \otimes \mathbb{Q}$. Let i^* denote the quotient map from V^* to its space $\bar{V}^* := V^*_\Gamma$ of Γ-coinvariants. From [1, §2], there is an embedding $\iota\colon \bar{V}^* \longrightarrow V^*$ with image $V^{*\Gamma}$ given by

$$\iota(\bar{v}) = \frac{1}{|\Gamma|}\sum_{\gamma\in\Gamma}\gamma v,$$

where v is any preimage in V^* of $\bar{v} \in \bar{V}^*$. Let $\bar{X}^*$ and $\bar{\Phi}$ denote the images of X^* and Φ under i^*. Then, $\bar{X}^*$ is X^*_Γ modulo torsion, where X^*_Γ denotes the module of Γ-coinvariants of X^*. It is straightforward to see that $\bar{X}^*$ and $\bar{X}_* := X_*^\Gamma$, and thus $\bar{V}^*$ and $\bar{V}_* := V_*^\Gamma$, are in duality via the pairing given by $\langle \bar{x}, \bar{\lambda}\rangle := \langle \iota\bar{x}, i_*\bar{\lambda}\rangle$, where $i_*\colon \bar{X}_* \longrightarrow X_*$ is the inclusion map. With respect to these pairings, i^* is the transpose of i_*.

For each $\beta \in \Phi$, let w_β denote the automorphism of X^* defined by

$$w_\beta(x) = x - \langle x, \beta^\vee\rangle\beta.$$

Let W denote the Weyl group of Ψ, i.e., the (finite) subgroup of $\mathrm{Aut}(X^*)$ generated by the w_β. Then, Γ acts naturally on W, and W acts on X^*. The group W^Γ of Γ-fixed elements of W acts on both $\bar{V}^*$ and $\bar{X}^*$ via the rule $w(i^*x) := i^*(w(x))$ for $w \in W^\Gamma$ and $x \in X^*$. Equivalently, for $\bar{x} \in \bar{X}^*$ and $w \in W^\Gamma$, we have $\iota(w(\bar{x})) = w(\iota\bar{x})$.

Lemma 3 (cf. [11, §1.32(a)]) *The natural action of W^Γ on $\bar{X}^*$ is faithful.*

Proof Let w be a nontrivial element of W^Γ. Then, there exists a positive root $\beta \in \Phi$ such that $w(\beta)$ is negative. Since Γ stabilizes Δ, it follows that $w(\gamma\cdot\beta) = \gamma\cdot(w\beta)$ is also negative for every $\gamma \in \Gamma$. Thus, $\iota(w(i^*\beta))$ is a linear combination of roots in Δ in which all of the coefficients are nonpositive, so $w(i^*\beta) \neq i^*\beta$. □

Notation 4 *For each root $\beta \in \Phi$, define a Γ-orbit Ξ_β in Φ as in* [1, §5]. That is, let $\Xi_\beta = \Gamma\cdot\beta$ if this is an orthogonal set. Otherwise, for each $\theta \in \Gamma\cdot\beta$, there exists a unique root $\theta' \neq \theta$ in $\Gamma \cdot \beta$ such that θ and θ' are not orthogonal. Moreover, $\theta + \theta'$ is a root in Φ and does not belong to $\Gamma \cdot \beta$. In this case, let $\Xi_\beta = \{\theta + \theta' \mid \theta \in \Gamma \cdot \beta\}$.

Remark 5 Thus, in all cases, Ξ_β is an orthogonal Γ-orbit of roots.

Lemma 6 *If $\alpha \in \bar{\Phi}$, then ${i^*}^{-1}(\alpha)$ is a Γ-orbit of roots in Φ.*

Proof This argument is similar to but more general than that given in the proof of [9, Lemma 10.3.2(ii)]. Suppose $\beta \in \Phi$ and $i^*(\beta) = \alpha$. Then, clearly $i^*\theta = \alpha$ for any $\theta \in \Gamma \cdot \beta$.

Now suppose $\beta' \in \Phi$, $\beta' \neq \beta$, and $i^*\beta' = \alpha$. Since $\iota(i^*(\beta' - \beta)) = 0$ and since Γ preserves Δ, when $\beta' - \beta$ is written as a linear combination of simple roots, the coefficients must sum to 0. In particular, $\beta' - \beta \notin \Phi$. Since β' cannot be a multiple of β, we have that $\langle \beta', \beta^\vee \rangle \leq 0$ by standard results about root systems. Similarly, $\langle \beta', \theta^\vee \rangle \leq 0$ for all $\theta \neq \beta'$ in $\Gamma \cdot \beta$. Therefore,

$$\Big\langle \beta' - \beta, \sum_{\theta \in \Gamma \cdot \beta} \theta^\vee \Big\rangle = \Big\langle \beta' - \beta, i_*\Big(\sum_{\theta \in \Gamma \cdot \beta} \theta^\vee\Big)\Big\rangle = \Big\langle i^*(\beta' - \beta), \sum_{\theta \in \Gamma \cdot \beta} \theta^\vee \Big\rangle,$$

and since $i^*(\beta' - \beta) = 0$, this pairing vanishes. Thus $\sum_{\theta \in \Gamma \cdot \beta} \langle \beta', \theta^\vee \rangle = \sum_{\theta \in \Gamma \cdot \beta} \langle \beta, \theta^\vee \rangle = 2$ or 1, depending on whether or not $\Gamma \cdot \beta$ is orthogonal. (This follows from the properties of root orbits discussed in [1, §5].) Since $\langle \beta', \theta^\vee \rangle \leq 0$ for all $\theta \neq \beta'$ in $\Gamma \cdot \beta$, it follows that $\beta' \in \Gamma \cdot \beta$. □

For each $\alpha \in \bar{\Phi}$, define

$$\alpha^\vee = \frac{|\Gamma \cdot \beta|}{|\Xi_\beta|} \sum_{\xi \in \Xi_\beta} \xi^\vee \in \bar{X}_*,$$

where β is any element of Φ such that $i^*\beta = \alpha$. The element of $\bar{X}_*$ defined by the above formula is independent of the particular choice of β by Lemma 6. Note that $\alpha^\vee$ does indeed lie in $\bar{X}_*$ since $|\Gamma \cdot \beta| / |\Xi_\beta| = 1$ or 2. Let $\bar{\Phi}^\vee = \{\alpha^\vee \mid \alpha \in \bar{\Phi}\}$.

Theorem 7 *With the above notation, $\bar{\Psi} := (\bar{X}^*, \bar{\Phi}, \bar{X}_*, \bar{\Phi}^\vee)$ is a root datum.*

Remark 8 If Ψ comes equipped with an action of $\mathrm{Gal}(k)$, and the action of Γ commutes with that of $\mathrm{Gal}(k)$, then it is clear that the action of $\mathrm{Gal}(k)$ preserves $\bar{\Psi}$.

Proof of Theorem 7 According to [8, §1.1], it suffices to show that

- $\bar{X}^*$ and $\bar{X}_*$ are in duality (which we have already observed),
- $\langle \alpha, \alpha^\vee \rangle = 2$ for all $\alpha \in \bar{\Phi}$, and
- The automorphisms w_α of $\bar{X}^*$ of the form

$$w_\alpha(\bar{x}) = \bar{x} - \langle \bar{x}, \alpha^\vee \rangle \alpha \qquad (\text{for } \alpha \in \bar{\Phi})$$

stabilize $\bar{\Phi}$ and generate a finite subgroup of $\mathrm{Aut}(\bar{X}^*)$.

Let $\alpha \in \bar{\Phi}$. Choose $\beta \in \Phi$ such that $i^*\beta = \alpha$, and choose $\xi_0 \in \Xi_\beta$. Then, we have

$$\begin{aligned}
\langle \alpha, \alpha^\vee \rangle &= \langle \iota\alpha, i_*\alpha^\vee \rangle \\
&= \Big\langle \frac{1}{|\Gamma \cdot \beta|} \sum_{\theta \in \Gamma\cdot\beta} \theta, \frac{|\Gamma \cdot \beta|}{|\Xi_\beta|} \sum_{\xi \in \Xi_\beta} \xi^\vee \Big\rangle \\
&= \frac{1}{|\Xi_\beta|} \Big\langle \sum_{\theta \in \Gamma\cdot\beta} \theta, \sum_{\xi \in \Xi_\beta} \xi^\vee \Big\rangle \\
&= \frac{1}{|\Xi_\beta|} \Big\langle \sum_{\xi' \in \Xi_\beta} \xi', \sum_{\theta \in \Xi_\beta} \xi^\vee \Big\rangle && \text{(by the definition of } \Xi_\beta) \\
&= \langle \xi_0, \xi_0^\vee \rangle && \text{(by Remark 5)} \\
&= 2,
\end{aligned}$$

as desired.

Now let $\bar{x} \in \bar{X}^*$, and choose $x \in X^*$ such that $i^*x = \bar{x}$. Then

$$\begin{aligned}
\langle \bar{x}, \alpha^\vee \rangle &= \langle x, i_*\alpha^\vee \rangle \\
&= \Big\langle x, \frac{|\Gamma \cdot \beta|}{|\Xi_\beta|} \sum_{\xi \in \Xi_\beta} \xi^\vee \Big\rangle \quad \text{(by Remark 5)} \\
&= \frac{|\Gamma \cdot \beta|}{|\Xi_\beta|} \sum_{\xi \in \Xi_\beta} \langle x, \xi^\vee \rangle.
\end{aligned} \tag{9}$$

It follows that

$$\begin{aligned}
w_\alpha(\bar{x}) &= \bar{x} - \langle \bar{x}, \alpha^\vee \rangle \alpha \\
&= i^*x - \langle \bar{x}, \alpha^\vee \rangle i^*\beta \\
&= i^*x - \frac{\langle \bar{x}, \alpha^\vee \rangle}{|\Gamma \cdot \beta|} i^*\Big(\sum_{\theta \in \Gamma\cdot\beta} \theta \Big) && \text{(by Lemma 6)} \\
&= i^*x - \frac{\langle \bar{x}, \alpha^\vee \rangle}{|\Gamma \cdot \beta|} i^*\Big(\sum_{\xi' \in \Xi_\beta} \xi' \Big) && \text{(by the definition of } \Xi_\beta) \\
&= i^*x - \frac{1}{|\Xi_\beta|} \sum_{\xi \in \Xi_\beta} \langle x, \xi^\vee \rangle i^*\Big(\sum_{\xi' \in \Xi_\beta} \xi' \Big) && \text{(by (9))} \\
&= i^*x - \sum_{\xi \in \Xi_\beta} \langle x, \xi^\vee \rangle i^*\Big(\frac{1}{|\Xi_\beta|} \sum_{\xi' \in \Xi_\beta} \xi' \Big).
\end{aligned}$$

But by Remark 5, for any $\xi \in \Xi_\beta$,

$$i^*\Big(\frac{1}{|\Xi_\beta|} \sum_{\xi' \in \Xi_\beta} \xi' \Big) = i^*\xi,$$

so we have

$$w_\alpha(\bar{x}) = i^*\Big(x - \sum_{\xi \in \Xi_\beta} \langle x, \xi^\vee \rangle \xi\Big). \tag{10}$$

Also by Remark 5, the reflections $w_\xi \in W$ for $\xi \in \Xi_\beta$ all commute with one another. If w denotes their product, then

$$x - \sum_{\xi \in \Xi_\beta} \langle x, \xi^\vee \rangle \xi = w(x),$$

so by (10), we have

$$w_\alpha(\bar{x}) = i^*(w(x)). \tag{11}$$

In particular, if $\alpha' \in \bar{\Phi}$, and $\beta' \in \Phi$ satisfies $i^*\beta' = \alpha'$, then

$$w_\alpha(\alpha') = i^*(w(\beta')) \in i^*(\Phi) = \bar{\Phi},$$

so $\bar{\Phi}$ is stable under the action of w_α, as desired.

It remains to show that the group $\bar{W} := \langle w_\alpha \mid \alpha \in \bar{\Phi} \rangle \subset \mathrm{Aut}(\bar{X}^*)$ is finite. To accomplish this, we show that $\bar{W}$ embeds naturally in the finite group W^Γ. By Lemma 3, there is a natural injection

$$W^\Gamma \longrightarrow \mathrm{Aut}(\bar{X}^*).$$

To construct an embedding $\bar{W} \longrightarrow W^\Gamma$, it is therefore enough to show that the image of this injection contains $\bar{W}$. Thus, given $\bar{w} \in \bar{W}$, we will show that there exists $w \in W^\Gamma$ whose action on $\bar{X}^*$ coincides with that of $\bar{w}$. It suffices to prove the existence of w only in the case in which $\bar{w}$ is a reflection w_α through a root $\alpha \in \bar{\Phi}$. In this case, let $w = \prod_{\xi \in \Xi_\beta} w_\xi$, where $\beta \in \Phi$ is such that $i^*\beta = \alpha$. It follows from Remark 5 that $w \in W^\Gamma$, and it follows from (11) that for any $x \in X^*$,

$$w_\alpha(i^*x) = i^*(w(x)) = w(i^*x).$$

This establishes the existence of the desired embedding. □

Remark 12 If Φ is reduced, then so is the root system $\bar{\Phi}$ constructed above, unless Φ has an irreducible factor of type A_{2n} whose stabilizer in Γ acts upon it nontrivially. To see this, it is easy to reduce to the case where Φ is irreducible and Γ is cyclic (see [1, Proposition 3.5]). The result then follows from [6, §1.3].

Remark 13 There is a way to choose a maximal reduced subsystem that we will later see is preferred. (See Lemma 28.) Specficially, take only the nondivisible (resp. nonmultipliable) roots of $\bar{\Phi}$ according as char k is not two (resp. two).

Lemma 14 *The map i^* induces a bijection between the set of Γ-invariant positive systems in Φ and the set of positive systems in $\bar{\Phi}$.*

Proof Let $\Pi \subseteq \Phi$ be a Γ-invariant positive system. Let $\bar{\Pi} = i^*(\Pi) \subseteq \bar{\Phi}$. Then there is some vector $v \in V_*$ such that for every root $\beta \in \Phi$, we have that $\langle \beta, v \rangle \neq 0$, and $\langle \beta, v \rangle > 0$ if and only if $\beta \in \Pi$. Since Π is Γ-invariant, we may replace v by $\sum_{\gamma \in \Gamma} \gamma v$, and thus assume that v is Γ-invariant, and so lies in $\bar{V}_*$. Suppose $\alpha \in \bar{\Phi}$. Then $\alpha = i^* \beta$ for some $\beta \in \Phi$, so $\langle \alpha, v \rangle = \langle \beta, v \rangle$. Thus, $\langle \alpha, v \rangle \neq 0$, and $\langle \alpha, v \rangle > 0$ if and only if $\alpha \in \bar{\Pi}$. This shows that $\bar{\Pi}$ is a positive system in $\bar{\Phi}$.

Conversely, suppose that $\bar{\Pi} \subseteq \bar{\Phi}$ is a positive system, and let $\Pi = {i^*}^{-1} \bar{\Pi}$. Then there is some vector $\bar{v} \in \bar{V}_*$ such that for every root $\alpha \in \bar{\Phi}$, we have that $\langle \alpha, \bar{v} \rangle \neq 0$, and $\langle \alpha, \bar{v} \rangle > 0$ if and only if $\alpha \in \bar{\Pi}$. For every root $\beta \in \Phi$, we have $\langle \beta, i_* v \rangle = \langle i^* \beta, v \rangle$, which is never zero, and is positive if and only if $\beta \in \Pi$. Thus, $\Pi \subset \Phi$ is a positive system. Since $i_* v$ is Γ-invariant, so is Π. □

Corollary 15 *Let $\bar{W}$ be the Weyl group of $\bar{\Psi}$. Then, the embedding of $\bar{W}$ into W^Γ in the proof of Theorem 7 is an isomorphism.*

Proof Since $\bar{W}$ acts simply transitively on the set of positive systems in $\bar{\Phi}$, and W^Γ acts simply transitively on the set of Γ-invariant positive systems in Φ, the result follows from Lemma 14. □

3 Automorphisms of Root Data and Automorphisms of Reductive Groups

Let $\Psi = (X^*, \Phi, X_*, \Phi^\vee)$ be a root datum on which a group Λ acts via automorphisms. Choosing a root basis Δ of Φ, we obtain a corresponding based root datum $\dot{\Psi}$. Then, $\dot{\Psi}$ also carries an action of Λ. Namely, for $\sigma \in \Lambda$, there exists a unique element $c(\sigma)$ in the Weyl group $W(\Psi)$ of Ψ such that $\sigma(\Delta) = c(\sigma)(\Delta)$. If we define $\sigma^\star$ to be the automorphism of $\dot{\Psi}$ given by

$$\sigma^\star \chi = c(\sigma)^{-1}(\sigma \chi) \tag{16}$$

for $\chi \in X^*$, then the action of Λ on $\dot{\Psi}$ is given by $\sigma \mapsto \sigma^\star$.

Since Λ acts on Ψ and $\dot{\Psi}$, it acts naturally on $\mathrm{Aut}(\Psi)$ and $\mathrm{Aut}(\dot{\Psi})$, as well as on the Weyl groups $W(\Psi) \subset \mathrm{Aut}(\Psi)$ and $W(\dot{\Psi}) \subset \mathrm{Aut}(\dot{\Psi})$. Just as the actions of Λ on $\dot{\Psi}$ and on Ψ differ, so the actions of Λ on $W(\dot{\Psi})$ and on $W(\Psi)$ differ, even though the Weyl groups themselves are equal. For $\sigma \in \Lambda$ and $w \in W(\dot{\Psi})$, let $(\sigma, w) \mapsto \sigma^\star(w)$ denote the action of Λ on $W(\dot{\Psi})$. Then, we have

$$\sigma w = c(\sigma)\,(\sigma^\star w)\,c(\sigma)^{-1}.$$

One can check readily that map $c\colon \Lambda \longrightarrow W(\dot{\Psi})$ is a cocycle in $Z^1(\Lambda, W(\dot{\Psi}))$.

We now turn our attention to based root data arising from reductive algebraic groups. Let G be a connected reductive k-group, B a Borel subgroup of G, and $T \subseteq B$ a maximal torus of G. Let $\dot{\Psi}(G, B, T)$ denote the corresponding based root datum.

Any map $\vartheta \in \mathrm{Aut}(G)$ determines an obvious isomorphism

$$\vartheta^*\colon \dot{\Psi}(G, B, T) \longrightarrow \dot{\Psi}(G, \vartheta(B), \vartheta(T)).$$

There is a natural homomorphism

$$\pi\colon \mathrm{Aut}(G) \longrightarrow \mathrm{Aut}(\dot{\Psi}(G, B, T))$$

defined as follows. For $\vartheta \in \mathrm{Aut}(G)$, choose $g_\vartheta \in G(k^{\mathrm{sep}})$ such that $\mathrm{Int}(g_\vartheta)$ takes $\vartheta(B)$ to B, and $\vartheta(T)$ to T. Then $\mathrm{Int}(g_\vartheta) \circ \vartheta$ stabilizes B and T, and we let $\pi(\vartheta)$ be the automorphism $(\mathrm{Int}(g_\vartheta) \circ \vartheta)^*$ of $\dot{\Psi}(G, B, T)$ (which is, in fact, independent of the choice of g_ϑ).

Now suppose that T is defined over k. Then, an element $\sigma \in \mathrm{Gal}(k)$ naturally determines an automorphism of $\Psi(G, T)$ hence an automorphism $\sigma^\star$ of $\dot{\Psi}(G, B, T)$ as defined in (16). We thus obtain an action of $\mathrm{Gal}(k)$ on $\dot{\Psi}(G, B, T)$, hence one on $\mathrm{Aut}(\dot{\Psi}(G, B, T))$ as above. These actions are independent of the particular choice of B and T in the sense that if $g \in G(k^{\mathrm{sep}})$ and $\sigma \in \mathrm{Gal}(k)$, then we have

$$\sigma^\star \circ \mathrm{Int}(g)^* = \mathrm{Int}(g)^* \circ \sigma^\star, \tag{17}$$

where we use the notation $\sigma^\star$ to denote both the action of σ on $\dot{\Psi}(G, B, T)$ and on $\dot{\Psi}(G, {}^gB, {}^gT)$.

There is a well-known exact sequence

$$1 \longrightarrow \mathrm{Inn}(G) \longrightarrow \mathrm{Aut}(G) \xrightarrow{\pi} \mathrm{Aut}(\dot{\Psi}(G, B, T)) \longrightarrow 1. \tag{18}$$

We note that the homomorphisms in (18) are $\mathrm{Gal}(k)$-equivariant.

Remark 19 Let Δ be the set of simple roots for (G, B, T). Let $\{X_\alpha\}_{\alpha\in\Delta} \subset \mathrm{Lie}(G)(k^{\mathrm{sep}})$ be a pinning. It is well known [8, Corollary 2.14] that $\{X_\alpha\}$ determines a unique splitting ψ of (18). Namely, if $f \in \mathrm{Aut}(\dot{\Psi}(G, B, T))$, define $\psi(f)$ to be the automorphism of G such that

- $\psi(f)$ stabilizes B and T,
- the restriction of $\psi(f)$ to T is determined by the automorphism of $X^*(T)$ given by f, and
- $\psi(f)(X_\alpha) = X_{f(\alpha)}$.

Thus, $\mathrm{Im}(\psi)$ lies in the subgroup $\mathrm{Aut}(G, B, T, \{X_\alpha\})$ of $\mathrm{Aut}(G)$ consisting of automorphisms that stabilize B, T, and the pinning $\{X_\alpha\}$. If B and T are defined over k, and $\{X_\alpha\}$ is $\mathrm{Gal}(k)$-stable, it follows from [3, §3.10] that ψ is $\mathrm{Gal}(k)$-equivariant.

4 Proofs of Theorems

Proof of Theorem 1 Consider an abstract root datum $\Psi = (X^*, \Phi, X_*, \Phi^\vee)$ with an action of $\mathrm{Gal}(k) \times \Gamma$. Suppose that Δ is a Γ-stable base for Ψ. Let $\dot{\Psi}$ be the corresponding based root datum. As discussed in Sect. 3, the action of $\mathrm{Gal}(k) \times \Gamma$ on Ψ determines one of $\mathrm{Gal}(k) \times \Gamma$ on $\dot{\Psi}$. Since Δ is Γ-stable, the actions of Γ on Ψ and $\dot{\Psi}$ coincide. In the notation of (16) with $\Lambda = \mathrm{Gal}(k)$, the elements $c(\sigma) \in W(\dot{\Psi})$ that arise from the action of $\mathrm{Gal}(k)$ on $\dot{\Psi}$ must lie in $W(\dot{\Psi})^\Gamma$ since this action commutes with that of Γ. Therefore, the map $c \colon \mathrm{Gal}(k) \longrightarrow W(\dot{\Psi})^\Gamma$ is a cocycle in $Z^1(k, W(\dot{\Psi})^\Gamma)$. We note that the $\mathrm{Gal}(k) \times \Gamma$-isomorphism class of $\dot{\Psi}$ depends only on that of Ψ.

By [10, Theorem 6.2.7], there exists a triple (G, B_0, T_0), unique up to k-isomorphism, consisting of a k-quasisplit connected reductive group G, a Borel k-subgroup B_0 of G, and a maximal k-torus T_0 of B_0, such that the associated based root datum $\dot{\Psi}(G, B_0, T_0)$ is $\mathrm{Gal}(k)$-isomorphic to $\dot{\Psi}$. We will identify $\dot{\Psi}$ and $\dot{\Psi}(G, B_0, T_0)$ via such an isomorphism.

Let $\{X_\alpha\}$ be a $\mathrm{Gal}(k)$-stable pinning for G relative to B_0 and T_0. The action of Γ on $\dot{\Psi}$ determines a homomorphism $\phi \colon \Gamma \longrightarrow \mathrm{Aut}(\dot{\Psi})$. Let φ be the composition

$$\varphi : \Gamma \xrightarrow{\phi} \mathrm{Aut}(\dot{\Psi}) = \mathrm{Aut}(\dot{\Psi}(G, B_0, T_0)) \xrightarrow{\psi} \mathrm{Aut}(G, B_0, T_0, \{X_\alpha\}),$$

where $\psi \colon \mathrm{Aut}(\dot{\Psi}(G, B_0, T_0)) \longrightarrow \mathrm{Aut}(G, B_0, T_0, \{X_\alpha\})$ is the homomorphism from Remark 19.

Let $\bar{G} = \left(G^{\varphi(\Gamma)}\right)^\circ$ and $\bar{T}_0 = \left(T_0^{\varphi(\Gamma)}\right)^\circ$. By Lemma 28, $\bar{G}$ is a k-quasisplit reductive group, $\bar{T}_0$ a maximal k-torus of $\bar{G}$, and

$$W(\dot{\Psi})^\Gamma = W(G, T_0)^{\varphi(\Gamma)} = W(\bar{G}, \bar{T}_0).$$

Thus we may view c as a cocycle in $Z^1(k, W(\bar{G}, \bar{T}_0))$.

By [7, Theorem 1.1], there is some $g \in \bar{G}(k^{\mathrm{sep}})$ such that for all $\sigma \in \mathrm{Gal}(k)$, $g^{-1}\sigma(g)$ lies in the normalizer $N_{\bar{G}}(\bar{T}_0)(k^{\mathrm{sep}})$, and the image of $g^{-1}\sigma(g)$ in $W(\bar{G}, \bar{T}_0)$ is equal to $c(\sigma)$. Let $T = {}^g T_0$ and $B = {}^g B_0$. Since g is $\varphi(\Gamma)$-fixed, T is a $\varphi(\Gamma)$-stable maximal k-torus of G, and B is a $\varphi(\Gamma)$-stable Borel subgroup of G containing T. We have therefore associated to Ψ a triple (G, T, φ) of the required type.

Suppose we vary the arbitrary choices made in the above construction of (G, T, φ). That is, suppose we choose

- another root datum Ψ' that is $\mathrm{Gal}(k) \times \Gamma$-isomorphic to Ψ, a based root datum $\dot{\Psi}'$ with underlying datum Ψ' and base Δ', and hence a cocycle c' in $Z^1(k, W(\dot{\Psi}')^\Gamma)$;
- another triple of k-groups (G', B_0', T_0') k-isomorphic to (G, B_0, T_0) and an identification of $\dot{\Psi}(G', B_0', T_0')$ with $\dot{\Psi}'$; and
- a $\mathrm{Gal}(k)$-stable pinning $\{X_\alpha'\}$ of G' relative to B_0' and T_0', along with the associated map $\psi' \colon \mathrm{Aut}(\dot{\Psi}(G', B_0', T_0')) \longrightarrow \mathrm{Aut}(G', B_0', T_0', \{X_\alpha'\})$ from Remark 19.

We will show that these choices lead to a triple (G', T', φ') that is equivalent to (G, T, φ).

By assumption, there exists a $\mathrm{Gal}(k) \times \Gamma$-isomorphism $\lambda\colon \Psi \longrightarrow \Psi'$. The unique element $w \in W(\Psi')$ mapping $\lambda(\Delta)$ to Δ' must lie in $W(\Psi')^\Gamma$ since $\lambda(\Delta)$ and Δ' are both Γ-stable. The map $\lambda(\dot\Psi) \longrightarrow \dot\Psi'$ induced by w is equivariant with respect to the actions of $\mathrm{Gal}(k)$ on these based root data. It follows that the composition $\dot\lambda = w\lambda$ is a $\mathrm{Gal}(k) \times \Gamma$-isomorphism $\dot\Psi \longrightarrow \dot\Psi'$. Via conjugation, $\dot\lambda$ induces Γ-equivariant isomorphisms $W(\dot\Psi) \longrightarrow W(\dot\Psi')$ and $\tau\colon Z^1(k, W(\dot\Psi)) \longrightarrow Z^1(k, W(\dot\Psi'))$.

The map $\tau(c)\colon \sigma \mapsto \dot\lambda c(\sigma)\dot\lambda^{-1}$ is a cocycle in $Z^1(k, W(\dot\Psi')^\Gamma)$, cohomologous to c'; more precisely, one can check that for $\sigma \in \mathrm{Gal}(k)$,

$$c'(\sigma) = w^{-1}(\dot\lambda c(\sigma)\dot\lambda^{-1})\sigma^{\star\prime}(w) = w^{-1}(\tau(c)(\sigma))\sigma^{\star\prime}(w), \tag{20}$$

where $\sigma^{\star\prime}$ denotes the result of the action of σ on w, viewed as an element of $W(\dot\Psi')$.

Use a particular k-isomorphism between (G', B_0', T_0') and (G, B_0, T_0) to identify these triples. Following the above construction, we obtain a homomorphism $\varphi'\colon \Gamma \longrightarrow \mathrm{Aut}(G, B_0, T_0, \{X'_\alpha\})$, as well as an element $g' \in \left(G^{\varphi'(\Gamma)}\right)^\circ$ and a k-torus $T' = {}^{g'}T_0$, analogous to g and T, respectively.

We have a unique element κ of $\mathrm{Aut}_k(\dot\Psi(G, B_0, T_0))$ that produces a commutative square of $\mathrm{Gal}(k)$-equivariant maps

$$\begin{array}{ccc} \dot\Psi & \longrightarrow & \dot\Psi(G, B_0, T_0) \\ \downarrow{\scriptstyle \dot\lambda} & & \downarrow{\scriptstyle \kappa} \\ \dot\Psi' & \longrightarrow & \dot\Psi(G, B_0, T_0) \end{array} \tag{21}$$

Here, the horizontal arrows are the identifications chosen in the respective constructions of φ and φ'. We therefore obtain a diagram

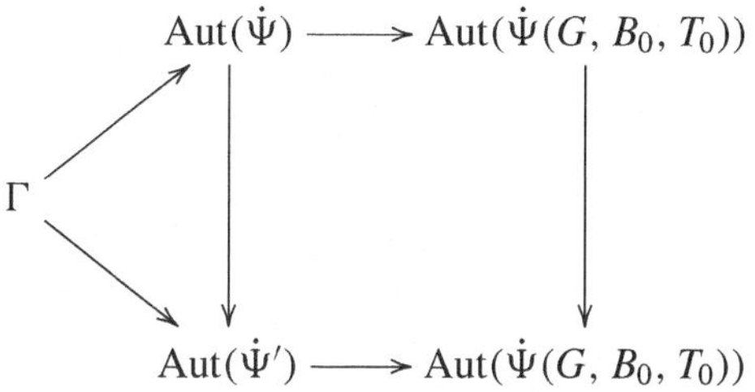

(22)

in which the square on the right is induced by (21) (and hence commutes), the vertical maps are given, respectively, by conjugation by $\dot\lambda$ and κ, the maps out of Γ are given by the actions of Γ on $\dot\Psi$ and $\dot\Psi'$, and the triangle commutes by the Γ-equivariance of $\dot\lambda$.

Identifying c and c', respectively, with cocycles in $Z^1(k, W(G, T_0)^{\varphi(\Gamma)})$ and $Z^1(k, W(G, T_0)^{\varphi'(\Gamma)})$ as in the above construction, it follows from (22) and (20) that

$$c'(\sigma) = w^{-1}(\kappa \circ c(\sigma) \circ \kappa^{-1})\sigma(w), \tag{23}$$

where $\sigma(w)$ here denotes the result of $\sigma \in \mathrm{Gal}(k)$ acting on the element $w \in W(\dot{\Psi}')$ via the identification of this group with the concrete Weyl group $W(G, T_0)$ in (22).

Let $n \in N_{\bar{G}}(\bar{T}_0)(k^{\mathrm{sep}})$ be a representative for w and set $\mu = \psi(\kappa) \in \mathrm{Aut}_k(G, B_0, T_0)$. Then by (23), $g'^{-1}\sigma(g')$ and $n^{-1}\mu(g^{-1}\sigma(g))\sigma(n)$ have the same image in $W(G, T_0)$. Rearranging terms and letting $h = g'n^{-1}\mu(g)^{-1}$, we obtain that $\sigma(h)$ and h have the same image modulo

$${}^{\sigma(\mu(g)n)}T_0 = \sigma({}^{\mu(g)n}T_0) = \sigma({}^{\mu(g)}T_0) = \sigma(\mu({}^{g}T_0)) = \sigma(\mu(T)) = \mu(T). \tag{24}$$

Let $\nu = \mathrm{Int}(h) \circ \mu$. Since

$$\begin{aligned}\nu(T) = \mathrm{Int}(h)(\mu(T)) &= \mathrm{Int}(g'n^{-1}\mu(g)^{-1})(\mu(T)) = \mathrm{Int}(g'n^{-1})(\mu({}^{g^{-1}}T))\\ &= \mathrm{Int}(g'n^{-1})(\mu(T_0)) = \mathrm{Int}(g'n^{-1})(T_0) = {}^{g'}T_0 = T',\end{aligned}$$

it follows from (24) that ν gives a k-isomorphism $T \longrightarrow T'$.

To show that (G', T', φ') is equivalent to (G, T, φ), it remains to show that ν is Γ-equivariant. It follows from the construction of φ that $\pi \circ \varphi$ is equal to the composition $\Gamma \longrightarrow \mathrm{Aut}(\dot{\Psi}) \longrightarrow \mathrm{Aut}(\dot{\Psi}(G, B_0, T_0))$ appearing in (22). Similarly, $\pi \circ \varphi'$ is equal to the analogous composition $\Gamma \longrightarrow \mathrm{Aut}(\dot{\Psi}') \longrightarrow \mathrm{Aut}(\dot{\Psi}(G, B_0, T_0))$. Thus for any $\gamma \in \Gamma$,

$$\pi(\varphi'(\gamma)) = \kappa \circ \pi(\varphi(\gamma)) \circ \kappa^{-1}.$$

Applying ψ to this equality and noting that $\psi \circ \pi \circ \varphi = \varphi$ by construction, we obtain

$$\psi(\pi(\varphi'(\gamma))) = \mu \circ \varphi(\gamma) \circ \mu^{-1}.$$

Note that by definition, $\psi(f)$ and $\psi'(f)$ agree on T_0 for any $f \in \mathrm{Aut}(\dot{\Psi}(G, B_0, T_0))$. Therefore, as automorphisms of T_0, we have

$$\varphi'(\gamma) = \psi'(\pi(\varphi'(\gamma))) = \psi(\pi(\varphi'(\gamma))) = \mu \circ \varphi(\gamma) \circ \mu^{-1}.$$

It follows that, as maps on T,

$$\begin{aligned}&\varphi'(\gamma) \circ \nu\\ &= \varphi'(\gamma) \circ \mathrm{Int}(h) \circ \mu\\ &= \varphi'(\gamma) \circ \mathrm{Int}(g'n^{-1}\mu(g)^{-1}) \circ \mu\\ &= \mathrm{Int}(g') \circ \varphi'(\gamma) \circ \mathrm{Int}(\mu(g)n)^{-1} \circ \mu\end{aligned}$$

$$
\begin{aligned}
&= \mathrm{Int}(g') \circ \mu \circ \varphi(\gamma) \circ \mu^{-1} \circ \mathrm{Int}(\mu(g)n)^{-1} \circ \mu \\
&= \mathrm{Int}(g') \circ \mu \circ \varphi(\gamma) \circ \mathrm{Int}(g\mu^{-1}(n))^{-1} \\
&= \mathrm{Int}(g') \circ \mu \circ \mathrm{Int}(g\mu^{-1}(n))^{-1} \circ \varphi(\gamma),
\end{aligned}
$$

where the last equality above comes from the fact that $g \in \bar{G}(k^{\mathrm{sep}})$ and $\mathrm{Int}(\mu^{-1}(n)) \in W(G, T_0)^{\varphi(\Gamma)}$. Thus $\varphi'(\gamma) \circ \nu$ is equal to

$$
\mathrm{Int}(g'n^{-1}\mu(g)^{-1}) \circ \mu \circ \varphi(\gamma) = \nu \circ \varphi(\gamma),
$$

showing that ν is Γ-equivariant. Therefore, (G', T', φ') is equivalent to (G, T, φ), and our construction induces a well-defined map $\mathrm{s}_\Gamma : \mathscr{R}_\Gamma \longrightarrow \mathscr{T}_\Gamma$.

We now show that $\mathrm{r}_\Gamma \circ \mathrm{s}_\Gamma$ is the identity map on $\mathscr{R}_\Gamma$. Let Ψ be a root datum representing some class in $\mathscr{R}_\Gamma$, and let (G, T, φ) be a triple representing the image of the class of Ψ under s_Γ. We need to show that $\Psi(G, T)$ is $\mathrm{Gal}(k) \times \Gamma$-isomorphic to Ψ. We will make free use of the notation developed in the construction of s_Γ.

The $\mathrm{Gal}(k)$-equivariant isomorphism of based root data $\dot{\Psi} \longrightarrow \dot{\Psi}(G, B_0, T_0)$ chosen in the definition of s_Γ is Γ-equivariant by construction (where the action of Γ on $\dot{\Psi}(G, B_0, T_0)$ is induced by φ). We may therefore identify $\dot{\Psi}$ and $\dot{\Psi}(G, B_0, T_0)$ as based root data with $\mathrm{Gal}(k) \times \Gamma$-action via this isomorphism. This allows us to identify Ψ and $\Psi(G, T_0)$ as root data with Γ-action (but not necessarily with $\mathrm{Gal}(k)$-action since the actions of $\mathrm{Gal}(k)$ on $\dot{\Psi}$ and Ψ differ in general).

Recall the element $g \in \bar{G}(k^{\mathrm{sep}})$ chosen in the definition of s_Γ. The map $\mathrm{Int}(g)^* \colon \Psi = \Psi(G, T_0) \longrightarrow \Psi(G, T)$ is Γ-equivariant since g is $\varphi(\Gamma)$-fixed. Furthermore, $\mathrm{Int}(g)^*$ is $\mathrm{Gal}(k)$-equivariant since for $\sigma \in \mathrm{Gal}(k)$ and $\chi \in X^*(T_0)$,

$$
\begin{aligned}
\mathrm{Int}(g)^*(\sigma\chi) &= \mathrm{Int}(g)^*\big(c(\sigma)(\sigma^\star\chi)\big) \\
&= {}^{gg^{-1}\sigma(g)}(\sigma^\star\chi) \\
&= {}^{\sigma(g)}(\sigma^\star\chi) \\
&= \sigma({}^{g}\chi) \\
&= \sigma(\mathrm{Int}(g)^*(\chi)).
\end{aligned}
$$

Thus $\Psi(G, T)$ is $\mathrm{Gal}(k) \times \Gamma$-isomorphic to Ψ, as desired.

Finally, we show that $\mathrm{s}_\Gamma \circ \mathrm{r}_\Gamma$ is the identity map on $\mathscr{T}_\Gamma$. Let (G, T, φ) represent a class in $\mathscr{T}_\Gamma$, and let (G', T', φ') represent the image of this class under $\mathrm{s}_\Gamma \circ \mathrm{r}_\Gamma$. Since $\mathrm{r}_\Gamma \circ (\mathrm{s}_\Gamma \circ \mathrm{r}_\Gamma) = (\mathrm{r}_\Gamma \circ \mathrm{s}_\Gamma) \circ \mathrm{r}_\Gamma = \mathrm{r}_\Gamma$, it follows that there is a $\mathrm{Gal}(k) \times \Gamma$ isomorphism $\Psi(G, T) \longrightarrow \Psi(G', T')$. By [8, Theorem 2.9], this isomorphism is induced by an isomorphism $\nu \colon G \longrightarrow G'$ that restricts to a Γ-equivariant k-isomorphism $T \longrightarrow T'$. Thus (G, T, φ) and (G', T', φ') are equivalent. □

Remark 25 Observe that in the definition of the map s_Γ above, the triple (G, T, φ) is constructed in such a way that G is k-quasisplit and φ fixes a $\mathrm{Gal}(k)$-invariant pinning of G. Thus, since $\mathrm{s}_\Gamma \circ \mathrm{r}_\Gamma$ is the identity map on $\mathscr{T}_\Gamma$, we see that every equivalence class in $\mathscr{T}_\Gamma$ contains such a triple.

Moreover, suppose that (G, T, φ) is a triple of this kind. Applying the construction of $\mathrm{s}_\Gamma \circ \mathrm{r}_\Gamma$ to this triple, we see that the triple we obtain is precisely (G, T, φ), provided that we make appropriate choices.

Remark 26 Recall that in the proof, it is shown that if (G, T, φ) and (G', T', φ') are two triples that arise by applying the s_Γ construction to a root datum Ψ, then (G, T, φ) and (G', T', φ') are equivalent. We note that the equivalence ν constructed in this case is of a special kind. Namely, ν is of the form $\mathrm{Int}(h) \circ \mu$, where $h \in G'(k^{\mathrm{sep}})$ and μ is a k-isomorphism from G to G'.

Now suppose that (G, T, φ) and (G', T', φ') are arbitrary equivalent triples with the properties that G and G' are k-quasisplit and φ and φ' fix $\mathrm{Gal}(k)$-invariant pinnings for G and G', respectively. Then combining the first part of this remark with Remark 25, it follows that there is an equivalence ν between (G, T, φ) and (G', T', φ') of the above special form.

Remark 27 Suppose that G' is k-quasisplit and T' is a maximal k-torus of G'. Suppose that the finite group Γ acts via $\mathrm{Gal}(k)$-equivariant automorphisms on $\Psi(G', T')$ preserving a base. Then, the equivalence class of $\Psi(G', T')$ lies in $\mathscr{R}_\Gamma$. Applying the construction in the definition of s_Γ to $\Psi(G', T')$, we obtain a triple (G, T, φ) where G is k-quasisplit. Since $\Psi(G', T')$ and $\Psi(G, T)$ are $\mathrm{Gal}(k)$-isomorphic, G' can be taken to equal G. Moreover, if $g \in G(k^{\mathrm{sep}})$ is chosen such that $T' = {}^{g}T_0$, then the cocycle c used to define T can be taken to be the image of $\sigma \mapsto g^{-1}\sigma(g)$ in $Z^1(k, W(G, T_0)^\Gamma)$. In particular, it follows from [1, Proposition 6.1] that T' is stably conjugate to T and that the $\mathrm{Gal}(k) \times \Gamma$-equivariant isomorphism between $\Psi(G, T')$ and $\Psi(G, T)$ can be given by an inner automorphism of G. (Of course, one could also construct another such triple involving a torus not necessarily stably conjugate to T' by taking the image of (G, T, φ) under a rational automorphism of G.)

We now turn our attention to Theorem 2. Let G be a quasisplit connected reductive k-group, B a Borel subgroup of G, and $T \subseteq B$ a maximal k-torus of G. Let $\{X_\alpha\}$ be a $\mathrm{Gal}(k)$-stable pinning of G with respect to (B, T). Suppose a group Γ acts on G via k-automorphisms, preserving B, T, and $\{X_\alpha\}$. Then Γ acts on the based root datum $\dot{\Psi}(G, B, T)$, and we will freely use the notation of Sect. 2 in the following.

In particular, we let $\bar{\Psi} = (\bar{X}^*, \bar{\Phi}, \bar{X}_*, \bar{\Phi}^\vee)$ be the restricted root datum associated to the action of Γ on $\Psi(G, B, T)$ by Theorem 7. Construct a new root datum $\bar{\Psi}'$ by replacing the root system $\bar{\Phi}$ of $\bar{\Psi}$ by a maximal reduced subsystem $\bar{\Phi}'$ as in Remark 13, and do likewise with the coroot system.

By [1, Proposition 3.5], $\bar{G} := (G^\Gamma)^\circ$ is a reductive k-group, and $\bar{T} := (T^\Gamma)^\circ$ is a maximal k-torus of $\bar{G}$. We may identify $\bar{X}^*$ with $X^*(\bar{T})$. Under this identification, the restriction β_{res} of a root $\beta \in \Phi(G, T)$ to $\bar{T}$ corresponds to $i^*\beta \in \bar{X}^*$.

Lemma 28 *Using the above notation, and under the above identification of $\bar{X}^*$ with $X^*(\bar{T})$, we have $\Psi(\bar{G}, \bar{T}) = \bar{\Psi}'$, and $W(\bar{G}, \bar{T}) = W(G, T)^\Gamma$.*

Proof Since the Weyl group of $\Psi(\bar{G}, \bar{T})$ is $W(\bar{G}, \bar{T})$, and $W(\bar{\Psi}) = W(\bar{\Psi}')$, and Corollary 15 implies that $W(\bar{\Psi}) = W(G, T)^\Gamma$, the claim about the Weyl groups will follow from the claim about root data.

To prove this, we reduce to the well-known case where Γ is cyclic. It is clear from the constructions that to show that $\bar{\Psi}' = \Psi(\bar{G}, \bar{T})$, it suffices to show $\bar{\Phi}' = \Phi(\bar{G}, \bar{T})$. This statement follows for G if it holds for a central quotient of G. Therefore, we may assume that (over k^{sep}) G is a direct product of simple groups. We can also reduce to the case where Γ acts transitively on the factors of G. As in the proof of [1, Proposition 3.5], we may identify the factors of G with each other, and replace Γ by a group $S \times \Gamma_1$, where S acts by permuting the factors in our product decomposition of G, and Γ_1 preserves each factor and acts in the same way on each, such that

- the action of $S \times \Gamma_1$ preserves $\{X_\alpha\}$,
- $\bar{G} = (G^{S\times\Gamma_1})^\circ$,
- $\bar{T} = (T^{S\times\Gamma_1})^\circ$, and
- $\bar{\Phi}$, hence $\bar{\Phi}'$, does not change when we replace the action of Γ by that of $S \times \Gamma_1$.

Working in stages, we may assume that Γ is simple. Thus, either Γ acts by permutation of the factors of G, or G has a single factor and thus is simple. In the former case, our result is trivial, so assume that G is simple. Then, G has a connected Dynkin diagram, whose automorphism group is solvable. Since Γ embeds in this automorphism group, Γ must be solvable hence cyclic. It is stated without proof in [6, §1.1] that in the cyclic case, $\Phi(\bar{G}, \bar{T}) = \bar{\Phi}'$ (and $W(\bar{G}, \bar{T}) = W(G, T)^\Gamma$). We include a proof below.

It follows from [11, §8.2(2'''')] that since Γ fixes a pinning (i.e., for each simple root $\beta \in \Phi(G, T)$, we have that $c_\beta = 1$ in the terminology *loc. cit.*), for each $\beta \in \Phi(G, T)$ such that $i^*\beta \in \bar{\Phi}'$, β_{res} belongs to $\Phi(\bar{G}, \bar{T})$. Since, by definition, $\bar{\Phi}' \subseteq i^*(\Phi(G, T))$, and since β_{res} corresponds to $i^*\beta$ under our identification of $X^*(\bar{T})$ with $\bar{X}^*$, this shows that $\bar{\Phi}' \subseteq \Phi(\bar{G}, \bar{T})$. On the other hand, by [1, Proposition 3.5(iv)], every root in $\Phi(\bar{G}, \bar{T})$ is the restriction of a root in $\Phi(G, T)$, so $\Phi(\bar{G}, \bar{T}) \subseteq i^*(\Phi(G, T)) = \bar{\Phi}$. But $\bar{\Phi}'$ is a maximal reduced subsystem of $\bar{\Phi}$, so $\Phi(\bar{G}, \bar{T}) = \bar{\Phi}'$, which concludes the proof. □

Proof of Theorem 2 Consider a class in $\mathscr{T}_\Gamma$. From Remark 25, we can represent this class by a triple (G, T, φ), where G is k-quasisplit and the action φ of Γ on G fixes a Gal(k)-invariant pinning. Let $\bar{G} = (G^{\varphi(\Gamma)})^\circ$ and $\bar{T} = (T^{\varphi(\Gamma)})^\circ$. In order to prove our result, we must show that the reduced root datum $\bar{\Psi}'$ contained in the restricted root datum $\bar{\Psi}$ of Theorem 7 is equivalent to $\Psi(\bar{G}, \bar{T})$. This is the content of Lemma 28. □

5 Remarks on Cohomological Parametrizations

In this section, we describe partitions of $\mathscr{R}_\Gamma$ and $\mathscr{T}_\Gamma$ into blocks, each of which can be parametrized cohomologically.

Let $\dot{\Psi}_0$ be a based root datum with Gal$(k) \times \Gamma$ action, i.e., the distinguished base is Gal$(k) \times \Gamma$-stable. Let Ψ_0 be the root datum underlying $\dot{\Psi}_0$. Let Ψ be a root datum with Gal$(k) \times \Gamma$-action such that

- Γ preserves a base of Ψ; and
- with respect to some (hence any) choice of Γ-invariant base of Ψ, the associated based root datum $\dot{\Psi}$ (with action $\sigma \mapsto \sigma^\star$ as in (16)) is $\mathrm{Gal}(k) \times \Gamma$-isomorphic to $\dot{\Psi}_0$.

The collection of such root data Ψ is closed under $\mathrm{Gal}(k) \times \Gamma$-isomorphism (the relation of equivalence from Sect. 1), and we let $\mathscr{R}_\Gamma^{\dot{\Psi}_0}$ denote the set of equivalence classes of data with the above properties.

Let ν be a $\mathrm{Gal}(k) \times \Gamma$-isomorphism $\dot{\Psi}_0 \longrightarrow \dot{\Psi}$ implementing the above equivalence. For $\sigma \in \mathrm{Gal}(k)$, let $\sigma(\nu)$ be the map $\sigma \circ \nu \circ \sigma^{-1}$, viewed as an isomorphism $\Psi_0 \longrightarrow \Psi$. In particular, the action of σ on Ψ here is via the original action of $\mathrm{Gal}(k)$, not the action $\sigma \mapsto \sigma^\star$ on $\dot{\Psi}$. Evidently, the map $\nu^{-1}\sigma(\nu)$ lies in the group $\mathrm{Aut}_\Gamma(\Psi_0)$ of Γ-equivariant automorphisms of Ψ_0. It is not difficult to show that in fact, $\nu^{-1}\sigma(\nu)$ lies in the subgroup $W(\Psi_0)^\Gamma$ of $\mathrm{Aut}_\Gamma(\Psi_0)$ and, moreover, that $\sigma \mapsto \nu^{-1}\sigma(\nu)$ determines a cocycle c in $Z^1(k, W(\Psi_0)^\Gamma)$.

Altering the choice of Ψ in its class or the choice of ν has the effect of replacing c by $\sigma \mapsto \kappa^{-1} \circ c(\sigma) \circ \sigma(\kappa)$ for some $\kappa \in \mathrm{Aut}_\Gamma(\Psi_0)$. Hence, the class of Ψ determines an element of

$$\mathscr{H}_\Gamma^{\dot{\Psi}_0} := \mathrm{Im}\big[H^1(k, W(\Psi_0)^\Gamma) \longrightarrow H^1(k, \mathrm{Aut}_\Gamma(\Psi_0))\big].$$

It is readily seen that, conversely, an element of this image determines a root datum Ψ as above, unique up to $\mathrm{Gal}(k) \times \Gamma$-isomorphism. Thus, we have defined a one-to-one correspondence $a_\Gamma : \mathscr{R}_\Gamma^{\dot{\Psi}_0} \longrightarrow \mathscr{H}_\Gamma^{\dot{\Psi}_0}$.

We now describe a corresponding parametrization for blocks in $\mathscr{T}_\Gamma$. Define $\mathscr{T}_\Gamma^{\dot{\Psi}_0}$ to be $s_\Gamma\big(\mathscr{R}_\Gamma^{\dot{\Psi}_0}\big)$, i.e., the set of classes in $\mathscr{T}_\Gamma$ containing triples whose associated based root data are $\mathrm{Gal}(k) \times \Gamma$-isomorphic to $\dot{\Psi}_0$.

Recall that the definition of s_Γ in Theorem 1 involves the choice of a triple (G, B_0, T_0) consisting of a k-quasisplit connected reductive group G, a Borel k-subgroup B_0 of G, and a maximal k-torus T_0 of B_0. We may identify the based root data (with $\mathrm{Gal}(k) \times \Gamma$-actions) $\dot{\Psi}(G, B_0, T_0)$ and $\dot{\Psi}_0$. We observe that the construction of the triple (G, T, φ) in the definition of s_Γ involves passing from the given root datum Ψ to a cocycle c in $Z^1(k, W(G, T_0)^{\varphi(\Gamma)}) = Z^1(k, W(\dot{\Psi}_0)^\Gamma)$. It can be checked that the passage from Ψ to c is precisely the one occurring in the above definition of the map a_Γ.

The proof of Theorem 1 proceeds to define (G, T, φ) in such a way that it depends only on c, and not directly on Ψ. Moreover, it follows from the proof that the equivalence class of (G, T, φ) depends only on the class of c in $H^1(k, \mathrm{Aut}_\Gamma(\Psi_0))$. Therefore, we obtain a map $b_\Gamma : \mathscr{H}_\Gamma^{\dot{\Psi}_0} \longrightarrow \mathscr{T}_\Gamma^{\dot{\Psi}_0}$ such that s_Γ factors through $\mathscr{H}_\Gamma^{\dot{\Psi}_0}$ according to the commutative diagram

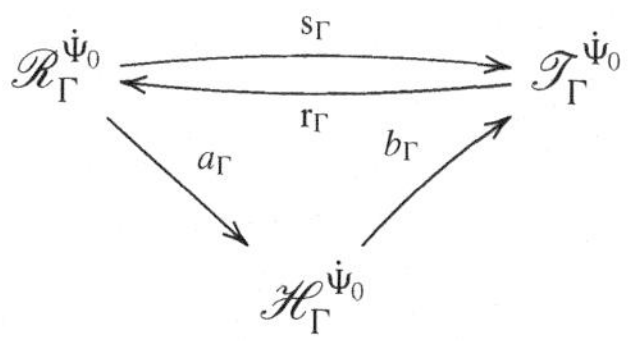

In particular, since a_Γ and s_Γ are bijections, b_Γ must also be a bijection.

References

1. J.D. Adler, J.M. Lansky, Lifting representations of finite reductive groups I: semisimple conjugacy classes. Canad. J. Math. **66**, 1201–1224 (2014). https://doi.org/10.4153/CJM-2014-013-6. Available at arXiv:1106.0786
2. J.D. Adler, J.M. Lansky, Lifting representations of finite reductive groups II: explicit conorm functions. Available at arxiv:1109.0794
3. M. Demazure, *Automorphismes des groupes réductifs*, Schémas en Groupes (Sém. Géométrie Algébrique, Inst. Hautes Études Sci., 1963/64), Inst. Hautes Études Sci., Paris, 1965, p. 87 (French). MR0228503 (37 #4083)
4. T.J. Haines, On Satake parameters for representations with parahoric fixed vectors. Int. Math. Res. Not. IMRN **20** (2015), 10367–10398. Available at arXiv:1402.3812. MR3455870
5. T.J. Haines, Dualities for root systems with automorphisms and applications to non-split groups. Represent. Theory **22**, 1–26 (2018). Available at arXiv:1604.01468
6. R.E. Kottwitz and D. Shelstad, Foundations of twisted endoscopy. Astérisque**255** (1999), vi+190 (English, with English and French summaries). MR1687096 (2000k:22024)
7. M.S. Raghunathan, Tori in quasi-split groups. J. Ramanujan Math. Soc. **19**(4), 281–287 (2004). MR2125504 (2005m:20114)
8. T.A. Springer, *Reductive groups*, Automorphic forms, representations, and L-functions, ed. by A. Borel, W. Casselman. Part 1, Proceedings of Symposia in Pure Mathematics, XXXIII, American Mathematical Society, Providence, R.I., 99 (1979), pp. 3–27. MR546587 (80h:20062)
9. T.A. Springer, in *Linear Algebraic Groups*. Progress in Mathematics, vol. 9 (Birkhäuser Boston Inc., Boston, MA, 1998). MR1642713 (99h:20075)
10. T.A. Springer, in *Linear Algebraic Groups*. Algebraic geometry IV, Encyclopedia of Mathematical Sciences (Springer, 1994), pp. 1–121. MR1100484 (92g:20061)
11. R. Steinberg, in *Endomorphisms of Linear Algebraic Groups*. Memoirs of the American Mathematical Society, No. 80, American Mathematical Society, Providence, R.I., 1968. MR0230728 (37 #6288)

The Character of a Simple Supercuspidal Representation of $SL(2, F)$

Moshe Adrian

Contents

1	Introduction	145
2	Notation	146
3	Background	146
4	Reduction Formula for the Character on the Split Torus	148
5	The Character Formula on the Split Torus	149
6	The Character Formula on Elliptic Tori	153
References		159

Abstract Let F be a non-Archimedean local field of characteristic zero with residual characteristic p. In this paper, we compute the character of a simple supercuspidal representation of $SL(2, F)$, when p is arbitrary.

1 Introduction

Let F be a non-Archimedean local field of characteristic zero, with residual characteristic p. Let $\mathbf{G}$ be a connected reductive group defined over F, and let $G = \mathbf{G}(F)$ be its F-points. If π is an irreducible admissible representation of G, we denote by θ_π its distribution character, which is a linear functional on $C_c^\infty(G)$, the locally constant, compactly supported complex-valued functions on G. Harish-Chandra showed that θ_π can be represented by a locally constant function on the regular semisimple set of G, which we will also denote by θ_π.

Suppose now that π is a supercuspidal representation. Much is known about θ_π. The first supercuspidal characters were computed by Sally and Shalika in [7], where they investigated the supercuspidal representations of $SL(2, F)$ when $p \neq 2$. Since then, there has been much work on understanding and computing supercuspidal characters, which has culminated in the work of Adler and Spice [1]. The computation

M. Adrian (✉)
Department of Mathematics, Queens College, CUNY,
65-30 Kissena Blvd., Queens, NY 11367-1597, USA
e-mail: moshe.adrian@qc.cuny.edu

A.-M. Aubert et al. (eds.), *Representations of Reductive p-adic Groups*,
Progress in Mathematics 328, https://doi.org/10.1007/978-981-13-6628-4_4

of these characters has yielded important consequences, including applications to the study of stability of supercuspidal L-packets constructed by Debacker and Reeder [2], Reeder [5], and Kaletha [4].

In this paper, we compute the character values of a simple supercuspidal representation of $SL(2, F)$ when p is arbitrary (see Theorem 5.3 and Theorem 6.6). Our computations give formulas that are geometric in nature: On the split torus case, we get a polynomial in q, whose powers are lengths of affine Weyl group elements that bound an explicit interval in the standard apartment of $SL(2, F)$ (see Remark 5.4). In the elliptic torus case, a similar polynomial arises, together with a Gauss sum.

We now briefly present an outline of the paper. In Sect. 2, we set up the notation that we will use throughout. In Sect. 3, we recall the definition of a simple supercuspidal representation as well as the Frobenius formula for the character of an induced representation. In Sect. 4, we present a reduction formula for the character of a simple supercuspidal representation of a split simply connected group on the split maximal torus. In Sect. 5, we compute the character of a simple supercuspidal representation of $SL(2, F)$ on the split torus. In Sect. 6, we compute the character on an elliptic torus.

2 Notation

Let F be a non-Archimedean local field of characteristic zero. Let $\mathfrak{o}$ be the ring of integers of F and $\mathfrak{p}$ its maximal ideal. Fix a uniformizer ϖ in F. If $p \neq 2$, we fix a non-square unit ϵ in $\mathfrak{o}^\times$. Let $|\cdot|$ denote the standard normalized absolute value on F, and val the associated valuation. We fix a Haar measure on F such that $\mathfrak{o}$ has volume 1, and we use the abbreviation vol for volume. A character of $\mathfrak{o}$ is said to be of *level* 1 if it is trivial on $\mathfrak{p}$, but nontrivial on $\mathfrak{o}$. If ψ is a character of F, we define the *twist* of ψ by $a \in F$ by $\psi_a(x) := \psi(ax)$.

3 Background

In this section, we recall the definition of a simple supercuspidal representation following Gross and Reeder [3], as well as the Frobenius formula for the character of a supercuspidal representation following [6].

Let $\mathbf{G}$ be a split, simply connected, almost simple, connected reductive group, and $\mathbf{T}$ a maximal F-split torus in $\mathbf{G}$. Associated to $\mathbf{T}$ we have the set of roots Φ of $\mathbf{T}$ in $\mathbf{G}$, a set of affine roots Ψ, and an apartment $\mathcal{A}$. Let $\mathbf{Z}$ be the center of $\mathbf{G}$, and set $G = \mathbf{G}(F)$, $T = \mathbf{T}(F)$, and $Z = \mathbf{Z}(F)$. Let $X^*(T)$ denote the character lattice of T, let T_0 be the maximal compact subgroup of T, and set

$$T_1 = \langle t \in T_0 : \lambda(t) \in 1 + \mathfrak{p} \; \forall \lambda \in X^*(T) \rangle.$$

Denoting the normalizer of T in G by N, we have the extended affine Weyl group $W^a := N/T_0$.

Now fix a Chevalley basis in the Lie algebra of $\mathbf{G}$. To each $\psi \in \Psi$, we have an associated affine root group U_ψ in G. Fix an alcove C in the apartment with corresponding simple affine and positive affine roots $\Pi \subset \Psi^+$. For $w \in W^a$, we let $\ell(w)$ denote the length of w. Set

$$I = \langle T_0, U_\psi : \psi \in \Psi^+ \rangle,$$

$$I_+ = \langle T_1, U_\psi : \psi \in \Psi^+ \rangle,$$

$$I_{++} = \langle T_1, U_\psi : \psi \in \Psi^+ \setminus \Pi \rangle.$$

Then I is the Iwahori subgroup corresponding to C and I_+ is its pro-unipotent radical. We set $H = ZI_+$.

The following results can be found in [3, §9].

Lemma 3.1 *The subgroup I_{++} is normal in I_+, with quotient*

$$I_+/I_{++} \cong \bigoplus_{\psi \in \Pi} U_\psi / U_{\psi+1}$$

as T_0-modules.

Definition 3.2 A character $\chi : H \to \mathbb{C}^*$ is called *affine generic* if
(i) χ is trivial on I_{++} and
(ii) χ is nontrivial on $U_\psi / U_{\psi+1}$ for every $\psi \in \Pi$ (see Lemma 3.1).

Theorem 3.3 *Let $\chi : H \to \mathbb{C}^*$ be an affine generic character. Then* $\mathrm{cInd}_H^G \chi$ *is an irreducible supercuspidal representation, where* cInd *denotes compact induction.*

Gross and Reeder have named the representations from Theorem 3.3 *simple supercuspidal representations.*

Now suppose that π is an irreducible smooth supercuspidal representation of G. Let K be a compact open subgroup of G, and suppose that σ is an irreducible representation of K such that

$$\pi = \mathrm{cInd}_K^G \sigma.$$

Let θ_σ denote the character of σ. The following is the Frobenius formula for the induced character θ_π.

Theorem 3.4 [6, Theorem 1.9] *Let g be a regular element of G. Then*

$$\theta_\pi(g) = \sum_{x \in K\backslash G/K} \sum_{y \in K \backslash KxK} \dot{\theta}_\sigma(ygy^{-1})$$

where

$$\dot{\theta}_\sigma(k) = \begin{cases} \theta_\sigma(k) & \text{if } k \in K, \\ 0 & \text{if } k \in G \setminus K. \end{cases}$$

4 Reduction Formula for the Character on the Split Torus

In this section, we show that the formula in Theorem 3.4 simplifies if G is simply connected, π is a simple supercuspidal representation, T is the split torus in G, and $g \in T$. Let us first recall the following basic fact about double coset decompositions. Suppose that K is a compact open subgroup of G, and let us choose any set of representatives $\{t_\alpha\}$ for the double cosets of $K\backslash G/K$. Then $Kt_\alpha K$ is the disjoint union of the cosets $Kt_\alpha s_1, Kt_\alpha s_2, \ldots, Kt_\alpha s_m$, where $s_1, s_2, \ldots, s_m$ is a set of representatives of $(K \cap t_\alpha^{-1}Kt_\alpha) \setminus K$. We will use this fact repeatedly in this paper.

Let $[W^a]$ denote any set of representatives for the group W^a.

Proposition 4.1 *Let* $\pi = \mathrm{cInd}_H^G \chi$ *be a simple supercuspidal representation of* G*, as in Theorem 3.3. If* $g \in T$ *and* g *is regular, then*

$$\theta_\pi(g) = |T_0/ZT_1| \sum_{x \in [W^a]} \sum_{y \in H\backslash HxH} \dot{\chi}(ygy^{-1}),$$

where

$$\dot{\chi}(h) = \begin{cases} \chi(h) & \text{if } h \in H \\ 0 & \text{if } h \in G \setminus H \end{cases}$$

Proof Our starting point is the well-known affine Bruhat decomposition $I\backslash G/I \longleftrightarrow W^a$. Since $I/I_+ \cong T_0/T_1$, the affine Bruhat decomposition descends to a decomposition

$$H\backslash G/H \longleftrightarrow N/ZT_1.$$

Therefore,

$$\theta_\pi(g) = \sum_{x \in N/ZT_1} \sum_{y \in H\backslash HxH} \dot{\chi}(ygy^{-1}). \tag{1}$$

We now analyze the inner sum. Write

$$\tau(x) := \sum_{y \in H\backslash HxH} \dot{\chi}(ygy^{-1}),$$

for $x \in N$. We claim that $\tau(x) = \tau(xt)\ \forall t \in T_0$. To see this, write

$$HxH = \bigcup_{i=1}^{m} Hxx_i,$$

where x_i are representatives of $H/(H \cap x^{-1}Hx)$. Then the map $Hxy \mapsto Hxty$ gives a bijection $H\backslash HxH \to H\backslash HxtH$, for $t \in T_0$. Therefore, in particular,

$$HxtH = \bigcup_{i=1}^{m} Hxtx_i.$$

We conclude that

$$\begin{aligned}\tau(xt) &= \sum_{y \in H\backslash HxtH} \dot{\chi}(ygy^{-1}) = \sum_{i=1}^{m} \dot{\chi}(xtx_i g x_i^{-1} t^{-1} x^{-1}) \\ &= \sum_{i=1}^{m} \dot{\chi}(xtx_i t^{-1} t g t^{-1} t x_i^{-1} t^{-1} x^{-1}) = \sum_{i=1}^{m} \dot{\chi}(xtx_i t^{-1} g t x_i^{-1} t^{-1} x^{-1}),\end{aligned}$$

the last equality coming from the fact that $g \in T$. Since $H/(H \cap x^{-1}Hx)$ is a disjoint union of certain spaces of the form $U_\gamma / U_{\gamma+n}$, and since conjugation by an element $t \in T_0$ preserves $U_\gamma / U_{\gamma+n}$, we get that

$$\sum_{i=1}^{m} \dot{\chi}(xtx_i t^{-1} g t x_i^{-1} t^{-1} x^{-1}) = \sum_{i=1}^{m} \dot{\chi}(xx_i g x_i^{-1} x^{-1}) = \tau(x).$$

We have therefore concluded that $\tau(xt) = \tau(x)\ \forall t \in T_0$.

Finally, note that we have the following short exact sequence

$$1 \to T_0/ZT_1 \to N/ZT_1 \to W^a = N/T_0 \to 1.$$

Since we just showed that τ is constant along fibers of the above exact sequence, together with (2), we now have that

$$\theta_\pi(g) = |T_0/ZT_1| \sum_{x \in [W^a]} \sum_{y \in H\backslash HxH} \dot{\chi}(ygy^{-1}).$$

□

5 The Character Formula on the Split Torus

In this section, we prove Theorem 5.3. In particular, we will compute the inner sum $\tau(x)$ in Proposition 4.1 for any set of representatives of W^a by decomposing HxH into a union of left cosets. Afterwards, we sum over all x to obtain θ_π.

By definition of affine generic, we have that $\chi|_{I_+}$ is given by

$$\chi|_{I_+} : I_+ \to \mathbb{C}^*$$

$$\begin{pmatrix} d_{11} & d_{12} \\ d_{21} & d_{22} \end{pmatrix} \mapsto \chi_1(d_{12})\chi_2(d_{21})$$

where χ_1 is some level 1 character of $\mathfrak{o}$ and where $\chi_2(d_{21}) = \chi_2'(\frac{1}{\varpi}d_{21})$, where χ_2' is some level 1 character of $\mathfrak{o}$. Extend both χ_1 and χ_2' to functions on F by setting both equal to 0 on $F \setminus \mathfrak{o}$.

Fix an element $g = \begin{pmatrix} a & 0 \\ 0 & a^{-1} \end{pmatrix} \in T_1$, set $r := val(a - a^{-1})$, and again define

$$\tau(x) := \sum_{y \in H \backslash HxH} \dot{\chi}(ygy^{-1}),$$

for $x \in N$.

Proposition 5.1 *Let* $x = \begin{pmatrix} b & 0 \\ 0 & b^{-1} \end{pmatrix}$.

(a) If $val(b) = n \geq 0$ *and* $2n < r$, *then* $\tau(x) = q^{2n}$.
(b) If $val(b) = n < 0$ *and* $-2n < r$, *then* $\tau(x) = q^{-2n}$.
(c) Otherwise, $\tau(x) = 0$.

Proof It will be convenient for the proof to introduce the following notation. For $r_1, r_2 \in \mathbb{N}$, we write

$$\begin{pmatrix} 1+\mathfrak{p} & \mathfrak{p}^{r_1} \\ \mathfrak{p}^{r_2} & 1+\mathfrak{p} \end{pmatrix} := \left\{ \begin{pmatrix} x_1 & x_2 \\ x_3 & x_4 \end{pmatrix} \in SL(2, F) : x_1, x_4 \in 1+\mathfrak{p}, x_2 \in \mathfrak{p}^{r_1}, x_3 \in \mathfrak{p}^{r_2} \right\}.$$

We prove (a). Let $val(b) = n \geq 0$. We first write the double coset HxH as a finite union of single right cosets. Since

$$H \cap x^{-1}Hx = Z \cdot \begin{pmatrix} 1+\mathfrak{p} & \mathfrak{o} \\ \mathfrak{p}^{2n+1} & 1+\mathfrak{p} \end{pmatrix},$$

we obtain an explicit disjoint union decomposition

$$H = \coprod_{z \in \mathfrak{p}/\mathfrak{p}^{2n+1}} Z \cdot \begin{pmatrix} 1+\mathfrak{p} & \mathfrak{o} \\ \mathfrak{p}^{2n+1} & 1+\mathfrak{p} \end{pmatrix} \cdot \begin{pmatrix} 1 & 0 \\ z & 1 \end{pmatrix}.$$

Therefore, we have the disjoint union

$$HxH = \coprod_{z \in \mathfrak{p}/\mathfrak{p}^{2n+1}} Hx \cdot \begin{pmatrix} 1 & 0 \\ z & 1 \end{pmatrix}.$$

Suppose that $y \in HxH$. We need to check when $ygy^{-1} \in H$ since $\dot{\chi}$ vanishes outside H. Using our above double coset decomposition, write $y = hxu$, where u is of the form $\begin{pmatrix} 1 & 0 \\ z & 1 \end{pmatrix}$ for some $z \in \mathfrak{p}/\mathfrak{p}^{2n+1}$ and for some $h \in H$. Then $ygy^{-1} \in H \Leftrightarrow xugu^{-1}x^{-1} \in H$. Moreover,

$$xugu^{-1}x^{-1} = \begin{pmatrix} a & 0 \\ b^{-2}z(a - a^{-1}) & a^{-1} \end{pmatrix}.$$

We now write $a - a^{-1} = \varpi^r u$ for some unit $u \in \mathfrak{o}^\times$. Then $b^{-2}z(a - a^{-1}) = \varpi^{-2n}\varpi^r z'$ for some $z' \in \mathfrak{p}/\mathfrak{p}^{2n+1}$.

Therefore,

$$\sum_{y \in H\backslash HxH} \dot{\chi}(ygy^{-1}) = \sum_{z' \in \mathfrak{p}/\mathfrak{p}^{2n+1}} \chi_2'(\varpi^{-2n+r-1}z').$$

Making a change of variables, we get

$$\sum_{z' \in \mathfrak{p}/\mathfrak{p}^{2n+1}} \chi_2'(\varpi^{-2n+r-1}z') = \sum_{v \in \mathfrak{p}^{-2n+r}/\mathfrak{p}^r} \chi_2'(v) = vol(\mathfrak{p}^r)^{-1} \int_{\mathfrak{p}^{-2n+r}\cap \mathfrak{o}} \chi_2'(v)\mathrm{d}v$$

since $\dot{\chi}$ vanishes outside H. If $\mathfrak{p}^{-2n+r} \supseteq \mathfrak{o}$, then this integral vanishes since the integral of a nontrivial character over a group vanishes. However, if $\mathfrak{p}^{-2n+r} \subsetneq \mathfrak{o}$, which is precisely the condition that $2n < r$, then

$$vol(\mathfrak{p}^r)^{-1} \int_{\mathfrak{p}^{-2n+r}\cap \mathfrak{o}} \chi_2'(v)\mathrm{d}v = vol(\mathfrak{p}^r)^{-1} \int_{\mathfrak{p}^{-2n+r}} \mathrm{d}v = vol(\mathfrak{p}^r)^{-1} vol(\mathfrak{p}^{-2n+r})$$

since χ_2' is trivial on $\mathfrak{p}$. By our choice of measure, we have that $vol(\mathfrak{p}^d) = q^{-d}$ for $d > 0$, thus proving part (a).

We now prove (b). So suppose that $val(b) = n < 0$. Then similarly as in (a), we get

$$H \cap x^{-1}Hx = Z \cdot \begin{pmatrix} 1+\mathfrak{p} & \mathfrak{p}^{-2n} \\ \mathfrak{p} & 1+\mathfrak{p} \end{pmatrix},$$

$$H = \coprod_{z \in \mathfrak{o}/\mathfrak{p}^{-2n}} Z \cdot \begin{pmatrix} 1+\mathfrak{p} & \mathfrak{p}^{-2n} \\ \mathfrak{p} & 1+\mathfrak{p} \end{pmatrix} \cdot \begin{pmatrix} 1 & z \\ 0 & 1 \end{pmatrix}.$$

Therefore, we have a disjoint union

$$HxH = \coprod_{z \in \mathfrak{o}/\mathfrak{p}^{-2n}} Hx \cdot \begin{pmatrix} 1 & z \\ 0 & 1 \end{pmatrix}.$$

Suppose that $y \in HxH$. By the above double coset decomposition, we may write $y = hxu$, where u is of the form $\begin{pmatrix} 1 & z \\ 0 & 1 \end{pmatrix}$ for some $z \in \mathfrak{o}/\mathfrak{p}^{-2n}$, and for some $h \in H$. A computation yields $xugu^{-1}x^{-1} = \begin{pmatrix} a & b^2 z(a^{-1} - a) \\ 0 & a^{-1} \end{pmatrix}$. Therefore,

$$\sum_{y \in H \backslash HxH} \dot{\chi}(ygy^{-1}) = \sum_{z \in \mathfrak{o}/\mathfrak{p}^{-2n}} \chi_1(b^2 z(a^{-1} - a)).$$

Rewriting $b^2 z(a^{-1} - a) = \varpi^{2n}\varpi^r z'$ where $z' \in \mathfrak{o}/\mathfrak{p}^{-2n}$, then after a change of variables, we obtain

$$\sum_{z \in \mathfrak{o}/\mathfrak{p}^{-2n}} \chi_1(b^2 z(a^{-1} - a)) = \sum_{z' \in \mathfrak{o}/\mathfrak{p}^{-2n}} \chi_1(\varpi^{2n}\varpi^r z')$$

$$= \sum_{v \in \mathfrak{p}^{2n+r}/\mathfrak{p}^r} \chi_1(v) = vol(\mathfrak{p}^r)^{-1} \int_{\mathfrak{p}^{2n+r} \cap \mathfrak{o}} \chi_1(v) dv.$$

If $\mathfrak{p}^{2n+r} \supseteq \mathfrak{o}$, then again this integral vanishes. However, if $\mathfrak{p}^{2n+r} \subsetneq \mathfrak{o}$, which is precisely the condition that $-2n < r$ then

$$vol(\mathfrak{p}^r)^{-1} \int_{\mathfrak{p}^{2n+r} \cap \mathfrak{o}} \chi_1(v) dv = vol(\mathfrak{p}^r)^{-1} \int_{\mathfrak{p}^{2n+r}} dv = vol(\mathfrak{p}^r)^{-1} vol(\mathfrak{p}^{2n+r}),$$

since χ_1 is trivial on $\mathfrak{p}$.

The proofs of parts (a) and (b) immediately imply part (c). □

Proposition 5.2 *Let* $x = \begin{pmatrix} 0 & c \\ -c^{-1} & 0 \end{pmatrix}$.

(a) If $val(c) = n \geq 0$ *and* $2n + 1 < r$, *then* $\tau(x) = q^{2n+1}$.
(b) If $val(c) = n < 0$ *and* $-2n - 1 < r$, *then* $\tau(x) = q^{-2n-1}$.
(c) Otherwise, $\tau(x) = 0$.

Proof The proof is completely analogous to that of Proposition 5.1. □

Theorem 5.3 *Let* π *be a simple supercuspidal representation of* $SL(2, F)$.

(a) Let $g = \begin{pmatrix} a & 0 \\ 0 & a^{-1} \end{pmatrix} \in T_1$ *and set* $r = val(a - a^{-1})$. *Let* $W^a(r) := \{w \in W^a : \ell(w) < r\}$. *Then*

$$\theta_\pi(g) = c_q \sum_{w \in W^a(r)} q^{\ell(w)},$$

where

$$c_q := \begin{cases} \frac{q-1}{2} & \text{if } p \neq 2 \\ q - 1 & \text{if } p = 2 \end{cases}$$

(b) If $g \in T \setminus ZT_1$, then $\theta_\pi(g) = 0$.

Proof We first prove (a). Let $\langle -1 + \mathfrak{p} \rangle$ denote the group generated by $\{-1 + z : z \in \mathfrak{p}\}$. Note that

$$|T_0/ZT_1| = |\mathfrak{o}^\times / \langle -1 + \mathfrak{p} \rangle| = \begin{cases} \frac{q-1}{2} & \text{if } p \neq 2 \\ q-1 & \text{if } p = 2 \end{cases}$$

Therefore, by Propositions 5.1 and 5.2, and Proposition 4.1, we have

$$\theta_\pi(g) = c_q(1 + 2q + 2q^2 + \cdots + 2q^{r-1}).$$

It is a straight forward calculation to show that if $x = \begin{pmatrix} b & 0 \\ 0 & b^{-1} \end{pmatrix}$ and $val(b) = k$, then $\ell(x) = |2k|$. Moreover, if $x = \begin{pmatrix} 0 & c \\ -c^{-1} & 0 \end{pmatrix}$ and $val(c) = k$, then $\ell(x) = |2k + 1|$. This concludes the proof of part (a).

We now prove (b). In the proof of Proposition 5.1 (a), we showed that

$$ygy^{-1} \in H \iff \begin{pmatrix} a & 0 \\ b^{-2}z(a - a^{-1}) & a^{-1} \end{pmatrix} \in H.$$

This trivially implies that $a \in \langle -1 + \mathfrak{p} \rangle$. Moreover, the condition $a \in \langle -1 + \mathfrak{p} \rangle$ is implied, for the same reason, when you compute the terms ygy^{-1} that appear in $\theta_\pi(g)$ for any other representative $x \in [W^a]$, as simple computations will show. Therefore, θ_π vanishes on $T \setminus ZT_1$. □

Remark 5.4

1. Up to central character, we have computed θ_π on all of the split torus (since we assumed above that $g \in T_1$). Since the central character is given by the data forming the simple supercuspidal representation, we have computed θ_π on the entire split torus.
2. The set $W^a(r)$ has geometric meaning. If C denotes the fundamental alcove in the standard apartment for $SL(2, F)$, then $W^a(r) \cdot C$ is the symmetric interval about C containing $2r - 1$ alcoves.
3. We note that the term $a - a^{-1}$ in Theorem 5.3 is, up to a sign, a canonical square root of the Weyl denominator. In particular, if D denotes the Weyl denominator, then $-D(g) = (a - a^{-1})^2$.

6 The Character Formula on Elliptic Tori

In this section, we compute θ_π on the elliptic tori of $SL(2, F)$. We first recall some relevant notions.

Let E/F be a quadratic extension. Then we may set $E = F(\sqrt{\Delta})$, with Δ a non-square in F, and we let $\delta = \sqrt{\Delta}$. Let ϖ_E be a uniformizer in E and set $\Gamma = Gal(E/F)$. If $w \in E^\times$, then we write $\overline{w}$ for the image of w under the non-trivial element of Γ. We have the norm map $N_{E/F} : E^\times \to F^\times$ given by $N_{E/F}(w) = w\overline{w}$. The elliptic tori of $SL(2, F)$ are of the form $E^1 := \{w \in E^\times : N_{E/F}(w) = 1\}$.

Lemma 6.1 *If* $w = a_0 + a_1\varpi_E \in E^1$, *then* $a_0, a_1 \in \mathfrak{o}$.

Proof Expanding the left-hand side of the equality $N_{E/F}(w) = 1$ in a power series gives the result. □

We now fix an embedding of the torus E^1 into $SL(2, F)$. Specifically, if we fix the basis $\Delta, 1$ of $F \oplus F$, then we have the embedding

$$E^1 \hookrightarrow SL(2, F)$$

$$a_0 + a_1\delta \mapsto \begin{pmatrix} a_0 & a_1 \\ a_1\Delta & a_0 \end{pmatrix}.$$

We now compute $\theta_\pi(g)$, where $g = a_0 + a_1\delta$, which we identify with the above 2×2 matrix. Recall that $\langle -1 + \mathfrak{p}\rangle$ denotes the group generated by $\{-1 + z : z \in \mathfrak{p}\}$. We set $r := val(a_1)$ and we assume that $a_0 \in 1 + \mathfrak{p}$. As we will see later, in order for $\theta_\pi(g) \neq 0$ to hold, we must have $a_0 \in \langle -1 + \mathfrak{p}\rangle$. To obtain the character formula in the $a_0 \in -1 + \mathfrak{p}$ case, we use the central character (see Remark 6.7).

First, we would like to use some of the ideas from the proof of Proposition 4.1. Recall that

$$\theta_\pi(g) = \sum_{x \in N/ZT_1} \sum_{y \in H\backslash HxH} \dot{\chi}(ygy^{-1}), \tag{2}$$

and that we have defined

$$\tau(x) := \sum_{y \in H\backslash HxH} \dot{\chi}(ygy^{-1}),$$

for $x \in N$. It is not necessarily the case anymore that $\tau(x) = \tau(xt)\ \forall t \in T_0$. If x satisfies $\tau(xt) = \tau(x)\ \forall t \in T_0$, then we can and will collect the $\tau(xt)$ terms together so that in the character formula we pick up a factor of $|T_0/ZT_1| = c_q$, as we did in Sect. 5. In the cases where $x \in N$ does not satisfy $\tau(x) = \tau(xt)$ for all $t \in T_0$, we need to sum over all of these elements xt, which are parameterized by $T_0/ZT_1 \cong \mathbb{F}_q/\langle -1\rangle$. These summations appear in Theorem 6.6. We have

Proposition 6.2 *Let* $x = \begin{pmatrix} b & 0 \\ 0 & b^{-1} \end{pmatrix}$. *Suppose first that* E/F *is ramified.*

(a) Let $val(b) = n > 0$.

(i) If $2n < r$, *then* $\tau(xt) = \tau(x) = q^{2n}\ \forall t \in T_0$.

(ii) *If* $2n = r$, *then* $\tau(x) = q^{2n}\chi_2(b^{-2}a_1\Delta)$.

(b) *Let* $val(b) = n < 0$.

(i) *If* $-2n < r$, *then* $\tau(xt) = \tau(x) = q^{-2n}\ \forall t \in T_0$.
(ii) *If* $-2n = r$, *then* $\tau(x) = q^{-2n}\chi_1(b^2a_1)$.

(c) *If* $val(b) = 0$, *then* $\tau(x) = \chi_1(b^2a_1)\chi_2(b^{-2}a_1\Delta)$.
(d) *If* x *does not satisfy the conditions of (a)-(i), (a)-(ii), (b)-(i), (b)-(ii), or (c), then* $\tau(x) = 0$.

Suppose now that E/F *is unramified.*

(a') *Let* $val(b) = n \geq 0$.

(i) *If* $2n + 1 < r$, *then* $\tau(xt) = \tau(x) = q^{2n}\ \forall t \in T_0$.
(ii) *If* $2n + 1 = r$, *then* $\tau(x) = q^{2n}\chi_2(b^{-2}a_1\Delta)$.

(b') *Let* $val(b) = n < 0$.

(i) *If* $-2n < r$, *then* $\tau(xt) = \tau(x) = q^{-2n}\ \forall t \in T_0$.
(ii) *If* $-2n = r$, *then* $\tau(x) = \sum_{z\in\mathfrak{o}/\mathfrak{p}^{-2n}} \chi_1(b^2a_1(1 - z^2\Delta))$.

(c') *If* x *does not satisfy the conditions of (a')-(i), (a')-(ii), (b')-(i), or (b')-(ii), then* $\tau(x) = 0$.

Proof We first prove (a), (c), and (a').

Let $val(b) = n \geq 0$. Recall from Proposition 5.1 the decomposition

$$HxH = \coprod_{z\in\mathfrak{p}/\mathfrak{p}^{2n+1}} Hx \cdot \begin{pmatrix} 1 & 0 \\ z & 1 \end{pmatrix}.$$

Suppose that $y \in HxH$. Write $y = hxu$, where u is of the form $\begin{pmatrix} 1 & 0 \\ z & 1 \end{pmatrix}$ for some $z \in \mathfrak{p}/\mathfrak{p}^{2n+1}$ and for some $h \in H$. Then $ygy^{-1} \in H \Leftrightarrow xugu^{-1}x^{-1} \in H$. Moreover,

$$xugu^{-1}x^{-1} = \begin{pmatrix} a_0 - za_1 & b^2a_1 \\ b^{-2}a_1(\Delta - z^2) & a_0 + za_1 \end{pmatrix}.$$

Suppose first that E/F is ramified, and recall that $\langle -1 + \mathfrak{p}\rangle$ was defined to be the group generated by $\{-1 + z : z \in \mathfrak{p}\}$. Then $xugu^{-1}x^{-1} \in H$ if and only if $a_0 \in \langle -1 + \mathfrak{p}\rangle$ and $b^{-2}a_1 \in \mathfrak{o}$, in other words, if and only if $a_0 \in \langle -1 + \mathfrak{p}\rangle$ and $2n \leq r$. Note that we used here that since $z \in \mathfrak{p}$, then the diagonal entries are indeed in $\langle -1 + \mathfrak{p}\rangle$. Now, we have $\chi(xugu^{-1}x^{-1}) = \chi_1(b^2a_1)\chi_2(b^{-2}a_1\Delta) = \chi_2(b^{-2}a_1\Delta)$ since $val(b) > 0$, $val(z) > 0$, and $val(a_1) \geq 0$ (see Lemma 6.1). We see now that if $2n < r$, then $\chi_2(b^{-2}a_1\Delta) = 1$ so that

$$\tau(x) = \sum_{y \in H \backslash HxH} \dot{\chi}(ygy^{-1}) = \sum_{z \in \mathfrak{p}/\mathfrak{p}^{2n+1}} \chi_2(b^{-2}a_1\Delta) = vol(\mathfrak{p}/\mathfrak{p}^{2n+1}) = q^{2n}.$$

If $2n = r$, then

$$\tau(x) = \sum_{y \in H \backslash HxH} \dot{\chi}(ygy^{-1}) = \sum_{z \in \mathfrak{p}/\mathfrak{p}^{2n+1}} \chi_2(b^{-2}a_1\Delta) = q^{2n}\chi_2(b^{-2}a_1\Delta).$$

Suppose now that E/F is unramified. Then $xugu^{-1}x^{-1} \in H$ if and only if $a_0 \in \langle -1 + \mathfrak{p} \rangle$ and $val(b^{-2}a_1) \geq 1$ (i.e., $r \geq 2n+1$) since $val(\Delta) = 0$. Indeed, we are also again using that $z \in \mathfrak{p}$, so that the diagonal entries are still in $\langle -1 + \mathfrak{p} \rangle$. Now, we have $\chi(xugu^{-1}x^{-1}) = \chi_2(b^{-2}a_1\Delta)$. If $r > 2n+1$, then $\chi_2(b^{-2}a_1\Delta) = 1$ so that $\tau(x) = q^{2n}$ as above. Otherwise, if $2n+1 = r$,

$$\tau(x) = \sum_{y \in H \backslash HxH} \dot{\chi}(ygy^{-1}) = \sum_{z \in \mathfrak{p}/\mathfrak{p}^{2n+1}} \chi_2(b^{-2}a_1\Delta) = q^{2n}\chi_2(b^{-2}a_1\Delta).$$

Let us now prove (b) and (b'). Suppose $val(b) = n < 0$. Then recall from Proposition 5.1, we have

$$HxH = \coprod_{z \in \mathfrak{o}/\mathfrak{p}^{-2n}} Hx \cdot \begin{pmatrix} 1 & z \\ 0 & 1 \end{pmatrix}.$$

Suppose that $y \in HxH$. Then we may write $y = hxu$, where u is of the form $\begin{pmatrix} 1 & z \\ 0 & 1 \end{pmatrix}$ for some $z \in \mathfrak{o}/\mathfrak{p}^{-2n}$, and for some $h \in H$. A computation yields $xugu^{-1}x^{-1} = \begin{pmatrix} a_0 + za_1\Delta & b^2a_1(1 - z^2\Delta) \\ b^{-2}a_1\Delta & a_0 - a_1z\Delta \end{pmatrix}$.

Suppose first that E/F is ramified. A similar analysis as in case (a) shows that $a_0 \in \langle -1 + \mathfrak{p} \rangle$ and $val(b^2a_1) \geq 0$, so that $-2n \leq r$. Note that we have used here that $\Delta \in \mathfrak{p}$, so that the diagonal entries are indeed in $\langle -1 + \mathfrak{p} \rangle$. Then, similarly as in case (a), if $-2n < r$, we have that $\tau(x) = q^{-2n}$. If $-2n = r$, then

$$\tau(x) = \sum_{y \in H \backslash HxH} \dot{\chi}(ygy^{-1}) = \sum_{z \in \mathfrak{o}/\mathfrak{p}^{-2n}} \chi_1(b^2a_1) = q^{-2n}\chi_1(b^2a_1).$$

Now suppose that E/F is unramified. Then we must have $a_0 \in \langle -1 + \mathfrak{p} \rangle$ and $val(b^2a_1(1 - z^2\Delta)) \geq 0$ (which implies in particular that we must have $val(a_1) \geq 1$), so that in particular we have $r \geq -2n$. Note here that $\Delta \in \mathfrak{o}$ and $z \in \mathfrak{o}$. But as just remarked, we must have $val(a_1) \geq 1$. Thus, the diagonal entries are indeed still in $\langle -1 + \mathfrak{p} \rangle$. Then, similarly as in case (a), if $r > -2n$, we have that $\tau(x) = q^{-2n}$. If $r = -2n$, then

$$\tau(x) = \sum_{y \in H \backslash HxH} \dot{\chi}(ygy^{-1}) = \sum_{z \in \mathfrak{o}/\mathfrak{p}^{-2n}} \chi_1(b^2a_1(1 - z^2\Delta)).$$

□

Remark 6.3

1. We note that the $val(b) > 0$ and $val(b) = 0$ cases needed to be separated for the E/F ramified case because of the b^2a_1 term. Specifically, if $val(a_1) = 0$, then we get a nontrivial contribution to $\tau(x)$ from b^2a_1. However, if $val(a_1) \geq 1$, then the two cases $val(b) > 0$ and $val(b) = 0$ can be combined to give $\tau(x) = q^{2n}$ in both of these cases. If E/F is unramified, however, this phenomenon does not occur. Namely, the b^2a_1 term does not contribute in the $val(b) = 0$ case. The reason is that the $b^{-2}a_1(\Delta - z^2)$ term forces $val(a_1) \geq 1$ if we assume that $val(b) = 0$. Therefore, we may combine the $val(b) > 0$ and $val(b) = 0$ cases into one case, as we have done.
2. In Proposition 6.2, it is a simple matter to compute $\tau(xt)$ in the cases where $\tau(xt) \neq \tau(x)$. For example, let $\gamma = \begin{pmatrix} s & 0 \\ 0 & s^{-1} \end{pmatrix}$ with $s \in \mathfrak{o}^\times$ so that $\gamma \in T_0$. In (a), we computed $\tau(x)$ when $2n = r$. Since $x\gamma$ falls into the same $2n = r$ case, we conclude that $\tau(x\gamma)$ has also been computed. In particular, we have that $\tau(x\gamma) = q^{2n}\chi_2((bs)^{-2}a_1\Delta)$.

Proposition 6.4 *Let* $x = \begin{pmatrix} 0 & c \\ -c^{-1} & 0 \end{pmatrix}$. *Suppose first that* E/F *is ramified.*

(a) Let $val(c) = n \geq 0$.

(i) if $2n + 1 < r$, *then* $\tau(xt) = \tau(x) = q^{2n+1}\ \forall t \in T_0$.
(ii) if $2n + 1 = r$, *then* $\tau(x) = q^{2n+1}\chi_2(-c^{-2}a_1)$.

(b) Let $val(c) = n < 0$.

(i) if $-2n - 1 < r$, *then* $\tau(xt) = \tau(x) = q^{-2n-1}\ \forall t \in T_0$.
(ii) if $-2n - 1 = r$, *then* $\tau(x) = q^{-2n-1}\chi_1(-c^2a_1\Delta)$.

(c) If x does not satisfy the conditions of (a)-(i), (a)-(ii), (b)-(i), or (b)-(ii), then $\tau(x) = 0$.

Suppose now that E/F *is unramified. Then*

(a') Let $val(c) = n \geq 0$.

(i) if $2n + 1 < r$, *then* $\tau(xt) = \tau(x) = q^{2n+1}\ \forall t \in T_0$.
(ii) if $2n + 1 = r$, *then* $\tau(x) = \sum_{z \in \mathfrak{o}/\mathfrak{p}^{2n+1}} \chi_2(-c^{-2}a_1(1 - z^2\Delta))$.

(b') Let $val(c) = n < 0$.

(i) if $-2n < r$, *then* $\tau(xt) = \tau(x) = q^{-2n-1}\ \forall t \in T_0$.
(ii) if $-2n = r$, *then* $\tau(x) = q^{-2n-1}\chi_1(-c^2a_1\Delta)$.

(c') If x does not satisfy the conditions of (a')-(i), (a')-(ii), (b')-(i), or (b')-(ii), then $\tau(x) = 0$.

Proof The proof is analogous to that of Proposition 6.2. □

Remark 6.5

1. In the case that E/F is unramified and $val(c) \geq 0$, one can compute that $val(a_1) \geq 1$ is forced for the character to not vanish. Thus, there is no need to distinguish between $val(c) > 0$ and $val(c) = 0$ cases, as we did in Remark 6.3.
2. The same remark as in Remark 6.3 can be made about elements $x\gamma$ where $\tau(x\gamma) \neq \tau(x)$.

We now present the character formula on elliptic tori. First recall the set $W^a(r) := \{w \in W^a : \ell(w) < r\}$ defined in Sect. 5. Finally, we note that if $g = a_0 + a_1\delta$ and $a_0 \notin \langle -1 + \mathfrak{p} \rangle$, then $\theta_\pi(g) = 0$ (this can be seen from the proofs of Propositions 6.2 and 6.4).

Theorem 6.6 *Let π be a simple supercuspidal representation of $SL(2, F)$ defined by the affine generic character of Sect. 5. Let $g = \begin{pmatrix} a_0 & a_1 \\ a_1\Delta & a_0 \end{pmatrix} \in E^1$, set $r = val(a_1)$, and assume that $a_0 \in 1 + \mathfrak{p}$. Let*

$$c_q := \begin{cases} \frac{q-1}{2} & \text{if } p \neq 2 \\ q-1 & \text{if } p = 2 \end{cases}$$

(a) Suppose that E/F is ramified.

(i) Suppose that r is even and $r > 0$. Then

$$\theta_\pi(g) = c_q \sum_{w \in W^a(r)} q^{\ell(w)} + q^r \sum_{s \in \mathbb{F}_q^\times/\langle -1 \rangle} (\chi_2(\varpi^{-r}s^2 a_1 \Delta) + \chi_1(\varpi^{-r}s^2 a_1)).$$

(ii) Suppose that r is odd. Then

$$\theta_\pi(g) = c_q \sum_{w \in W^a(r)} q^{\ell(w)} + q^r \sum_{s \in \mathbb{F}_q^\times/\langle -1 \rangle} (\chi_2(-\varpi^{-r+1}s^2 a_1) + \chi_1(-\varpi^{-r-1}s^2 a_1 \Delta)).$$

(iii) Suppose that $r = 0$. Then

$$\theta_\pi(g) = \sum_{s \in \mathbb{F}_q^\times/\langle -1 \rangle} \chi_1(s^2 a_1)\chi_2(s^{-2} a_1 \Delta).$$

(b) Suppose that E/F is unramified.

(i) Suppose that r is even and $r > 0$. Then

$$\theta_\pi(g) = c_q \left(1 + 2q + 2q^2 + \cdots + 2q^{r-2} + q^{r-1}\right) + \sum_{s \in \mathbb{F}_q^\times/\langle -1 \rangle} \sum_{z \in \mathfrak{o}/\mathfrak{p}^r} \chi_1(\varpi^{-r}s^2 a_1(1 - z^2\Delta)) + q^{r-1} \sum_{s \in \mathbb{F}_q^\times/\langle -1 \rangle} \chi_1(-\varpi^{-r}s^2 a_1 \Delta).$$

(ii) *Suppose that r is odd. Then*

$$\theta_\pi(g) = c_q\left(1 + 2q + 2q^2 + \cdots + 2q^{r-2} + q^{r-1}\right) + \sum_{s\in\mathbb{F}_q^\times/\langle -1\rangle}\sum_{z\in\mathfrak{o}/\mathfrak{p}^r}\chi_2(-\varpi^{-r+1}s^2a_1(1-z^2\Delta)) + q^{r-1}\sum_{s\in\mathbb{F}_q^\times/\langle -1\rangle}\chi_2(\varpi^{-r+1}s^2a_1\Delta).$$

(iii) *Suppose that $r = 0$. Then $\theta_\pi(g) = 0$.*

Remark 6.7

1. Up to central character, we have computed θ_π on all of the elliptic torus (since we assumed above that $a_0 \in 1 + \mathfrak{p}$). Since the central character is given by the data forming the simple supercuspidal representation, we have computed θ_π on the entire elliptic torus.
2. The set $W^a(r)$ again arises naturally in the computation of the character on elliptic tori. The geometric meaning of this set was discussed in Remark 5.4.
3. The term $\left(1 + 2q + 2q^2 + \cdots + 2q^{r-2} + q^{r-1}\right)$ is precisely $\left(\sum_{w\in W^a(r)} q^{\ell(w)}\right) - q^{r-1}$.

Acknowledgements It is a great pleasure to have been able to discuss this work with Paul Sally, Jr. This paper has benefited from conversations with Paul Sally, Jr., Gordan Savin, and Loren Spice.

References

1. J. Adler, L. Spice, Supercuspidal characters of reductive, p-adic groups. Amer. J. Math. **131**(4), 1137–1210 (2009)
2. S. DeBacker, M. Reeder, Depth-zero supercuspidal L-packets and their stability. Ann. Math. **169**(3), 795–901 (2009)
3. B. Gross, M. Reeder, Arithmetic invariants of discrete Langlands parameters. Duke Math. J. **154**, 431–508 (2010)
4. T. Kaletha, Epipelagic L-packets and rectifying characters. Inventiones Mathematicae **202**(1), 1–89 (2015)
5. M. Reeder, Supercuspidal L-packets of positive depth and twisted coxeter elements. J. Reine Angew. Math. **620**, 1–33 (2008)
6. P.J. Sally, Jr., Some remarks on discrete series characters for reductive p-adic groups, in *Representations of Lie groups, Kyoto, Hiroshima, 1986*, (91g:22026) (1988), pp. 337–348
7. P. Sally, J. Shalika, Characters of the discrete series of representations of $SL(2)$ over a local field. Proc. Nat. Acad. Sci. U.S.A. **61**, 1231–1237 (1968)

Classification of Strongly Positive Representations of Even General Unitary Groups

Yeansu Kim and Ivan Matić

Contents

1 Introduction ... 162
2 Notation and Preliminaries ... 162
  2.1 Notation ... 162
3 The Tadić's Structure Formula: General Unitary Groups ... 163
4 Classification of Strongly Positive Representations of Even General Unitary Groups ... 166
  4.1 Construction of the Map ... 167
  4.2 Classification of Strongly Positive Representations: $D(\rho; \sigma_{cusp})$ ... 167
  4.3 Classification of Strongly Positive Representations ... 169
5 Appendix: Even Unitary Case ... 171
  5.1 Notation for Even Unitary Groups ... 172
  5.2 Tadić's Structure Formula for Unitary Groups ... 173
  5.3 Strongly Positive Representations for Even Unitary Groups ... 173
References ... 174

Abstract We explicitly identify the structure of Jacquet modules of parabolically induced representations of even unitary groups and even general unitary groups over a p-adic field F of characteristic different than two. As an application, we obtain a classification of the strongly positive discrete series representations of those groups.

Keywords Tadić's structure formula · Jacquet modules · Strongly positive representations

Mathematics Subject Classification (2000) 20C11 · 11F70

The first author was supported by a National Research Foundation of Korea (NRF) grant funded by the Korean government (MSIP) (No. $2017R1C1B2010081$). The second author was supported by the Croatian Science Foundation under project 9364.

Y. Kim
Department of Mathematics Education, Chonnam National University, Yongbong-ro 77, Gwangju City, Korea
e-mail: ykim@jnu.ac.kr

I. Matić
Department of Mathematics, University of Osijek, Trg Ljudevita Gaja 6, Osijek, Croatia
e-mail: imatic@mathos.hr

A.-M. Aubert et al. (eds.), *Representations of Reductive p-adic Groups*,
Progress in Mathematics 328, https://doi.org/10.1007/978-981-13-6628-4_5

1 Introduction

The first purpose of this paper is to explicitly construct Tadić's structure formula for even unitary groups and even general unitary groups. The structure formula explores the Jacquet modules of parabolically induced representations. In the case of general linear groups, the Jacquet modules of parabolically induced representations are studied in [2, 17]. The case of classical groups is very different due to the different Weyl groups and how they act on Levi subgroups. In [14], Tadić explicitly describes the structure of Jacquet modules in the cases of Sp_{2n}, GSp_{2n}, and SO_{2n+1}, later generalized to the cases of SO_{2n}, metaplectic groups, and $GSpin$ groups in [1, 3–6]. Using Tadić's structure formula, one can determine all Jacquet modules of certain classes of representations [9, 10].

The Tadić's structure formula also happens to be extremely useful for the study of reducibility and composition series of certain induced representations which happen to be important for understanding of the unitary dual, such as standard representations and generalized principal series [15].

As an application of Tadić's structure formula, the second purpose of this paper is to obtain a classification of strongly positive representations of even unitary groups and even general unitary groups. We note that the strongly positive representations serve as basic building blocks in the classification of the discrete series of classical groups, including unitary ones, obtained in [11, 12], and in the classification of discrete series representations of odd $GSpin$ groups, recently provided in [7].

This paper is organized as follows. In Sect. 2, we outline standard notation. In Sect. 3, we obtain Tadić's structure formula for even general unitary groups, which describes the explicit structure of the Jacquet modules of parabolically induced representations of general unitary groups. In Sect. 4, we obtain a classification of strongly positive representations of even general unitary groups. In the appendix, we also discuss the even unitary group case.

2 Notation and Preliminaries

2.1 Notation

Let F be a non-Archimedean local field of characteristic different than two and let E/F be a quadratic extension. Let $\Gamma = Gal(E/F)$ and write $x \to \bar{x}$ for its non-trivial element. Choose an element $\beta \in E$ such that $E = F(\beta)$ and $\bar{\beta} = -\beta$. To define the unitary groups, we set

$$J_n = \begin{pmatrix} & \beta I_n \\ -\beta I_n & \end{pmatrix}.$$

We let $\mathbf{H}_n = \mathbf{GU}(n, n)$ be the quasi-split general unitary group in $2n$ variables defined with respect to E/F and J_n. Its F-points are

$$\mathbf{H}_n(F) = \{g \in GL_{2n}(E) | {}^t\bar{g}J_n g = \lambda J_n, \lambda \in E^\times\}$$

We fix λ throughout the paper.

Let $\mathbf{s}$ and $\mathbf{M_s}$ be as in Remark 3.1. For a parabolic subgroup $\mathbf{P_s} = \mathbf{M_s}\mathbf{N_s}$ of $\mathbf{H}_n$, we denote the induced representation $\mathrm{Ind}_{\mathbf{P_s}}^{\mathbf{H}_n}(\rho_1 \otimes \cdots \otimes \rho_k \otimes \tau)$ by

$$\rho_1 \times \cdots \times \rho_k \rtimes \tau$$

where each ρ_i (resp. τ) is a representation of some $\mathbf{GL}_{n_i}(E)$ (resp. $\mathbf{H}_{n-n'}(F)$). In particular, $\mathrm{Ind}_{\mathbf{P_s}}^{\mathbf{H}_n}$ is a functor from admissible representations of $\mathbf{M_s}(F)$ to admissible representations of $\mathbf{H}_n(F)$ that sends unitary representations to unitary representations. We also denote the normalized Jacquet module with respect to $\mathbf{P_s}$ by $r_{\mathbf{s}}(\tau)$. In particular, $r_{\mathbf{s}}$ is a functor from admissible representations of $\mathbf{H}_n(F)$ to admissible representations of $\mathbf{M_s}(F)$.

The Grothendieck group of the category of all admissible representations of finite length of $\mathbf{H}_n(F)$, i.e. a free abelian group over the set of all irreducible representations of $\mathbf{H}_n(F)$ (resp. $\mathbf{GL}_n(E)$) is denoted by $R_{GU}(n)$ (resp. $R_{GL}(n)$) and set $R_{GU} = \bigoplus_{n \geq 0} R_{GU}(n)$, $R_{GL} = \bigoplus_{n \geq 0} R_{GL}(n)$.

In the case of $\mathbf{GL}$, we denote the induced representation $\mathrm{Ind}_{\mathbf{P'}}^{\mathbf{GL}_n}(\rho_1 \otimes \cdots \otimes \rho_k)$ by

$$\rho_1 \times \cdots \times \rho_k$$

such that $\mathbf{P'} = \mathbf{M'}\mathbf{N'}$ is the standard parabolic subgroup of $\mathbf{GL}_n$ where $\mathbf{M'} \cong \mathbf{GL}_{n_1} \times \mathbf{GL}_{n_2} \times \cdots \times \mathbf{GL}_{n_k}$ and each ρ_i is a representation of $\mathbf{GL}_{n_i}(E)$ for $i = 1, \ldots, k$. We also follow the notation in [2]. Let ρ be an irreducible unitary cuspidal representation of some $\mathbf{GL}_p(E)$. We define the segment, $\Delta := [\nu^a\rho, \nu^{a+k}\rho] = \{\nu^a\rho, \nu^{a+1}\rho, \ldots \nu^{a+k}\rho\}$ where $a \in \mathbb{R}$ and $k \in \mathbb{Z}_{\geq 0}$. If $a > 0$, we call the segment Δ strongly positive.

3 The Tadić's Structure Formula: General Unitary Groups

We fix the F-Borel subgroup $\mathbf{B}$ of upper triangular matrices in $\mathbf{H}_n$. Then, $\mathbf{B} = \mathbf{TU}$, where $\mathbf{T}$ is a maximal torus of diagonal elements in $\mathbf{H}_n$ and let $\mathbf{A}_0$ be the maximal F-split subtorus of $\mathbf{T}$. Then,

$$\mathbf{T}(F) = \left\{ \begin{pmatrix} x_1 & & & & & \\ & \ddots & & & & \\ & & x_n & & & \\ & & & \lambda\bar{x}_1^{-1} & & \\ & & & & \ddots & \\ & & & & & \lambda\bar{x}_n^{-1} \end{pmatrix} | x_i \in E^\times, \lambda \in F^\times \right\}$$

and

$$\mathbf{A}_0(F) = \left\{ \begin{pmatrix} x_1 & & & & & \\ & \ddots & & & & \\ & & x_n & & & \\ & & & \lambda x_1^{-1} & & \\ & & & & \ddots & \\ & & & & & \lambda x_n^{-1} \end{pmatrix} | x_i, \lambda \in F^{\times} \right\}$$

For simplicity, we let $a(x_1, \ldots, x_n; \lambda)$ be an element of the form $diag(x_1, \ldots, x_n, \lambda x_1^{-1}, \ldots, \lambda x_n^{-1})$ in $\mathbf{A}_0(F)$. Let $\Phi(\mathbf{H}_n(F), \mathbf{A}_0(F))$ be the restricted roots of $\mathbf{H}_n(F)$ with respect to $\mathbf{A}_0(F)$ and let $\Delta := \{\alpha_i\}_{i=1}^n$ be the set of simple roots, where $\alpha_i = e_i - e_{i+1}, 1 \leq i \leq n-1,\ \alpha_n = 2e_n - e_0$.

The Weyl group $W(\mathbf{H}_n(F)/\mathbf{A}_0(F))$ is isomorphic to $S_n \rtimes \{\pm 1\}^n$, where S_n is the permutation group of n letters. More precisely, for $(ij) \in S_n$,

$$(ij) \cdot a(x_1, \ldots, x_n; \lambda) = a(x_1, \ldots, x_{i-1}, x_j, x_{i+1}, \ldots, x_{j-1}, x_i, x_{j+1}, \ldots, x_n; \lambda)$$

and for $\epsilon = (\epsilon_1, \ldots, \epsilon_n) \in \{\pm 1\}^n$,

$$\epsilon_i \cdot a(x_1, \ldots, x_n; \lambda) = a(x_1, \ldots, x_{i-1}, \lambda x_i^{\epsilon_i}, x_{i+1}, \ldots, x_n; \lambda).$$

Remark 3.1 Let $\mathbf{s} = (n_1, n_2, \ldots, n_k)$ be an ordered partition of some n' such that $n' \leq n$ and let $\Theta = \Delta \backslash \{\alpha_{n_1}, \alpha_{n_1+n_2}, \ldots, \alpha_{n_1+\ldots+n_k}\}$. Let $\mathbf{A_s}$ be the subtorus of $\mathbf{A}_0$ that corresponds to Θ and let $\mathbf{M_s}$ be the centralizer of $\mathbf{A_s}$. Then, its F-points is of the form

$$\mathbf{M_s}(F) = \left\{ \begin{pmatrix} g_1 & & & & & & \\ & \ddots & & & & & \\ & & g_k & & & & \\ & & & g & & & \\ & & & & \lambda^t \bar{g}_1^{-1} & & \\ & & & & & \ddots & \\ & & & & & & \lambda^t \bar{g}_n^{-1} \end{pmatrix} | g_i \in \mathbf{GL}_{n_i}(E), g \in \mathbf{H}_{n-n'}(F), \lambda \in F^{\times} \right\}.$$

Therefore, $\mathbf{M_s}(F) \cong \mathbf{GL}_{n_1}(E) \times \mathbf{GL}_{n_2}(E) \times \cdots \times \mathbf{GL}_{n_k}(E) \times \mathbf{H}_{n-n'}(F)$ and for simplicity, the element $diag(g_1, g_2, \ldots, g_k, g, \lambda^t \bar{g}_1^{-1}, \lambda^t \bar{g}_2^{-1}, \ldots, \lambda^t \bar{g}_k^{-1})$ in $\mathbf{M_s}(F)$ is denoted by $(g_1, g_2, \ldots, g_k, g)$.

Then, for an element $(g_1, g_2, \ldots, g_k, g) \in \mathbf{M_s}(F)$, the Weyl group $W(\mathbf{H}_n(F)/\mathbf{A_s}(F))$ is a subgroup of $S_k \rtimes \{\pm 1\}^k$. In particular, for $(ij) \in W(\mathbf{H}_n(F)/\mathbf{A_s}(F)) \subset S_k$,

$$(ij) \cdot (g_1, g_2, \ldots, g_k, g) = (g_1, \ldots, g_{i-1}, g_j, g_{i+1}, \ldots, g_{j-1}, g_i, g_{j+1}, \ldots, g_k, g),$$

and for $\epsilon = (\epsilon_1, \ldots, \epsilon_k) \in \{\pm 1\}^k \subset W(\mathbf{H}_n(F)/\mathbf{A}_s(F))$ with $\epsilon_i = -1, \epsilon_k = 1$ for $k \neq i$,

$$\epsilon \cdot (g_1, \ldots, g_i, \ldots, g_k, g) = (g_1, \ldots, \lambda^t \bar{g}_i^{-1}, \ldots, g_k, g). \tag{3.1}$$

Therefore, the Weyl group action on the maximal F-split torus $\mathbf{A}_0(F)$ and Levi subgroup $\mathbf{M}_s(F)$ of $\mathbf{H}_n$ is similar to that for general symplectic groups (Note that the main difference is the Weyl group action on the Levi subgroup (3.1)). In [14, Sect. 4], Tadić characterizes the representative element of the set $[W_{\Delta\backslash\alpha}\backslash W/W_{\Delta\backslash\beta}]$ and its explicit action on the simple roots for $\mathbf{GSp}_{2n}$. We also get the same results, i.e. from Lemmas 4.1 through 4.8 of [14] in the case of even general unitary groups, since those lemmas only depend on the simple roots, Weyl group and its action on the simple roots and we also know that simple roots for $\mathbf{H}_n$ is same as those for $\mathbf{GSp}_{2n}$.

We now explain the Tadić's structure formula for $\mathbf{H}_n$ and follow the notation in [14] for simplicity. Let i_1, i_2 be integers which satisfy $1 \leq i_1, i_2 \leq n$. Take an integer d such that $0 \leq d \leq \min\{i_1, i_2\}$. Suppose that an integer k satisfies $\max\{0, (i_1 + i_2 - n) - d\} \leq k \leq \min\{i_1, i_2\} - d$. Let $p_n(d, k)_{i_1,i_2} \in S_n$ be defined by

$$p_n(d,k)_{i_1,i_2}(j) = \begin{cases} j & \text{for } 1 \leq j \leq k; \\ j + i_1 - k & \text{for } k+1 \leq j \leq i_2 - d; \\ (i_1 + i_2 - d + 1) - j & \text{for } i_2 - d + 1 \leq j \leq i_2; \\ j - i_2 + k & \text{for } i_2 + 1 \leq j \leq i_1 + i_2 - d - k; \\ j & \text{for } i_1 + i_2 - d - k + 1 \leq j \leq n. \end{cases}$$

Let $q_n(d,k)_{i_1,i_2}$ be $(p_n(d,k)_{i_1,i_2}, (\mathbf{1}_{i_2-d}, -\mathbf{1}_d, \mathbf{1}_{n-i_2}))$ where $\mathbf{1}_i = 1, \ldots, 1$ (1 appears i times). Let $w = q_n(d,k)_{i_1,i_2}$. Then, for $(g_1, g_2, g_3, g_4, h) \in \mathbf{GL}_k(E) \times \mathbf{GL}_{i_2-d-k}(E) \times \mathbf{GL}_d(E) \times \mathbf{GL}_{i_1-d-k}(E) \times \mathbf{H}_{n-i_1-i_2+d+k}(F)$, we have $w \cdot (g_1, g_2, g_3, g_4, h) = (g_1, g_4, \lambda\, {}^t\bar{g}_3^{-1}, g_2, h)$.

Let π_i be an irreducible smooth representation of $\mathbf{GL}_{n_i}(E)$ for $i = 1, 2, 3, 4$ and let σ be an irreducible smooth representation of $\mathbf{H}_m$. We have,

$$w^{-1} \cdot (\pi_1 \otimes \pi_2 \otimes \pi_3 \otimes \pi_4 \otimes \sigma) = \pi_1 \otimes \pi_4 \otimes \check{\pi}_3 \otimes \pi_2 \otimes \omega_{\pi_3}\sigma. \tag{3.2}$$

where $\check{\pi}(g) := \pi({}^t\bar{g}^{-1})$.

Set

$$(\pi_1 \otimes \pi_2 \otimes \pi_3) \tilde{\rtimes} (\pi_4 \otimes \sigma) = \check{\pi}_1 \times \pi_2 \times \pi_4 \otimes \pi_3 \rtimes \omega_{\pi_1}\sigma. \tag{3.3}$$

Applying (3.2) and (3.3), we get

Theorem 3.1 (Tadić's structure formula for general unitary groups.)
For $\pi \in R_{GL}(i)$ and $\sigma \in R_{GU}(n-i)$, the following structure formula holds

$$\mu^*(\pi \rtimes \sigma) = \mathfrak{M}^*(\pi) \tilde{\rtimes} \mu^*(\sigma).$$

Lemma 3.2 *Let ρ be an irreducible cuspidal representation of $\mathbf{GL}_k(E)$ and $a, b \in \mathbb{R}$ be such that $b - a \in \mathbb{Z}_{\geq 0}$. Let σ be an admissible representation of finite length* *of* $\mathbf{H}_n(F)$*. Write* $\mu^*(\sigma) = \sum_{\pi',\sigma'} \pi' \otimes \sigma'$*. Then* $\mathfrak{M}^*(\delta([\nu^a\rho, \nu^b\rho])) = \sum_{i=a-1}^{b}\sum_{j=i}^{b} \delta([\nu^a\rho, \nu^i\rho]) \otimes \delta([\nu^{j+1}\rho, \nu^b\rho]) \otimes \delta([\nu^{i+1}\rho, \nu^j\rho])$ *and* $\mu^*(\delta([\nu^a\rho, \nu^b\rho]) \rtimes \sigma) = \sum_{i=a-1}^{b}\sum_{j=i}^{b} \sum_{\pi',\sigma'} \delta([\nu^{-i}\ {}^t\bar{\rho}^{-1}, \nu^{-a}\ {}^t\bar{\rho}^{-1}]) \times \delta([\nu^{j+1}\rho, \nu^b\rho]) \times \pi' \otimes \delta([\nu^{i+1}\rho, \nu^j\rho]) \rtimes \omega\sigma'$*. We omit* $\delta([\nu^x\rho, \nu^y\rho])$ *if* $x > y$*.*

We recall the definition of the strongly positive representations of $GSpin$ groups.

Definition 3.3 An irreducible representation σ of $\mathbf{H}_n(F)$ is called strongly positive if for every embedding

$$\sigma \hookrightarrow \nu^{s_1}\rho_1 \times \nu^{s_2}\rho_2 \times \cdots \times \nu^{s_k}\rho_k \rtimes \sigma_{cusp}$$

where $\rho_i, i = 1, 2, \ldots, k$ are irreducible unitary cuspidal representations of GL, σ_{cusp} is an irreducible cuspidal representation of $\mathbf{H}_{n'}(F)$ and $s_i \in \mathbb{R}, i = 1, 2, \ldots, k$, then we have $s_i > 0$ for each i.

The following lemma is also useful when we explicitly calculate Jacquet modules:

Lemma 3.4 *Let* ρ *be a cuspidal representation of* $\mathbf{GL}_k(E)$ *and let* σ_{cusp} *be a cuspidal representation of* $\mathbf{H}_n(F)$*. Write* $\rho = \nu^{e(\rho)}\rho^u$*, where* $e(\rho) \in \mathbb{R}$ *and* ρ^u *is a unitary cuspidal representation. If* $\rho \rtimes \sigma_{cusp}$ *has a strongly positive discrete series subrepresentation, then we have*

(i) $\widetilde{\rho^u} \cong \bar{\rho^u}$*, i.e.* ρ^u *is conjugate self-dual.*
(ii) $\omega_\rho\sigma_{cusp} \cong \sigma_{cusp}$*.*

Proof Let σ be a strongly positive subrepresentation of $\nu^{e(\rho)}\rho^u \rtimes \sigma_{cusp}$. Then, $e(\rho) > 0$ since σ is strongly positive. If (i) or (ii) does not hold, then due to Lemma 2.1 in [16], $\nu^\alpha\rho^u \rtimes \sigma_{cusp}$ is irreducible for every α. Then, we have the following embedding:

$$\sigma \hookrightarrow \nu^{e(\rho)}\rho^u \rtimes \sigma_{cusp} \cong \nu^{-e(\rho)}\widetilde{\bar{\rho^u}} \rtimes \sigma_{cusp}.$$

Since $-e(\rho) < 0$, this contradicts the strong positivity of σ.

4 Classification of Strongly Positive Representations of Even General Unitary Groups

In this section, we classify the strongly positive representations of even general unitary groups. We mostly follow the arguments in [8] and appendix to [5] and generalize those to our case.

4.1 Construction of the Map

In this section, we construct the map from the set of strongly positive representations into certain induced representations. We consider the induced representations of the following form

$$\delta(\Delta_1) \times \delta(\Delta_2) \times \cdots \times \delta(\Delta_k) \rtimes \sigma_{cusp} \tag{4.1}$$

where $\Delta_1, \Delta_2, \ldots, \Delta_k$ is a sequence of strongly positive segments (See Notation 2.1 for the definition of strongly positive segments) satisfying $0 < e(\Delta_1) \leq e(\Delta_2) \leq \cdots \leq e(\Delta_k)$ (we allow $k = 0$ here), σ_{cusp} an irreducible cuspidal representation of $\mathbf{H}_m(F)$.

Then, we show that

Theorem 4.1

(i) The induced representation $\delta(\Delta_1) \times \delta(\Delta_2) \times \cdots \times \delta(\Delta_k) \rtimes \sigma_{cusp}$ *of the form (4.1) has a unique irreducible subrepresentation which we denote by* $\delta(\Delta_1, \ldots, \Delta_k; \sigma_{cusp})$.

(ii) The strongly positive representation can be embedded into induced representation of the form (4.1).

Proof (i) and (ii) are GU analogue of Theorem 3.3 and Theorem 3.4 in [8], respectively. Since the idea of their proofs depends on the behaviour of GL parts of Jacquet modules, we apply those in [8] to the case of even general unitary groups and we do not repeat here.

4.2 Classification of Strongly Positive Representations: $D(\rho; \sigma_{cusp})$

Let ρ be a conjugate self-dual irreducible cuspidal representation of $\mathbf{GL}_{n_\rho}(E)$ and σ_{cusp} be an irreducible cuspidal representation of $\mathbf{H}_m(F)$. Let $D(\rho; \sigma_{cusp})$ be the set of strongly positive representations whose cuspidal supports are the representation σ_{cusp} and twists of the representation ρ by positive valued characters. Let $a \geq 0$ be the unique non-negative real number such that $\nu^a \rho \rtimes \sigma_{cusp}$ reduces [13]. Furthermore, we assume that this reducibility point a is in $\frac{1}{2}\mathbb{Z}$ (see (HI) of [12], page 771). Let k_ρ denote $\lceil a \rceil$, the smallest integer which is not smaller than a. In this section, we obtain the classification of strongly positive representations in $D(\rho; \sigma_{cusp})$.

In a previous section, Theorem 4.1 implies that every strongly positive representation can be viewed as the unique irreducible subrepresentation of induced representation of the form (4.1). Therefore, there exists an mapping from the set of strongly positive representations of $\mathbf{H}_n(F)$ into the set of induced representations of the form (4.1).

Now, we further refine the image of this mapping when we restrict the mapping to $D(\rho; \sigma_{cusp})$.

Theorem 4.2 *Let σ_{sp} be an irreducible strongly positive representation in $D(\rho; \sigma_{cusp})$ and consider it as the unique irreducible subrepresentation of induced representation of the form (4.1). Write $\Delta_i = [\nu^{a_i}\rho, \nu^{b_i}\rho]$. Then,*

$$a_i = a - k + i,\ b_1 < \cdots < b_k \text{ and } k \leq \lceil a \rceil.$$

Proof We only consider the Theorem when $a > 0$. We use induction as in [8]. The cases $k = 0$ and $k = 1$ are exactly as in [8] and we skip the proof. We now consider the case when $k = 2$. Now, we have

$$\sigma_{sp} \hookrightarrow \delta([\nu^{a_1}\rho, \nu^{b_1}\rho]) \times \delta([\nu^{a_2}\rho, \nu^{b_2}\rho]) \rtimes \sigma_{cusp}$$

As in the case $k = 1$, we easily show that $a_2 = a$. Since σ_{sp} is the unique irreducible subrepresentation of $\delta([\nu^{a_1}\rho, \nu^{b_1}\rho]) \times \delta([\nu^{a}\rho, \nu^{b_2}\rho]) \rtimes \sigma_{cusp}$ we also have $\sigma_{sp} \hookrightarrow \delta([\nu^{a_1}\rho, \nu^{b_1}\rho]) \rtimes \delta([\nu^{a}\rho, \nu^{b_2}\rho]; \sigma_{cusp})$. This embedding gives us the following embedding

$$\sigma_{sp} \hookrightarrow \delta([\nu^{a_1+1}\rho, \nu^{b_1}\rho]) \times \nu^{a_1}\rho \rtimes \delta([\nu^{a}\rho, \nu^{b_2}\rho]; \sigma_{cusp})$$

If $\nu^{a_1}\rho \rtimes \delta([\nu^{a}\rho, \nu^{b_2}\rho]; \sigma_{cusp})$ is irreducible, we have the embedding $\sigma_{sp} \hookrightarrow \delta([\nu^{a_1+1}\rho, \nu^{b_1}\rho]) \times \nu^{-a_1}\rho \rtimes \delta([\nu^{a}\rho, \nu^{b_2}\rho]; \sigma_{cusp})$ and this contradicts the strong positivity of σ_{sp}. Therefore, $\nu^{a_1}\rho \rtimes \delta([\nu^{a}\rho, \nu^{b_2}\rho]; \sigma_{cusp})$ is reducible.

GU analogue of Proposition 4.3 [5] implies that $a_1 \in \{a-1, a, b_2+1\}$. Let us first consider the case when $a_1 = a > 1/2$. Similarly as in Proposition 3.1 in [14], we use the following calculation of Jacquet modules:

$$r_{GL}(\nu^{a}\rho \rtimes \delta([\nu^{a}\rho, \nu^{b_2}\rho]; \sigma_{cusp})) = \nu^{-a}\tilde{\rho} \times \delta([\nu^{a}\rho, \nu^{b_2}\rho]) \otimes \sigma_{cusp}$$

$$+ \nu^{a}\rho \times \delta([\nu^{a}\rho, \nu^{b_2}\rho]) \otimes \sigma_{cusp} \tag{4.2}$$

$$\mu^*(\nu^{a}\rho \rtimes \delta([\nu^{a}\rho, \nu^{b_2}\rho]; \sigma_{cusp})) \geq \delta([\nu^{a+1}\rho, \nu^{b_2}\rho]) \otimes \nu^{a}\rho \rtimes \delta(\nu^{a}\rho, \sigma_{cusp}) \tag{4.3}$$

Due to GU analogue of Lemma 4.1 in [5], $\nu^{a}\rho \rtimes \delta(\nu^{a}\rho, \sigma_{cusp})$ is irreducible. Therefore, the irreducible subquotient of $\nu^{a}\rho \rtimes \delta([\nu^{a}\rho, \nu^{b_2}\rho]; \sigma_{cusp})$ that contains right-hand side of (4.3) in its Jacquet modules must also contain both terms in (4.2). This implies that $\nu^{a}\rho \rtimes \delta([\nu^{a}\rho, \nu^{b_2}\rho]; \sigma_{cusp})$ is irreducible, which is a contradiction. Now, we consider $a_1 = a = 1/2$. In this case, GU analogue of Appendix of [5] (or Lemma 5.7 of [5]) implies that irreducible subrepresentation of $\delta([\nu^{1/2}\rho, \nu^{b_1}\rho]) \times \delta([\nu^{1/2}\rho, \nu^{b_2}\rho]) \rtimes \sigma_{cusp})$ is not strongly positive, which is a contradiction. Similarly as in [8], we have a contradiction in the case $a_1 = b_2 + 1$. The remaining case is when $a_1 = a - 1$, which is possible only if $a > 1$. In that case, we also have $b_1 < b_2$. Completing argument of induction on k is also exactly as in [8] and we skip the proof.

We also show that the mapping from $D(\rho; \sigma_{cusp})$ to the set of induced representations of the form (4.1) is well defined in the following theorem:

Theorem 4.3 *Let σ_{sp} be an irreducible strongly positive representation in $D(\rho; \sigma_{cusp})$. Then, there exist a unique set of strongly positive segments $\Delta_1, \Delta_2, \ldots, \Delta_k$, with $0 < e(\Delta_1) < e(\Delta_2) < \cdots < e(\Delta_k)$, and a unique irreducible cuspidal representation $\sigma' \in R$ such that $\sigma_{sp} \simeq \delta(\Delta_1, \Delta_2, \ldots, \Delta_k; \sigma')$.*

Proof The proof is similar to [8] and we, therefore, omit the proof in this case since we constructed all the tools that we need in Sect. 3.

In Theorem 4.2 and Theorem 4.3, we construct an injective mapping from $D(\rho; \sigma_{cusp})$ into the set of induced representations of the form (4.1) with refinement on the unitary exponents as in Theorem 4.2. More precisely, let $Jord_{(\rho,a)}$ stand for the set of all increasing sequences $b_1, b_2, \ldots, b_{k_\rho}$, where $b_i \in \mathbb{R}$, $b_i - a + k_\rho - i \in \mathbb{Z}_{\geq 0}$ for $i = 1, \ldots, k_\rho$ and $-1 < b_1 < b_2 < \cdots < b_{k_\rho}$. So far, we construct the following injective mapping:

$$D(\rho; \sigma_{cusp}) \hookrightarrow Jord_{(\rho,a)}$$

Now, it remains to show that this map is surjective. Let $b_1, b_2, \ldots, b_{k_\rho}$ denote an increasing sequence appearing in $Jord_{(\rho,a)}$. We showed in Sect. 4.1 that the induced representation

$$\delta([\nu^{a-k_\rho+1}\rho, \nu^{b_1}\rho]) \times \delta([\nu^{a-k_\rho+2}\rho, \nu^{b_2}\rho]) \times \cdots \times \delta([\nu^{a}\rho, \nu^{b_{k_\rho}}\rho]) \rtimes \sigma_{cusp} \quad (4.4)$$

has a unique irreducible subrepresentation, which we denote by $\sigma_{(b_1,\ldots,b_{k_\rho};a)}$.

We apply the induction argument in [8] to show that the above subrepresentation is strongly positive and we do not repeat the argument here.

Theorem 4.4 *The representation $\sigma_{(b_1,\ldots,b_{k_\rho};a)}$ is strongly positive.*

4.3 Classification of Strongly Positive Representations

Let ρ_i be a conjugate self-dual irreducible cuspidal representation of $\mathbf{GL}_{n_{\rho_i}}(E)$ for $i = 1, \ldots, k$ and σ_{cusp} is an irreducible cuspidal representation of $\mathbf{H}_m(F)$. Let $D(\rho_1, \rho_2, \ldots, \rho_k; \sigma_{cusp})$ be the set of strongly positive representations whose cuspidal supports are the representation σ_{cusp} and the twists of the representations ρ_i by positive-valued characters for $i = 1, \ldots, k$. Let $a_{\rho_i} \geq 0$ be the unique non-negative real number such that $\nu^{a_{\rho_i}}\rho_i \rtimes \sigma_{cusp}$ reduces for each $i = 1, \ldots, k$ [13]. Furthermore, we assume that this reducibility point a_{ρ_i} is in $\frac{1}{2}\mathbb{Z}$ (see (HI) of [12], page 771).

With Theorem 4.1, we use induction to prove the following two theorems as in [8]:

Theorem 4.5 *Let σ_{sp} be a strongly positive representation in $D(\rho_1, \rho_2, \ldots, \rho_k; \sigma_{cusp})$. Then σ_{sp} can be considered the unique irreducible subrepresentation of the following induced representation:*

$$(\prod_{i=1}^{k}\prod_{j=1}^{k_i}\delta([\nu^{a_{\rho_i}-k_i+j}\rho_i, \nu^{b_j^{(i)}}\rho_i])) \rtimes \sigma_{cusp}, \tag{4.5}$$

where $k_i \in \mathbb{Z}_{\geq 0}$, $k_i \leq \lceil a_{\rho_i} \rceil$, $b_j^{(i)} > 0$ such that $b_j^{(i)} - a_{\rho_i} \in \mathbb{Z}_{\geq 0}$, for $i = 1, \ldots, k$ $j = 1, \ldots, k_i$. Also, $b_j^{(i)} < b_{j+1}^{(i)}$ for $1 \leq j \leq k_i - 1$.

Theorem 4.5 implies that we construct the mapping from $D(\rho_1, \rho_2, \ldots, \rho_k; \sigma_{cusp})$ to the set of induced representations of the form (4.1). We now show that this mapping is well defined and injective.

Theorem 4.6 *Suppose that the representation $\sigma_{sp} \in D(\rho_1, \rho_2, \ldots, \rho_k; \sigma_{cusp})$ can be embedded as the unique irreducible subrepresentations of both representations $(\prod_{i=1}^{k}\prod_{j=1}^{k_i}\delta([\nu^{a_{\rho_i}-k_i+j}\rho_i, \nu^{b_j^{(i)}}\rho_i])) \rtimes \sigma_{cusp}$ and $(\prod_{i=1}^{k'}\prod_{j=1}^{k_i'}\delta([\nu^{a_{\rho_i'}-k_i'+j}\rho_i', \nu^{c_j^{(i)}}\rho_i'])) \rtimes \sigma_{cusp}'$ as in Theorem 4.5. Then we have $k = k'$, $\sigma_{cusp} \cong \sigma_{cusp}'$ and $\{\prod_{j=1}^{k_i}\delta([\nu^{a_{\rho_i}-k_i+j}\rho_i, \nu^{b_j^{(i)}}\rho_i]) | i = 1, \ldots, k\}$ is a permutation of $\{\prod_{j=1}^{k_i'}\delta([\nu^{a_{\rho_i'}-k_i'+j}\rho_i', \nu^{c_j^{(i)}}\rho_i']) | i = 1, \ldots, k\}$.*

Proof Since $\sigma_{sp} \in D(\rho_1, \rho_2, \ldots, \rho_k; \sigma_{cusp})$, $\sigma_{cusp}' \cong \sigma_{cusp}$ and $\{\rho_i' | i = 1, \ldots, k'\} \subset \{\rho_i | i = 1, \ldots, k\}$. Then, comparing the Jacquet modules, we easily see that $k = k'$ and $\{\prod_{j=1}^{k_i}\delta([\nu^{a_{\rho_i}-k_i+j}\rho_i, \nu^{b_j^{(i)}}\rho_i]) | i = 1, \ldots, k\}$ is a permutation of $\{\prod_{j=1}^{k_i'}\delta([\nu^{a_{\rho_i'}-k_i'+j}\rho_i', \nu^{c_j^{(i)}}\rho_i']) | i = 1, \ldots k\}$.

Now, we extend the above mapping to the set of all strongly positive representations of $\mathbf{H}(F)$. We first show the uniqueness of partial cuspidal support of strongly positive representation.

Proposition 4.7 *Let σ_{sp} denote a strongly positive representation of $\mathbf{H}_n(F)$. Then there is a unique, up to isomorphism, cuspidal representation σ_{cusp} of $\mathbf{H}_m(F)$ such that σ_{sp} is a subrepresentation of $\pi \rtimes \sigma_{cusp}$, for an irreducible representation π of $GL_{n-m}(E)$.*

Proof Suppose that there are non-isomorphic irreducible cuspidal representations σ_1 of $\mathbf{H}_{m_1}(F)$ and σ_2 of $\mathbf{H}_{m_2}(F)$, such that $\sigma_{sp} \hookrightarrow \pi_1 \rtimes \sigma_1$ and $\sigma_{sp} \hookrightarrow \pi_2 \rtimes \sigma_2$ for appropriate irreducible representations π_1 and π_2.

Thus, there are cuspidal representations $\rho_1, \rho_2, \ldots, \rho_k$ of general linear groups such that

$$\sigma_{sp} \hookrightarrow \nu^{x_1}\rho_1 \times \nu^{x_2}\rho_2 \times \cdots \times \nu^{x_k}\rho_k \rtimes \sigma_1.$$

Strong positivity of σ_{sp} implies $x_i > 0$ for all i.

Also, Frobenius reciprocity implies $\mu^*(\sigma_{sp}) \geq \pi_2 \otimes \sigma_2$, which implies that

$$\mu^*(\nu^{x_1}\rho_1 \times \nu^{x_2}\rho_2 \times \cdots \times \nu^{x_k}\rho_k \rtimes \sigma_1) \geq \pi_2 \otimes \sigma_2.$$

Repeated application of Lemma 3.2 implies that π_2 is an irreducible subquotient of $\rho_1' \times \rho_2' \times \cdots \times \rho_k'$, where $\rho_i' \in \{\nu^{x_i}\rho_i, \nu^{-x_i}{}^t\overline{\rho_i}^{-1}\}$, for $i = 1, 2, \ldots, k$. Since σ_1 is not isomorphic to σ_2, using Lemma 4.2 with obtain that there is an $i \in \{1, 2, \ldots, k\}$ such that $\rho_i' \cong \nu^{-x_i}{}^t\overline{\rho_i}^{-1}$. Since $x_i > 0$, this contradicts strong positivity of σ_{sp} and the proposition is proved.

Furthermore, by comparing Jacquet modules as in the proof of Theorem 4.6, we also show the uniqueness of GL cuspidal supports of strongly positive representation. Therefore, for any strongly positive representation σ_{sp} of $\mathbf{H}(F)$, there exists unique set of $\rho_1, \rho_2, \ldots, \rho_k$ and σ_{cusp} such that σ_{sp} can be considered to be the element in $D(\rho_1, \rho_2, \ldots, \rho_k; \sigma_{cusp})$.

Let SP be the set of all strongly positive representations of $\mathbf{H}(F)$. To see this mapping explicitly, let us collect the data from the induced representations of the form (4.5). Let LJ be the set of $(Jord, \sigma')$ where $Jord = \bigcup_{i=1}^{k}\bigcup_{j=1}^{k_i}\{(\rho_i, b_j^{(i)})\}$ and σ' be an irreducible cuspidal representation in R such that

(i) $\{\rho_1, \rho_2, \ldots, \rho_k\}$ is a (possibly empty) set of mutually non-isomorphic irreducible conjugate self-dual cuspidal unitary representations of GL such that $\nu^{a'_{\rho_i}}\rho_i \rtimes \sigma'$ reduces for $a'_{\rho_i} > 0$ (this defines a'_{ρ_i}),

(ii) $k_i = \lceil a'_{\rho_i} \rceil$,

(iii) for each $i = 1, 2, \ldots, k$, $b_1^{(i)}, b_2^{(i)}, \ldots, b_{k_i}^{(i)}$ is a sequence of real numbers such that $a'_{\rho_i} - b_j^{(i)} \in \mathbb{Z}$, for $j = 1, 2, \ldots, k_i$, and $-1 < b_1^{(i)} < b_2^{(i)} < \cdots < b_{k_i}^{(i)}$.

Now, the last step is to show that this mapping is surjective onto LJ. Following [8], we have

Theorem 4.8 *The maps described above give a bijective correspondence between the sets SP and LJ.*

5 Appendix: Even Unitary Case

We shortly derive the structural formula for unitary groups, which is also mentioned in Section 15 of [12], for the purpose of obtaining a classification of strongly positive representations of even unitary groups, following the same approach as in the previous sections. The strongly positive representations of even unitary groups have also been classified in Sections 7 and 15 of [12], using a different approach.

5.1 Notation for Even Unitary Groups

We let $\mathbf{G}_n = \mathrm{U}(n, n)$ be the quasi-split unitary group in $2n$ variables defined with respect to E/F and J_n and let $R_U(n)$ be the Grothendieck group of the category of all admissible representations of finite length of $\mathbf{G}_n(F)$ and set $R_U = \bigoplus_{n \geq 0} R_U(n)$. As in the even general unitary groups. Let also $\mathbf{B} = \mathbf{TU}$, $\mathbf{A}_0$, $\mathbf{T}$ for $\mathbf{G}_n$ be defined as in Sect. 3.

Then,

$$\mathbf{T}(F) = \left\{ \begin{pmatrix} x_1 & & & & & \\ & \ddots & & & & \\ & & x_n & & & \\ & & & \bar{x}_1^{-1} & & \\ & & & & \ddots & \\ & & & & & \bar{x}_n^{-1} \end{pmatrix} | x_i \in E^\times \right\}$$

and

$$\mathbf{A}_0(F) = \left\{ \begin{pmatrix} x_1 & & & & & \\ & \ddots & & & & \\ & & x_n & & & \\ & & & x_1^{-1} & & \\ & & & & \ddots & \\ & & & & & x_n^{-1} \end{pmatrix} | x_i \in F^\times \right\}$$

The F-points of Levi subgroups in $\mathbf{G}_n$ that corresponds to $\mathbf{s} = (n_1, n_2, \ldots, n_k)$ is of the form

$$\mathbf{M}_\mathbf{s}(F) = \left\{ \begin{pmatrix} g_1 & & & & & & \\ & \ddots & & & & & \\ & & g_k & & & & \\ & & & g & & & \\ & & & & {}^t\bar{g}_1^{-1} & & \\ & & & & & \ddots & \\ & & & & & & {}^t\bar{g}_n^{-1} \end{pmatrix} | g_i \in \mathbf{GL}_{n_i}(E), g \in \mathbf{G}_{n-n'}(F) \right\}.$$

5.2 *Tadić's Structure Formula for Unitary Groups*

Note that the Weyl group for unitary groups is isomorphic to general unitary groups. Therefore, we use the same notation for $q_n(d,k)_{i_1,i_2}$ $(=\omega)$ as in the general unitary group case. Then, for $(g_1, g_2, g_3, g_4, h) \in \mathbf{GL}_k(E) \times \mathbf{GL}_{i_2-d-k}(E) \times \mathbf{GL}_d(E) \times \mathbf{GL}_{i_1-d-k}(E) \times \mathbf{G}_{n-i_1-i_2+d+k}(F)$, we have $w \cdot (g_1, g_2, g_3, g_4, h) = (g_1, g_4, {}^t\bar{g}_3^{-1}, g_2, h)$.

Let π_i be an irreducible smooth representation of $\mathbf{GL}_{n_i}(E)$ for $i = 1, 2, 3, 4$ and let σ be an irreducible smooth representation of $\mathbf{G}_m$. By our previous calculation,

$$w^{-1} \cdot (\pi_1 \otimes \pi_2 \otimes \pi_3 \otimes \pi_4 \otimes \sigma) = \pi_1 \otimes \pi_4 \otimes \check{\pi}_3 \otimes \pi_2 \otimes \sigma. \tag{5.1}$$

Set

$$(\pi_1 \otimes \pi_2 \otimes \pi_3)\widetilde{\rtimes}'(\pi_4 \otimes \sigma) = \check{\pi}_1 \times \pi_2 \times \pi_4 \otimes \pi_3 \rtimes \sigma. \tag{5.2}$$

We follow argument in Sect. 3 by replacing (3.3) by (5.2), we have

Theorem 5.1 *(Tadić's structure formula for unitary groups.) For $\pi \in R_{GL}(i)$ and $\sigma \in R_U(n-i)$, the following structure formula holds*

$$\mu^*(\pi \rtimes \sigma) = \mathfrak{M}^*(\pi)\widetilde{\rtimes}'\mu^*(\sigma).$$

5.3 *Strongly Positive Representations for Even Unitary Groups*

With Tadić's structure formula for unitary groups (Sect. 5.2), we apply the arguments as in Sect. 4 to obtain the analogous results for even unitary groups. In this subsection, we only state the main result for even unitary groups and skip the proof since we already go through the similar arguments in Sect. 4.

Let SP' be the set of all strongly positive representations of $\mathbf{G}(F)$ and LJ' be the set of $(Jord, \sigma')$ where σ' be an irreducible cuspidal representation in R_U and *Jord* be exactly as in the case of even general unitary groups. Then, one can repeat the same arguments as before to obtain the bijective correspondence between SP' and LJ'.

Acknowledgements The first author would like to thank the organizers of the workshop on Representation theory of p-adic groups at IISER Pune, Professors Anne-Marie Aubert, Manish Mishra, Alan Roche, Steven Spallone for their invitation and hospitality. The authors would also like to thank the referee for his/her careful reading of the paper and for helpful comments and suggestions.

References

1. D. Ban, Parabolic induction and Jacquet modules of representations of O(2n, F). Glas. Mat. Ser. III **34**(54) 2, 147–185 (1999)
2. J. Bernstein, A.V. Zelevinsky, Induced representations of reductive p-adic groups I. Ann. Sci. École Norm. Sup. **10**, 4411–472 (1977)
3. M. Hanzer, G. Muić, Parabolic induction and Jacquet functors for metaplectic groups. J. Algebra **323**, 241–260 (2010)
4. C. Jantzen, Jacquet modules of induced representations for p-adic special orthogonal groups. J. Algebra **305**, 802–819 (2006)
5. Y. Kim, Strongly positive representations of $GSpin_2n + 1$ and the Jacquet module method. with an appendix, "Strongly positive representations in an exceptional rank-one reducibility case" by Ivan Matic. Math. Z. **279**, 271–296 (2015)
6. Y. Kim, Strongly positive representations of even *GSpin* groups. Pacific J. Math. **280**, 69–88 (2016)
7. Y. Kim, I. Matić, *Discrete series of odd GSpin groups*. Preprint, available at arXiv:1706.01111
8. I. Matić, Strongly positive representations of metaplectic groups. J. Algebra **334**, 255–274 (2011)
9. I. Matić, Jacquet modules of strongly positive representations of the metaplectic groups. Trans. Amer. Math. Soc. **365**(5), 2755–2788 (2013)
10. I. Matić, M. Tadić, On Jacquet modules of representations segment type. Manuscripta Math. **147**, 437–476 (2015)
11. C. Mœglin, Sur la classification des séries discrètes des groupes classiques p-adiques: paramètres de Langlands et exhaustivité. J. Eur. Math. Soc. **4**, 143–200 (2002)
12. C. Mœglin, M. Tadić, Construction of discrete series for classical groups. J. Am. Math. Soc. **15**, 715–786 (2002)
13. A. Silberger, Special representations of reductive p-adic groups are not integrable. Ann. of Math. **111**, 571–587 (1980)
14. M. Tadić, Structure arising from induction and Jacquet modules of representations of classical p-adic groups. J. Algebra **177**, 1–33 (1995)
15. M. Tadić, On reducibility of parabolic induction. Israel J. Math. **107**, 29–91 (1998)
16. M. Tadić, On regular square integrable representations of p-adic groups. Amer. J. Math. **120**, 159–210 (1998)
17. A.V. Zelevinsky, Induced representations of reductive p-adic groups. II. On irreducible representations of GL(n)

On the Unicity of Types for Toral Supercuspidal Representations

Peter Latham and Monica Nevins

Contents

1 Introduction 175
2 Notation 178
3 The Construction of Toral Supercuspidal Representations 179
4 Results on Equivalence 181
5 Stabilizers 182
6 Mackey Decomposition and Strategies for Identifying Types 183
7 Unicity Results 184
8 The Inequivalence of Unicity and Strong Unicity 186
References 190

Abstract For tame arbitrary-length toral, also called positive regular, supercuspidal representations of a simply connected and semisimple p-adic group G, constructed as per Adler-Yu, we determine which components of their restriction to a maximal compact subgroup are types. We give conditions under which there is a unique such component and then present a class of examples for which there is not, disproving the strong version of the conjecture of unicity of types on maximal compact open subgroups. We restate the unicity conjecture and prove it holds for the groups and representations under consideration under a mild condition on depth.

1 Introduction

Let G be a connected reductive p-adic group. In [2], J. Bernstein gives a decomposition of the category of smooth representations of G into indecomposable full subcategories, called *blocks*, that are indexed by the inertial support of the irreducible

The second author's research is supported by NSERC Canada Discovery Grant 06294.

P. Latham
Department of Mathematics, King's College London, Strand, London, UK
e-mail: peter.latham@kcl.ac.uk

M. Nevins (✉)
Department of Mathematics and Statistics, University of Ottawa, Ottawa, Canada
e-mail: mnevins@uottawa.ca

A.-M. Aubert et al. (eds.), *Representations of Reductive p-adic Groups*,
Progress in Mathematics 328, https://doi.org/10.1007/978-981-13-6628-4_6

representations they contain. Given an irreducible supercuspidal representation π of G, a *type* for π is a pair (K, λ) consisting an irreducible representation λ of a compact open subgroup K of G such that $\mathrm{Ind}_K^G \lambda$ is a projective generator of the block containing π. In this case, λ occurs as a subrepresentation of $\pi|_K$, and we say π *contains* the type (K, λ). Types are now known to exist for many classes of supercuspidal representations. In particular, the work of Adler [1], generalized by Yu [24], shows that every essentially tame supercuspidal representation contains a type; the work of Kim [11] and Fintzen [7] assure us that under mild assumptions on the residual characteristic, then all supercuspidal representations of G are of this form.

Given a supercuspidal representation π of G containing a type (K, λ), it is simple to produce additional types: Any G-conjugate of (K, λ) is a type, as is any pair (K', τ) where K' is a compact open subgroup of G containing K and τ is an irreducible representation of K' that contains λ upon restriction to K. A natural question to ask is whether π can contain any additional types, specifically on *maximal* compact open subgroups, that are not related to (K, λ) in this way. It is expected that this should never happen, and the conjecture that this is the case is known as the *unicity of types*. The name is due to Paškūnas [21], whose thesis extended G. Henniart's appendix "Sur l'unicité des types pour GL_2" of the article [3]. The goal of this paper is to establish the unicity of types for a class of essentially tame supercuspidal representations which we call toral, defined below.

The unicity conjecture is a theorem for $G = \mathrm{GL}_n(F)$ [21], $G = \mathrm{SL}_2(F)$ [12], for essentially tame representations of $G = \mathrm{SL}_n(F)$ [14], and for depth-zero supercuspidal representations of any connected reductive p-adic group G [13]. In each of these cases, it was seen that a stronger property holds, namely that if $K \subset G$ is any maximal compact open subgroup, then there exists, up to G-conjugacy, at most one type defined on K for each supercuspidal representation π of G. We will refer to this as the *strong unicity property*. While strong unicity implies unicity, in Sect. 8 we provide counterexamples to prove that they are in fact inequivalent in general.

From now on, let us specialize to the case that G is a simply connected semisimple p-adic group. Essentially tame supercuspidal representations are constructed from sequences of twisted Levi subgroups that split over a tamely ramified Galois extension, together with characters of these subgroups and a representation of the smallest twisted Levi subgroup. We restrict our attention here to those sequences for which the smallest twisted Levi subgroup is an anisotropic (also called elliptic) maximal torus of G. For the purposes of this paper, we call these *toral* supercuspidal representations, though we caution the reader that some authors reserve "toral" to mean the more restrictive case that the twisted Levi sequence has length $d = 1$. Relating to work of T. Kaletha [10], F. Murnaghan calls our representations "positive regular" (as justified in [18]).

The strategy for proving the unicity conjecture for supercuspidal representations of G is as follows: Let π be an essentially tame supercuspidal representation and (K, κ) a type as arising from the above construction. Since G is semisimple and simply connected, we have both that $\text{c-Ind}_{\mathsf{K}}^G \kappa$ is irreducible (hence equivalent to π) and that every maximal compact open subgroup of G is the stabilizer G_y of a vertex

y in the Bruhat-Tits building of G. Thus for any such G_y containing K, it follows directly that we have an induced type $(G_y, \mathrm{Ind}_{\mathsf{K}}^{G_y} \kappa)$.

The conjecture of unicity of types, therefore, amounts to the statement that, up to G-conjugacy, all types for π on a maximal compact open subgroup arise in this way. Strong unicity is the statement that furthermore any two types for π on G_y are conjugate by an element of $N_G(G_y)$; this is equivalent to the statement that (K, κ) is not contained in two distinct but conjugate maximal compact open subgroups.

The restriction of π to a maximal compact open subgroup G_y decomposes as an infinite direct sum of irreducible representations of G_y. Describing these branching rules is a difficult open problem of interest in its own right. Here, it suffices to note that with G as above, the Bernstein block corresponding to π is generated by π. Therefore, the types of π supported on G_y are exactly those irreducible components of $\pi|_{G_y}$ that do not occur in $\pi'|_{G_y}$ for any other (inequivalent) irreducible representation π' of G. Proving this, in turn, is made possible by the major work of Hakim and Murnaghan [8] which establishes the equivalences among essentially tame supercuspidal representations entirely in terms of the Adler-Yu data used to construct them.

To state our main result (Theorem 7.3), let T be an anisotropic maximal torus of G and let $\mathcal{B}^T$ denote the fixed point set of T acting on the Bruhat-Tits building $\mathcal{B}(G)$ of G. This set contains in particular a point x which is the image of the building of T in the building of G. Let $c_T \geq 0$ be the *simplicial radius* of $\mathcal{B}^T$, relative to x, as defined in Sect. 5.

Theorem *Let T be a tame elliptic maximal torus of G and suppose π is a supercuspidal representation of G built from a datum containing T. If the character ϕ_0 of T appearing in the datum has depth greater than $2c_T$, then π satisfies the unicity conjecture relative to any maximal compact open subgroup of G. Moreover, in this case, π satisfies the strong unicity property if and only if $\mathcal{B}^T$ consists precisely of the closure of a single facet in $\mathcal{B}(G)$.*

We note that the theorem holds without any hypothesis on the depth of ϕ_0 when T is *unramified*; see Corollary 7.4.

An essential ingredient of the proof is the analysis of the fixed points of $\mathcal{B}(G)$ under the action of both the torus T and of the inducing subgroup K, using particularly Lemma 5.1. As we discuss in Sect. 8, while the hypothesis for strong unicity (e.g., that $\mathcal{B}^T = \overline{\{x\}}$) holds in many cases (notably, for $G = \mathrm{SL}_n$, for unramified tori and for purely ramified Coxeter tori), it fails for many classes of anisotropic tori in general. In these cases, the number of inequivalent types of the form (G_y, λ) can grow arbitrarily large as the rank of G increases; see Example 8.4.

This paper is organized as follows: We establish our notation in Sect. 2, including particularly of the building $\mathcal{B}(G)$ and of Moy–Prasad filtration subgroups, and recall the definition of the essentially tame toral supercuspidal representations we study in Sect. 3. In Sect. 4, we prove a proposition about the inequivalence of toral supercuspidal representations under certain twists, based on [8, Cor 6.10] (recalled here as Lemma 4.1) and some results of F. Murnaghan in [17]. In Sect. 5, we establish a key result relating Moy-Prasad filtration subgroups to stabilizers of subsets of an apartment, generalizing a proposition in [20], and we define the notion of the simplicial radius of a bounded subset of $\mathcal{B}(G)$. In Sect. 6, we recall the Mackey decomposition of $\pi|_{G_y}$ for a vertex y and prove two general results about components that can contain types. Our main results on unicity are proven in Sect. 7. We discuss the success and failure of strong unicity and provide examples where it fails, in Sect. 8.

Several interesting problems remain open. The hypothesis on the depth of the character of T given in the statement of Theorem 7.3 arises as a result of our method of proof. There is no such restriction for the unicity theorems on $G = \mathrm{GL}_n$ or on $G = \mathrm{SL}_n$, where a different argument (particular to type A) was employed by V. Paškūnas to address the small-depth case. Finally, not all essentially tame supercuspidal representations arise from toral data; the consideration of general twisted Levi sequences is necessary to completely resolve the unicity conjecture in these cases.

2 Notation

Let F be a non-Archimedean local field of residual characteristic p, with integer ring $\mathcal{O}$ and maximal ideal $\mathfrak{p}$. We normalize the valuation ν on F so that $\nu(F^\times) = \mathbb{Z}$; if E is an extension field of F, then we also denote by ν the unique extension of this valuation to E. Fix an additive character Λ of F that is nontrivial on $\mathcal{O}$ but trivial on $\mathfrak{p}$. If $H \subset G$ are groups and $g \in G$, let ${}^gH = \{ghg^{-1} \mid h \in H\}$ and for any representation τ of H let ${}^g\tau$ denote the corresponding representation of ${}^gH$. If σ is a representation of G, we write $\mathrm{Res}_H\sigma$ for its restriction to H.

Let $\mathbf{G}$ be a semisimple simply connected linear algebraic group defined over F and let $G = \mathbf{G}(F)$. Let $\mathcal{B}(G) = \mathcal{B}(\mathbf{G}, F)$ denote the reduced building of $\mathbf{G}$ over F; since $\mathbf{G}$ is semisimple, it coincides with the enlarged building. Since $\mathbf{G}$ is simply connected, the stabilizer of a point x in the building, G_x, coincides with the parahoric subgroup $G_{x,0}$ associated with the facet containing x. Both hypotheses together imply that the maximal compact open subgroups of G are exactly the maximal parahoric subgroups of G, that is, G_y for each vertex y of the building.

Let $\mathbf{S}$ be a maximal F-split torus in $\mathbf{G}$ defined over F. Fix a maximal unramified extension F^{un} of F, and let $\mathbf{S}'$ be a maximal F^{un}-split torus of $\mathbf{G}$ defined over F and containing $\mathbf{S}$. Let $\mathbf{Z}$ be the centralizer of $\mathbf{S}'$ in $\mathbf{G}$, which is a maximal torus of $\mathbf{G}$ defined over F. Denote by $\Phi = \Phi(\mathbf{G}, \mathbf{S})$, the root system of $\mathbf{G}$ relative to $\mathbf{S}$ and by $\mathcal{A} = \mathcal{A}(\mathbf{S}, F)$ the apartment of $\mathcal{B}(\mathbf{G}, F)$ corresponding to $\mathbf{S}$. Let Ψ_Φ be the set of affine roots, corresponding to our choice of valuation ν; these are functions on $\mathcal{A}$. Denote the root subgroup of $\mathbf{G}$ corresponding to $\alpha \in \Phi$ by $\mathbf{G}_\alpha$. For $\alpha \in \Phi$, we set $G_\alpha = \mathbf{G}_\alpha(F)$; this group admits a filtration by compact open subgroups G_ψ, as ψ runs over the affine roots with gradient $\dot{\psi} = \alpha$. For more details, see [5, §4] or the careful exposition in [6, §2].

Let $S = \mathbf{S}(F)$, $S' = \mathbf{S}'(F)$ and $Z = \mathbf{Z}(F)$. Let $Z_\mathfrak{b}$ be the maximal bounded subgroup of Z. Recall that any torus $T = \mathbf{T}(F)$ admits a natural filtration by subgroups

T_r for $r \geq 0$, and its Lie algebra $\mathfrak{t}$ is filtered by lattices $\mathfrak{t}_r$ for $r \in \mathbb{R}$. In particular, Z admits a natural filtration by subgroups Z_r for $r \geq 0$, with $Z_0 \subseteq Z_{\mathrm{b}}$.

For any $x \in \mathcal{A}$ and $r \geq 0$, Moy and Prasad [16] defined $G_{\alpha,x,r} = \langle G_\psi \mid \psi(x) \geq r, \dot{\psi} = \alpha \rangle$ and thus filtration subgroups

$$G_{x,r} = \langle Z_r, G_{\alpha,x,r} \mid \alpha \in \Phi \rangle$$

of G_x. We set $G_{x,r+} = \cup_{s>r} G_{x,s}$. They similarly defined lattices $\mathfrak{g}_{x,r}$ in $\mathfrak{g} = \mathrm{Lie}(G)(F)$ and $\mathfrak{g}^*_{x,r}$ in $\mathfrak{g}^*$, indexed by $r \in \mathbb{R}$. Conjugation by G allows us to extend these definitions to any $x \in \mathcal{B}(G)$.

We say that a group $\mathbf{G}'$ (or its set of F-points $G' = \mathbf{G}'(F)$) is a *twisted Levi subgroup* of $\mathbf{G}$ if there is an extension E of F such that $\mathbf{G}'$ is an E-Levi subgroup of $\mathbf{G}$ defined over F. We say $\mathbf{G}'$ is tamely ramified if $\mathbf{G}'$ (and thus $\mathbf{G}$) splits over a tamely ramified extension.

Suppose now that $\mathbf{G}'$ is a (tamely ramified) twisted Levi subgroup of $\mathbf{G}$ and let $\mathbf{T}$ be a maximal torus of $\mathbf{G}'$; let E be a tamely ramified splitting field of $\mathbf{T}$ over F. Let $\mathfrak{g}'$ denote the F-points of $\mathrm{Lie}(\mathbf{G}')$, and denote by $\mathfrak{z}'^*$ the F-points of the dual of the center of $\mathrm{Lie}(\mathbf{G}')$.

Let $r > 0$. Following [8, Definition 3.7], an element $X^* \in \mathfrak{z}'^*_{-r}$ is called *G-generic of depth -r* if it satisfies the conditions **GE1** and **GE2** of [24, §8]. By [24, Lemma 8.1], if p is not a torsion prime for the root datum (in the sense of [22]) of $(\mathbf{G}, \mathbf{T})$, then these conditions reduce to the requirement that $\nu(\langle X^*, H_a \rangle) = -r$, for each $a \in \Phi(\mathbf{G}, \mathbf{T}) \setminus \Phi(\mathbf{G}', \mathbf{T})$, where $H_a \in \mathrm{Lie}(\mathbf{G})(E)$ is the coroot associated with a.

Fix also a Moy–Prasad isomorphism [8, §2.6]

$$e\colon G'_{x,r}/G'_{x,r+} \to \mathfrak{g}'_{x,r}/\mathfrak{g}'_{x,r+}.$$

A character ϕ of G' of depth r is said to be *realized by an element* $X^* \in \mathfrak{z}'^*_{-r}$ on $G'_{x,r}$ if for every $Y \in \mathfrak{g}'_{x,r}$,

$$\phi(e(Y + \mathfrak{g}'_{x,r+})) = \Lambda(\langle X^*, Y \rangle).$$

By [8, Definition 3.9], ϕ is *G-generic (relative to x) of depth* $r > 0$ if ϕ is trivial on $G'_{x,r+}$, nontrivial on $G'_{x,r}$, and is realized on $G'_{x,r}$ by an element $X^* \in \mathfrak{z}'^*_{-r}$ that is G-generic of depth $-r$. A particular consequence of the G-genericity of a character ϕ of G' is that for any $g \in G$, we have that ${}^g\phi$ and ϕ coincide on $G'_{x,r}$ if and only if $g \in G'$ [24, Lemma 8.3].

3 The Construction of Toral Supercuspidal Representations

Fix G semisimple and simply connected, and retain all the notations above. Following [24, §3] and [8, §3.1], we define a *positive-depth generic toral supercuspidal datum of length d* of G (hereafter: *toral supercuspidal datum*) to be a pair $(\vec{G}, \vec{\phi})$, where

- $\vec{G} = (G^0 \subsetneq G^1 \subsetneq \cdots \subsetneq G^d = G)$, for some $d \geq 1$, is a sequence of tamely ramified twisted Levi subgroups, such that in particular $G^0 = T$ is an anisotropic maximal torus in G;
- $\vec{\phi} = (\phi_0, \phi_1, \ldots, \phi_{d-1})$ where ϕ_i is a G^{i+1}-generic character of G^i of depth r_i, and these real numbers satisfy $0 < r_0 < r_1 < \cdots < r_{d-1}$.

For convenience, we write $s_i = r_i/2$ for each $i = 0, \ldots, d-1$. Given a tamely ramified twisted Levi subgroup $\mathbf{G}'$ of $\mathbf{G}$ that is defined over F, one may embed $\mathcal{B}(\mathbf{G}', F)$ into $\mathcal{B}(\mathbf{G}, F)$ with canonical image. Since T is anisotropic, the image of $\mathcal{B}(T)$ in $\mathcal{B}(G)$ consists of a single point x, and it is this point relative to which the characters ϕ_i are G^{i+1}-generic. Thus, a toral supercuspidal datum implies the datum $(\vec{G}, \vec{\phi}, d, \vec{r}, \vec{s}, x)$, and we will take this extra data for granted where there is no possibility of confusion.

Let $\Psi = (\vec{G}, \vec{\phi})$ be a toral supercuspidal datum. Set $\mathsf{K}^0 = G^0_x = T$. For each $1 \leq i \leq d$, define

$$\mathsf{K}^i := G^0_x G^1_{x,s_0} \cdots G^i_{x,s_{i-1}} \subseteq T G_{x,s_0} \subseteq G_x.$$

The groups in these products have large pairwise intersections, so we next define subgroups J^i and J^i_+ which will have the property that $\mathsf{K}^i \mathsf{J}^{i+1} = \mathsf{K}^{i+1}$ and $\mathsf{K}^i \cap \mathsf{J}^{i+1} = \mathsf{K}^i \cap \mathsf{J}^{i+1}_+ = G^i_{x,r_i}$.

Let E be a tamely ramified Galois extension of F over which $\mathbf{T}$ splits. For each $0 \leq i \leq d$, let $\Phi_i = \Phi(\mathbf{G}^i, \mathbf{T})$ be the corresponding root system. For $0 \leq i < d$, let $\mathsf{J}(E)^{i+1}$ be the group generated by $\mathbf{T}(E)_{r_i}$ together with the root subgroups $\mathbf{G}_a(E)_{x,r_i}$ for $a \in \Phi_i$ and $\mathbf{G}_b(E)_{x,s_i}$ for $b \in \Phi_{i+1} \setminus \Phi_i$. Similarly, let

$$\mathsf{J}(E)^{i+1}_+ = \langle \mathbf{T}(E)_{r_i}, \mathbf{G}_a(E)_{x,r_i}, \mathbf{G}_b(E)_{x,s_i+} \mid a \in \Phi_i, b \in \Phi_{i+1} \setminus \Phi_i \rangle.$$

These subgroups of $\mathbf{G}^{i+1}(E)$ are invariant under $\mathrm{Gal}(E/F)$, and we set $\mathsf{J}^{i+1} = \mathsf{J}(E)^{i+1} \cap G^{i+1}$, $\mathsf{J}^{i+1}_+ = \mathsf{J}(E)^{i+1}_+ \cap G^{i+1}$.

Next for each $0 \leq i < d$, the subdatum $((G^i, G^{i+1}), \phi_i)$ is used to construct a representation ϕ'_i of K^{i+1}. The first step is to extend the restriction of ϕ_i to G^i_{x,r_i} trivially to a character $\hat{\phi}_i$ of J^{i+1}_+ [24, §4]. Next, when $\mathsf{J}^{i+1}_+ = \mathsf{J}^{i+1}$, define a character ϕ'_i of K^{i+1} by

$$\phi'_i(kj) = \phi_i(k)\hat{\phi}_i(j), \tag{3.1}$$

for each $k \in \mathsf{K}^i$, $j \in \mathsf{J}^{i+1}$. On the other hand, when $\mathsf{J}^{i+1}_+ \neq \mathsf{J}^{i+1}$, then instead one uses $\hat{\phi}_i$ to construct a Heisenberg–Weil representation ω of $\mathsf{K}^i \ltimes \mathsf{J}^{i+1}$, and then define the representation ϕ'_i on $k \in \mathsf{K}^i$, $j \in \mathsf{J}^{i+1}$ by

$$\phi'_i(kj) = \phi_i(k)\omega(k, j). \tag{3.2}$$

Now set $\mathsf{K} = \mathsf{K}^d$. There is a well-defined inflation process extending each ϕ'_i trivially across the remaining subgroups to give a representation of K, which we denote $\kappa_i = \inf_{\mathsf{K}^{i+1}}^{\mathsf{K}} \phi_i'$ [8, §3.4]. Note that this representation κ_i is independent of any other characters ϕ_j, $j \neq i$, in the datum [8, Proposition 3.26].

Putting these together, we obtain a representation $\kappa(\Psi) = \kappa_0 \otimes \cdots \otimes \kappa_{d-1}$ of K such that $\pi(\Psi) = \text{c-Ind}_{\mathsf{K}}^{G} \kappa(\Psi)$ is an irreducible supercuspidal representation of G of depth $r = r_{d-1}$ [24, Theorem 15.1]. It then follows that $(\mathsf{K}, \kappa(\Psi))$ is a type for $\pi(\Psi)$.

4 Results on Equivalence

We begin by noting when two toral supercuspidal data give rise to equivalent representations, from [8, Cor 6.10].

Lemma 4.1 [Hakim-Murnaghan] *Let $\Psi = (\vec{G}, \vec{\phi})$ and $\dot{\Psi} = (\dot{\vec{G}}, \dot{\vec{\phi}})$ be two toral supercuspidal data. Then $\pi(\Psi) \cong \pi(\dot{\Psi})$ if and only if there exists some $g \in G$ such that $G^0 = {}^g\dot{G}^0$ and $\mathrm{Res}_{G^0}(\phi_0\phi_1 \ldots \phi_{d-1}) = \mathrm{Res}_{G^0}{}^g(\dot{\phi}_0\dot{\phi}_1 \ldots \dot{\phi}_{d-1})$. In this case, $G^i = {}^g\dot{G}^i$ for all $i \geq 0$.*

Now suppose $\Psi = (\vec{G}, \vec{\phi})$ is a toral supercuspidal datum of length d with $T = G^0$. Let ξ be a character of T of depth strictly less than r_0. The G^1-genericity of the character ϕ_0 of G^0 of depth r_0 depends only on $\mathrm{Res}_{G^0_{x,r_0}} \phi$. Since $\xi\phi_0$ and ϕ_0 coincide with G^0_{x,r_0}, we conclude that $\xi\phi_0$ is also a G^1-generic character of G^0 of depth r_0. Thus

$$\Psi_\xi = (\vec{G}, (\xi\phi_0, \phi_1, \ldots, \phi_{d-1}))$$

is another toral supercuspidal datum.

Lemma 4.2 *The character ξ inflates to a character $\overline{\xi}$ of K such that*

$$\kappa(\Psi_\xi) = \overline{\xi}\, \kappa(\Psi).$$

Proof Since ξ is a character of $T = G^0 = \mathsf{K}^0$ of depth less than r_0, it is trivial on G^0_{x,r_0}. Since $\mathsf{K}^1 = \mathsf{K}^0\mathsf{J}^1$ and $\mathsf{K}^0 \cap \mathsf{J}^1 = G^0_{x,r_0}$, ξ may be uniquely inflated to a character $\xi' = \inf_{\mathsf{K}^0}^{\mathsf{K}^1} \xi$ of K^1 that is trivial on J^1. In particular, for any $k \in \mathsf{K}^0$, $j \in \mathsf{J}^1$, we have $\xi'(kj) = \xi(k)$.

Since ϕ_0 and $\xi\phi_0$ coincide on G^0_{x,r_0}, we may denote the extension to J^1_+ of their common restriction by $\hat{\phi}_0$ and, if applicable, the corresponding Heisenberg–Weil representation of $\mathsf{K}^0 \ltimes \mathsf{J}^1$ by ω. Using now (3.1) and (3.2), as applicable, to construct $(\xi\phi_0)'$ and ϕ_0', we deduce that in both cases $(\xi\phi_0)' = \xi'\phi_0'$.

Set $\overline{\xi} = \inf_{\mathsf{K}^1}^{\mathsf{K}} \xi'$; then $\inf_{\mathsf{K}^1}^{\mathsf{K}} \xi'\phi_0' = \overline{\xi}\kappa_0$. Since for $i > 0$, κ_i depends only on the character ϕ_i, we conclude that $\overline{\xi}$ factors out of the tensor product to yield $\kappa(\Psi_\xi) = \overline{\xi}\kappa(\Psi)$, as desired. □

Proposition 4.3 *Let $\Psi = (\vec{G}, \vec{\phi})$ be a toral supercuspidal datum and suppose ξ is a character of $T = G^0$ of depth less than that of ϕ_0. If $\Psi_\xi = (\vec{G}, (\xi\phi_0, \phi_1, \ldots, \phi_{d-1}))$, then $\pi(\Psi_\xi)$ is an irreducible supercuspidal representation of G and it is equivalent to $\pi(\Psi)$ if and only if ξ is trivial.*

Proof Suppose that $\pi(\Psi) \cong \pi(\Psi_\xi)$. In the setting of Lemma 4.1, this means there is some $g \in N_G(T)$ such that $\mathrm{Res}_T(\phi_0\phi_1 \ldots \phi_{d-1}) = \mathrm{Res}_{T^g}(\xi\phi_0\phi_1 \ldots \phi_{d-1})$. Then ${}^g(\xi\phi_0){}^g\phi_1 \ldots {}^g\phi_{d-1}$ is a refactorization of $\phi_0 \ldots \phi_{d-1}$, in the sense of [17, Definition 5.3], and so by [17, Proposition 5.4], for each $i > 0$, $\prod_{j=i}^{d} \phi_j$ and $\prod_{j=i}^{d} {}^g\phi_j$ coincide with $G^i_{x,r_{i-1}+}$. The genericity of the characters ϕ_j recursively implies, as in the proof of [17, Proposition 5.6], that $g \in G^1$, and so ${}^g\phi_i = \phi_i$ for all $i \geq 1$. Therefore, our equality reduces to $\mathrm{Res}_{T_{r_0}} \phi_0 = \mathrm{Res}_{T_{r_0}} {}^g(\xi\phi_0) = \mathrm{Res}_{T_{r_0}} {}^g\phi_0$, whence $g \in T$ by genericity. Returning to the first equality, we conclude that ξ is the trivial character of T. □

5 Stabilizers

Let $S, Z, \mathcal{A} = \mathcal{A}(S), \Phi$ be as in Sect. 2. Given a subset $\Omega \subseteq \mathcal{A}$, the subgroup of G that fixes Ω pointwise is generated by Z_{b} (the maximal bounded subgroup of the maximal torus Z) and those G_ψ satisfying $\psi(z) \geq 0$ for each $z \in \Omega$ [4, §6.4]. For $x, y \in \mathcal{A}$, we write $[x, y]$ for the geodesic from x to y; then $G_{[x,y]} = G_x \cap G_y$.

The following result is a generalization of [20, Prop 3.3], and it allows us to relate Moy–Prasad filtration subgroups at x to stabilizers of its neighborhoods in any apartment $\mathcal{A}$ containing x. Note that $\mathcal{A}$ is the affine space under $X_*(S) \otimes_{\mathbb{Z}} \mathbb{R}$. Given two points $x, z \in \mathcal{A}$, we identify $z - x$ with a vector in $X_*(S) \otimes_{\mathbb{Z}} \mathbb{R}$, and then for each $\alpha \in \Phi$, $\alpha(z - x)$ is a well-defined real number.

Lemma 5.1 *Let $\mathcal{A} = \mathcal{A}(\mathbf{S}, F)$ be an apartment in $\mathcal{B}(\mathbf{G}, F)$ containing x and let $\Phi = \Phi(\mathbf{G}, \mathbf{S})$ be the corresponding root system. Let $s \geq 0$ and define*

$$\Omega_{\mathcal{A}}(x, s) = \{z \in \mathcal{A} \mid \forall \alpha \in \Phi, \alpha(z - x) \leq s\}.$$

Then $G_{x,s} \subseteq G_{\Omega_{\mathcal{A}}(x,s)} = G_{\overline{\Omega_{\mathcal{A}}(x,s)}}$, where $\overline{\Omega_{\mathcal{A}}(x, s)}$ denotes the simplicial closure of $\Omega_{\mathcal{A}}(x, s)$ in $\mathcal{A}$.

Proof That $G_{\Omega_{\mathcal{A}}(x,s)} = G_{\overline{\Omega_{\mathcal{A}}(x,s)}}$ follows from [4, 2.4.13], since G is semisimple and simply connected. Let $z \in \Omega_{\mathcal{A}}(x, s)$. For each affine root ψ such that $\psi(x) \geq s$, let α denote its gradient. Then $\psi(x) - \psi(z) = \alpha(x - z) \leq s$ so $\psi(z) \geq \psi(x) - s \geq 0$. Thus for all affine roots ψ, if $G_\psi \subset G_{x,s}$ then $G_\psi \subset G_{\Omega_{\mathcal{A}}(x,s)}$. Since also $Z_s \subseteq Z_{\mathrm{b}}$, we conclude $G_{x,s} \subseteq G_{\Omega_{\mathcal{A}}(x,s)}$.

More generally, we may define

$$\Omega(x, s) = \bigcup_{\mathcal{A} \ni x} \Omega_{\mathcal{A}}(x, s)$$

which is the union over all apartments of $\mathcal{B}(G)$ containing x. By local compactness, this is reduced to a finite union. Let $\overline{\Omega(x,s)}$ denote its simplicial closure.

Definition 5.2 Let $\Omega \subset \mathcal{B}(G)$ be a bounded convex set and suppose $x \in \Omega$. For each apartment $\mathcal{A} = \mathcal{A}(\mathbf{S}, F)$ containing x define

$$r(\Omega, \mathcal{A}, x) = \sup\{\alpha(z - x) \mid z \in \Omega \cap \mathcal{A}, \alpha \in \Phi(\mathbf{G}, \mathbf{S})\}.$$

Then, the *simplicial radius* of Ω with respect to x is defined to be

$$c(\Omega, x) = \sup\{r(\Omega, \mathcal{A}, x) \mid x \in \mathcal{A}\}.$$

As one motivating example, note that the simplicial radius of $\overline{\Omega(x,s)}$ with respect to x is $c(\overline{\Omega(x,s)}, x) = s$. For another, letting $\overline{\{x\}}$ denote the simplicial closure of $\{x\}$ in $\mathcal{B}(G)$, we have $c(\overline{\{x\}}, x) < 1$ since $\overline{\{x\}}$ is constrained between adjacent affine root hyperplanes in any apartment containing x. Note that in each of these examples, we have the equality $c(\Omega, x) = r(\Omega, \mathcal{A}, x)$ for each apartment $\mathcal{A}$ containing x.

One can be slightly more precise about $c(\overline{\{x\}}, x)$ when x arises as the point identified with $\mathcal{B}(T)$ from a datum Ψ. In this case, x is an *optimal point* of $\mathcal{B}(G)$, in the sense of [16, §6.1], whence the family of such values $c_x = c(\overline{\{x\}}, x)$ could be computed for any G. For example, if $G = \mathrm{SL}_n(F)$, then the optimal points are among the barycentres of the facets, whence if x lies in a k-dimensional facet $\mathcal{F}$ then $c_x = 1 - \frac{1}{k}$.

6 Mackey Decomposition and Strategies for Identifying Types

Let y be a vertex of $\mathcal{B}(G)$ and G_y the corresponding maximal compact open subgroup of G. Let $\Psi = (\vec{G}, \vec{\phi})$ be a toral supercuspidal datum with $T = G^0$ and let $\pi = \pi(\Psi)$. We are interested in the irreducible representations of G_y occurring in $\mathrm{Res}_{G_y}\pi$.

Mackey theory gives a decomposition

$$\mathrm{Res}_{G_y}\pi \cong \bigoplus_{g \in G_y \backslash G / \mathsf{K}} \mathrm{Ind}_{G_y \cap {}^g\mathsf{K}}^{G_y} {}^g\kappa,$$

where each *Mackey component* $\tau(g) := \mathrm{Ind}_{G_y \cap {}^g\mathsf{K}}^{G_y} {}^g\kappa$ is a finite-dimensional representation of G_y. Note that since $G_y = N_G(G_y)$, the Mackey components are parametrized by a subset of the G-orbit of the vertex y in $\mathcal{B}(G)$. We emphasize that these Mackey components are not, in general, irreducible; a first strategy for identifying those that contain types is the following.

Proposition 6.1 *If $\mathsf{K} \subseteq G_{g^{-1}y}$ then $(G_y, \tau(g))$ is a type for π.*

Proof Consider instead the twisted Mackey component

$$^{g^{-1}}\tau(g) = {}^{g^{-1}}(\mathrm{Ind}_{G_y \cap {}^g\mathsf{K}}^{G_y} {}^g\kappa) \cong \mathrm{Ind}_{G_{g^{-1}y} \cap \mathsf{K}}^{G_{g^{-1}y}} \kappa$$

which is a representation of $G_{g^{-1}y}$. If $\mathsf{K} \subseteq G_{g^{-1}y}$, then $^{g^{-1}}\tau(g) \cong \mathrm{Ind}_{\mathsf{K}}^{G_{g^{-1}y}} \kappa$. By the transitivity of compact induction, c-$\mathrm{Ind}_{G_{g^{-1}y}}^{G} \tau' = \pi$, whence by Frobenius reciprocity $(G_{g^{-1}y}, {}^{g^{-1}}\tau(g))$ is a type for π, and the result follows.

On the other hand, the key strategy to discern Mackey components that cannot contain a type of π is the following.

Theorem 6.2 *Given y, Ψ, g as above, suppose that there exists a nontrivial character ξ of T of depth less than r_0 such that its inflation $\overline{\xi}$ to K is trivial on $\mathsf{K} \cap G_{g^{-1}y}$. Then no irreducible subrepresentation of the Mackey component*

$$\mathrm{Ind}_{G_y \cap {}^g\mathsf{K}}^{G_y} {}^g\kappa(\Psi)$$

is a type.

Proof The twisted Mackey component $^{g^{-1}}\tau(g)$ depends only $\mathrm{Res}_{G_{g^{-1}y} \cap \mathsf{K}} \kappa(\Psi)$. By Lemma 4.2, $\kappa(\Psi_\xi) = \overline{\xi}\kappa(\Psi)$; by hypothesis $\mathrm{Res}_{G_{g^{-1}y} \cap \mathsf{K}} \overline{\xi}$ is trivial. Thus, $\mathrm{Ind}_{G_y \cap {}^g\mathsf{K}}^{G_y} {}^g\kappa(\Psi)$ is a common component of both $\pi(\Psi)$ and $\pi(\Psi_\xi)$. By Proposition 4.3, $\pi(\Psi) \not\cong \pi(\Psi_\xi)$. Since these are inequivalent irreducible supercuspidal representations, they have distinct inertial support, and thus, no irreducible representation of G_y occurring in $\mathrm{Ind}_{G_y \cap {}^g\mathsf{K}}^{G_y} {}^g\kappa(\Psi)$ can be a type.

7 Unicity Results

We first identify the obvious types occurring in $\mathrm{Res}_{G_y} \pi$.

Lemma 7.1 *Suppose y is a vertex of the facet containing x. Then $(G_y, \tau(1))$ is a type.*

Proof Let $\mathcal{F}$ be the facet containing x. We have $\mathsf{K} \subseteq G_x$. Since $y \in \overline{\mathcal{F}}$ by hypothesis, $G_x \subseteq G_y$. This implies that $\mathsf{K} \subseteq G_y$, whence the first statement by Proposition 6.1.

Now, continuing with the notation of the previous section, we identify $g \in G$ for which the associated Mackey components do not contain types.

Proposition 7.2 *Let $g \in G$. Suppose that there exists $z \in [x, g^{-1}y] \cap \overline{\Omega(x, s_0)}$ such that $T \not\subseteq G_z$. Then no irreducible subrepresentation of $\tau(g)$ is a type.*

Proof Let $\mathcal{A}$ be an apartment of $\mathcal{B}(G)$ containing x and $g^{-1}y$; then $\mathcal{A}$ contains the geodesic $[x, g^{-1}y]$. Let $z \in [x, g^{-1}y] \cap \overline{\Omega_{\mathcal{A}}(x, s_0)}$ as in the proposition. Since z is on the line $[x, g^{-1}y]$, whose pointwise stabilizer is $G_x \cap G_{g^{-1}y}$, we have that $G_x \cap G_{g^{-1}y} \subseteq G_z$. At the same time, Lemma 5.1 implies that $G_{x,s_0} \subseteq G_{\overline{\Omega_{\mathcal{A}}(x,s_0)}} \subseteq G_z$. Thus

$$\mathsf{K} \cap G_{g^{-1}y} \subseteq TG_{x,s_0} \cap G_x \cap G_{g^{-1}y} \subseteq TG_{x,s_0} \cap G_z = (T \cap G_z)G_{x,s_0}.$$

Noting that $T_{s_0} \subseteq T \cap G_z$, we deduce that any character of T that is trivial on $T \cap G_z$ has depth strictly less than s_0.

By hypothesis $T \cap G_z \subsetneq T$; so let ξ be a nontrivial character of T that is trivial on $T \cap G_z$. Since its depth is less than s_0, its inflation $\overline{\xi}$ is trivial on $(T \cap G_z)G_{x,s_0}$. We deduce that ξ satisfies the hypotheses of Theorem 6.2, whence the proposition.

Let

$$\mathcal{B}^T = \{z \in \mathcal{B}(G) \mid T \subseteq G_z\}$$

be the set of fixed points of T acting on $\mathcal{B}(G)$. This is a convex subset of $\mathcal{B}(G)$ containing x, and it is compact since T is an anisotropic maximal torus. Let c_T denote the simplicial radius of $\mathcal{B}^T$ with respect to x, as per Definition 5.2.

Now let $\mathcal{A}$ be an apartment containing x. The hypothesis of Proposition 7.2 will be satisfied for all $g^{-1}y \in \mathcal{A}$ if $\mathcal{A}^T = \mathcal{A} \cap \mathcal{B}^T$ is contained in the interior of the simplicial closure $\overline{\Omega_{\mathcal{A}}(x, s_0)}$ of $\Omega_{\mathcal{A}}(x, s_0)$. In particular this holds if $s_0 > c_T \geq r(\mathcal{B}^T, \mathcal{A}, x)$.

Putting these geometric ideas together yields our main theorem.

Theorem 7.3 *Let T be a tamely ramified anisotropic maximal torus of G and let $s_0 > c_T$. Let $\Psi = (\vec{G}, \vec{\phi})$ be any toral supercuspidal datum such that $G^0 = T$ and such that the depth of ϕ_0 is at least $2s_0$. Then $\pi(\Psi)$ satisfies the conjecture of unicity of types relative to any maximal compact open subgroup of G. If moreover $\mathcal{B}^T$ is the closure of a single facet of $\mathcal{B}(G)$ then $\pi(\Psi)$ has the property of strong unicity.*

Proof Let $g \in G_y \backslash G / \mathsf{K}$. Suppose first that $g^{-1}y \in \mathcal{B}^T$, so that $T \subseteq G_{g^{-1}y}$. Since $s_0 > c_T$, we have $g^{-1}y \in \Omega(x, s_0)$ so $G_{x,s_0} \subseteq G_{g^{-1}y}$ as well. Thus $\mathsf{K} \subseteq TG_{x,s_0} \subseteq G_{g^{-1}y}$ and by the same argument as Lemma 7.1, we conclude that $(G_y, \tau(g))$ is a type for π.

Now suppose that $g^{-1}y \notin \mathcal{B}^T$, so that $T \not\subset G_{g^{-1}y}$. Since $s_0 > c_T$, $\Omega(x, s_0)$ contains an open neighborhood of $\mathcal{B}^T$, so the line $[x, g^{-1}y]$ must meet $\Omega(x, s_0) \setminus \mathcal{B}^T$ in at least one point z. Using now Proposition 7.2, we conclude that the corresponding Mackey component contains no types for $\pi(\Psi)$.

Finally, we note that if $\mathcal{B}^T = \overline{\mathcal{F}}$ for a facet $\mathcal{F} \subset \mathcal{B}(G)$, then since G is semisimple simply connected, each orbit of a vertex y in $\mathcal{B}(G)$ meets $\mathcal{B}^T$ at most once. By the above arguments, this implies strong unicity.

By [23, 3.6.1], if T is an anisotropic maximal torus which splits over an unramified extension, then $\mathcal{B}^T$ consists of a single vertex, namely $\{x\}$, whence $c_T = 0$. Since all our toral supercuspidal representations have positive depth, we have the following immediate corollary.

Corollary 7.4 *Let T be an* unramified *anisotropic maximal torus of G. Let $\Psi = (\vec{G}, \vec{\phi})$ be any toral supercuspidal datum such that $G^0 = T$. Then $\pi(\Psi)$ satisfies strong unicity of types relative to any maximal compact open subgroup of G.*

8 The Inequivalence of Unicity and Strong Unicity

Recall that strong unicity of types is the statement that $\mathrm{Res}_K \pi$ should contain at most one type, for each choice of maximal compact open subgroup K. We have the following converse to the strong unicity statement in Theorem 7.3. Note that this result is without a condition on depth.

Lemma 8.1 *Let T be a tamely ramified anisotropic maximal torus. If $\mathcal{B}^T$ contains two distinct but G-conjugate vertices y and y' of alcoves whose closure contain x, then for any toral supercuspidal datum $\Psi = (\vec{G}, \vec{\phi})$ such that $G^0 = T$, the strong unicity property fails to hold for $\pi(\Psi)$, that is, there exist at least two nonisomorphic types in* $\mathrm{Res}_{G_y} \pi(\Psi)$.

Proof Let $x \in \mathcal{B}(T) \subseteq \mathcal{B}(G)$ and set $\pi = \pi(\Psi)$. Let $y \neq y'$ be G-conjugate vertices of two distinct alcoves $\mathcal{C}, \mathcal{C}'$ of $\mathcal{B}(G)$ such that $x \in \overline{\mathcal{C}} \cap \overline{\mathcal{C}'}$. Then for any $s_0 > 0$, we have that $y, y' \in \overline{\Omega(x, s_0)}$. Let $g \in G$ be such that $y' = g^{-1}y$. If both y and $g^{-1}y$ lie in $\mathcal{B}^T$, then by the proof of Theorem 7.3, we conclude that both $(G_y, \tau(1))$ and $(G_y, \tau(g))$ are types for π occurring in $\mathrm{Res}_{G_y}\pi$. Note that although these types are induced from the G-conjugate types (K, κ) and $({}^g\mathsf{K}, {}^g\kappa)$, they are not themselves G-conjugate since $g \notin N_G(G_y) = G_y$. Neither are these types isomorphic, since $\pi \cong \text{c-Ind}_{G_y}^G \tau(1)$ implies

$$\mathbb{C} \cong \mathrm{Hom}_G(\pi, \pi) \cong \mathrm{Hom}_{G_y}(\tau(1), \mathrm{Res}_{G_y}\pi)$$

by Frobenius reciprocity.

We next prove the existence of pairs (G, T) satisfying the hypotheses of Lemma 8.1 with an example such that G has rank 2. We thank Jeff Adler for providing this instructive example of a torus that stabilizes more than the closure of a single facet.

Example 8.2 Consider $G = \mathrm{Sp}_4(F)$, given in matrix form relative to $J = \left[\begin{smallmatrix} 0 & I \\ -I & 0 \end{smallmatrix}\right] \in M_4(F)$ as the set

$$G = \{g \in \mathrm{GL}_4(F) \mid {}^t g J g = J\}$$

where ${}^t g$ denotes the transpose. Note that G contains a generalized Levi subgroup G' isomorphic to $\mathrm{SL}_2(F) \times \mathrm{SL}_2(F)$, defined by the long roots.

Let ϖ be a uniformizer of F. For $i = 1, 2$, we choose an anisotropic torus T_i of $\mathrm{SL}_2(F)$ that isomorphic to the norm-one elements of some quadratic ramified extension $E_i = F[\sqrt{\gamma_i \varpi}]$ with $\nu(\gamma_i) = 0$. Let $T = T_1 \times T_2$ be the corresponding anisotropic torus of G'; explicitly we embed T into G as the subgroup

$$T = \left\{ \begin{bmatrix} a & 0 & b & 0 \\ 0 & c & 0 & d \\ b\gamma_1\varpi & 0 & a & 0 \\ 0 & d\gamma_2\varpi & 0 & c \end{bmatrix} \;\middle|\; \begin{array}{l} a, b, c, d \in \mathcal{O}, \\ a^2 - b^2\gamma_1\varpi = 1, \\ c^2 - d^2\gamma_2\varpi = 1 \end{array} \right\}.$$

For each i, the point in the building $\mathcal{B}(\mathrm{SL}_2, F)$ corresponding to T_i is the midpoint of the fundamental alcove. Thus, the point x representing $\mathcal{B}(T) \subset \mathcal{B}(G)$ is the midpoint of the diagonal facet $\mathcal{F}$ in the closure of the fundamental alcove $\mathcal{C}$ in the standard apartment $\mathcal{A}$ of $\mathcal{B}(G)$, as illustrated in Fig. 1. Evidently $\overline{\mathcal{F}} \subseteq \mathcal{A}^T$.

Let us now prove that $\mathcal{A}^T \neq \overline{\mathcal{F}}$. Let y denote the (non-special) vertex of $\overline{\mathcal{C}}$ opposite $\mathcal{F}$, and $w \in G$ a reflection in the wall of $\mathcal{A}$ containing $\mathcal{F}$ (viewed as a representative of the corresponding element of the affine Weyl group). Adopting the convention that a matrix ring stands for its intersection with $\mathrm{Sp}_4(F)$, we compute directly that

$$G_{\mathcal{C}} = \begin{bmatrix} \mathcal{O} & \mathcal{O} & \mathcal{O} & \mathcal{O} \\ \mathfrak{p} & \mathcal{O} & \mathcal{O} & \mathcal{O} \\ \mathfrak{p} & \mathfrak{p} & \mathcal{O} & \mathfrak{p} \\ \mathfrak{p} & \mathfrak{p} & \mathcal{O} & \mathcal{O} \end{bmatrix} \quad \text{and} \quad G_{w\mathcal{C}} = \begin{bmatrix} \mathcal{O} & \mathfrak{p} & \mathcal{O} & \mathcal{O} \\ \mathcal{O} & \mathcal{O} & \mathcal{O} & \mathcal{O} \\ \mathfrak{p} & \mathfrak{p} & \mathcal{O} & \mathcal{O} \\ \mathfrak{p} & \mathfrak{p} & \mathfrak{p} & \mathcal{O} \end{bmatrix}, \tag{8.3}$$

each of which contains T as a subgroup. It, therefore, follows that $\{y, wy\} \subset \overline{\mathcal{C}} \cup \overline{w\mathcal{C}} \subseteq \mathcal{A}^T$, even though neither y nor wy lie in $\overline{\mathcal{F}}$. Applying now Lemma 8.1, we deduce that strong unicity fails for any supercuspidal representation $\pi(\Psi)$ constructed from $G^0 = T$. □

Remark 8.3 In the setting of the preceding example, we can conclude slightly more. First note that if $z \in \mathcal{A}$ and $T \subseteq G_z$, then for each positive long root α, we must have

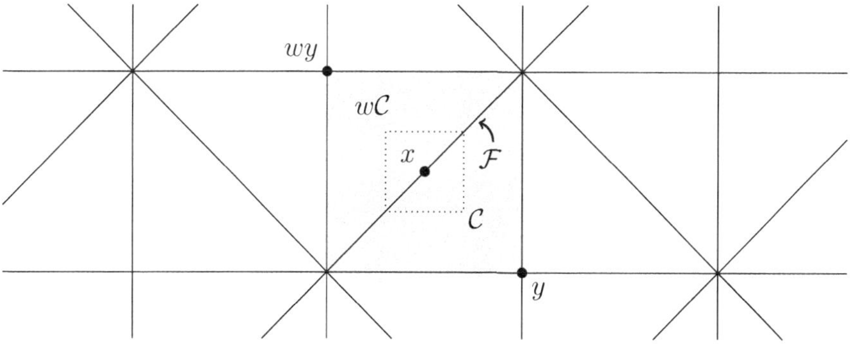

Fig. 1 A portion of the standard apartment $\mathcal{A}$ of $\mathrm{Sp}_4(F)$, identifying an alcove $\mathcal{C}$, the point x that is the image of $\mathcal{B}(T)$ in $\mathcal{B}(G)$, the facet $\mathcal{F}$ containing x and the vertex y of $\mathcal{C}$ not in $\overline{\mathcal{F}}$. The images of y and of $\mathcal{C}$ under a reflection w in the Weyl group are also indicated. The subset $\mathcal{A}^T$ of $\mathcal{A}$ fixed by T is the closed region shaded in gray. For sake of example, the set $\Omega_{\mathcal{A}}(x, s_0)$, with $s_0 = \frac{1}{10}$, is indicated with a dotted line; its closure is $\mathcal{A}^T$

$0 \leq \alpha(z) \leq 1$. It follows that $\mathcal{A}^T = \overline{\mathcal{C}} \cup \overline{w\mathcal{C}}$, as indicated in Fig. 1. We claim that in fact $\mathcal{B}^T = \mathcal{A}^T$.

Since $G_x \cap G_y = G_{\mathcal{C}}$, the orbit of y is parametrized by the set $G_x/G_{\mathcal{C}} \simeq \mathrm{GL}_2(\mathcal{O})/\mathcal{I}$, where $\mathcal{I}$ denotes its Iwahori subgroup. A set of representatives is

$$\left\{ \overline{a} = \begin{bmatrix} 1 & a & & \\ 0 & 1 & & \\ & & 1 & 0 \\ & & -a & 1 \end{bmatrix}, w = \begin{bmatrix} 0 & 1 & & \\ -1 & 0 & & \\ & & 0 & 1 \\ & & -1 & 0 \end{bmatrix} \;\middle|\; a \in \mathcal{O}/\mathfrak{p} \right\}.$$

One can verify directly that for each $a \in \mathcal{O}^\times, {}^{\overline{a}}T \nsubseteq G_y$. Thus since the fixed point set $\mathcal{B}^T$ is closed and convex but contains no vertices of chambers adjacent to x outside of $\mathcal{A}$, we conclude that $\mathcal{B}^T = \overline{\mathcal{C}} \cup \overline{w\mathcal{C}} = \mathcal{A}^T$, which is contained entirely in the standard apartment. Therefore, we can compute the simplicial radius of $\mathcal{B}^T$ with respect to x, yielding $c_T = \frac{1}{2}$.

We had seen that $\tau(1)$ and $\tau(w)$ gave types. For $a \in \mathcal{O}^\times$, since the choice $z = \overline{a}^{-1}y$ satisfies the hypotheses of Proposition 7.2 for any $s_0 > 0$, we deduce that the corresponding Mackey components $\tau(\overline{a})$ do not contain types. For the remaining Mackey components, $g^{-1}y$ is not in the closure of an alcove adjacent to x, and Theorem 7.3 implies that if $s_0 > c_T = \frac{1}{2}$, any corresponding toral supercuspidal representation contains exactly the two types on G_y identified above.

Example 8.2 generalizes immediately, to give families of supercuspidal representations for which the number of distinct types supported on a given non-special maximal compact subgroup grows exponentially with the rank of G.

Example 8.4 Let $n \geq 2$ and consider the subgroup $G' \cong \mathrm{SL}_2(F)^n$ of $G = \mathrm{Sp}_{2n}(F)$ generated by the root subgroups corresponding to long roots. To allow us to be explicit, let the roots of Sp_{2n} with respect to the diagonal torus $\mathbf{S}$ be the set $\Phi = \{\varepsilon_i \pm \varepsilon_j, \pm 2\varepsilon_i \mid 1 \leq i \neq j \leq n\}$, with simple system $\Delta = \{\varepsilon_i - \varepsilon_{i+1}, 2\varepsilon_n \mid 1 \leq i < n\}$. With respect to the basis $\{e_1, \ldots, e_n\}$ of $X_*(\mathbf{S}) \otimes_{\mathbb{Z}} \mathbb{R}$ (whose affine space is $\mathcal{A}$), dual to $\{\varepsilon_1, \ldots, \varepsilon_n\}$, the vertices of the fundamental alcove $\mathcal{C}$ are $v_i = \sum_{j=1}^{i} \frac{1}{2} e_j$, for $0 \leq i \leq n$. Let $W = W_{\mathbf{G}}$ denote the Weyl group of $\mathbf{G}$ relative to $\mathbf{S}$.

In $\mathrm{SL}_2(F)$, there are two conjugacy classes of unramified anisotropic tori, attached to the distinct conjugacy classes of vertices in $\mathcal{B}(\mathrm{SL}_2, F)$. There are between 2 and 4 conjugacy classes of ramified anisotropic tori, attached to the midpoint of facets; see [19], for example. With respect to the coordinates above, the roots of each $\mathrm{SL}_2(F)$ subgroup are $\pm 2\varepsilon_i$, and thus up to conjugacy we can arrange that the vertices have e_i-coordinates in $\{0, \frac{1}{2}\}$, whereas the midpoints have e_i-coordinate $\frac{1}{4}$.

Let each of $T_1, \ldots, T_n$ represent an anisotropic torus of $\mathrm{SL}_2(F)$, ordered so that: for $1 \leq i \leq m$, T_i is unramified and attached to $\frac{1}{2}$; for $m+1 \leq i \leq m+\ell$, T_i is ramified and attached to $\frac{1}{4}$; and for $i > m+\ell$, T_i is again unramified, but attached to 0. Then $T = T_1 \times \cdots \times T_n$ embeds as an anisotropic maximal torus of G, and $\{x\} = \mathcal{B}(T) \subseteq \mathcal{B}(G)$ has coordinates

$$x = \sum_{j=1}^{m} \frac{1}{2} e_j + \sum_{j=m+1}^{m+\ell} \frac{1}{4} e_j.$$

This is the midpoint of the 1-dimensional facet $\mathcal{F}$ whose closure is the geodesic $[v_m, v_{m+\ell}]$. Let $\Omega = \{\sum_{j=1}^{m} \frac{1}{2} e_j + \sum_{j=m+1}^{m+\ell} a_j e_j \mid 0 \leq a_j \leq \frac{1}{2}\}$; then $\mathcal{F} \subseteq \Omega$ and one can verify directly that $\Omega = \{z \in \mathcal{A} \mid T \subseteq G_z\} = \mathcal{A}^T$ using the argument of Remark 8.3. Note that since every maximal facet of Ω has x in its closure, $\Omega \subseteq \overline{\Omega_{\mathcal{A}}(x, s)}$ for any $s > 0$.

Note that $\overline{\mathcal{F}} \neq \Omega$ if and only if $\ell > 1$. In this case, let $W^{\Omega} = \{w \in W \mid w\Omega \subseteq \Omega\}$ and let $W_{\Omega} = \{w \in W \mid \forall z \in \Omega, wz = z\}$. The elements of $W(T) := W^{\Omega}/W_{\Omega}$ permute the vertices of Ω. Let $\mathbf{G}' = \mathrm{Sp}_{2m} \times \mathrm{SL}_2^{\ell} \times \mathrm{Sp}_{2n-2m-2\ell} \subseteq \mathbf{G}$; then we can identify $W(T)$ with $W_{\mathbf{G}}/W_{\mathbf{G}'}$. The vertices of $\Omega \cap \mathcal{C}$ are $v_m, \ldots, v_{m+\ell}$. The vertices v_m and $v_{m+\ell}$ are fixed by each element of $W(T)$, but the orbit of $y = v_{m+t}$ under $W(T)$, for $0 < t < \ell$, contains $\binom{\ell}{t}$ distinct vertices of Ω.

Let Ψ be any toral supercuspidal datum such that $G^0 = T$, and $\pi = \pi(\Psi)$. By the argument of the proof of Lemma 8.1, for each vertex $y = v_{m+t}$ with $0 < t < \ell$, $\mathrm{Res}_{G_y} \pi$ contains $\binom{\ell}{t} > 1$ inequivalent types. □

Remark 8.5 The quadratic tori of Examples 8.2 and 8.4 are the smallest of a broad class of tori to which the preceding arguments apply. For example, using L. Morris's classification of anisotropic maximal principal tori of $\mathrm{Sp}_{2n}(F)$ in [15] one can explicitly construct products of tori of arbitrary rank, obtaining analogous results. Similar constructive arguments may be made for groups of type B_n and D_n.

The question of determining the fixed points of an anisotropic maximal torus T acting on $\mathcal{B}(G)$ was partially addressed by F. Hurst in his thesis [9]. Under the hypotheses that $\mathbf{G}$ is simple, connected and split over F (but not of type E_7 or E_8), that T splits over a purely tamely ramified cyclic Galois extension, and some mild assumptions on F, Hurst proves that $\mathcal{B}^{T_{0+}} = \overline{\{x\}}$ if and only if T is a Coxeter torus (in which case x lies in an alcove) [9, Satz 13.14]. For other T, he shows that x lies on the wall of an alcove. Then it follows that $\mathcal{B}^{T_{0+}}$ contains the closure of the G_x-orbit of this alcove in $\mathcal{B}(G)$, since $T_{0+} \subseteq G_{x,0+} \subseteq G_{\mathcal{C}}$ for all alcoves $\mathcal{C}$ adjacent to x. (This is deduced via more explicit arguments in [9, Satz 13.15].)

F. Hurst computes a range of examples in [9, §13], including of particular interest one in type F_4 (labeled $6_p 6_p$) where T_{0+} fixes pointwise an alcove that is *not* adjacent to x. Hence, this example may yield a torus T with a more interesting set of fixed points $\mathcal{B}^T$ and consequently may be an unusual example to explore.

Acknowledgements The second author warmly thanks Anne-Marie Aubert, Manish Mishra, Alan Roche and Steven Spallone for the invitation to the excellent conference *Representation theory of p-adic groups* at IISER Pune, India. The stimulating environment of the workshop contributed significantly to this article; in particular, fellow participant Jeff Adler provided invaluable insight into tori and buildings, and he offered up the torus of Example 8.2.

References

1. J.D. Adler, Refined anisotropic K-types and supercuspidal representations. Pacific J. Math. **185**(1), 1–32 (1998)
2. J. Bernstein, *Le "centre" de Bernstein, Representations of reductive groups over a local field*, ed. by P. Deligne (Travaux en Cours, Hermann, Paris, 1984), pp. 1–32
3. C. Breuil, A. Mézard, Multiplicités modulaires et représentations de $\mathrm{GL}_2(\mathbf{Z}_p)$ et de $\mathrm{Gal}(\overline{\mathbf{Q}}_p/\mathbf{Q}_p)$ en $l = p$, Duke Math. J. **115**(2), 205–310 (2002). With an appendix by Guy Henniart
4. F. Bruhat, J. Tits, Groupes réductifs sur un corps local. Inst. Hautes Études Sci. Publ. Math. **41**, 5–251 (1972)
5. F. Bruhat, J. Tits, Groupes réductifs sur un corps local. II. Schémas en groupes. Existence d'une donnée radicielle valuée. Inst. Hautes Études Sci. Publ. Math. **60**, 197–376 (1984)
6. J. Fintzen, On the Moy-Prasad filtration. Preprint arXiv:1511.00726v3 [math.RT] (2017)
7. J. Fintzen, Types for tame p-adic groups. Preprint arXiv:1810.04198 [math.TR] (2018)
8. J. Hakim, F. Murnaghan, Distinguished tame supercuspidal representations. Int. Math. Res. Pap. IMRP **2**, Art. ID rpn005, 166 (2008)
9. F. Hurst, Primitive Tori in einfachen spaltenden Gruppen und ihre Fixpunkte im Bruhat-Tits Gebäude, Inaugural-Dissertation zur Erlangung der Doktorwürde der Naturwissenschaftlich-Mathematischen Gesamtfakultät der Ruprecht-Karls-Universität Heidelberg (2005), 164pp
10. T. Kaletha, Regular supercuspidal representations, Preprint arXiv:1602.03144v2 (2016)
11. J.-L. Kim, Supercuspidal representations: an exhaustion theorem. J. Amer. Math. Soc. **20**(2), 273–320 (electronic) (2007)
12. P. Latham, Unicity of types for supercuspidal representations of p-adic SL_2. J. Number Theor. **162**, 376–390 (2016)
13. P. Latham, The unicity of types for depth-zero supercuspidal representations. Represent. Theor. **21**, 590–610 (2017)
14. P. Latham, On the unicity of types in special linear groups. Manuscripta Math. **157**(3–4), 445–465 (2018)
15. L. Morris, Some tamely ramified supercuspidal representations of symplectic groups. Proc. London Math. Soc. **63**(3), 519–551 (1991)
16. A. Moy, G. Prasad, Unrefined minimal K-types for p-adic groups. Invent. Math. **116**(1–3), 393–408 (1994)
17. F. Murnaghan, Parametrization of tame supercuspidal representations, in *On certain L-functions*. Clay Mathematical Proceedings 13, American Mathematical Society, Providence, RI, 2011, pp. 439–469
18. F. Murnaghan, Distinguished positive regular representations. Bull. Iranian Math. Soc. **43**(4), 291–311 (2017)
19. M. Nevins, Branching rules for supercuspidal representations of $SL_2(k)$, for k a p-adic field. J. Algebra **377**, 204–231 (2013)
20. M. Nevins, On branching rules of depth-zero representations. J. Algebra **408**, 1–27 (2014)
21. V. Paskunas, Unicity of types for supercuspidal representations of GL_N. Proc. Lond. Math. Soc. **91**(3), 623–654 (2005)
22. T.A. Springer, *Reductive groups, automorphic forms, representations and L-functions, in (Proceedings Symposium Pure Mathematical, Oregon State University, Corvallis, Ore, 1977)*, Part 1 (XXXIII, American Mathematical Society, Providence, R.I, Proceeding Symposium Pure Mathematical, 1979), pp. 3–27
23. J. Tits, Reductive groups over local fields, automorphic forms, representations and L-functions (Oregon State University, Corvallis, Ore, 1977), *Part 1, Proceedings Symposium Pure Mathematical, XXXIII*, American Mathematical Society, Providence, R.I., 1979, pp. 29–69
24. J.-K. Yu, Construction of tame supercuspidal representations. J. Amer. Math. Soc. **14**(3), 579–622 (2001) (electronic)

Local Gamma Factors, Converse Theorems and Related Problems

Chufeng Nien

Contents

1 Introduction 192
  1.1 Notation 192
  1.2 Gamma Factors in the p-Adic Case 193
2 Local Converse Problem and Jacquet's Conjecture 194
  2.1 Local Converse Problem Over p-Adic Fields 194
  2.2 Local Converse Problem Over Finite Fields 195
  2.3 Sharpness on the Bound $[\frac{n}{2}]$ 196
3 Explicit Computation of Gamma Factors Over Finite Fields 197
  3.1 Bessel Functions 197
  3.2 $n \times 1$ Gamma Factor and Gauss Sums 198
  3.3 Some $n \times 1$ Local Converse Theorem for $\mathrm{GL}_n(\mathbb{F}_p)$ 199
4 Gamma Factors for Level Zero Cuspidals 201
5 Converse Theorems and Distinction 202
  5.1 p-Adic Results 202
  5.2 Finite Field Results 204
References 204

Abstract This paper reviews the development of Local Converse Theorems and related results on distinction for representations of general linear groups over finite and p-adic fields.

Keywords Generic representation · Whittaker models · Gamma factors · Local Converse Theorem · Distinguished representation · Gauss sum

2000 Mathematics Subject Classification. Primary 20C33 · Secondary 11L05

C. Nien (✉)
Key Laboratory of Computing and Stochastic Mathematics (Ministry of Education),
School of Mathematics and Statistics, Hunan Normal University, Changsha,
Hunan 410081, China
e-mail: nienpig@hotmail.com

A.-M. Aubert et al. (eds.), *Representations of Reductive p-adic Groups*,
Progress in Mathematics 328, https://doi.org/10.1007/978-981-13-6628-4_7

1 Introduction

We will give a survey on local gamma factors, Local Converse Theorems and related problems. For a more complete picture on these subjects, please see some other articles. For example, the references [9, 10] contain nice and substantial introductions.

1.1 Notation

Let $\mathbb{F}_q$ be the finite field of q elements and F be a p-adic field. We will write K for such a field, finite or p-adic, depending on context. Denote by B_n the standard Borel subgroup of upper-triangular matrices in GL_n. Let U_n be its unipotent radical and write A_n for the Levi subgroup of B_n consisting of diagonal matrices. Denote by Z_n the center of GL_n.

Let ψ be a fixed nontrivial additive character of K. The **standard non-degenerate** character ψ_n is given by

$$\psi_n(u) = \psi(\sum_{i=1}^{n-1} u_{i,i+1}), \text{ for } u = (u_{i,j}) \in \mathrm{U}_n.$$

A character ψ' of U_n is called **non-degenerate** if

$$\psi'(u) = \psi(\sum_{i=1}^{n-1} a_i u_{i,i+1}), \text{ for } u = (u_{i,j}) \in \mathrm{U}_n,$$

with $a_i \in K^\times$ for all i. We call an irreducible representation π of GL_n ψ'-**generic** if

$$\dim \mathrm{Hom}_{\mathrm{GL}_n}(\pi, \mathrm{Ind}_{\mathrm{U}_n}^{\mathrm{GL}_n} \psi') \neq 0$$

for ψ' non-degenerate. By transitivity of the conjugation action of A_n on the set of generic characters, if π is ψ'-generic, then it is also ψ''-generic for any other non-degenerate character ψ'' of U_n. Hence, we often say "generic" rather than "ψ'-generic."

For π irreducible, the image of π under an embedding in $\mathrm{Ind}_{\mathrm{U}_n}^{\mathrm{GL}_n} \psi_n$ is unique.

Theorem 1.1 [Uniqueness of Whittaker models]

$$\dim \mathrm{Hom}_{\mathrm{U}_n}(\pi|_{\mathrm{U}_n}, \psi_n) = \dim \mathrm{Hom}_{\mathrm{GL}_n}(\pi, \mathrm{Ind}_{\mathrm{U}_n}^{\mathrm{GL}_n} \psi_n) \leq 1,$$

for any irreducible representation π *of* $\mathrm{GL}_n(K)$.

When π is generic, the above Hom-space has dimension one. Moreover, any irreducible cuspidal representation of $\mathrm{GL}_n(K)$ is generic.

Let $\ell_{\psi_n} \in \mathrm{Hom}_{\mathrm{U}_n}(\pi|_{\mathrm{U}_n}, \psi_n)$ be nonzero. Such an element is called a Whittaker functional of π. Let $W_v(g) = \ell_{\psi_n}(\pi(g)v)$, for $v \in V_\pi$ and $g \in \mathrm{GL}_n(K)$. The function $W_v \in \mathrm{Ind}_{\mathrm{U}_n}^{\mathrm{GL}_n} \psi_n$ is called the Whittaker function attached to the vector v. By Theorem 1.1, the subspace generated by all Whittaker functions W_v is unique. We denote it by $\mathcal{W}(\pi, \psi)$. This space is called the Whittaker model of π. For $W_v \in \mathcal{W}(\pi, \psi)$, denote by $\widetilde{W}_v$ the function on GL_n given by

$$\widetilde{W}_v(g) = W_v(w_n({}^t g^{-1})), \ g \in \mathrm{GL}_n(K)$$

where w_n is the element of $\mathrm{GL}_n(K)$ with 1's on the anti-diagonal and zeros elsewhere (the longest Weyl element). Let $\tilde{\pi}$ be the contragredient of π. Then,

$$\widetilde{W}_v \in \mathcal{W}(\tilde{\pi}, \psi^{-1}).$$

1.2 Gamma Factors in the p-Adic Case

Let $n > t \geq 1$ be integers. Let π be an irreducible generic representation of $\mathrm{GL}_n(F)$ with central character ω_π and let τ be an irreducible generic representation of $\mathrm{GL}_t(F)$ with central character ω_τ. Let $W_\pi \in \mathcal{W}(\pi, \psi_n)$ and $W_\tau \in \mathcal{W}(\tau, \psi_t^{-1})$ be Whittaker functions of π and τ. For $g \in \mathrm{GL}_n(F)$, we denote by R_g the action by right translation of g on $\mathbb{C}$-valued functions on $\mathrm{GL}_n(F)$ to $\mathbb{C}$. Let $w_{n,t} = \begin{pmatrix} \mathrm{I}_t & 0 \\ 0 & w_{n-t} \end{pmatrix}$.

For $n - t - 1 \geq j \geq 0$, a local zeta integral for $W_\pi \in \mathcal{W}(\pi, \psi)$, $W_\tau \in \mathcal{W}(\tau, \psi^{-1})$ is defined by

$$\mathcal{Z}(W_\pi, W_\tau, s; j) := \int_g \int_x W_\pi \begin{pmatrix} g & 0 & 0 \\ x & \mathrm{I}_{n-t-1-j} & 0 \\ 0 & 0 & \mathrm{I}_{j+1} \end{pmatrix} W_\tau(g) |\det g|^{s-\frac{n-t}{2}} dx dg,$$

where integration in the variable g is over $\mathrm{U}_t(F)\backslash\mathrm{GL}_t(F)$ and integration in the variable x is over $\mathrm{Mat}_{(n-t-j-1)\times t}(F)$.

In the celebrated paper [18], Jacquet, Piatetski-Shapiro, and Shalika established the following functional equations.

Theorem 1.2 [18, Section 2.7]

(1) Each integral $\mathcal{Z}(W_\pi, W_\tau, s; j)$ is absolutely convergent for $\Re(s)$ sufficiently large and is a rational function of q^{-s}. More precisely, for fixed j, the integrals $\mathcal{Z}(W_\pi, W_\tau, s; j)$ span a fractional ideal (independent of j)

$$\mathbb{C}[q^s, q^{-s}]L(s, \pi \times \tau)$$

of the ring $\mathbb{C}[q^s, q^{-s}]$, where the local L-factor $L(s, \pi \times \tau)$ has the form $P(q^s)^{-1}$, with $P \in \mathbb{C}[x]$ and $P(0) = 1$.

(2) *For $n-t-1 \geq j \geq 0$, there is a factor $\epsilon(s, \pi \times \tau, \psi)$ independent of j, such that*

$$\frac{\mathcal{Z}(R_{w_{n,r}} \widetilde{W}_\pi, \widetilde{W}_\tau, 1-s; n-t-j-1)}{L(1-s, \tilde{\pi} \times \tilde{\tau})} = \omega_\tau(-1)^{n-1} \epsilon(s, \pi \times \tau, \psi) \frac{\mathcal{Z}(W_\pi, W_\tau, s; j)}{L(s, \pi \times \tau)}.$$

The local Gamma factor[1] attached to a pair of representations π and τ is defined by

$$\Gamma(s, \pi \times \tau, \psi) = \epsilon(s, \pi \times \tau, \psi) \frac{L(1-s, \tilde{\pi} \times \tilde{\tau})}{L(s, \pi \times \tau)}. \tag{1.1}$$

The functional equation in Theorem 1.2 (2) can be rewritten as

$$\begin{aligned} &\mathcal{Z}(R_{w_{n,t}} \widetilde{W}_\pi, \widetilde{W}_\tau, 1-s; n-t-j-1) \\ &= \omega_\tau(-1)^{n-1} \Gamma(s, \pi \times \tau, \psi) \mathcal{Z}(W_\pi, W_\tau, s; j). \end{aligned} \tag{1.2}$$

2 Local Converse Problem and Jacquet's Conjecture

2.1 Local Converse Problem Over p-Adic Fields

It is natural to ask if an irreducible generic representation is determined by sufficiently many of its twisted Gamma factors. More precisely, we ask the following.

Question 2.1 [$n \times m$ Local Converse Problem for $\mathrm{GL}_n(F)$] *Let π_1 and π_2 be irreducible generic representations of $\mathrm{GL}_n(F)$. If the (local) Gamma factors $\Gamma(s, \pi_1 \times \tau, \psi)$ and $\Gamma(s, \pi_2 \times \tau, \psi)$ agree for any irreducible generic representation τ of $\mathrm{GL}_t(F)$, with $t = 1, 2, \ldots, m$, can we deduce that π_1 and π_2 are isomorphic?*

In 1993, in [14], Henniart proved $n \times (n-1)$ Local Converse Theorem for $\mathrm{GL}_n(F)$ and Jiang-Ping (Jeff) Chen proved $n \times (n-2)$ Local Converse Theorem [8] in 1996. However, a conjecture credited to Jacquet predicts (essentially) that $n \times [\frac{n}{2}]$ twisted Gamma factors should be enough.

Conjecture 2.2 [Jacquet's conjecture] *Let π_1 and π_2 be irreducible smooth* **generic** *representations of $\mathrm{GL}_n(F)$ with the* **same central character**. *Assume that the local Gamma factors $\Gamma(s, \pi_1 \times \tau, \psi)$ and $\Gamma(s, \pi_2 \times \tau, \psi)$ are equal for all irreducible* **generic** *representations τ of $\mathrm{GL}_t(F)$, $1 \leq t \leq [\frac{n}{2}]$. Then, π_1 and π_2 are isomorphic.*

For a while, no further progress was made toward the conjecture and people turned to consider finite field analogue for inspiration.

[1] In this paper, we use Gamma or Γ-factors to indicate the gamma factors in the p-adic case and use gamma or γ-factors in the case of finite fields.

2.2 Local Converse Problem Over Finite Fields

Roditty in her thesis considered the finite field analogue of zeta integrals and gamma factors.

Theorem 2.3 [26] *Let π be an irreducible* **cuspidal** *representation of* $\mathrm{GL}_n(\mathbb{F}_q)$ *and τ an irreducible* **generic** *representation of* $\mathrm{GL}_t(\mathbb{F}_q)$ *with $n > t$. Then, there exists a complex number $\gamma(\pi \times \tau, \psi)$ such that*

$$\gamma(\pi \times \tau, \psi)q^{tj} \sum_{m\in \mathrm{U}_t\backslash \mathrm{GL}_t(\mathbb{F}_q)} \sum_{x\in \mathrm{M}_{n-t-j-1,t}} W_\pi \begin{pmatrix} m & 0 & 0 \\ x & \mathrm{I}_{n-t-j-1} & 0 \\ 0 & 0 & \mathrm{I}_{j+1} \end{pmatrix} W_\tau(m)$$
$$= \sum_{m\in \mathrm{U}_j\backslash \mathrm{GL}_t(\mathbb{F}_q)} \sum_{y\in \mathrm{M}_{t,j}} W_\pi(\begin{pmatrix} 0 & \mathrm{I}_{n-t-j} & 0 \\ 0 & 0 & \mathrm{I}_j \\ m & 0 & y \end{pmatrix}) W_\tau(m),$$

for all $0 \le j \le n-t-1$, $W_\pi \in \mathcal{W}(\pi, \psi)$ and $W_\tau \in \mathcal{W}(\tau, \psi^{-1})$.

Due to lack of uniqueness of some auxiliary models, the above functional equations can only be built when $n > t$ and the irreducible representation π of $\mathrm{GL}_n(\mathbb{F}_q)$ is cuspidal. As in the p-adic case, we can consider the $n \times m$ Local Converse Problem.

Question 2.4 [$n \times m$ Local Converse Problem for $\mathrm{GL}_n(\mathbb{F}_q)$] *Let π_1 and π_2 be irreducible cuspidal representations of* $\mathrm{GL}_n(\mathbb{F}_q)$ *with the same central character.*[2] *If the (local) gamma factors $\gamma(\pi_1 \times \tau, \psi)$ and $\gamma(\pi_2 \times \tau, \psi)$ agree for any irreducible generic representation τ of* $\mathrm{GL}_t(\mathbb{F}_q)$, *with $t = 1, 2, \ldots, m$, can we deduce that π_1 and π_2 are isomorphic?*

In [26], Roditty verified $n \times (n-1)$ and $n \times (n-2)$ Local Converse Theorem for $\mathrm{GL}_n(\mathbb{F}_q)$. In 2014, the author proved the following finite field analogue of Jacquet's conjecture.

Theorem 2.5 [$n \times [\frac{n}{2}]$ Local Converse Theorem, [19]] *Let π_1 and π_2 be irreducible cuspidal representations of* $\mathrm{GL}_n(\mathbb{F}_q)$ *with the same central character. If*

$$\gamma(\pi_1 \times \tau, \psi) = \gamma(\pi_2 \times \tau, \psi),$$

for all irreducible generic representation of $\mathrm{GL}_r(\mathbb{F}_q)$, $1 \le r \le [\frac{n}{2}]$, *then $\pi_1 \cong \pi_2$.*

Denote by P_n the mirabolic subgroup of GL_n, consisting of matrices in GL_n with last row equal to $(0, \ldots, 0, 1)$. Let $\alpha = \begin{pmatrix} 0 & \mathrm{I}_{n-1} \\ 1 & 0 \end{pmatrix}$.

The proof of the Theorem 2.5 is based on the following geometric decomposition of GL_n.

---

[2]This condition can be removed if we assume that $\gamma(\pi_1 \times \tau, \psi) = \gamma(\pi_2 \times \tau, \psi)$ for any $\tau \in \mathbb{F}_q^\times$. Refer to [22].

Proposition 2.6 [19]

$$\mathrm{A}_n W(\mathrm{GL}_n) \subset \cup_{0 \leq r \leq [\frac{n}{2}], n-[\frac{n}{2}] \leq k \leq n} \alpha^r \mathrm{P}_n \alpha^k \mathrm{Z}_n.$$

Or equivalently, $\mathrm{GL}_n = \mathrm{U}_n(\cup_{0 \leq r \leq [\frac{n}{2}], n-[\frac{n}{2}] \leq k \leq n} \alpha^r \mathrm{P}_n \alpha^k)\mathrm{Z}_n \mathrm{U}_n$.

2.3 In [17], a project with Dihua Jiang and Shaun Stevens, we removed the criterion on the agreement of central characters in Conjecture 2.2 by appealing to the stability of Gamma factors.

Corollary 2.7 [17] *Let π_1 and π_2 be irreducible generic representations of* $\mathrm{GL}_n(F)$. *If their local Gamma factors $\Gamma(s, \pi_1 \times \chi, \psi_F)$ and $\Gamma(s, \pi_2 \times \chi, \psi_F)$ are equal (as functions in the complex variable s) for any character χ of $F^\times$, then they possess the same central character.*

Applying the multiplicativity of Gamma factors, we reduce Conjecture 2.2 to the following, which allows us checking the agreement of twisted Gamma factors against irreducible cuspidals only instead of all irreducible generic representations.

Conjecture 2.8 *Let π_1 and π_2 be irreducible smooth* **generic** *representations of* $\mathrm{GL}_n(F)$. *Assume that the local Gamma factors $\Gamma(s, \pi_1 \times \tau, \psi)$ and $\Gamma(s, \pi_2 \times \tau, \psi)$ are equal (as functions in the complex variable s) for all irreducible* **cuspidal** *representations τ of* $\mathrm{GL}_t(F)$, $1 \leq t \leq [\frac{n}{2}]$. *Then, π_1 and π_2 are isomorphic.*

We further set up a scheme for verifying Conjecture 2.8 and used this to confirm many cases. Based on our formulation, Adrian et al. [1] verified Conjecture 2.2 for $\mathrm{GL}_p(F)$ for p prime. Finally, Jacquet and Liu [16], using an analytic approach, and Chai [7], using p-adic Bessel functions, independently settled this longstanding conjecture.

2.3 Sharpness on the Bound $[\frac{n}{2}]$

After confirmation of Jacquet's conjecture, it was natural to consider sharpness of the bound $[\frac{n}{2}]$. The paper [3] showed that $[\frac{n}{2}]$ is the sharp bound for necessary twisting in the Local Converse Theorem for a pair of cuspidal representations of $\mathrm{GL}_n(F)$ for n prime. However, for special families of cuspidal representations, you may wonder if it is possible to lower the upper bound for necessary twisting? For instance, for simple cuspidal representations(i.e., cuspidal representations of minimal positive depth), the upper bound may be lowered to 1: See [2, 27] when the residue characteristic is coprime to n and [5] for general case.

Question 2.9 *In the case of depth zero cuspidal representations, can we lower the upper bound for necessary twisting to 1?*

Since depth zero cuspidal representations of $\mathrm{GL}_n(F)$ are constructed from cuspidal representations of $\mathrm{GL}_n(\mathbb{F}_q)$, we start by studying the finite field analogue.

3 Explicit Computation of Gamma Factors Over Finite Fields

3.1 Bessel Functions

Bessel functions were introduced by Gelfand in [11]. They play an important role in explicit computations of gamma factors and in Converse Theorems over finite fields.

Proposition 3.1 *Let π be an irreducible generic representation of $\mathrm{GL}_n(\mathbb{F}_q)$, and let χ_π be its character. The genericity of π guarantees the existence of a function $\mathcal{B} \in \mathcal{W}(\pi, \psi)$ satisfying*

$$\mathcal{B}(u_1 g u_2) = \psi_n(u_1)\psi_n(u_2), \textit{ for all } u_1, u_2, \in \mathrm{U}_n, \ g \in \mathrm{GL}_n(\mathbb{F}_q). \tag{3.1}$$

By Theorem 1.1, this function is unique up to a scalar.

Proposition 3.2 [[11, Proposition 4.5] or [26, Lemma 6.1.1]] *Let π be an irreducible generic representation of $\mathrm{GL}_n(\mathbb{F}_q)$, and let χ_π be its character. Define*

$$\mathcal{B}(g) = |\mathrm{U}_n(\mathbb{F}_q)|^{-1} \sum_{u \in \mathrm{U}_n} \psi_n(u^{-1})\chi_\pi(gu), \textit{ for } g \in \mathrm{GL}_n(\mathbb{F}_q).$$

Then, $\mathcal{B}$ satisfies (3.1). *Moreover, $\mathcal{B}(\mathrm{I}_n) = 1$.*

We call $\mathcal{B}$ the Bessel function of π in terms of ψ and denote it by $\mathcal{B}_{\pi,\psi}$ to indicate the dependence on π and ψ. We then have the following nice expression for twisted gamma factors in terms of the associated Bessel functions.

Proposition 3.3 [26] *Let π be an irreducible cuspidal representation of $\mathrm{GL}_n(\mathbb{F}_q)$ and τ an irreducible generic representation of $\mathrm{GL}_r(\mathbb{F}_q)$ with $r < n$. Then,*

$$\gamma(\pi \times \tau, \psi) = \sum_{\mathrm{U}_r \backslash \mathrm{GL}_r} \mathcal{B}_{\pi,\psi_n}\begin{pmatrix} 0 & \mathrm{I}_{n-r} \\ m & 0 \end{pmatrix} \mathcal{B}_{\tau,\psi_r^{-1}}(m). \tag{3.2}$$

Bessel functions have nice symmetry properties. These together with the geometric decomposition, Lemma 2.6, are the key to the proof of Theorem 2.5. Furthermore, this symmetry also contributes to the proof of the converse theorem on distinction, which is to be discussed in later sections.

Proposition 3.4 [Symmetry of Bessel functions, [26]] *Let π be an irreducible generic representation of $\mathrm{GL}_n(\mathbb{F}_q)$. Then,*

$$\mathcal{B}_{\pi,\psi}(g^{-1}) = \overline{\mathcal{B}_{\pi,\psi}(g)}, \ g \in \mathrm{GL}_n(\mathbb{F}_q).$$

3.2 $n \times 1$ Gamma Factor and Gauss Sums

Let $\overline{\mathbb{F}_q}$ denote the algebraic closure of $\mathbb{F}_q$ and $\mathbb{F}_{q^d}$ be the unique field extension of $\mathbb{F}_q$ with index d such that $\mathbb{F}_q \subset \mathbb{F}_{q^d} \subset \overline{\mathbb{F}_q}$. In [12], Green showed that the character of an irreducible representation of $\mathrm{GL}_n(\mathbb{F}_q)$ can be computed in terms of a suitable associated character of

$$\mathbb{F}_{q^{n_1}}^\times \times \mathbb{F}_{q^{n_2}}^\times \times \cdots \times \mathbb{F}_{q^{n_k}}^\times$$

where $n_1 + \cdots + n_k = n$.

Let η be a character of $\mathbb{F}_{q^n}^\times$ satisfying the following properties:

$$\eta^{q^m-1} \neq 1, \text{ for all positive } m < n. \tag{3.3}$$

Such a character η is called **regular**.

Theorem 3.5 (Green's construction, [12]) *A regular character η of $\mathbb{F}_{q^n}^\times$ correspond to irreducible cuspidal representation π_η of $\mathrm{GL}_n(\mathbb{F}_q)$. Moreover, any irreducible cuspidal representation corresponds to some regular character.*

Regular characters η_1 and η_2 correspond to the same cuspidal π if and only if

$$\eta_1 = \eta_2^{q^t} \text{ for some } t \geq 1. \tag{3.4}$$

Let $\mathrm{tr}_n : \mathbb{F}_{q^n} \to \mathbb{F}_q$ and $\mathrm{Nr}_n : \mathbb{F}_{q^n}^\times \to \mathbb{F}_q^\times$ be the trace map and norm map (resp.) for the extension $\mathbb{F}_{q^n} / \mathbb{F}_q$. Using Green's formula, we can relate $n \times 1$ gamma factors over finite fields and abelian Gauss sums.

Theorem 3.6 [20, Theorem 1.1] *Let π be an irreducible cuspidal representation of $\mathrm{GL}_n(\mathbb{F}_q)$, $n \geq 2$ and $\chi \in \widehat{\mathbb{F}_q^\times}$. Then,*

$$\gamma(\pi \times \chi, \psi) = (-q^{-1}\chi(-1))^{n-1} \sum_{\sigma \in \mathbb{F}_{q^n}^\times} \psi_n(\mathrm{tr}_n \sigma^{-1}) \eta_\pi(\sigma) \chi(\mathrm{Nr}_n(\sigma)),$$

where η_π is the regular character of $\mathbb{F}_{q^n}^\times$ corresponding to π in Green's construction.

The above theorem and [6, Theorem 9.3] imply that the gamma factors for $\mathrm{GL}_n(\mathbb{F}_q)$ defined by Roditty coincide with the ones constructed through cohomology by Braverman and Kazhdan in [6].

The next result is an analogue of a well known property of abelian Gauss sums.

Corollary 3.7 (absolute values of gamma factors, [21, Corollary 4.1]) *Let π be an irreducible cuspidal representation of $\mathrm{GL}_n(\mathbb{F}_q)$ and τ an irreducible generic representation of $\mathrm{GL}_t(\mathbb{F}_q)$ with $n > t$. Then,*

$$|\gamma(\pi \times \tau, \psi)| = q^{\frac{-t(n-t-1)}{2}}.$$

Proof Use

$$\gamma(\pi \times \tau, \psi)\gamma(\tilde{\pi} \times \tilde{\tau}, \psi^{-1}) = q^{-t(n-t-1)}$$

and

$$\mathcal{B}_{\tilde{\pi},\psi^{-1}} = \overline{\mathcal{B}_{\pi,\psi}}.$$

Theorem 3.6 and Corollary 3.7 suggest the following question.

Question 3.8 *Let π be an irreducible cuspidal representation of $\mathrm{GL}_n(\mathbb{F}_q)$ and τ an irreducible, cuspidal representation of $\mathrm{GL}_n(\mathbb{F}_q)$. Can we express the gamma factor $\gamma(\pi \times \tau, \psi)$ in terms of the characters η_π and η_τ? Moreover, if this is possible, can we view the resulting expression as some sort of generalization of a Gauss sum?*

3.3 *Some $n \times 1$ Local Converse Theorem for* $\mathrm{GL}_n(\mathbb{F}_p)$

Let $\mathrm{Nr}_{n:d}$ denote the norm map from $\mathbb{F}_{q^n}^\times$ to $\mathbb{F}_{q^d}^\times$ where $d \mid n$. Guided by Langlands functoriality, there should exist a tensor lifting taking an irreducible representation $\pi \otimes \tau$ of $\mathrm{GL}_n(\mathbb{F}_q) \times \mathrm{GL}_m(\mathbb{F}_q)$ to an irreducible representation Π of $\mathrm{GL}_{mn}(\mathbb{F}_q)$ such that Π corresponds to the character $\eta_\pi \circ \mathrm{Nr}_{mn:n} \cdot \eta_\tau \circ \mathrm{Nr}_{mn:m}$ of $\mathbb{F}_{q^{mn}}^\times$. Note that the character $\eta_\pi \circ \mathrm{Nr}_{mn:n} \cdot \eta_\tau \circ \mathrm{Nr}_{mn:m}$ is possibly not regular so that the corresponding representation Π need not be cuspidal. We expect that Π corresponds to the unique generic subrepresentation of

$$\mathrm{Ind}_{\mathrm{GL}_{n_1} \times \cdots \times \mathrm{GL}_{n_k}}^{\mathrm{GL}_{mn}} \pi_1 \otimes \cdots \otimes \pi_k,$$

where $n_1 + \cdots + n_k = mn$ and each π_i is cuspidal. If we extend Roditty's gamma factors to generic representations through multiplicativity,[3] then $\gamma(\Pi \times 1, \psi) := \prod_{i=1}^k \gamma(\pi_i \times 1, \psi)$ and we expect that

$$\gamma(\Pi \times 1, \psi) = q^{\frac{2-m^2-m}{2}} \gamma(\pi \times \tau, \psi).$$

Based on the known case $m = 1$ (Theorem 3.6), we propose the following conjectural formula for $n \times m$ gamma factors in terms of the corresponding characters of $\mathbb{F}_{q^{mn}}^\times$.

Conjecture 3.9 *For $n > m$, let π and τ be irreducible cuspidal representations of $\mathrm{GL}_n(\mathbb{F}_q)$ and $\mathrm{GL}_m(\mathbb{F}_q)$ respectively, and χ and η be the corresponding regular characters of $\mathbb{F}_{q^n}^\times$ and of $\mathbb{F}_{q^m}^\times$. Then,*

$$\gamma(\pi \times \tau, \psi) = c \cdot \chi(-1)^{m-1} \eta(-1)^{n-1} G(\chi \circ \mathrm{Nr}_{mn:n} \cdot \eta \circ \mathrm{Nr}_{mn:m}, \psi), \tag{3.5}$$

where

[3]Multiplicativity of gamma factors of $\mathrm{GL}_n(\mathbb{F}_q)$ has been established by Soudry in an unpublished note.

$$G(\beta^{-1}, \psi) := \sum_{a \in \mathbb{F}_{q^N}^{\times}} \beta^{-1}(a)\psi(\mathrm{tr}_N a) = \sum_{a \in \mathbb{F}_{q^N}^{\times}} \beta(a)\psi(\mathrm{tr}_N a^{-1})$$

is the classical Gauss sum for a character β^{-1} *of* $\mathbb{F}_{q^N}^{\times}$, *and*

$$c = (-1)^{m(n-1)} q^{-mn + \frac{m^2+m}{2}}.$$

The above conjecture interprets representation theoretic matters in terms of number theoretic objects. For the case $m = 1$, this conjectural formula has been verified in Theorem 3.6. When m divides n, Conjecture 3.9 and the multiplicativity of gamma factors imply the Hasse-Davenport relation for Gauss sums.

The Gauss sums involved in $n \times 1$ gamma factors over finite fields are algebraic integers, and the results developed for Gauss sums over $\mathbb{F}_p$ enable us to extract the invariants for characterizing our target representations. Using Theorem 3.6 and known results about Gauss sums developed in number theory, the author and Lei Zhang translated the $n \times 1$ Local Converse Problem for $\mathrm{GL}_n(\mathbb{F}_p)$ into a combinatorial statement and obtained the following results.

Theorem 3.10 [22] *Let* π_1 *and* π_2 *be irreducible cuspidal representations of* $\mathrm{GL}_n(\mathbb{F}_p)$ *where* $n \leq 5$ *and* p *is prime.*[4] *If*

$$\gamma(\pi_1 \times \chi, \psi) = \gamma(\pi_2 \times \chi, \psi)$$

for all $\chi \in \widehat{\mathbb{F}_p^{\times}}$, *then* $\pi_1 \cong \pi_2$.

Recall that **Mersenne primes** are primes of the form $2^n - 1$. In this case, n must be a prime and all non-trivial characters of $\mathbb{F}_{2^n}^{\times}$ are regular.

Proposition 3.11 [22] *If* $2^n - 1$ *is a prime (Mersenne prime), then the* $n \times 1$ *Local Converse Theorem holds for* $\mathrm{GL}_n(\mathbb{F}_2)$.

The 50th Mersenne prime, $M_{50} = 2^{77232917} - 1$, was found on December 26, 2017, and it became the largest prime number known to mankind. Thus, the 77232917×1 Local Converse Theorem holds for $\mathrm{GL}_{77232917}(\mathbb{F}_2)$. Moreover, in the appendix to [22], Zhiwei Yun gives a series of examples confirming the $n \times 1$ Local Converse Theorem by transferring the questions on Gauss sums to ones concerning Kloosterman sheaves.

Theorem 3.12 (Appendix of [22] by Zhiwei, Yun) *Let* $n \geq 1$ *be an integer satisfying* $n < \frac{q-1}{2\sqrt{q}} + 1$. *Let* $\chi_1, \chi_2 : \mathbb{F}_{q^n}^{\times} \to \mathbb{C}^{\times}$ *be characters. Suppose that*

$$G(\chi_1 \cdot \eta, \psi) = G(\chi_2 \cdot \eta, \psi),$$

[4]Here, the cardinality of $\mathbb{F}_q$ is required to be a prime to make use of certain properties of Gauss sums.

for any character $\eta : \mathbb{F}_q^\times \to \mathbb{C}^\times$. *Then,* χ_1 *and* χ_2 *are in the same Frobenius orbit; i.e., there exists* $j \in \mathbb{Z}$ *such that* $\chi_1 = \chi_2^{q^j}$, *where* $G(\chi, \psi) = \sum_{x \in \mathbb{F}_{q^n}^\times} \chi(x)\psi(\mathrm{tr}_n x)$.

However, we also found counterexamples to the 6×1 Local Converse Problem.

Theorem 3.13 [22] *There exist non-isomorphic irreducible cuspidal representations* π_1 *and* π_2 *of* $\mathrm{GL}_6(\mathbb{F}_3)$ *such that*

$$\gamma(\pi_1 \times \chi, \psi) = \gamma(\pi_2 \times \chi, \psi)$$

for all $\chi \in \widehat{\mathbb{F}_3^\times}$.

To interpret these different outcomes to $n \times 1$ Local Converse Problems, we introduced a notion of **primitive form** for cuspidal representations and made the conjecture below, drawing inspiration from Arthur's work [4]. Let r be a prime divisor of n. For each regular character χ of $\mathbb{F}_{q^n}^\times$, denote by π_χ the cuspidal representation of $\mathrm{GL}_r(\mathbb{F}_{q^{n/r}})$ corresponding to χ which is called a **primitive representation** associated to χ.

Conjecture 3.14 *[22] For* $i = 1, 2$, *let* χ_i *be a regular character of* $\mathbb{F}_{q^n}^\times$ *and* r *be a prime divisor of* n. *If* $\gamma(\pi_{\chi_1} \times \eta, \psi) = \gamma(\pi_{\chi_2} \times \eta, \psi)$ *for all characters* η *of* $\mathbb{F}_{q^{n/r}}^\times$. *Then* $\pi_{\chi_1} \cong \pi_{\chi_2}$.

4 Gamma Factors for Level Zero Cuspidals

Let $\mathfrak{o}$ be the ring of integers in F and $\mathfrak{p}$ the prime ideal in $\mathfrak{o}$. Denote by $\mathfrak{k}$ the residue field of F. Let cInd denote the functor of compact induction and $\mathrm{Z}_n(F)$ the center of $\mathrm{GL}_n(F)$. Let $\mathrm{K}_n = \mathrm{GL}_n(\mathfrak{o})$, the unique maximal compact subgroup of $\mathrm{GL}_n(F)$ (up to conjugacy), and set $\mathfrak{P}_n := \mathrm{I}_n + \mathrm{Mat}_n(\mathfrak{p})$. Let ψ_F be an additive complex character of F such that ψ_F is nontrivial on $\mathfrak{o}$ and trivial on $\mathfrak{p}$. Let ψ be the character of $\mathfrak{o}/\mathfrak{p}$ induced by ψ_F.

Let τ be an irreducible cuspidal representation of $\mathrm{GL}_n(\mathfrak{k})$. We write $\overline{\tau}$ for the representation of K_n obtained by composing τ with the natural map $^- : K_n \to \mathrm{GL}_n(\mathfrak{k})$ given by reduction modulo $\mathfrak{p}$.

Let $\pi \cong \mathrm{cInd}_{Z_n(F)K_n(F)}^{\mathrm{GL}_n(F)} \omega\bar{\tau}$, where ω is a character of $\mathrm{Z}_n(F)$ such that $\omega|_{K_n \cap \mathrm{Z}_n(F)} = \bar{\tau}|_{K_n \cap \mathrm{Z}_n(F)}$.

Theorem 4.1 *If* τ *is an irreducible cuspidal representation of* $\mathrm{GL}_n(\mathfrak{k})$, *then* π *is an irreducible cuspidal representation of* $\mathrm{GL}_n(F)$.

The representations π are the **level zero cuspidal** (or depth zero cuspidal) representations of $\mathrm{GL}_n(F)$.

By using (generalized) Bessel functions of cuspidal representations of $\mathrm{GL}_n(F)$ introduced by V. Paškūnas and Stevens in [25], the author and Lei Zhang were able to establish the following relation between Gamma factors for a pair of level zero cuspidal representations of $\mathrm{GL}_n(F)$ and $\mathrm{GL}_t(F)$ and gamma factors for the corresponding cuspidal representations of $\mathrm{GL}_n(\mathbb{F}_q)$ and $\mathrm{GL}_t(\mathbb{F}_q)$.

Theorem 4.2 [22] *For $n \geq 2$, let π_1 and π_2 be irreducible* **level zero cuspidal** *representations of $\mathrm{GL}_n(F)$ and $\mathrm{GL}_t(F)$) (resp.) with $t < n$ such that $\pi \cong \mathrm{cInd}_{Z_n(F)K_n(F)}^{\mathrm{GL}_n(F)} \omega_1 \overline{\tau_1}$ and $\pi' \cong \mathrm{cInd}_{Z_t(F)K_t(F)}^{\mathrm{GL}_t(F)} \omega_2 \overline{\tau_2}$. Then,*

$$\omega_{\tau_2}(-1)^{n-1} \Gamma(s, \pi \times \pi', \psi_F) = q^{\frac{t(n-t-1)}{2}} \gamma(\tau_1 \times \tau_2, \psi).$$

Given two level zero cuspidal representations of $\mathrm{GL}_n(F)$,

$$\pi_i \cong \mathrm{cInd}_{Z_n(F)K_n(F)}^{\mathrm{GL}_n(F)} \omega \tau_i,$$

for $i = 1, 2$, where τ_1(resp. τ_2) is an irreducible cuspidal representation of $\mathrm{GL}_n(\mathbb{F}_q)$ (resp. $\mathrm{GL}_t(\mathbb{F}_q)$), and q is the cardinality of the residue field of F. Assume that

$$\Gamma(s, \pi_1 \times \chi, \psi_F) = \Gamma(s, \pi_2 \times \chi, \psi_F), \text{ for all } \chi \in \widehat{F^\times}.$$

Then, π_1 and π_2 possess the same central character ω, by Corollary 2.7. Therefore, $n \times 1$ Local Converse Theorem for level zero cuspidals for $n \leq 5$, q prime or $n < \frac{q-1}{2\sqrt{q}} + 1$ follows the results of Theorem 3.10, 3.12 and 4.2.

Theorem 4.3 ([22] and the appendix by Zhiwei Yu) *Let π be an irreducible,* **level zero cuspidal**, *unitary representations of $\mathrm{GL}_n(F)$, where the cardinality of the residue field q of F is either a prime, and $n \leq 5$ or $n < \frac{q-1}{2\sqrt{q}} + 1$. Then, π is uniquely determined by the set of twisted Gamma factors $\{\Gamma(s, \pi \times \chi, \psi_F) \mid \chi \in \widehat{F^\times}\}$.*

Theorem 4.4 [22] *Let $\mathcal{F}_2$ be the 2-adic field with residue field of two elements. Assume that $2^n - 1$ is a Mersenne prime. Let π_1 and π_2 be a pair of irreducible, unitarizable, level zero supercuspidal representations of $\mathrm{GL}_n(\mathcal{F}_2)$. Suppose that*

$$\Gamma(s, \pi_1 \times \chi, \psi_{\mathcal{F}_2}) = \Gamma(s, \pi_2 \times \chi, \psi_{\mathcal{F}_2}), \textit{ for all } \chi \in \widehat{\mathcal{F}_2^\times},$$

then π_1 and π_2 are isomorphic.

5 Converse Theorems and Distinction

5.1 p-Adic Results

Definition 5.1 Let H be a subgroup of G. A representation (π, V) of G is called H-distinguished if

$$\mathrm{Hom}_G(\pi, \mathrm{Ind}_H^G 1) \cong \mathrm{Hom}_H(\pi|_H, 1) \neq 0.$$

Let E be a quadratic extension of a p-adic field F such that the characteristic of the residue field of F is odd. Let ψ_E be a non-trivial character of E whose restriction to F is trivial. Let $G_n = \mathrm{GL}_n(E)$ and $H_n = \mathrm{GL}_n(F)$. Hakim studied distinction for generic representations of G_2 using special values of Gamma factors.

Theorem 5.2 [13] *Let π be an irreducible unitary* **generic** *representation of G_2 with central character whose restriction to $F^\times$ is trivial. If*

$$\Gamma(\frac{1}{2}, \pi \times \tau, \psi_E) = 1$$

for all $F^\times$-distinguished character τ of $E^\times$, then π is H_2-distinguished.

In 1997, under the supervision of Jacquet, Youngbin Ok in his thesis proved the $n \times (n-1)$ Local Converse Theorem on distinction.

Theorem 5.3 [24] *Let π be an irreducible unitary* **cuspidal** *representation of* $\mathrm{GL}_n(E)$ *with central character whose restriction to $F^\times$ is trivial. Then, the following are equivalent:*

(1) π is H_n-distinguished.
(1) $\Gamma(\frac{1}{2}, \pi \times \tau, \psi_E) = 1$ for any H_r-distinguished unitary generic representation τ of G_r, for $r = 1, \ldots, n-1$.

In 2011, Omer Offen gave a criterion for representation of G_n to be H_n-distinguished in terms of special values of twisted Γ and ϵ factors.

Theorem 5.4 [23] *Let π (resp. τ) be an irreducible* **generic** *and H_n-distinguished (resp. H_t-distinguished) representation of G_n (resp. G_t). Then,*

$$\epsilon(\frac{1}{2}, \pi \times \tau) = \Gamma(\frac{1}{2}, \pi \times \tau, \psi_E) = 1.$$

In 2015, Hakim and Offen established $n \times (n-2)$ Local Converse Theorem on distinction over p-adic fields.

Theorem 5.5 [15] *Let π be a* **cuspidal** *representation of G_n, $n \geq 3$. Then, π is H_n-distinguished if and only if*

$$\Gamma(\frac{1}{2}, \pi \times \tau, \psi_E) = 1$$

for all irreducible generic H_r-distinguished representations τ of G_r, for $r = 1, \ldots, n-2$.

Imitating Jacquet's conjecture on the Local Converse Problem, we proposed the following conjecture on distinction.

Conjecture 5.6 [$n \times [\frac{n}{2}]$ Local Converse Problem on distinction for $\mathrm{GL}_n(F)$] *The $n \times [\frac{n}{2}]$ Local Converse Theorem for distinction hold for* $\mathrm{GL}_n(F)$.

As a first step, we worked on the finite field analogue to seek supporting evidence for a positive answer.

5.2 Finite Field Results

Definition 5.7 Let π be an irreducible representation of $\mathrm{GL}_n(\mathbb{F}_q)$ and τ irreducible representation of $\mathrm{GL}_r(\mathbb{F}_q)$, $1 \le r \le n$. We call τ exceptional to π or (π, τ) an exceptional pair if there exists an irreducible representation ρ of $\mathrm{GL}_{n-r}(\mathbb{F}_q)$ such that

$$\dim \mathrm{Hom}(\pi, \ \mathrm{Ind}_P^{\mathrm{GL}_n} \tau \times \rho) \neq 0,$$

where P is the standard parabolic subgroup of GL_n, corresponding to partition $(r, n-r)$. Here, we allow $r = n$ and the only irreducible representation of $\mathrm{GL}_n(\mathbb{F}_q)$ which is exceptional to a cuspidal π is π itself.

Definition 5.8 Let π be an irreducible generic representation of $\mathrm{GL}_n(\mathbb{F}_q)$ and τ be an irreducible generic representation of $\mathrm{GL}_r(\mathbb{F}_q)$, $r < n$. Then, we define

$$\gamma(\pi \times \tau, \ \psi) = \sum_{\mathrm{U}_r \backslash \mathrm{GL}_r} \mathcal{B}_{\pi, \psi_n} \begin{pmatrix} 0 & \mathrm{I}_{n-r} \\ m & 0 \end{pmatrix} \mathcal{B}_{\tau, \psi_r^{-1}}(m). \tag{5.1}$$

When π is cuspidal, by Eq. (3.2), $\gamma(\pi \times \tau, \ \psi)$ in this definition coincides with the one defined in Theorem 2.3. However, the functional equations in Theorem 2.3 may fail for generic non-cuspidal π.

Let q be a power of an odd prime and ψ_0 be a non-trivial additive character of $\mathbb{F}_{q^2}$, whose restriction to $\mathbb{F}_q$ is trivial.

Theorem 5.9 [20] *Let π be an irreducible generic representation of $\mathrm{GL}_2(\mathbb{F}_{q^2})$ with central character ω_π whose restriction to $\mathbb{F}_q^\times$ is trivial. Then, π is $\mathrm{GL}_2(\mathbb{F}_q)$-distinguished if and only if $\gamma(\pi \times \chi^{-1}, \psi_0) = 1$ for all $\mathbb{F}_q^\times$-distinguished $\chi \in \widehat{\mathrm{GL}}_1((\mathbb{F}_{q^2})$ which are non-exceptional to π.*

Theorem 5.10 [[21], $n \times [\frac{n}{2}]$ Local Converse Theorem on distinction over finite fields] *For $n \geq 3$, let π be an irreducible cuspidal representation of $\mathrm{GL}_n(\mathbb{F}_{q^2})$ with central character ω_π whose restriction to $\mathbb{F}_q^\times$ is trivial. If $\gamma(\pi \times \rho, \ \psi_0) = q^{-r(n-r-1)}$ for all generic $\mathrm{GL}_r(\mathbb{F}_q)$-distinguished representations ρ of $\mathrm{GL}_r(\mathbb{F}_{q^2})$ for $r = 1, \ldots, [\frac{n}{2}]$, then π is $\mathrm{GL}_n(\mathbb{F}_q)$-distinguished.*

Acknowledgements This work was supported by the Ministry of Science and Technology, Taiwan 105-2115-M-006-010-MY2.

References

1. M. Adrian, B. Liu, S. Stevens, P. Xu, On the Jacquet conjecture on the local converse problem for p-adic GL_N. Represent. Theor. **20**, 1–13 (2016)
2. M. Adrian, B. Liu, Some results on simple supercuspidal representations of $\mathrm{GL}_n(F)$. J. Number Theor. **160**, 117–147 (2016)

3. M. Adrian, B. Liu, S. Stevens, K.-F. Tam, On the sharpness of the bound for the Local Converse Theorem of p-adic GL(prime). To appear in Proc. Amer. Math. Soc. (2017)
4. J. Arthur, *Functoriality and the Trace Formula.* preprint
5. C.J. Bushnell, G. Henniart, Langlands parameters for epipelagic representations of GL_n. Math. Ann. **358**(1–2), 433–463 (2014)
6. A. Braverman, D. Kazhdan, γ-functions of representations and lifting. With an appendix by V. Vologodsky. GAFA 2000 (Tel Aviv, 1999). Geometric Functional Analysis, Special Volume, Part I, pp. 237–278 (2000)
7. J. Chai, Bessel functions and local converse conjecture of Jacquet. To appear in J. Eur. Math. Soc. (JEMS)
8. J.-P.J. Chen, The $n \times (n-2)$ Local Converse Theorem for $GL(n)$ over a p-adic field. J. Number Theor. **120**(2), 193–205 (2006)
9. J. Cogdell, I. Piatetski-Shapiro, Converse theorems for GL_n. Publications Mathématiques de l'IHÉS (79), 157–214 (1994)
10. J. Cogdell, I. Piatetski-Shapiro, Converse theorems for GL_n. II. J. Reine Angew. Math. **507**, 165–188 (1999)
11. S.I. Gelfand, Representation of the full linear group over a finite field. Math. USSR Sbornik **12**(1), 13–39 (1970)
12. J.A. Green, The characters of the finite general linear groups. Trans. Amer. Math. Soc. **80**, 402–447 (1955)
13. J. Hakim, Distinguished p-adic representations. Duke Math. J. **62**(1), 1–22 (1991)
14. G. Henniart, Characterization of the local Langlands correspondence by ϵ-factors of pairs. Invent. Math. **113**(2), 339–350 (1993)
15. J. Hakim, O. Offen, Distinguished representations of $GL(n)$ and Local Converse Theorems. Manuscripta Math. **148**(1–2), 1–27 (2015)
16. H. Jacquet, B. Liu, On the Local Converse Theorem for p-adic GL_n. To appear in Amer. J. Math
17. D. Jiang, C. Nien, S. Stevens, Towards the Jacquet conjecture on the Local Converse Problem for p-adic GL_n. J. Eur. Math. Soc. (JEMS) **17**(4), 991–1007 (2015)
18. H. Jacquet, I. Piatetski-Shapiro, J. Shalika, Rankin-Selberg convolutions. Amer. J. Math. **105**, 367–464 (1983)
19. C. Nien, A proof of finite field analogue of Jacquet's conjecture. Amer J. Math. **136**(3), 653–674 (2014)
20. C. Nien, $n \times 1$ Local Gamma factors and Gauss Sum. Finite Fields Appl. **46**, 255–270 (2017)
21. C. Nien, Gamma factors and quadratic extension over finite fields. manuscripta math. **158**(1–2), 31–54 (2019)
22. C. Nien, L. Zhang, Converse Theorem meets Gauss sums, with an appendix by Zhiwei Yun. submitted
23. O. Offen, On local root numbers and distinction. J. Reine Angew. Math. **652**, 165–205 (2011)
24. Y. Ok, Distinction and Gamma factors at $\frac{1}{2}$: supercuspidal case. Ph.D. Thesis, Columbia University (1997)
25. V. Paškūnas, S. Stevens, On the realization of maximal simple types and epsilon factors of pairs. Amer. J. Math. **130**, 1211–1261 (2008)
26. E.-A. Roditty, On gamma factors and Bessel functions for representations of general linear groups over finite field. M.Sc. Thesis, Tel-Aviv University (2010)
27. P. Xu, A remark on the simple cuspidal representations of $GL(n, F)$. Preprint, available at arxiv.org/pdf/1310.3519v2.pdf

On Completions of Hecke Algebras

Maarten Solleveld

Contents

1 Introduction 208
2 Affine Hecke Algebras 212
  2.1 The Schwartz and C^*-completions 217
  2.2 The Space of Irreducible Representations 224
3 Reductive p-adic Groups 228
  3.1 The Plancherel Isomorphism 231
  3.2 The Space of Irreducible Representations 236
4 Morita Equivalences 238
  4.1 Conditions and First Consequences 238
  4.2 Preservation of Temperedness 243
  4.3 Comparison of Completions 249
5 Hecke Algebras from Bushnell–Kutzko Types 253
6 Hecke Algebras from Bernstein's Progenerators 258
References 260

Abstract Let G be a reductive p-adic group and let $\mathcal{H}(G)^{\mathfrak{s}}$ be a Bernstein block of the Hecke algebra of G. We consider two important topological completions of $\mathcal{H}(G)^{\mathfrak{s}}$: a direct summand $\mathcal{S}(G)^{\mathfrak{s}}$ of the Harish-Chandra–Schwartz algebra of G and a two-sided ideal $C_r^*(G)^{\mathfrak{s}}$ of the reduced C^*-algebra of G. These are useful for the study of all tempered smooth G-representations. We suppose that $\mathcal{H}(G)^{\mathfrak{s}}$ is Morita equivalent to an affine Hecke algebra $\mathcal{H}(\mathcal{R}, q)$ – as is known in many cases. The latter algebra also has a Schwartz completion $\mathcal{S}(\mathcal{R}, q)$ and a C^*-completion $C_r^*(\mathcal{R}, q)$, both defined in terms of the underlying root datum $\mathcal{R}$ and the parameters q. We prove that, under some mild conditions, a Morita equivalence $\mathcal{H}(G)^{\mathfrak{s}} \sim_M \mathcal{H}(\mathcal{R}, q)$ extends to Morita equivalences $\mathcal{S}(G)^{\mathfrak{s}} \sim_M \mathcal{S}(\mathcal{R}, q)$ and $C_r^*(G)^{\mathfrak{s}} \sim_M C_r^*(\mathcal{R}, q)$. We also check that our conditions are fulfilled in all known cases of such Morita equivalences between Hecke algebras. This is applied to compute the topological K-theory of the reduced C^*-algebra of a classical p-adic group.

The author was supported by an NWO Vidi grant "A Hecke algebra approach to the local Langlands correspondence" (nr. 639.032.528).

M. Solleveld (✉)
IMAPP, Radboud Universiteit, Heyendaalseweg 135, 6525AJ Nijmegen, The Netherlands
e-mail: m.solleveld@science.ru.nl

A.-M. Aubert et al. (eds.), *Representations of Reductive p-adic Groups*,
Progress in Mathematics 328, https://doi.org/10.1007/978-981-13-6628-4_8

2010 Mathematics Subject Classification 20C08 · 22E50 · 22E35

1 Introduction

Let G be a connected reductive group over a non-Archimedean local field. Let Rep(G) be the category of smooth G-representations on complex vector spaces. To study such representations, it is often useful to consider various group algebras of G. Most fundamentally, there is the Hecke algebra $\mathcal{H}(G)$, the convolution algebra of locally constant, compactly supported functions $f : G \to \mathbb{C}$. The category Rep(G) is equivalent to the category $\mathrm{Mod}(\mathcal{H}(G))$ of nondegenerate $\mathcal{H}(G)$-modules.

For the purposes of harmonic analysis, and in particular for the study of tempered smooth G-representations, the Harish-Chandra–Schwartz algebra $\mathcal{S}(G)$ [14] can be very convenient. This is a topological completion of $\mathcal{H}(G)$, consisting of locally constant functions $f : G \to \mathbb{C}$ that decay rapidly in a specified sense. By [43, §III.7], an admissible smooth G-representation is tempered if and only if it is naturally an $\mathcal{S}(G)$-module. For larger representations, it is best to define the category of tempered smooth G-representations as the category $\mathrm{Mod}(\mathcal{S}(G))$ of nondegenerate $\mathcal{S}(G)$-modules [32, Appendix].

Further, from the point of view of operator algebras or noncommutative geometry, the reduced C^*-algebra $C_r^*(G)$ may be the most suitable. The modules of $C_r^*(G)$ are Banach spaces, so they are usually not smooth as G-representations. But $\mathcal{S}(G) \subset C_r^*(G)$ and the smooth vectors in any $C_r^*(G)$-module do form a nondegenerate $\mathcal{S}(G)$-module and hence a smooth G-representation. Moreover, this operation provides a bijection between the irreducible representations of $C_r^*(G)$ and those of $\mathcal{S}(G)$. This feature distinguishes $C_r^*(G)$ from other Banach group algebras like $L^1(G)$ or the maximal C^*-algebra of G.

Let $\mathrm{Rep}(G)^{\mathfrak{s}}$ be a Bernstein block of Rep(G) [5]. It is well known that in many cases (see Sects. 5 and 6) $\mathrm{Rep}(G)^{\mathfrak{s}}$ is equivalent to the category of modules of an affine Hecke algebra $\mathcal{H}(\mathcal{R}, q)$. Here $\mathcal{R}$ is a root datum and q is a parameter function for $\mathcal{R}$. In such cases, it would be useful if one could detect, in terms of $\mathcal{H}(\mathcal{R}, q)$, whether a G-representation in $\mathrm{Rep}(G)^{\mathfrak{s}}$

(i) is tempered;
(ii) is unitary;
(iii) admits a continuous extension to a $C_r^*(G)$-module.

The structure needed to make sense of this is available for (extended) affine Hecke algebras with positive parameters. They have a natural $*$-operation, so unitarity is defined. Temperedness of finite-dimensional $\mathcal{H}(\mathcal{R}, q)$-modules can be defined conveniently either in terms of growth of matrix coefficients or by means of weights for a large commutative subalgebra of $\mathcal{H}(\mathcal{R}, q)$ [25, §2.7].

Furthermore, there exists a Schwartz completion $\mathcal{S}(\mathcal{R}, q)$ of $\mathcal{H}(\mathcal{R}, q)$ [25, §6.2] that parallels the completion $\mathcal{S}(G)$ of $\mathcal{H}(G)$ [12]. By [25, Corollary 6.7], a finite dimensional $\mathcal{H}(\mathcal{R}, q)$-module is tempered if and only if it extends continuously to an $\mathcal{S}(\mathcal{R}, q)$-module. As in the group case, we define the category of tempered $\mathcal{H}(\mathcal{R}, q)$-modules to be the module category of $\mathcal{S}(\mathcal{R}, q)$.

The algebra $\mathcal{H}(\mathcal{R}, q)$ is a Hilbert algebra, so it acts by multiplication on its own Hilbert space completion. Then one can define the reduced C^*-completion $C_r^*(\mathcal{R}, q)$ as the closure of $\mathcal{H}(\mathcal{R}, q)$ in the algebra of bounded linear operators on that Hilbert space. It is reasonable to expect that this algebra plays a role analogous to $C_r^*(G)$.

Let $\mathcal{H}(G)^{\mathfrak{s}}$ (resp. $\mathcal{S}(G)^{\mathfrak{s}}$ and $C_r^*(G)^{\mathfrak{s}}$) be the direct summand of $\mathcal{H}(G)$ (resp. $\mathcal{S}(G)$ and $C_r^*(G)$) corresponding to $\mathrm{Rep}(G)^{\mathfrak{s}}$ via the Bernstein decomposition

$$\mathrm{Mod}(\mathcal{H}(G)) \cong \mathrm{Rep}(G) = \prod\nolimits_{\mathfrak{s} \in \mathfrak{B}(G)} \mathrm{Rep}(G)^{\mathfrak{s}}.$$

In view of the above, it is natural to ask whether an equivalence of categories

$$\mathrm{Rep}(G)^{\mathfrak{s}} \cong \mathrm{Mod}(\mathcal{H}(G)^{\mathfrak{s}}) \xrightarrow{\sim} \mathrm{Mod}(\mathcal{H}(\mathcal{R}, q)) \tag{1}$$

extends to Morita equivalences

$$\mathcal{S}(G)^{\mathfrak{s}} \sim_M \mathcal{S}(\mathcal{R}, q) \quad \text{and} \quad C_r^*(G)^{\mathfrak{s}} \sim_M C_r^*(\mathcal{R}, q). \tag{2}$$

That would solve the issues (i) and (iii) completely, and would provide a partial answer to (ii). Namely, since irreducible tempered representations are unitary, it would imply that (1) matches the unitary tempered representations on both sides. (It is not clear what it could say about unitary non-tempered representations.) Furthermore, (2) would make $\mathrm{Mod}(\mathcal{S}(G)^{\mathfrak{s}})$ and $\mathrm{Mod}(C_r^*(G)^{\mathfrak{s}})$ amenable to much more explicit calculations, in terms of the generators and relations from $\mathcal{H}(\mathcal{R}, q)$. That is important for topological K-theory, where one deals with finitely generated projective $C_r^*(G)$-modules.

While (2) looks fairly plausible, it is not automatic. To prove it, we impose three conditions on the Morita equivalence $\mathcal{H}(G)^{\mathfrak{s}} \sim_M \mathcal{H}(\mathcal{R}, q)$, Conditions 4.1 and 4.2 (described in Sect. 4.1) and Condition 2.1 (described in Sect. 2).

- Condition 4.1 is about compatibility with parabolic induction and restriction.
- Condition 4.2 says roughly that under this Morita equivalence every (suitable) parabolic subgroup of G should give rise to a parabolic subalgebra of $\mathcal{H}(\mathcal{R}, q)$. In all such instances, the relation between a parabolic subgroup and a parabolic subalgebra should respect the appropriate notions of positivity.
- Sometimes we obtain, instead of $\mathcal{H}(\mathcal{R}, q)$, an extended affine Hecke algebra $\mathcal{H}(\mathcal{R}, q) \rtimes \Gamma$, where Γ is a finite group. Then we require Condition 2.1, which says that Γ respects all the relevant structure.

Here Condition 4.1 has little to do with affine Hecke algebras. If it holds, then it does so based on general functoriality principles. Condition 4.2 is needed to bring affine Hecke algebras into play. If it does not hold, then our results simply cannot be formulated in such terms. The positivity part is innocent and can usually be achieved by a good choice of a standard minimal parabolic subgroup of G. Condition 2.1 is more technical. It serves to rule out some phenomena that could happen for arbitrary Γ but not for reductive groups.

Notice that the above conditions do not say anything about $*$-homomorphisms between $\mathcal{H}(\mathcal{R}, q)$ and $\mathcal{H}(G)^{\mathfrak{s}}$. Instead, our conditions are chosen so that they will hold true for affine Hecke algebras arising from reductive p-adic groups via the two main methods: Bushnell–Kutzko types and Bernstein's progenerators.

Theorem 1 (see Theorem *4.12*)
Suppose that there is a Morita equivalence between $\mathcal{H}(G)^{\mathfrak{s}}$ and an extended affine Hecke algebra $\mathcal{H}(\mathcal{R}, q) \rtimes \Gamma$ with positive parameters, such that Conditions 4.1, 4.2 and 2.1 hold. Then it induces Morita equivalences $\mathcal{S}(G)^{\mathfrak{s}} \sim_M \mathcal{S}(\mathcal{R}, q)$ and $C_r^*(G)^{\mathfrak{s}} \sim_M C_r^*(G)$.

Hitherto this was proved only for the Schwartz completions in the case of Iwahori–spherical representations of split groups [12, Theorem 10.2]. In all cases where a Morita equivalence on the Hecke algebra level is known (to the author), we check that the conditions of Theorem 1 are fulfilled. This includes principal series representations of F-split groups (F is any non-Archimedean local field), level-zero representations, inner forms of $\mathrm{GL}_n(F)$, inner forms of $\mathrm{SL}_n(F)$, symplectic groups (not necessarily split), and special orthogonal groups (also possibly non-split).

In all these cases, we obtain a pretty good picture of $C_r^*(G)^{\mathfrak{s}}$, up to Morita equivalence. This can, for instance, be used to compute the topological K-theory of these algebras. Indeed, in [41] the author determined the K-theory of $C_r^*(\mathcal{R}, q)$ for many root data $\mathcal{R}$ (it does not depend on q). These calculations, together with Theorem 1 for classical groups, lead to a result which is useful in relation to the Baum–Connes conjecture.

Theorem 2 (see Theorem *6.3*)
Let G be a special orthogonal or a symplectic group over F (not necessarily split), or an inner form of $\mathrm{GL}_n(F)$. Then $K_*(C_r^*(G))$ is a free abelian group. For every Bernstein block $\mathrm{Rep}(G)^{\mathfrak{s}}$, the rank of $K_*(C_r(G)^{\mathfrak{s}})$ is finite and can be computed explicitly.

Let us also discuss other approaches to the topics (i), (ii), and (iii) from page 208. In most cases where a Morita equivalence $\mathcal{H}(G)^{\mathfrak{s}} \sim_M \mathcal{H}(\mathcal{R}, q) \rtimes \Gamma$ is known, these issues are not discussed in the literature. When the Morita equivalence comes from a type (K, λ) in the sense of Bushnell–Kutzko [9], several relevant techniques are available. Let $e_\lambda \in \mathcal{H}(G)$ and $\mathcal{H}(G, J, \lambda) = \mathrm{End}_G(\mathrm{ind}_J^G \lambda)$ be the idempotent and the algebra associated with the type (K, λ) in [9, §2]. In this setting, the Morita equivalence can be implemented by injective algebra homomorphisms

$$\mathcal{H}(\mathcal{R}, q) \rtimes \Gamma \xrightarrow{\Upsilon_\lambda} \mathcal{H}(G, J, \lambda) \to e_\lambda \mathcal{H}(G) e_\lambda \to \mathcal{H}(G)^{\mathfrak{s}}, \tag{3}$$

where we assume that Υ_λ is an isomorphism. It follows quickly from the definition of types that the last two maps in (3) induce Morita equivalences [9, (2.12) and Theorem

4.3]. Both algebras $\mathcal{H}(G, J, \lambda)$ and $e_\lambda \mathcal{H}(G) e_\lambda$ are endowed with a natural trace and $*$-operation, which are preserved by the injections

$$\mathcal{H}(G, J, \lambda) \to e_\lambda \mathcal{H}(G) e_\lambda \to \mathcal{H}(G)^{\mathfrak{s}}. \tag{4}$$

Considerations with Hilbert algebras show that the maps (4) induce equivalences between the associated categories of finite length tempered representations [8]. Moreover, after the first version of this paper appeared, Ciubotaru showed that (4) also induces equivalences between the respective subcategories of unitary modules [11]. Hence whenever Υ_λ is a $*$-isomorphism, the categories of unitary modules of all the algebras in (3) are equivalent. Earlier, this had been proved under certain additional hypotheses [3, 4].

Notice that in (3), all algebras are endowed with extra structure, which on the left-hand side comes from affine Hecke algebras and for the other terms from the embedding in $\mathcal{H}(G)$. In particular, both $\mathcal{H}(G)$ and $\mathcal{H}(\mathcal{R}, q) \rtimes \Gamma$ are endowed with a canonical trace, which stems from evaluation of functions at the unit element of G. Usually, (3) will transfer the trace on $\mathcal{H}(G)^{\mathfrak{s}}$ to a positive scalar multiple of the trace on $\mathcal{H}(\mathcal{R}, q) \rtimes \Gamma$. If that is the case and Υ_λ is a $*$-isomorphism, then by [12, Theorem 10.1], (3) induces an equivalence between the category of finite length tempered G-representations in $\mathrm{Rep}(G)^{\mathfrak{s}}$ and the category of finite dimensional tempered modules of $\mathcal{H}(\mathcal{R}, q) \rtimes \Gamma$. This relies on properties of the Plancherel measures of G and of $\mathcal{H}(\mathcal{R}, q) \rtimes \Gamma$, and it uses that Υ_λ preserves these Plancherel measures, up to a scalar multiple.

When the Morita equivalence $\mathcal{H}(G)^{\mathfrak{s}} \sim_M \mathcal{H}(\mathcal{R}, q) \rtimes \Gamma$ does not arise from a type, fewer techniques for (i), (ii), and (iii) were known. Heiermann established such Morita equivalences for symplectic and special orthogonal groups and for inner forms of $\mathrm{GL}_n(F)$ [15], by using Bernstein's progenerators of $\mathrm{Rep}(G)^{\mathfrak{s}}$. In [16], he showed that these equivalences preserve temperedness of finite length modules. Here, it is unknown whether the $*$-operation and the trace are preserved by the Morita equivalence. Since maps like (3) are lacking, it is even unclear how such a statement could be formulated in this setting.

Summarizing, in the literature, already several results about the behaviour of finite length modules under a Morita equivalence between a Bernstein block $\mathrm{Rep}(G)^{\mathfrak{s}}$ and the module category of an (extended) affine Hecke algebra can be found, but there is so far almost nothing about the Schwartz completions and the C^*-completions.

Let us briefly describe the contents of the paper. In the first section, we recall the definitions of affine Hecke algebras and their topological completions. We formulate the Plancherel isomorphism for these completions, from [12], and we establish suitable versions for affine Hecke algebras extended by finite groups. In addition, we analyse the space of irreducible representations and the subspace of irreducible tempered representations, mainly relying on [39]. This is formulated in terms of the Langlands classification and induction from discrete series representations of parabolic subalgebras.

After that, we look at the aforementioned group algebras for a reductive p-adic group G. We recall the Plancherel isomorphism for the Schwartz algebra of G [14, 43] and for the reduced C^*-algebra of G [26]. In parallel to our discussion of affine Hecke algebras, we analyse the space of irreducible smooth G-representation in terms of the Langlands classification and parabolic induction of square-integrable representations, following [37].

This forms the setup for the proof of Theorem 1, our main result, which occupies Sect. 4. The crucial idea behind our argument is that in the Plancherel isomorphisms for $\mathcal{S}(G)^{\mathfrak{s}}$ and $\mathcal{S}(\mathcal{R}, q)$, very similar algebras appear. In both settings, one encounters a bundle of matrix algebras over a compact torus, one takes C^∞-sections of those, and then invariants with respect to a finite group acting via intertwining operators. We compare the resulting algebras on both sides, analyzing the data used to describe the Plancherel isomorphisms. First, we prove that a Morita equivalence between $\mathcal{H}(G)^{\mathfrak{s}}$ and $\mathcal{H}(\mathcal{R}, q)$, plus the mild extra conditions listed on page 209, imply that the two necessary sets of data, for $\mathcal{S}(G)^{\mathfrak{s}}$ and for $\mathcal{S}(\mathcal{R}, q)$, become equivalent after some manipulations. The most problematic part is to prove that the Morita equivalences preserve temperedness and match square-integrable G-representations with discrete series $\mathcal{H}(\mathcal{R}, q)$-modules. For that, we use very specific information about the spaces of irreducible representations and their subspaces of tempered representations. When we have compared all the data needed for the Plancherel isomorphisms on both sides, we establish the desired Morita equivalences between topological algebras.

In Sect. 5, we check that the conditions from Sect. 4 are fulfilled in (most) known cases of Morita equivalences coming from types. In the Sect. 6, we do the same for Heiermann's Morita equivalences constructed with the use of projective generators, and we derive Theorem 2.

2 Affine Hecke Algebras

Let $\mathfrak{a}$ be a finite dimensional real vector space and let $\mathfrak{a}^*$ be its dual. Let $Y \subset \mathfrak{a}$ be a lattice and $X = \mathrm{Hom}_{\mathbb{Z}}(Y, \mathbb{Z}) \subset \mathfrak{a}^*$ the dual lattice. Let

$$\mathcal{R} = (X, R, Y, R^\vee, \Delta). \tag{5}$$

be a based root datum. Thus R is a reduced root system in X, $R^\vee \subset Y$ is the dual root system, Δ is a basis of R and the set of positive roots is denoted R^+. Furthermore, a bijection $R \to R^\vee$, $\alpha \mapsto \alpha^\vee$ is given, such that $\langle \alpha, \alpha^\vee \rangle = 2$ and such that the corresponding reflections $s_\alpha : X \to X$ (resp. $s_\alpha^\vee : Y \to Y$) stabilize R (resp. $R^\vee$). We do not assume that R spans $\mathfrak{a}^*$. The reflections s_α generate the Weyl group $W = W(R)$ of R, and $S_\Delta := \{s_\alpha \mid \alpha \in \Delta\}$ is the collection of simple reflections.

We have the affine Weyl group $W^{\mathrm{aff}} = \mathbb{Z}R \rtimes W$ and the extended (affine) Weyl group $W^e = X \rtimes W$. Both can be considered the groups of affine transformations of $\mathfrak{a}^*$. We denote the translation corresponding to $x \in X$ by t_x. As is well known,

W^{aff} is a Coxeter group, and the basis Δ of R gives rise to a set S^{aff} of simple (affine) reflections. More explicitly, let $\Delta_M^\vee$ be the set of maximal elements of $R^\vee$, with respect to the dominance ordering coming from Δ. Then

$$S^{\text{aff}} = S_\Delta \cup \{t_\alpha s_\alpha \mid \alpha^\vee \in \Delta_M^\vee\}.$$

The length function ℓ of the Coxeter system $(W^{\text{aff}}, S^{\text{aff}})$ extends naturally to W^e. The elements of length zero form a subgroup $\Omega \subset W^e$ and $W^e = W^{\text{aff}} \rtimes \Omega$.

A complex parameter function for $\mathcal{R}$ is a map $q : S^{\text{aff}} \to \mathbb{C}^\times$ such that $q(s) = q(s')$ if s and s' are conjugate in W^e. This extends naturally to a map $q : W^e \to \mathbb{C}^\times$ which is 1 on Ω and satisfies

$$q(ww') = q(w)q(w') \quad \text{if} \quad \ell(ww') = \ell(w) + \ell(w').$$

Equivalently (see [20, §3.1]), one can define q as a W-invariant function

$$q : R \cup \{2\alpha : \alpha \in R, \alpha^\vee \in 2Y\} \to \mathbb{C}^\times. \tag{6}$$

We speak of equal parameters if $q(s) = q(s')\ \forall s, s' \in S^{\text{aff}}$ and of positive parameters if $q(s) \in \mathbb{R}_{>0}\ \forall s \in S^{\text{aff}}$. We fix a square root $q^{1/2} : S^{\text{aff}} \to \mathbb{C}^\times$.

The affine Hecke algebra $\mathcal{H} = \mathcal{H}(\mathcal{R}, q)$ is the unique associative complex algebra with basis $\{N_w \mid w \in W^e\}$ and multiplication rules

$$\begin{aligned} N_w N_{w'} &= N_{ww'} && \text{if } \ell(ww') = \ell(w) + \ell(w'), \\ \big(N_s - q(s)^{1/2}\big)\big(N_s + q(s)^{-1/2}\big) &= 0 && \text{if } s \in S^{\text{aff}}. \end{aligned} \tag{7}$$

In the literature, one also finds this algebra defined in terms of the elements $q(s)^{1/2}N_s$, in which case the multiplication can be described without square roots. This explains why $q^{1/2}$ does not appear in the notation $\mathcal{H}(\mathcal{R}, q)$. For $q = 1$ (7) just reflects the defining relations of W^e, so $\mathcal{H}(\mathcal{R}, 1) = \mathbb{C}[W^e]$.

The set of dominant elements in X is

$$X^+ = \{x \in X : \langle x, \alpha^\vee \rangle \geq 0\ \forall \alpha \in \Delta\}.$$

The subset $\{N_{t_x} : x \in X^+\} \subset \mathcal{H}(\mathcal{R}, q)$ is closed under multiplication and isomorphic to X^+ as a semigroup. For any $x \in X$ we put

$$\theta_x = N_{t_{x_1}} N_{t_{x_2}}^{-1}, \text{ where } x_1, x_2 \in X^+ \text{ and } x = x_1 - x_2.$$

This does not depend on the choice of x_1 and x_2, so $\theta_x \in \mathcal{H}(\mathcal{R}, q)^\times$ is well defined.

We write $T = \text{Hom}_{\mathbb{Z}}(X, \mathbb{C}^\times)$, so that $\mathcal{O}(T) \cong \mathbb{C}[X]$. The Bernstein presentation of $\mathcal{H}(\mathcal{R}, q)$ [20, §3] says that:

- $\{\theta_x : x \in X\}$ forms a $\mathbb{C}$-basis of a subalgebra of $\mathcal{H}(\mathcal{R}, q)$ isomorphic to $\mathbb{C}[X]$, which we identify with $\mathcal{O}(T)$.

- $\mathcal{H}(W, q) := \mathbb{C}\{N_w : w \in W\}$ is a finite dimensional subalgebra of $\mathcal{H}(\mathcal{R}, q)$ (known as the Iwahori–Hecke algebra of W).
- The multiplication map $\mathcal{O}(T) \otimes \mathcal{H}(W, q) \to \mathcal{H}(\mathcal{R}, q)$ is a $\mathbb{C}$-linear bijection.
- There are explicit cross relations between $\mathcal{H}(W, q)$ and $\mathcal{O}(T)$, deformations of the standard action of W on $\mathcal{O}(T)$.

To define parabolic subalgebras of affine Hecke algebras, we associate some objects with any set of simple roots $Q \subset \Delta$. Let R_Q be the root system they generate, $R_Q^\vee$ the root system generated by $Q^\vee$, and W_Q their Weyl group. We also define

$$\begin{array}{ll} X_Q = X\big/\big(X \cap (Q^\vee)^\perp\big) & X^Q = X/(X \cap \mathbb{Q}Q), \\ Y_Q = Y \cap \mathbb{Q}Q^\vee & Y^Q = Y \cap Q^\perp, \\ T_Q = \mathrm{Hom}_{\mathbb{Z}}(X_Q, \mathbb{C}^\times) & T^Q = \mathrm{Hom}_{\mathbb{Z}}(X^Q, \mathbb{C}^\times), \\ \mathfrak{a}_Q = Y_Q \otimes_{\mathbb{Z}} \mathbb{R} & \mathfrak{a}^Q = Y^Q \otimes_{\mathbb{Z}} R, \\ \mathcal{R}_Q = (X_Q, R_Q, Y_Q, R_Q^\vee, Q) & \mathcal{R}^Q = (X, R_Q, Y, R_Q^\vee, Q), \\ \mathcal{H}_Q = \mathcal{H}(\mathcal{R}_Q, q_Q) & \mathcal{H}^Q = \mathcal{H}(\mathcal{R}^Q, q^Q). \end{array}$$

Here, q_Q and q^Q are derived from q via (6). Both $\mathcal{H}_Q$ and $\mathcal{H}^Q$ are called parabolic subalgebras of $\mathcal{H}$. The quotient map $X \mapsto X_Q$ yields a natural projection

$$\mathcal{H}^Q \to \mathcal{H}_Q \; : \; \theta_x N_w \mapsto \theta_{x_Q} N_w. \tag{8}$$

In this way, one can regard $\mathcal{H}_Q$ as a "semisimple" quotient of $\mathcal{H}^Q$. The algebra $\mathcal{H}^Q$ is embedded in $\mathcal{H}$ via the Bernstein presentation, as the image of $\mathcal{O}(T) \otimes \mathcal{H}(W_Q, q) \to \mathcal{H}$. Any $t \in T^Q$ and any $u \in T^Q \cap T_Q$ give rise to algebra automorphisms

$$\begin{array}{ll} \psi_u : \mathcal{H}_Q \to \mathcal{H}_Q, \; \theta_{x_Q} N_w \mapsto u(x_Q)\theta_{x_Q} N_w, \\ \psi_t : \mathcal{H}^Q \to \mathcal{H}^Q, \; \theta_x N_w \;\; \mapsto t(x)\theta_x N_w. \end{array} \tag{9}$$

Let Γ be a finite group acting on $\mathcal{R}$, i.e. it acts $\mathbb{Z}$-linearly on X and preserves R and Δ. We also assume that Γ acts on T by affine transformations, whose linear part comes from the action on X. Thus Γ acts on $\mathcal{O}(T) \cong \mathbb{C}[X]$ by

$$\gamma(\theta_x) = z_\gamma(x)\theta_{\gamma x}, \tag{10}$$

for some $z_\gamma \in T$. We suppose throughout that $q^{1/2}$ is Γ-invariant, and that Γ acts on $\mathcal{H}(\mathcal{R}, q)$ by the algebra automorphisms

$$\mathrm{Ad}(\gamma) : \sum_{w \in W, x \in X} c_{x,w}\theta_x N_w \;\mapsto\; \sum_{w \in W, x \in X} c_{x,w} z_\gamma(x)\theta_{\gamma(x)} N_{\gamma w \gamma^{-1}}. \tag{11}$$

This being a group action, the multiplication relations in $\mathcal{H}(\mathcal{R}, q)$ imply that we must have $z_\gamma \in T^W$. We build the crossed product algebra

$$\mathcal{H}(\mathcal{R}, q) \rtimes \Gamma. \tag{12}$$

In [39], we considered a slightly less general action of Γ on $\mathcal{H}(\mathcal{R}, q)$, where the elements $z_\gamma \in T^W$ from (10) were all equal to 1. But the relevant results from [39] do not rely on Γ fixing the unit element of T, so they are also valid for the actions as in (11). In this paper, we will tacitly use some results from [39] in the generality of (11). We note that nontrivial $z_\gamma \in T^W$ are sometimes needed to describe Hecke algebras coming from p-adic groups, for example in [30, §4].

Since $\mathcal{H}(\mathcal{R}, q)$ is of finite rank as a module over its commutative subalgebra $\mathcal{O}(T)$, all irreducible $\mathcal{H}(\mathcal{R}, q)$-modules have finite dimension. The set of $\mathcal{O}(T)$-weights of a $\mathcal{H}(\mathcal{R}, q)$-module V will be denoted by $\mathrm{Wt}(V)$.

We regard $\mathfrak{t} = \mathfrak{a} \oplus i\mathfrak{a}$ as the polar decomposition of $\mathfrak{t}$, with associated real part map $\Re : \mathfrak{t} \to \mathfrak{a}$. The vector space $\mathfrak{t}$ can be interpreted as the Lie algebra of the complex torus $T = \mathrm{Hom}_{\mathbb{Z}}(X, \mathbb{C}^\times)$. The latter has a polar decomposition $T = T_{\mathrm{rs}} \times T_{\mathrm{un}}$, where $T_{\mathrm{rs}} = \mathrm{Hom}_{\mathbb{Z}}(X, \mathbb{R}_{>0})$ and $T_{\mathrm{un}} = \mathrm{Hom}_{\mathbb{Z}}(X, S^1)$. The polar decomposition of an element $t \in T$ is written as $t = |t|\,(t\,|t|^{-1})$. The exponential map $\exp : \mathfrak{t} \to T$ becomes bijective when restricted to $\mathfrak{a} \to T_{\mathrm{rs}}$. We denote its inverse by $\log : T_{\mathrm{rs}} \to \mathfrak{a}$.

We write

$$
\begin{aligned}
\mathfrak{a}^+ &= \{\mu \in \mathfrak{a} : \langle \alpha , \mu \rangle \geq 0 \,\forall \alpha \in \Delta\},\\
\mathfrak{a}^{*+} &= \{x \in \mathfrak{a}^* : \langle x , \alpha^\vee \rangle \geq 0 \,\forall \alpha \in \Delta\},\\
\mathfrak{a}^- &= \{\lambda \in \mathfrak{a} : \langle x , \lambda \rangle \leq 0 \,\forall x \in \mathfrak{a}^{*+}\} = \Big\{ \sum\nolimits_{\alpha \in \Delta} \lambda_\alpha \alpha^\vee : \lambda_\alpha \leq 0 \Big\}.
\end{aligned}
$$

The interior $\mathfrak{a}^{--}$ of $\mathfrak{a}^-$ equals $\big\{\sum_{\alpha\in\Delta} \lambda_\alpha \alpha^\vee : \lambda_\alpha < 0\big\}$ if Δ spans $\mathfrak{a}^*$, and is empty otherwise. We write

$$
T^- = \exp(\mathfrak{a}^-) \subset T_{\mathrm{rs}} \quad \text{and} \quad T^{--} = \exp(\mathfrak{a}^{--}) \subset T_{\mathrm{rs}}.
$$

We say that a module V for $\mathcal{H}(\mathcal{R}, q)$ (or for $\mathcal{H}(\mathcal{R}, q) \rtimes \Gamma$) is tempered if $|\mathrm{Wt}(V)| \subset T^-$, and that it is discrete series if $|\mathrm{Wt}(V)| \subset T^{--}$. The latter is only possible if R spans $\mathfrak{a}$, for otherwise $\mathfrak{a}^{--}$ and T^{--} are empty. We alleviate these notions by calling a $\mathcal{H} \rtimes \Gamma$-module essentially discrete series if its restriction to $\mathcal{H}_\Delta$ is discrete series. Equivalently, essentially discrete series means that $\mathrm{Wt}(V) \subset T^{--} T_{\mathrm{un}} T^\Delta$. Such a representation is tempered if and only if $\mathrm{Wt}(V) \subset T^{--} T_{\mathrm{un}}$. We denote the set of (equivalence classes of) irreducible tempered essentially discrete series representations by $\mathrm{Irr}_{L^2}(\mathcal{H}(\mathcal{R}, q) \rtimes \Gamma)$.

It follows from the Bernstein presentation [20, §3] that

$$
\text{the centre of } \mathcal{H}(\mathcal{R}, q) \rtimes \Gamma \text{ contains } \mathcal{O}(T)^{W\Gamma} = \mathcal{O}(T/W\Gamma), \tag{13}
$$

with equality if $W\Gamma$ acts faithfully on T. By Schur's Lemma, $\mathcal{O}(T)^{W\Gamma}$ acts on every irreducible $\mathcal{H} \rtimes \Gamma$-representation π by a character. Such a character can be identified with a $W\Gamma$-orbit $W\Gamma t \subset T$. We will just call $W\Gamma t$ the central character of π. Then $W\Gamma |t| \subset T_{\mathrm{rs}}$ and $cc(\pi) := W\Gamma \log |t|$ is a single $W\Gamma$-orbit in $\mathfrak{a}$. We fix a

$W\Gamma$-invariant inner product on $\mathfrak{a}$ and we define

$$\|cc(\pi)\| = \|\log|t|\|. \tag{14}$$

In some cases that we will encounter, the appropriate parabolic subalgebras of $\mathcal{H}(\mathcal{R}, q) \rtimes \Gamma$ are not $\mathcal{H}(\mathcal{R}^Q, q^Q)$, but $\mathcal{H}(\mathcal{R}^Q, q^Q) \rtimes \Gamma_Q$ for some subgroup $\Gamma_Q \subset \Gamma$. To make this work well, we need some assumptions on the groups Γ_Q for $Q \subset \Delta$.

Condition 2.1

(1) $\Gamma_Q \subset \Gamma_{Q'}$ if $Q \subset Q'$;
(2) the action of Γ_Q on T stabilizes T_Q and T^Q;
(3) Γ_Q acts on T^Q by multiplication with elements of K_Q.

Notice that Γ_Q is a subgroup of $\Gamma(Q, Q) = \{\gamma \in \Gamma : \gamma(Q) = Q\}$, but that we do not require these two groups to be equal.

We also note that Conditions 2.1 entail that $\Gamma_\emptyset$ acts trivially on $\mathcal{O}(T) = \mathcal{H}^\emptyset$, so

$$\mathrm{Irr}(\mathcal{H}^\emptyset \rtimes \Gamma_\emptyset) \cong T \times \mathrm{Irr}(\Gamma_\emptyset). \tag{15}$$

Remark 2.2 Often there is a larger root system $\tilde{R} \supset R$ in X, such that $W_Q \Gamma_Q$ is contained in the parabolic subgroup of $W(\tilde{R})$ associated with $\tilde{R} \cap \mathbb{Q}Q$. Then parts (2) and (3) of Condition 2.1 are automatically satisfied (and part (1) is usually obvious).

Under Conditions 2.1 Γ_Q commutes with K_Q. The conditions also entail that the projection $\mathcal{H}^Q \to \mathcal{H}_Q$ and the isomorphisms $\phi_t : \mathcal{H}^Q \to \mathcal{H}^Q$ ($t \in T^Q$) are Γ_Q-equivariant, so they extend to algebra homomorphisms

$$\mathcal{H}^Q \rtimes \Gamma_Q \to \mathcal{H}_Q \rtimes \Gamma_Q \quad \text{and} \quad \phi_t : \mathcal{H}^Q \rtimes \Gamma_Q \to \mathcal{H}^Q \rtimes \Gamma_Q \; (t \in T^Q). \tag{16}$$

Via the first map of (16), we can inflate any representation of $\mathcal{H}_Q \rtimes \Gamma_Q$ to $\mathcal{H}^Q \rtimes \Gamma_Q$, which we often do tacitly. For any representation π of $\mathcal{H}^Q \rtimes \Gamma_Q$ and any $t \in T^Q$ we write

$$\pi \otimes t = \pi \circ \phi_t \in \mathrm{Mod}(\mathcal{H}^Q \rtimes \Gamma_Q).$$

Lemma 2.3 *(a) Every irreducible $\mathcal{H}^Q \rtimes \Gamma_Q$-representation is of the form $\pi_Q \otimes t^Q$ for some $\pi_Q \in \mathrm{Irr}(\mathcal{H}_Q \rtimes \Gamma_Q)$ and $t^Q \in T^Q$.*
(b) $\pi_Q \otimes t^Q$ is tempered if and only if π is tempered and $t^Q \in T^Q_{\mathrm{un}}$.
(c) $\pi_Q \otimes t^Q$ is essentially discrete series if and only if π_Q is discrete series.

Proof. (a) First we consider the situation without Γ_Q. Let $\pi \in \mathrm{Irr}(\mathcal{H}^Q)$ with central character $W_Q t \in T/W_Q$. The group $W_Q = W(R_Q)$ acts trivially on T^Q, so $W_Q t = t^Q W_Q t_Q$ for some $t^Q \in T^Q$, $t_Q \in T_Q$. Then $\pi \otimes (t^Q)^{-1}$ factors through $\mathcal{H}^Q \to \mathcal{H}_Q$ (say as π_Q), and $\pi = \pi_Q \otimes t^Q$.

To include Γ_Q, we use Clifford theory [27, Theorem A.6]. It says that every irreducible $\mathcal{H}^Q \rtimes \Gamma_Q$-representation is of the form

$$\pi \rtimes \rho := \mathrm{ind}_{\mathcal{H}^Q \rtimes \Gamma_{Q,\pi}}^{\mathcal{H}^Q \rtimes \Gamma_Q} (\pi \otimes \rho).$$

Here $\Gamma_{Q,\pi}$ is the stabilizer of $\pi \in \mathrm{Irr}(\mathcal{H}^Q)$ in Γ_Q and (ρ, V_ρ) is an irreducible representation of a twisted group algebra of $\Gamma_{Q,\pi}$. If $\mathcal{O}(T)$ acts by $t^Q t_1$ on a vector subspace $V_1 \subset V_\pi$, then for $\gamma \in \Gamma_Q$ it acts by the character $\gamma^{-1}(t^Q t_1)$ on $N_\gamma(V_1 \otimes V_\rho)$. By Condition 2.1(3) $\gamma^{-1}(t^Q t_1) \in t^Q K_Q \gamma^{-1}(t_1)$. Hence $(\pi \rtimes \rho) \otimes (t^Q)^{-1}$ factors through $\mathcal{H}^Q \rtimes \Gamma_Q \to \mathcal{H}_Q \rtimes \Gamma_Q$ as $\pi_Q \rtimes \rho$, and

$$\pi \rtimes \rho = (\pi_Q \rtimes \rho) \otimes t^Q.$$

(b) As $T = T^Q T_Q$ with $T^Q \cap T_Q \subset T_{\mathrm{un}}$, there is a factorization $T_{rs} = T^Q_{rs} \times T_{Q,rs}$ and (with respect to $\mathcal{R}^Q$) $T^-_{rs} = \{1\} \times T^-_{Q,rs}$. Also $|\mathrm{Wt}(\pi \otimes t^Q)| = |t^Q|\,|\mathrm{Wt}(\pi)|$. These observations imply the result.
(c) This is obvious from $\mathrm{Wt}(\pi \otimes t^Q) = t^Q \mathrm{Wt}(\pi)$.

2.1 The Schwartz and C*-completions

To get nice completions of $\mathcal{H}(\mathcal{R}, q)$, we assume from now on that q is a positive parameter function for $\mathcal{R}$. As a topological vector space, the Schwartz completion of $\mathcal{H}(\mathcal{R}, q)$ will consist of rapidly decreasing functions on W^e, with respect to a suitable length function $\mathcal{N}$. For example, we can take a W-invariant norm on $X \otimes_{\mathbb{Z}} \mathbb{R}$ and put $\mathcal{N}(w t_x) = \|x\|$ for $w \in W$ and $x \in X$. Then we can define, for $n \in \mathbb{N}$, the following norm on $\mathcal{H}$:

$$p_n\Big(\sum\nolimits_{w \in W^e} h_w N_w\Big) = \sup\nolimits_{w \in W^e} |h_w| (\mathcal{N}(w) + 1)^n.$$

The completion of $\mathcal{H}$ with respect to these norms is the Schwartz algebra $\mathcal{S} = \mathcal{S}(\mathcal{R}, q)$. It is known from [25, Sect. 6.2] that it is a Fréchet algebra. The Γ-action on $\mathcal{H}$ extends continuously to $\mathcal{S}$, so the crossed product algebra $\mathcal{S}(\mathcal{R}, q) \rtimes \Gamma$ is well defined. By [25, Lemma 2.20], a finite dimensional $\mathcal{H} \rtimes \Gamma$-representation is tempered if and only if it extends continuously to an $\mathcal{S} \rtimes \Gamma$-representation.

We define a *-operation and a trace on $\mathcal{H}(\mathcal{R}, q)$ by

$$\Big(\sum\nolimits_{w \in W^e} c_w N_w\Big)^* = \sum\nolimits_{w \in W^e} \overline{c_w} N_{w^{-1}},$$
$$\tau\Big(\sum\nolimits_{w \in W^e} c_w N_w\Big) = c_e.$$

Since $q(s_\alpha) > 0$, * preserves the relations (7) and defines an anti-involution of $\mathcal{H}(\mathcal{R}, q)$. The set $\{N_w : w \in W^e\}$ is an orthonormal basis of $\mathcal{H}(\mathcal{R}, q)$ for the inner product

$$\langle h_1 \,,\, h_2 \rangle = \tau(h_1^* h_2).$$

This gives $\mathcal{H}(\mathcal{R}, q)$ the structure of a Hilbert algebra. The Hilbert space completion $L^2(\mathcal{R})$ of $\mathcal{H}(\mathcal{R}, q)$ is a module over $\mathcal{H}(\mathcal{R}, q)$, via left multiplication. Moreover, every $h \in \mathcal{H}(\mathcal{R}, q)$ acts as a bounded linear operator [25, Lemma 2.3]. The reduced C^*-algebra of $\mathcal{H}(\mathcal{R}, q)$ [25, §2.4], denoted $C_r^*(\mathcal{R}, q)$, is defined as the closure of $\mathcal{H}(\mathcal{R}, q)$ in the algebra of bounded linear operators on $L^2(\mathcal{R})$. By [25, Theorem 6.1]

$$\mathcal{H}(\mathcal{R}, q) \subset \mathcal{S}(\mathcal{R}, q) \subset C_r^*(\mathcal{R}, q).$$

As in (12), we can extend this to a C^*-algebra $C_r^*(\mathcal{R}, q) \rtimes \Gamma$, provided that q is Γ-invariant.

Let us recall some background about $C_r^*(\mathcal{R}, q) \rtimes \Gamma$, mainly from [25, 39]. It follows from [12, Corollary 5.7] that it is a finite type I C^*-algebra and that $\mathrm{Irr}(C_r^*(\mathcal{R}, q))$ is precisely the tempered part of $\mathrm{Irr}(\mathcal{H}(\mathcal{R}, q))$. According to [25, Theorem 4.23] all irreducible $\mathcal{S}(\mathcal{R}, q) \rtimes \Gamma$-representations extend continuously to $C_r^*(\mathcal{R}, q) \rtimes \Gamma$. Hence we can regard the representation theory of $C_r^*(\mathcal{R}, q) \rtimes \Gamma$ as the tempered unitary representation theory of $\mathcal{H}(\mathcal{R}, q) \rtimes \Gamma$).

The structure of $C_r^*(\mathcal{R}, q) \rtimes \Gamma$ is described in terms of parabolically induced representations. As induction data, we use triples (Q, δ, t) where $Q \subset \Delta$, $\delta \in \mathrm{Irr}_{L^2}(\mathcal{H}_Q)$ and $t \in T^Q$. We regard two triples (Q, δ, t) and (Q', δ', t') as equivalent if $Q = Q'$, $t = t'$, and $\delta \cong \delta'$. Notice that $\mathcal{H}_Q$ comes from a semisimple root datum, so it can have discrete series representations. We inflate such a representation to $\mathcal{H}^Q$ via the projection (8). To a triple (Q, δ, t), we associate the $\mathcal{H} \rtimes \Gamma$-representation

$$\pi^\Gamma(Q, \delta, t) = \mathrm{ind}_{\mathcal{H}^Q}^{\mathcal{H} \rtimes \Gamma}(\delta \circ \psi_t). \tag{17}$$

(When $\Gamma = 1$, we often suppress it from these and similar notations.) For $t \in T^Q_{\mathrm{un}} = T^Q \cap T_{\mathrm{un}}$, these representations extend continuously to the respective C^*-completions of the involved algebras. Let Ξ_{un} be the set of triples (Q, δ, t) as above, such that, moreover, $t \in T_{\mathrm{un}}$. Considering Q and δ as discrete variables, we regard Ξ_{un} as a disjoint union of finitely many compact real tori (of different dimensions).

Let $\mathcal{V}_\Xi^\Gamma$ be the vector bundle over Ξ_{un}, whose fibre at $\xi = (Q, \delta, t)$ is the vector space underlying $\pi^\Gamma(Q, \delta, t)$. That vector space is independent of t, so the vector bundle is trivial. Let $\mathrm{End}(\mathcal{V}_\Xi^\Gamma)$ be the algebra bundle with fibres $\mathrm{End}_{\mathbb{C}}\big(\pi^\Gamma(Q, \delta, t)\big)$. These data give rise to a canonical map

$$\begin{array}{ccc} \mathcal{H}(\mathcal{R}, q) \rtimes \Gamma & \to & \mathcal{O}\big(\Xi; \mathrm{End}(\mathcal{V}_\Xi^\Gamma)\big) \\ h & \mapsto & \big(\xi \mapsto \pi^\Gamma(\xi)(h)\big) \end{array} \tag{18}$$

which we refer to as the Fourier transform. By [25, Lemma 2.22], every discrete series representation is unitary, so V_δ carries an $\mathcal{H}_Q$-invariant inner product and $\mathrm{End}_{\mathbb{C}}(V_\delta)$ has a natural *-operation. For any $t \in T^Q$, this becomes an $\mathcal{H}^Q$-invariant nondegenerate pairing between $\delta \circ \phi_t$ and $\delta \circ \phi_{t|t|^{-2}}$. By [25, Proposition 4.19], this extends canonically to an inner product on the vector space

$$\pi^\Gamma(Q, \delta, t) = \Gamma \ltimes \mathcal{H}(W, q) \otimes_{\mathcal{H}(W_Q,q)} V_\delta. \tag{19}$$

That yields an anti-involution on $\mathrm{End}_\mathbb{C}(\pi^\Gamma(Q, \delta, t))$ and a nondegenerate $\mathcal{H} \rtimes \Gamma$-invariant pairing between $\pi^\Gamma(Q, \delta, t)$ and $\pi^\Gamma(Q, \delta, t\,|t|^{-2})$.

The algebra $\mathcal{O}\big(\Xi; \mathrm{End}(\mathcal{V}_\Xi^\Gamma)\big)$ is endowed with the anti-involution

$$(f^*)(Q, \delta, t) = f(Q, \delta, t\,|t|^{-2})^*. \tag{20}$$

With respect to this anti-involution, (18) is a *-homomorphism.

To administer the upcoming intertwining operators, we use a finite groupoid $\mathcal{G}$ which acts on $\mathrm{End}(\mathcal{V}_\Xi^\Gamma)$. It is made from the elements of $W \rtimes \Gamma$ and of $K_Q := T_Q \cap T^Q$. More precisely, its base space is the power set of Δ, and for $Q, Q' \subseteq \Delta$ the collection of arrows from Q to Q' is

$$\mathcal{G}_{QQ'} = \{(g, u) : g \in \Gamma \ltimes W, u \in K_Q, g(Q) = Q'\}. \tag{21}$$

Whenever it is defined, the multiplication in $\mathcal{G}$ is

$$(g', u') \cdot (g, u) = (g'g, g^{-1}(u')u).$$

In particular, writing $W\Gamma(Q, Q) = \{w \in W\Gamma : w(Q) = Q\}$, we have the group

$$\mathcal{G}_{QQ} = W\Gamma(Q, Q) \rtimes K_Q. \tag{22}$$

Usually, we will write elements of $\mathcal{G}$ simply as gu. There is an analogous groupoid $\mathcal{G}^Q$ for $\mathcal{H}^Q \rtimes \Gamma_Q$, which under Conditions 2.1 satisfies $\mathcal{G}^Q_{QQ} = \Gamma_Q \times K_Q$.

For $\gamma \in \Gamma W$ with $\gamma(Q) = Q' \subset \Delta$, there are algebra isomorphisms

$$\begin{aligned} \psi_\gamma &: \mathcal{H}_Q \to \mathcal{H}_{Q'},\ \theta_{x_Q} N_w \mapsto \theta_{\gamma(x_Q)} N_{\gamma w \gamma^{-1}}, \\ \psi_\gamma &: \mathcal{H}^Q \to \mathcal{H}^{Q'},\ \theta_x N_w \ \mapsto \theta_{\gamma x} N_{\gamma w \gamma^{-1}}. \end{aligned} \tag{23}$$

The groupoid $\mathcal{G}$ acts from the left on Ξ_{un} by

$$(g, u) \cdot (Q, \delta, t) := (g(Q), \delta \circ \psi_u^{-1} \circ \psi_g^{-1}, g(ut)), \tag{24}$$

the action being defined if and only if $g(Q) \subset \Delta$.

Suppose that $g(Q) = Q' \subset \Delta$ and $\delta' \cong \delta \circ \psi_u^{-1} \circ \psi_g^{-1}$. By [25, Theorem 4.33] and [39, Theorem 3.1.5], there exists an intertwining operator

$$\pi^\Gamma(gu, Q, \delta, t) \in \mathrm{Hom}_{\mathcal{H}(\mathcal{R},q) \rtimes \Gamma}\big(\pi^\Gamma(Q, \delta, t), \pi^\Gamma(Q', \delta', g(ut))\big), \tag{25}$$

which depends algebraically on $t \in T^Q_{\mathrm{un}}$. This implies that, for all $\xi \in \Xi$ and $g \in \mathcal{G}$ such that $g\xi$ is defined, $\pi^\Gamma(\xi)$ and $\pi^\Gamma(g\xi)$ have the same irreducible constituents, counted with multiplicity [39, Lemma 3.1.7].

The action of $\mathcal{G}$ on the continuous sections $C(\Xi_{un}; \mathrm{End}(\mathcal{V}_\Xi^\Gamma))$ is given by

$$(g \cdot f)(g\xi) = \pi^\Gamma(g, \xi) f(\xi) \pi^\Gamma(g, \xi)^{-1} \qquad g \in \mathcal{G}_{QQ}, \xi = (Q, \delta, t). \tag{26}$$

The next result is the Plancherel isomorphism for affine Hecke algebras, proven in [12, Theorem 5.3 and Corollary 5.7] and [39, Theorem 3.2.2].

Theorem 2.4 *The Fourier transform* (18) *induces *-homomorphisms*

$$\begin{aligned} \mathcal{H}(\mathcal{R}, q) \rtimes \Gamma &\to \mathcal{O}\big(\Xi; \mathrm{End}(\mathcal{V}_\Xi^\Gamma)\big)^{\mathcal{G}}, \\ \mathcal{S}(\mathcal{R}, q) \rtimes \Gamma &\to C^\infty\big(\Xi_{un}; \mathrm{End}(\mathcal{V}_\Xi^\Gamma)\big)^{\mathcal{G}}, \\ C_r^*(\mathcal{R}, q) \rtimes \Gamma &\to C\big(\Xi_{un}; \mathrm{End}(\mathcal{V}_\Xi^\Gamma)\big)^{\mathcal{G}}. \end{aligned}$$

The first is injective, the second is an isomorphism of Fréchet algebras, and the third is an isomorphism of C^*-algebras.

For $q = 1$, these simplify to the well-known isomorphisms

$$\begin{aligned} \mathcal{H}(\mathcal{R}, 1) \rtimes \Gamma &= \mathcal{O}(T) \rtimes W\Gamma \to \mathcal{O}\big(T; \mathrm{End}_{\mathbb{C}}(\mathbb{C}[W\Gamma])\big)^{W\Gamma}, \\ \mathcal{S}(\mathcal{R}, 1) \rtimes \Gamma &= C^\infty(T_{un}) \rtimes W\Gamma \to C^\infty\big(T_{un}; \mathrm{End}_{\mathbb{C}}(\mathbb{C}[W\Gamma])\big)^{W\Gamma}, \\ C_r^*(\mathcal{R}, 1) \rtimes \Gamma &= C(T_{un}) \rtimes W\Gamma \to C\big(T_{un}; \mathrm{End}_{\mathbb{C}}(\mathbb{C}[W\Gamma])\big)^{W\Gamma}. \end{aligned} \tag{27}$$

Unfortunately, the bookkeeping in Theorem 2.4 is not entirely suitable for our purposes, because sometimes the parabolic subalgebras need to be extended by diagram automorphisms. In those cases, we should rather use induction data based on $\mathrm{Irr}_{L^2}(\mathcal{H}^Q \rtimes \Gamma_Q)$ than based on $\mathrm{Irr}_{L^2}(\mathcal{H}_Q)$ or $\mathrm{Irr}_{L^2}(\mathcal{H}^Q)$.

We fix a system of subgroups $\Gamma_Q \subset \Gamma$ ($Q \subset \Delta$), satisfying Condition 2.1. With Lemma 2.3 in mind, we define new induction data. They are triples (Q, σ, t) with $Q \subset \Delta$, $t \in T^Q$, and $\sigma \in \mathrm{Irr}_{L^2}(\mathcal{H}_Q \rtimes \Gamma_Q)$. We regard another such triple (Q', σ', t') as equivalent if and only if $Q' = Q$, $t' = t$, and $\sigma' \cong \sigma$. We keep the same groupoid $\mathcal{G}$ as before, and it also acts on the new triples via (24). To such a triple, we associate the representation

$$\pi(Q, \sigma, t) = \mathrm{ind}_{\mathcal{H}^Q \rtimes \Gamma_Q}^{\mathcal{H} \rtimes \Gamma} (\sigma \otimes t). \tag{28}$$

The vector space underlying $\pi(Q, \sigma, t)$ does not depend on t, we denote it by $V_{Q,\sigma}$. There is a natural homomorphism

$$\begin{aligned} \mathcal{H}(\mathcal{R}, q) \rtimes \Gamma &\to \mathcal{O}(T^Q) \otimes \mathrm{End}_{\mathbb{C}}(V_{Q,\sigma}) \\ h \quad &\mapsto \big(t \mapsto \pi(Q, \sigma, t)(h)\big). \end{aligned} \tag{29}$$

We refer to the system of these maps, for all Q and σ, as the Fourier transform for $\mathcal{H}(\mathcal{R}, q) \rtimes \Gamma$. The recipe for the intertwining operators from [25, §4] and [39, Theorem 3.1.5] remains valid, so we get

$$\pi(gu, Q, \sigma, t) \in \mathrm{Hom}_{\mathcal{H}\rtimes\Gamma}\big(\pi(Q, \sigma, t), \pi(g(Q), \sigma', g(ut))\big) \tag{30}$$

with the same properties as in (25). In particular, $\pi(Q, \sigma, t)$ and $\pi(g(Q), \sigma', g(ut))$ have the same irreducible constituents, counted with multiplicity. With these notions, we can vary on the Plancherel isomorphism (Theorem 2.4).

To do so, we first consider essentially discrete series representations of $\mathcal{H}^Q \rtimes \Gamma_Q$. Pick $\delta_1 \in \mathrm{Irr}_{L^2}(\mathcal{H}_Q)$ and $t_1 \in T^Q_{\mathrm{un}}$. We note that $\mathrm{ind}_{\mathcal{H}^Q}^{\mathcal{H}^Q\rtimes\Gamma_Q}(\delta \otimes t_1)$ is unitary and essentially discrete series, because Γ_Q stabilizes Q. Write $\mathcal{G}^Q_{QQ}\delta_1 = \{\delta_i\}_i$. The summand of $C^\infty\big(\Xi_{Q,\mathrm{un}}; \mathrm{End}(\mathcal{V}^{\Gamma_Q}_{\Xi_Q})\big)^{\mathcal{G}^Q}$ associated with (Q, δ_1) is

$$\Big(\bigoplus\nolimits_i C^\infty\big(T^Q_{\mathrm{un}}; \mathrm{End}_{\mathbb{C}}(\pi^{\Gamma_Q}(Q, \delta_i, t_i))\big)\Big)^{\mathcal{G}^Q_{QQ}}. \tag{31}$$

Let $\{\sigma_j\}_j$ be the members of $\mathrm{Irr}_{L^2}(\mathcal{H}_Q \rtimes \Gamma_Q)$ contained in $\mathrm{ind}_{\mathcal{H}_Q}^{\mathcal{H}_Q\rtimes\Gamma_Q}(\delta_1 \circ \psi_u)$ for some $u \in K_Q$. This set is stable under $\mathcal{G}^Q_{QQ} = \Gamma_Q \times K_Q$. The summand of $\Big(\bigoplus_\sigma C^\infty\big(T^Q_{\mathrm{un}}; \mathrm{End}_{\mathbb{C}}(V_\sigma)\big)\Big)^{\mathcal{G}^Q}$ corresponding to the σ_j is

$$\Big(\bigoplus\nolimits_j C^\infty\big(T^Q_{\mathrm{un}}; \mathrm{End}_{\mathbb{C}}(V_{\sigma_j})\big)\Big)^{\mathcal{G}^Q_{QQ}}. \tag{32}$$

Lemma 2.5 *The algebras* (31) *and* (32) *are naturally isomorphic.*

Proof. For $\sigma \in \mathrm{Irr}_{L^2}(\mathcal{H}^Q \rtimes \Gamma_Q)$ we write $(\sigma, t) > (\delta_1, t_1)$ if

$$\mathrm{Hom}_{\mathcal{H}^Q\rtimes\Gamma_Q}\big(\sigma \otimes t, \mathrm{ind}_{\mathcal{H}^Q}^{\mathcal{H}^Q\rtimes\Gamma_Q}(\delta \otimes t_1)\big) \cong \mathrm{Hom}_{\mathcal{H}^Q}(\sigma \otimes t, \delta \otimes t_1)$$

is nonzero. Since Γ_Q is finite, the set of such (σ, t) is finite. Hence the map

$$\bigoplus_{(\sigma,t)>(\delta_1,t_1)} \sigma \circ \psi_t \, : \, \mathcal{H}^Q \rtimes \Gamma_Q \to \bigoplus_{(\sigma,t)>(\delta_1,t_1)} \mathrm{End}_{\mathbb{C}}(V_\sigma)$$

is surjective. The specialization of (31) at $\mathcal{G}^Q_{QQ}(Q, \delta_1, t_1)$ is also $\bigoplus_{(\sigma,t)>(\delta_1,t_1)} \mathrm{End}_{\mathbb{C}}(V_\sigma)$, for that specialization is just

$$\mathrm{ind}_{\mathcal{H}^Q}^{\mathcal{H}^Q\rtimes\Gamma_Q}(\delta \circ \psi_{t_1})(\mathcal{S}(\mathcal{R}^Q, q^Q) \rtimes \Gamma_Q).$$

Similarly, specializing the algebra (32) at all $(\sigma, t) > (\delta_1, t_1)$ gives a surjection from (32) to $\bigoplus_{(\sigma,t)>(\delta_1,t_1)} \mathrm{End}_{\mathbb{C}}(V_\sigma)$.

Now we can explicitly compare (32) with (31). Both are algebras of smooth sections of (trivial) algebra bundles, and specialization at the points associated with (δ_1, t_1) yields the same algebra in both cases. This holds for any $t_1 \in T^Q_{\mathrm{un}}$ and that accounts for all base points of these algebra bundles, so (31) and (32) are isomorphic.

Moreover, the isomorphism is canonical: It is the composition of the inverse of the map in Theorem 2.4 and the map induced by (29) (both for $\mathcal{H}^Q \rtimes \Gamma_Q$). □

Next we formulate our variation on the Plancherel isomorphism.

Proposition 2.6 *The Fourier transform from* (29) *induces isomorphisms of Fréchet *-algebras*

$$\mathcal{S}(\mathcal{R}, q) \rtimes \Gamma \to \Big(\bigoplus\nolimits_{Q,\sigma} C^\infty \big(T^Q_{\text{un}}; \mathrm{End}_{\mathbb{C}}(V_{Q,\sigma})\big)\Big)^{\mathcal{G}},$$
$$C^*_r(\mathcal{R}, q) \rtimes \Gamma \to \Big(\bigoplus\nolimits_{Q,\sigma} C\big(T^Q_{\text{un}}; \mathrm{End}_{\mathbb{C}}(V_{Q,\sigma})\big)\Big)^{\mathcal{G}}.$$

Proof. We will analyse the right-hand side of Theorem 2.4 (for the Schwartz algebras) and compare it with the current setting.

For every $Q \subset \Delta$, Lemma 2.5 yields a canonical isomorphism

$$\Big(\bigoplus\nolimits_{\sigma \in \mathrm{Irr}_{L^2}(\mathcal{H}_Q \rtimes \Gamma_Q)} C^\infty \big(T^Q_{\text{un}}; \mathrm{End}_{\mathbb{C}}(V_\sigma)\big)\Big)^{\mathcal{G}^Q_{QQ}} \cong$$
$$\Big(\bigoplus\nolimits_{\delta \in \mathrm{Irr}_{L^2}(\mathcal{H}_Q)} C^\infty \big(T^Q_{\text{un}}; \mathrm{End}_{\mathbb{C}}(\mathbb{C}[\Gamma_Q] \otimes V_\delta)\big)\Big)^{\mathcal{G}^Q_{QQ}}. \tag{33}$$

To obtain the right-hand side of Theorem 2.4 from (33), we must apply $\mathrm{ind}_{\mathcal{H}^Q \rtimes \Gamma_Q}^{\mathcal{H} \rtimes \Gamma}$ to $\mathbb{C}[\Gamma_Q] \otimes V_\delta \cong \mathrm{ind}_{\mathcal{H}^Q}^{\mathcal{H}^Q \rtimes \Gamma_Q} (\delta \otimes t)$ and then take invariants with respect to the larger groupoid $\mathcal{G} \supset \mathcal{G}^Q$. The formula [39, (3.12)] is the same for $\mathcal{G}^Q$ and for $\mathcal{G}$, so the intertwiners associated with elements of $\mathcal{G}^Q$ need not be adjusted in this process.

With exactly the same procedure, we can turn (33) into the right-hand side of the current proposition. The intertwining operators associated with elements of $\mathcal{G}$ agree under the isomorphisms obtained from (33) by applying $\mathrm{ind}_{\mathcal{H}^Q \rtimes \Gamma_Q}^{\mathcal{H} \rtimes \Gamma}$, because in both settings they were constructed with [39, (3.12) and Theorem 3.1.5]. Consequently

$$\Big(\bigoplus\nolimits_{Q,\sigma} C^\infty \big(T^Q_{\text{un}}; \mathrm{End}_{\mathbb{C}}(V_{Q,\sigma})\big)\Big)^{\mathcal{G}} \cong C^\infty \big(\Xi_{\text{un}}; \mathrm{End}(\mathcal{V}^\Gamma_\Xi)\big)^{\mathcal{G}},$$

proving the proposition for the Schwartz algebras. For $C^*_r(\mathcal{R}, q) \rtimes \Gamma$, one can use the same argument, with everywhere C^∞ replaced by continuous functions. □

Choose representatives Q for $\mathcal{P}(\Delta)$ modulo $W\Gamma$-association. For every such Q, we choose representatives σ for the action of $\mathcal{G}_{QQ} = W\Gamma(Q, Q) \times K_Q$ on $\mathrm{Irr}_{L^2}(\mathcal{H}_Q \rtimes \Gamma_Q)$. By Lemma 2.3, these σ also form representatives for the action of $\mathcal{G}_{QQ} \ltimes T^Q_{\text{un}}$ on $\mathrm{Irr}_{L^2}(\mathcal{H}^Q \rtimes \Gamma_Q)$. We denote the resulting set of representatives of pairs by $(Q, \sigma)/\mathcal{G}$. Let $\mathcal{G}_{Q,\sigma}$ be the setwise stabilizer of $(Q, \sigma, T^Q_{\text{un}})$ in the group $\mathcal{G}_{QQ}$. Proposition 2.6 can be rephrased as isomorphisms

$$\begin{aligned} \mathcal{S}(\mathcal{R}, q) \rtimes \Gamma &\to \bigoplus\nolimits_{(Q,\sigma)/\mathcal{G}} C^\infty \big(T^Q_{\text{un}}; \mathrm{End}_{\mathbb{C}}(V_{Q,\sigma})\big)^{\mathcal{G}_{Q,\sigma}}, \\ C^*_r(\mathcal{R}, q) \rtimes \Gamma &\to \bigoplus\nolimits_{(Q,\sigma)/\mathcal{G}} C\big(T^Q_{\text{un}}; \mathrm{End}_{\mathbb{C}}(V_{Q,\sigma})\big)^{\mathcal{G}_{Q,\sigma}}. \end{aligned} \tag{34}$$

Sometimes we have to consider the opposite algebra $(\mathcal{H}(\mathcal{R}, q) \rtimes \Gamma)^{\text{op}}$ and its completions. It is, morally, clear that all the previous results can also be developed for right $\mathcal{H} \rtimes \Gamma$-modules, that is, for $(\mathcal{H} \rtimes \Gamma)^{\text{op}}$-modules. However, none of that has been written down, so we prefer more steady ground.

For every $\mathcal{H} \rtimes \Gamma$-representation (π, V_π), the full linear dual V_π^* becomes a $(\mathcal{H} \rtimes \Gamma)^{\text{op}}$-representation π^* by

$$\pi^*(h^{\text{op}})\lambda = \lambda \circ \pi(h).$$

This sets up a bijection between finite dimensional left and right modules of $\mathcal{H} \rtimes \Gamma$. In view of the canonical inner products from on the spaces (19), this bijection commutes with induction from parabolic subalgebras.

For infinite dimensional representations, there is often some choice for which dual space of V_π we use here. In particular, when V_π is a Hilbert space we can use V_π also as dual space. With this convention, one checks easily that π is unitary if and only if π^* is unitary.

The $\mathcal{O}(T)$-weights of π^* are the same as for π, so π^* is tempered or (essentially) discrete series if and only if π is so. Thus the pairs (Q, σ) with $\sigma \in \text{Irr}_{L^2}(\mathcal{H}_Q \rtimes \Gamma_Q)$ are in natural bijection with the pairs (Q, σ^*) in

$$\bigcup\nolimits_{Q \subset \Delta} \text{Irr}_{L^2}\big((\mathcal{H}(\mathcal{R}_Q, q_Q) \rtimes \Gamma_Q)^{\text{op}}\big). \tag{35}$$

The bijection is $\mathcal{G}$-equivariant for the $\mathcal{G}$-action on (35) as in (24). Hence $\mathcal{G}_{Q,\sigma} = \mathcal{G}_{Q,\sigma^*}$, and we can take $(Q, \sigma^*)/\mathcal{G}$ to be the image of $(Q, \sigma)/\mathcal{G}$.

Lemma 2.7 *The Fourier transform for right $\mathcal{H}(\mathcal{R}, q) \rtimes \Gamma$-modules induces isomorphisms of Fréchet *-algebras*

$$\begin{aligned}
(\mathcal{S}(\mathcal{R}, q) \rtimes \Gamma)^{\text{op}} &\to \bigoplus\nolimits_{(Q,\sigma^*)/\mathcal{G}} C^\infty\big(T_{\text{un}}^Q; \text{End}_{\mathbb{C}}(V_{Q,\sigma^*})\big)^{\mathcal{G}_{Q,\sigma^*}}, \\
(C_r^*(\mathcal{R}, q) \rtimes \Gamma)^{\text{op}} &\to \bigoplus\nolimits_{(Q,\sigma^*)/\mathcal{G}} C\big(T_{\text{un}}^Q; \text{End}_{\mathbb{C}}(V_{Q,\sigma^*})\big)^{\mathcal{G}_{Q,\sigma^*}}.
\end{aligned}$$

Proof. The opposite algebra of

$$\text{End}_{\mathbb{C}}(V_{Q,\sigma}) = \text{End}_{\mathbb{C}}\big(\text{ind}_{\mathcal{H}^Q \rtimes \Gamma_Q}^{\mathcal{H} \rtimes \Gamma} V_\delta\big)$$

is naturally isomorphic to $\text{End}_{\mathbb{C}}\big(V_{Q,\sigma}^*\big)$, which by [25, Proposition 4.19] is canonically isomorphic with

$$\text{End}_{\mathbb{C}}\big(\text{ind}_{\mathcal{H}^Q \rtimes \Gamma_Q}^{\mathcal{H} \rtimes \Gamma}(V_\sigma^*)\big) = \text{End}_{\mathbb{C}}(V_{Q,\sigma^*}).$$

For $g \in \mathcal{G}_{Q,\sigma}$, we take $\pi(g, Q, \sigma^*, t|t|^{-2})$ to be the transpose inverse of $\pi(g, Q, \sigma, t)$. Thus an element of $C\big(T_{\text{un}}^Q; \text{End}_{\mathbb{C}}(V_{Q,\sigma})\big)$ is $\mathcal{G}_{Q,\sigma}$-invariant if and only if its transpose in $C\big(T_{\text{un}}^Q; \text{End}_{\mathbb{C}}(V_{Q,\sigma^*})\big)$ is $\mathcal{G}_{Q,\sigma^*}$-invariant for the action

$$(g \cdot f)(g(Q, \sigma^*, t)) = \pi(g, Q, \sigma^*, t) f(Q, \sigma^*, t) \pi(g, Q, \sigma^*, t)^{-1}.$$

Now we take the opposite algebras in Theorem 2.4 and we find the desired isomorphisms.

The implementing algebra homomorphisms are given by transpose, the Fourier transform from (29) and again transpose, which works out to the Fourier transform for $(\mathcal{H}(\mathcal{R}, q) \rtimes \Gamma)^{\mathrm{op}}$-modules. Since the correspondence between left and right $\mathcal{H}(\mathcal{R}, q) \rtimes \Gamma$-modules preserves unitarity, the latter Fourier transform is still a *-homomorphism. □

2.2 The Space of Irreducible Representations

We compare the irreducible representations of $\mathcal{H} \rtimes \Gamma$ to its representations induced from proper parabolic subalgebras (i.e., the algebras $\mathcal{H}^Q \rtimes \Gamma_Q$ with $Q \subsetneq \Delta$). Let $\mathrm{Gr}(\mathcal{H} \rtimes \Gamma)$ be the Grothendieck group of the category of finite length $\mathcal{H} \rtimes \Gamma$-representations and write $\mathrm{Gr}_{\mathbb{Q}}(\mathcal{H} \rtimes \Gamma) = \mathbb{Q} \otimes_{\mathbb{Z}} \mathrm{Gr}(\mathcal{H} \rtimes \Gamma)$. Then (parabolic) induction induces a $\mathbb{Q}$-linear map $\mathrm{Gr}_{\mathbb{Q}}(\mathcal{H}_Q \rtimes \Gamma_Q) \to \mathrm{Gr}_{\mathbb{Q}}(\mathcal{H} \rtimes \Gamma)$.

Theorem 2.8 *(a) The collection of irreducible $\mathcal{H} \rtimes \Gamma$-representations whose image in $\mathrm{Gr}_{\mathbb{Q}}(\mathcal{H} \rtimes \Gamma)$ is not a $\mathbb{Q}$-linear combination of representations induced from proper parabolic subalgebras is a nonempty union of $T^{W\Gamma}$-orbits in $\mathrm{Irr}(\mathcal{H} \rtimes \Gamma)$.*

(b) Suppose that, for every $w \in W\Gamma \setminus \bigcup_{Q \subsetneq \Delta} W(R_Q)\Gamma_Q$, the set T^w_Δ is finite. Then the set in part (a) is a finite union of $T^{W\Gamma}$-orbits.

Proof. (a) Recall from (13) that $\mathcal{O}(T)^{W(R_Q)\Gamma_Q} \subset Z(\mathcal{H}^Q \rtimes \Gamma_Q)$. Hence all irreducible representations of $\mathcal{H}^Q \rtimes \Gamma_Q$ come in families parametrized by $T^{W(R_Q)\Gamma_Q}$. Since $T^{W\Gamma} \subset T^{W(R_Q)\Gamma_Q}$ for all $Q \subset \Delta$, the set of irreducible representations under consideration is a union of $T^{W\Gamma}$-orbits.

By [40, Lemma 2.3], $\mathrm{Irr}(\mathcal{H} \rtimes \Gamma)$ can be parametrized by the extended quotient

$$T//W\Gamma = \Big(\bigcup_{w \in W\Gamma} \{w\} \times T^w \Big) / W\Gamma,$$

where $W\Gamma$ acts on the union by $w' \cdot (w, t) = (w' w w'^{-1}, w'(t))$. This parametrization respects central characters, up to a twists which are constant on connected components of $T//W\Gamma$ [40, Theorem 2.6]. In this parametrization of $\mathrm{Irr}(\mathcal{H} \rtimes \Gamma)$, almost all elements of a piece $\{w\} \times T^w$ with $w \in W(R_Q)\Gamma_Q$ come from representations induced from $\mathcal{H}^Q \rtimes \Gamma_Q$, and such w account for all representations induced from proper parabolic subalgebras. Hence the set considered in the statement can be parametrized by

$$\Big(\bigcup_{w \in W\Gamma : w \notin W(R_Q)\Gamma_Q \forall Q \subsetneq \Delta} \{w\} \times T^w \Big) / W\Gamma \tag{36}$$

This set is nonempty because every Coxeter element of $W = W(R)$ contributes at least $(w, 1)$ to it.

(b) This is obvious from (36). □

The induction data from Ξ give rise to a partition $\mathrm{Irr}(\mathcal{H} \rtimes \Gamma)$ into finite packets.

Theorem 2.9 [39, Theorem 3.3.2(b)]
For every $\pi \in \mathrm{Irr}(\mathcal{H} \rtimes \Gamma)$ there exists a unique $\mathcal{G}$-association class $\mathcal{G}(Q, \delta, t) \in \Xi/\mathcal{G}$ such that π is a constituent of $\pi^\Gamma(Q, \delta, t)$ and the invariant $\|cc(\delta)\|$ from (14) *is maximal for this property.*

With the new induction data (Q, σ, t) from (28), we can vary on Theorem 2.9. (Now $\sigma \in \mathrm{Irr}_{L^2}(\mathcal{H}_Q \rtimes \Gamma_Q)$, whereas the above δ was a representation of $\mathcal{H}_Q$.)

Theorem 2.10 *(a) For every $\pi \in \mathrm{Irr}(\mathcal{H} \rtimes \Gamma)$ there exists a triple (Q, σ, t) as above, such that π is a constituent of $\pi(Q, \sigma, t)$ and $\|cc(\sigma)\|$ is maximal for this property. In this situation we say that π is a* Langlands constituent *of $\pi(Q, \sigma, t)$.*
(b) In the setting of part (a), the restriction of $\sigma \otimes t$ to $\mathcal{H}^Q$ is a direct sum of irreducible representations in one Γ_Q-orbit, say $\Gamma_Q(\delta \otimes t) \subset \mathrm{Irr}_{L^2}(\mathcal{H}^Q)$. Then (Q, δ, t) is uniquely determined by π, up to the action of $\mathcal{G}$.
(c) Let (Q, σ, t) be any induction datum as in (28)*. Every constituent of $\pi(Q, \sigma, t)$ is either a Langlands constituent or a constituent of $\pi(Q', \sigma', t')$ for some induction datum with $\|cc(\sigma')\| > \|cc(\sigma)\|$.*
(d) π is tempered if and only if $t \in T_{\mathrm{un}}$, where $t \in T^Q$ comes from part (a).

Proof. (a) Let (Q, δ, t) be as in Theorem 2.9. Thus π is a constituent of

$$\pi^\Gamma(Q, \delta, t) = \mathrm{ind}_{\mathcal{H}^Q \rtimes \Gamma_Q}^{\mathcal{H} \rtimes \Gamma} \big(\mathrm{ind}_{\mathcal{H}^Q}^{\mathcal{H}^Q \rtimes \Gamma_Q} (\delta \otimes t)\big),$$

and the norm of the central character of δ is maximal for this property. Let T^{--Q} be the subset T^{--} of T, but computed with respect to Q. Every $\mathcal{A}$-weight of $\mathrm{ind}_{\mathcal{H}^Q}^{\mathcal{H}^Q \rtimes \Gamma_Q}(\delta \otimes t)$ lies in one of the Γ_Q-orbits of weights of $\delta \otimes t$, which are entirely contained in $T^{--Q} T_{\mathrm{un}}$ because Γ_Q stabilizes Q. In other words, $\mathrm{ind}_{\mathcal{H}^Q}^{\mathcal{H}^Q \rtimes \Gamma_Q}(\delta \otimes t)$ is a direct sum of finitely many irreducible essentially discrete series representations of $\mathcal{H}^Q \rtimes \Gamma$, all with central characters in the same $W_Q \rtimes \Gamma_Q$-orbit. By Lemma 2.3, all these summands are of the form $\sigma_i \otimes t$ with $\sigma_i \in \mathrm{Irr}_{L^2}(\mathcal{H}_Q \rtimes \Gamma_Q)$. Hence

$$\pi^\Gamma(Q, \delta, t) = \bigoplus\nolimits_i \pi(Q, \sigma_i, t) \tag{37}$$

and π is a constituent of (at least) one $\pi(Q, \sigma_i, t)$.

The central characters of the σ_i and of δ have the same norm, and that of δ was maximal given π. Hence the norm of the central character of σ_i is also maximal, given π.

(b) Suppose that (Q, σ, t) satisfies the requirements of part (a), that is, π is a Langlands constituent of $\pi(Q, \sigma, t)$. In the proof of Lemma 2.3, we observed that the restriction of $\sigma \otimes t$ to $\mathcal{H}^Q$ is a direct sum of representations of the form $\gamma((\delta \otimes k^{-1}) \otimes kt)$ with $\gamma \in \Gamma_Q$ and $k \in K_Q$. For all $\gamma \in \Gamma_Q$, π is a constituent of

$$\pi^\Gamma(Q, \delta, t) = \pi^\Gamma(Q, \gamma(\delta \otimes k^{-1}), \gamma(kt)).$$

The norm of the central character of $\gamma(\delta \otimes k^{-1})$ is maximal for this property, by the assumption on σ. By Theorem 2.9, all $(Q, \gamma(\delta \otimes k^{-1}), \gamma(kt))$ lie in a unique $\mathcal{G}$-association class in Ξ determined by π.
(c) In view of (37), it suffices to consider the constituents of $\pi^\Gamma(Q, \delta, t)$. Then the statement is implicit in [39], we make it more explicit here. By [39, Lemma 3.1.7 and Theorem 3.3.2], we may, furthermore, assume that $\xi = (Q, \delta, t)$ is in positive position, that is, $|t| \in T^{Q+} = \exp(\mathfrak{a}^Q \cap \mathfrak{a}^+)$. Write $P(\xi) = \{\alpha \in \Delta : |\alpha(t)| = 1\}$ and consider the representation $\mathrm{ind}_{\mathcal{H}^Q}^{\mathcal{H}^{P(\xi)} \rtimes \Gamma(P(\xi),P(\xi))}(\delta \otimes t)$. By [39, Proposition 3.1.4] that representation is completely reducible, and all its irreducible summands are of the form $\delta' \otimes t'$ where $\delta' \in \mathrm{Irr}\big(\mathcal{H}_{P(\xi)} \rtimes \Gamma(P(\xi), P(\xi))\big)$ is tempered and $t' \in T^{P(\xi)} \cap tT_{P(\xi)}$ with

$$|t'| \in T^{P(\xi)++} = \exp\big(\{\mu \in \mathfrak{a}^{P(\xi)} : \langle \alpha\, , \, \mu \rangle > 0 \;\forall \alpha \in \Delta \setminus P(\xi)\}\big).$$

Thus every constituent $\pi(Q, \sigma, t)$ is a constituent of

$$\mathrm{ind}_{\mathcal{H}^{P(\xi)} \rtimes \Gamma(P(\xi),P(\xi))}^{\mathcal{H} \rtimes \Gamma}(\delta' \otimes t') \tag{38}$$

for such a $\delta' \otimes t'$. By [39, Theorem 3.3.2], the Langlands constituents of $\pi(Q, \sigma, t)$ (or $\pi^\Gamma(Q, \delta, t)$) are precisely those constitutents which are quotients of a representation (38). The Langlands classification for extended affine Hecke algebras [39, Corollary 2.2.5] says that the latter representation has a unique irreducible quotient (called the Langlands quotient, hence our terminology Langlands constituents). Moreover, by [39, Lemma 2.2.6(b)], all other constituents of (38) are Langlands quotients for data where the norm of the central character is bigger than $\|cc(\sigma)\|$. Then [39, Proposition 3.1.4 and Theorem 3.3.2] entail that those other irreducible representations occur as Langlands constituents of a $\pi^\Gamma(Q', \delta', t')$ with $\big\|cc(\delta')\big\| > \|cc(\sigma)\| = \|cc(\delta)\|$.
(d) As observed in the proof of part (c), in the construction for part (a), and for [39, Theorem 3.3.2], we may assume that (Q, σ, t) is positive and that π is the unique Langlands quotient of (38), for one of the above $\delta' \otimes t'$. By the uniqueness in the Langlands classification for extended affine Hecke algebras [39, Corollary 2.2.5], π is tempered if and only if $P(\xi) = \Delta$ and $t' \in T^\Delta_{\mathrm{un}}$.

Here $P(\xi) = \Delta$ implies $|t| = |t'| \in T^\Delta$, and then $t' \in T^\Delta_{\mathrm{un}}$ says that $|t| = 1$. Conversely, $t \in T_{\mathrm{un}}$ implies $P(\xi) = \Delta$ and $t = t' \in T^\Delta$. □

With Theorem 2.10, one can express the structure of $\mathrm{Irr}(\mathcal{H} \rtimes \Gamma)$ in terms of its subset of irreducible tempered representations. In essence, the former is the complexification of the latter (which is something like a real algebraic variety with multiplicities). We do not need the full strength of this. For our purposes, it suffices to consider half-lines in the parameter space, such that Q, σ and the unitary part of t are fixed and the absolute value of t can be scaled by a positive factor.

Proposition 2.11 *Let (Q, σ, t) be an induction datum with $\sigma \in \mathrm{Irr}_{L^2}(\mathcal{H}^Q \rtimes \Gamma_Q)$.*

(a) For $r \in \mathbb{R}_{>-1}$, the number of inequivalent Langlands constituents of $\pi(Q, \sigma, t\,|t|^r)$ does not depend on r.

(b) *For all but finitely many $r \in \mathbb{R}_{>-1}$, $\pi(Q, \sigma, t\,|t|^r)$ is completely reducible. Then all its irreducible subquotients are Langlands constituents.*

Proof. (a) Let $\xi = (Q, \delta, t) \in \Xi$ be as in Theorem 2.10.

Notice that the intertwining operators $\pi(gu, Q, \sigma, t)$ depend algebraically on $t \in T^Q$. This implies that, for every $gu \in \mathcal{G}$, $\pi(Q, \sigma, t)$ and $\pi(gu(Q, \sigma, t))$ have the same irreducible subquotients (counted with multiplicity), see [39, Lemma 3.1.7] or [38, Lemma 3.4]. Since every $\mathcal{G}$-orbit in Ξ contains an element in positive position, we may assume that (Q, δ, t) is positive. Then $(Q, \delta, t\,|t|^r)$ is positive for $r \geq -1$ and for $r > -1$ its stabilizer in $\mathcal{G}$ does not depend on r.

Now the statement for $\pi^\Gamma(Q, \delta, t)$ is an instance of [39, Proposition 3.4.1]. Together with (37) and Theorem 2.10(c), this implies the statement for $\pi(Q, \sigma, t)$.

(b) By [39, Proposition 3.1.4(a)], the representation

$$\mathrm{ind}_{\mathcal{H}^Q}^{\mathcal{H}^{P(\xi)} \rtimes \Gamma(P(\xi),P(\xi))}(\delta \otimes t\,|t|^r) = \bigoplus_i \mathrm{ind}_{\mathcal{H}^Q \rtimes \Gamma_Q}^{\mathcal{H}^{P(\xi)} \rtimes \Gamma(P(\xi),P(\xi))}(\sigma_i \otimes t\,|t|^r) \tag{39}$$

is completely reducible. Therefore, it suffices to consider all irreducible direct summands of (39) separately. This brings us to representations of the form

$$\mathrm{ind}_{\mathcal{H}^{P(\xi)} \rtimes \Gamma(P(\xi),P(\xi))}^{\mathcal{H} \rtimes \Gamma}(\delta' \otimes t'|t|^r). \tag{40}$$

Then $(P(\xi), \delta', t'|t|^r)$ is a datum for the Langlands classification for extended affine Hecke algebras [39, Corollary 2.2.5]. This result says that such a representation has a unique irreducible quotient, so (40) is irreducible as soon as it is completely reducible. In that case, its Langlands constituent obviously is the whole of (40). That implies the claim about the constituents of $\pi(Q, \sigma, t\,|t|^r)$ when that representation is completely reducible.

Next we show that (40) is irreducible for almost all $r \in \mathbb{R}_{>-1}$. From [39, Lemma 2.2.6(a)] we know that the space of $\mathcal{H} \rtimes \Gamma$-endomorphisms of (40) is just $\mathbb{C}\mathrm{Id}$.

For all $z \in \mathbb{C}$, the isotropy group $\mathcal{G}_{(P(\xi),\delta',t')}$ also fixes $(P(\xi), \delta', t'|t|^z)$. Since $\mathcal{G}_{P(\xi),\delta'}$ acts on T^Q by group automorphisms and translations, $\mathcal{G}_{(P(\xi),\delta',t'|t|^z)}$ equals $\mathcal{G}_{(P(\xi),\delta',t')}$ for almost all $z \in \mathbb{C}$. In particular, this happens for some $z \in -1 + i\mathbb{R}$. Then $t'|t|^z \in T_{\mathrm{un}}^{P(\xi)}$ and by [39, Corollary 3.1.3] the representation $\pi^\Gamma(P(\xi), \delta', t'|t|^z)$ is unitary (and in particular completely reducible). By [39, Theorem 3.3.1(b)]

$$\mathrm{End}_{\mathcal{H} \rtimes \Gamma}\big(\mathrm{ind}_{\mathcal{H}^{P(\xi)} \rtimes \Gamma(P(\xi),P(\xi))}^{\mathcal{H} \rtimes \Gamma}(\delta' \otimes t'|t|^z)\big) \tag{41}$$

is spanned by intertwining operators coming from $\mathcal{G}_{P(\xi),\delta',t'}$, just like for (40). Therefore, (41) consists only of $\mathbb{C}\mathrm{Id}$, which implies that the representation is irreducible for that $z \in -1 + i\mathbb{R}$. In the above algebraic family of finite dimensional representations parametrized by $z \in \mathbb{C}$, irreducibility is an open condition: Slightly varying z cannot destroy irreducibility. Hence the locus of z's where the representation is reducibile is a Zariski-closed subset of $\mathbb{C}$, that is, it is finite. In particular, (40) is irreducible for all but finitely many $r \in \mathbb{R}_{>-1}$. □

3 Reductive p-adic Groups

Let F be a non-Archimedean local field and let G be the group of F-points of a connected reductive algebraic group defined over F. We endow G with the topology coming from the metric on F and we fix a Haar measure. Let $\mathcal{H}(G)$ be the Hecke algebra of G, the vector space of locally constant compactly supported functions $f : G \to \mathbb{C}$ endowed with the convolution product (with respect to the Haar measure). Let $\mathcal{S}(G)$ be the Harish-Chandra–Schwartz algebra of G, as defined in [14] and [43, §III.6]. By definition, a smooth G-representation (by default on a complex vector space) is tempered if and only if it extends continuously to a module for $\mathcal{S}(G)$. Let $C_r^*(G)$ be the reduced C^*-algebra of G, the completion of $\mathcal{H}(G)$ in the algebra of bounded linear operators on the Hilbert space $L^2(G)$. By [42, Theorem 29], there are dense inclusions

$$\mathcal{H}(G) \subset \mathcal{S}(G) \subset C_r^*(G).$$

The *-operation and the trace on these algebras are

$$f^*(g) = \overline{f(g^{-1})} \quad \text{and} \quad \tau(f) = f(1_G).$$

Fix a minimal parabolic F-subgroup $P_0 = M_0 U_0$ and let $W(G, M_0)$ be the Weyl group of G with respect to the maximal F-split torus A_0 in the centre of M_0. Write $\mathfrak{a}_0 = X^*(A_{M_0}) \otimes_{\mathbb{Z}} \mathbb{R}$ and endow this vector space with a $W(G, M_0)$-invariant inner product.

Every Levi F-subgroup of G is conjugate to a standard Levi subgroup, that is, one that contains M_0. Let M be such a standard Levi subgroup of G, and let A_M be the maximal F-split torus in $Z(M)$. There is a canonical decomposition

$$\mathfrak{a}_0 = \mathfrak{a}_M \oplus \mathfrak{a}^M, \tag{42}$$

where $\mathfrak{a}_M = X^*(A_M) \otimes_{\mathbb{Z}} \mathbb{R}$ and $\mathfrak{a}^M = \{\chi \in X^*(A_0) : \chi|_{A_M} = 1\} \otimes_{\mathbb{Z}} \mathbb{R}$. Let $R(G, M)$ be the set of roots of G with respect to A_M. For a parabolic subgroup $P = MU$ of G, we let $R(P, M)$ be the set of roots of (G, M) that appear in (the Lie algebra of) P.

When $M_1 \subset M$ is another standard Levi subgroup of G, we write $\mathfrak{a}_{M_1}^M = \mathfrak{a}_{M_1} \cap \mathfrak{a}^M$. Every parabolic subgroup $P_1 = M_1 U_1$ of M determines an obtuse cone in $\mathfrak{a}_{M_1}^M$:

$${}^+\mathfrak{a}_{P_1}^M = \big\{ \sum\nolimits_{\alpha \in R(P_1, M_1)} c_\alpha \alpha\big|_{A_{M_1}} : c_\alpha > 0 \; \forall \alpha \big\}.$$

Here we would obtain the same cone if we used only the simple roots for (P_1, M_1). The closure of ${}^+\mathfrak{a}_{P_1}^M$ in $\mathfrak{a}_{M_1}^M$ is

$$\overline{{}^+\mathfrak{a}_{P_1}^M} = \big\{ \sum\nolimits_{\alpha \in R(P_1, M_1)} c_\alpha \alpha\big|_{A_{M_1}} : c_\alpha \geq 0 \; \forall \alpha \big\}.$$

It is easy to see that the normalized Jacquet restriction functor

$$J_P^G : \mathrm{Rep}(G) \to \mathrm{Rep}(M)$$

does not preserve temperedness. Fortunately, the normalized parabolic induction functor $I_P^G : \mathrm{Rep}(M) \to \mathrm{Rep}(G)$ does and also respects non-temperedness:

Proposition 3.1 *Let $\pi \in \mathrm{Rep}(M)$ be of finite length. Then $I_P^G(\pi)$ is tempered if and only if π is tempered.*

Proof. The if-part is well known, see [43, Lemme III.2.3] or [28, Lemme VII.2.2].

By conjugating P, M and π, we can achieve that $P \supset P_0$ and $M \supset M_0$. Recall from [43, Proposition III.2.2] that π is tempered if and only if, for every parabolic subgroup $P_1 = M_1 U_1$ of M with $M_1 \supset M_0$ and every A_{M_1}-weight χ of $J_{P_1}^M(\pi)$:

$$\log|\chi| \in \overline{{}^+\mathfrak{a}_{P_1}^M}. \tag{43}$$

Moreover, it is equivalent to impose this condition for all P_1 such that $P_1 = M$ or P_1 is a standard maximal parabolic subgroup of M.

To show that I_P^G preserves non-temperedness, it suffices to consider the case that P is a standard maximal parabolic subgroup of G. Namely, there exists a chain of parabolic subgroups

$$P \subset P_1 = M_1 U_1 \subset \cdots \subset P_n = M_n U_n \subset G = M_{n+1}$$

such that every $P_{i-1} \cap M_i$ is a maximal parabolic subgroup of M_i. If we can prove that each $I_{P_{i-1}\cap M_i}^{M_i}$ preserves non-temperedness, the transitivity of parabolic induction [28, Lemme VI.1.4] implies that I_P^G does so as well.

Since I_P^G is an exact functor [28, Théorème VI.1.1], we may, furthermore, assume that π is irreducible. So, we suppose that π is irreducible and not tempered, and (contrary to what we want to prove) that $I_P^G(\pi)$ is tempered. Let us consider the $Z(G)$-character of π. It is also the $Z(G)$-character of $I_P^G(\pi)$. Since $I_P^G(\pi)$ is tempered, its central character is unitary [28, Corollaire VII.2.6].

We claim that the A_M-character ζ of π must also be unitary.

Suppose it is not, and consider its absolute value $|\zeta| \in X_{\mathrm{nr}}(M) \setminus \{1\}$. Let $\alpha \in R(G, M_0)$ be the unique simple root of (G, M) and let $s_\alpha \in W(G, M_0)$ be the associated reflection. The length of s_α in $W(G, M_0)$ is one, so it is a minimal length representative for a double coset in $W(M, M_0) \setminus W(G, M_0) \backslash W(M, M_0)$. By the Geometric Lemma [28, Théorème VI.5.1], both ζ and $s_\alpha\zeta$ occur as A_M-weights of $J_P^G(I_P^G(\pi))$. Since M is a maximal Levi subgroup of G and $|\zeta|_{A_G} = |\zeta|_{Z(G)} = 1$, both $\log|\zeta|$ and $\log|s_\alpha\zeta|$ lie in the one-dimensional vector space $\mathfrak{a}_M^G$. The reflection s_α acts as -1 on $\mathfrak{a}_M^G$, so $|s_\alpha\zeta| = |\zeta|^{-1}$. As $|\zeta| \neq 1$, it is not possible that both $\log|\zeta|$ and $\log|s_\alpha\zeta|$ lie in the cone $\overline{{}^+\mathfrak{a}_P^G}$. From (43), we see that this contradicts the temperedness of $I_P^G(\pi)$. Consequently, $|\zeta|$ must be 1, and ζ must be unitary.

Now we invoke the non-temperedness of π. By [43, Proposition III.2.2(iii)], there exists a standard parabolic subgroup $P' = M'U'$ of G such that:

- $M' = M$ or M' is a maximal Levi subgroup of M;

- $J^M_{P' \cap M}(\pi)$ has an $A_{M'}$-weight χ with $\log|\chi| \in \mathfrak{a}_{M'}$ not in $\overline{{}^+\mathfrak{a}^M_{P' \cap M}}$.

When $M' = M$, then $\chi = \zeta$ and the above claim says that $|\chi| = 1$. That would be in contradiction with the second bullet.

Hence $M' \subsetneq M$. As $\chi|_{A_M} = \zeta_{A_M}$ is unitary, χ is of the form $c_\beta \beta$, where $\beta \in R(M, M_0)$ is the unique simple root for (M, M'). Then the second bullet says that $c_\beta < 0$. By the Geometric Lemma [28, Théorème VI.5.1] χ also an $A_{M'}$-weight of $J^G_{P'}(I^G_P(\pi))$. As $\beta \in \overline{{}^+\mathfrak{a}^G_{P'}}$ (a potentially larger cone than $\overline{{}^+\mathfrak{a}^M_{P' \cap M}}$), $c_\beta \beta \notin \overline{{}^+\mathfrak{a}^G_{P'}}$. Together with (43), this shows that $I^G_P(\pi)$ cannot be tempered. □

Let $L = \mathcal{L}(F)$ be a Levi subgroup G and let $(\sigma, V_\sigma) \in \mathrm{Irr}(L)$ be an irreducible tempered supercuspidal L-representation. Let $X_{\mathrm{nr}}(L)$ be the group of unramified characters $L \to \mathbb{C}^\times$ and let $X_{\mathrm{unr}}(L)$ be the subgroup of unitary unramified characters. Recall that the inertial equivalence class of the pair (L, σ) consists of all pairs of the form

$$(gLg^{-1}, (g \cdot \sigma) \otimes \chi), \text{ where } g \in G \text{ and } \chi \in X_{\mathrm{nr}}(gLg^{-1}).$$

We write $\mathfrak{s} = [L, \sigma]_G$ and call this an inertial equivalence class for G. It gives rise to a subset $\mathrm{Irr}(G)^{\mathfrak{s}} \subset \mathrm{Irr}(G)$, namely all those irreducible smooth G-representations whose supercuspidal support lies in $\mathfrak{s}$. This in turn is used to define a subcategory $\mathrm{Rep}(G)^{\mathfrak{s}}$ of $\mathrm{Rep}(G)$, namely those smooth G-representations all whose irreducible constituents lie in $\mathrm{Irr}(G)^{\mathfrak{s}}$. The Bernstein blocks $\mathrm{Rep}(G)^{\mathfrak{s}}$ have better finiteness properties than $\mathrm{Rep}(G)$:

Theorem 3.2 *[18, §1.6]*

(a) Let $M \subset G$ be a Levi subgroup containing L. There exist tempered $\pi_{M,i} \in \mathrm{Irr}(M)^{[L,\sigma]_M}$ $(i = 1, \ldots, \kappa_M)$, such that

$$\{\pi_{M,i} \otimes \chi_M : i = 1, \ldots, \kappa_M, \chi_M \in X_{\mathrm{unr}}(M)\}$$

is the collection of irreducible tempered representations in $\mathrm{Rep}(M)^{[L,\sigma]_M}$ that are not isomorphic to the normalized parabolic induction of representation of a proper Levi subgroup of M.

(b) Let M run through a set of representatives for the conjugacy classes of Levi subgroups of G containing L. Then the set

$$\bigcup_M \{I^G_P(\pi_{M,i} \otimes \chi_M) : i = 1, \ldots, \kappa_M, \chi_M \in X_{\mathrm{nr}}(M)\}$$

spans the Grothendieck group of the category of finite length representations in $\mathrm{Rep}(G)^{\mathfrak{s}}$.

Let $\mathfrak{B}(G)$ be the set of all inertial equivalence classes $\mathfrak{s}$ for G. By [5, Corollaire 3.9], it is countably infinite (unless $G = 1$). The Bernstein decomposition [5, Theorem 2.10] says that

$$\begin{aligned} \mathrm{Rep}(G) &= \prod_{\mathfrak{s}\in\mathfrak{B}(G)} \mathrm{Rep}(G)^{\mathfrak{s}}, \\ \mathcal{H}(G) &= \bigoplus_{\mathfrak{s}\in\mathfrak{B}(G)} \mathcal{H}(G)^{\mathfrak{s}}, \end{aligned} \tag{44}$$

where $\mathcal{H}(G)^{\mathfrak{s}}$ is the largest two-sided ideal of $\mathcal{H}(G)$ which annihilates $\mathrm{Rep}(G)^{\mathfrak{s}'}$ for all $\mathfrak{s}' \neq \mathfrak{s}$. Alternatively, it is characterized by $\mathcal{H}(G)^{\mathfrak{s}} \cdot V = V$ for all $V \in \mathrm{Rep}(G)^{\mathfrak{s}}$.

Let $\mathcal{S}(G)^{\mathfrak{s}}$ (resp. $C_r^*(G)^{\mathfrak{s}}$) be the two-sided ideal of $\mathcal{S}(G)$ (resp. $C_r^*(G)$) generated by $\mathcal{H}(G)^{\mathfrak{s}}$. Upon completion, (44) yields further Bernstein decompositions

$$\begin{aligned} \mathcal{S}(G) &= \bigoplus_{\mathfrak{s}\in\mathfrak{B}(G)} \mathcal{S}(G)^{\mathfrak{s}}, \\ C_r^*(G) &= \bigoplus_{\mathfrak{s}\in\mathfrak{B}(G)} C_r^*(G)^{\mathfrak{s}}. \end{aligned} \tag{45}$$

The latter must be interpreted as a direct sum in the Banach algebra sense: It is the completion of the algebraic direct sum with respect to the operator norm of $C_r^*(G)$.

For a compact open subgroup K of G, we let $\langle K\rangle$ be the corresponding idempotent of $\mathcal{H}(G)$. Then

$$\mathcal{H}(G, K) := \langle K\rangle \mathcal{H}(G) \langle K\rangle \tag{46}$$

is the subalgebra of K-bi-invariant functions in $\mathcal{H}(G)$. We define $\mathcal{S}(G, K)$ and $C_r^*(G, K)$ analogously. For every compact open subgroup K of G, $\mathcal{S}(G, K)$ is a Fréchet algebra [42, Theorem 29]. The Schwartz algebra $\mathcal{S}(G)$ is their union (over all possible K), so it is an inductive limit of Fréchet algebras.

We will focus on one Bernstein block $\mathrm{Rep}(G)^{\mathfrak{s}}$ of $\mathrm{Rep}(G)$. By [5, Corollaire 3.9], there exists a compact open subgroup $K_{\mathfrak{s}}$ of G such that every representation in $\mathrm{Rep}(G)^{\mathfrak{s}}$ is generated by its $K_{\mathfrak{s}}$-fixed vectors. This leads to Morita equivalences

$$\begin{aligned} \mathcal{H}(G)^{\mathfrak{s}} &\sim_M \mathcal{H}(G, K_{\mathfrak{s}})^{\mathfrak{s}} := \mathcal{H}(G)^{\mathfrak{s}} \cap \mathcal{H}(G, K_{\mathfrak{s}}) \\ \mathcal{S}(G)^{\mathfrak{s}} &\sim_M \mathcal{S}(G, K_{\mathfrak{s}})^{\mathfrak{s}} := \mathcal{S}(G)^{\mathfrak{s}} \cap \mathcal{S}(G, K_{\mathfrak{s}}) \\ C_r^*(G)^{\mathfrak{s}} &\sim_M C_r^*(G, K_{\mathfrak{s}})^{\mathfrak{s}} := C_r^*(G)^{\mathfrak{s}} \cap C_r^*(G, K_{\mathfrak{s}}). \end{aligned} \tag{47}$$

3.1 The Plancherel Isomorphism

We will describe the structure of $\mathcal{S}(G)^{\mathfrak{s}}$ and $\mathcal{S}(G, K_{\mathfrak{s}})^{\mathfrak{s}}$ in more detail. Let $[L, \sigma]_L = T_{\mathfrak{s}} \subset \mathrm{Irr}(L)$ be the set of isomorphism classes of L-representations of the form $\sigma \otimes \chi$ with $\chi \in X_{\mathrm{nr}}(L)$. Thus there is a finite covering of complex varieties

$$X_{\mathrm{nr}}(L) \to T_{\mathfrak{s}} : \chi \mapsto \sigma \otimes \chi. \tag{48}$$

Let $T_{\mathfrak{s},\mathrm{un}}$ be the subset of unitary representations in $T_{\mathfrak{s}}$, it is covered by $X_{\mathrm{unr}}(L)$ via (48). We write

$$X_{\mathrm{nr}}(L, \sigma) = \{\chi \in X_{\mathrm{nr}}(L) : \sigma \otimes \chi \cong \sigma\}.$$

This is a finite subgroup of $X_{\mathrm{unr}}(L)$. The map (48) induces an isomorphism of algebraic varieties $X_{\mathrm{nr}}(L)/X_{\mathrm{nr}}(L, \sigma) \to T_{\mathfrak{s}}$.

The group $W(G,L) = N_G(L)/L$ acts on $\mathrm{Irr}(L)$ by

$$(gL \cdot \pi)(l) = \pi(glg^{-1}). \tag{49}$$

(The representation $gL \cdot \pi$ is only determined up to isomorphism.) This action stabilizes $X_{\mathrm{nr}}(L)$, the unitary representations in $\mathrm{Irr}(L)$, and the supercuspidal L-representations. Let $W_{\mathfrak{s}}$ be the stabilizer of $T_{\mathfrak{s}}$ in $N_G(L)/L$. This group will play the same role as $W\Gamma$ did in Sect. 2. The theory of the Bernstein centre [5, Théorème 2.13] says that the centre of $\mathcal{H}(G, K_{\mathfrak{s}})^{\mathfrak{s}}$ is naturally isomorphic with $\mathcal{O}(T_{\mathfrak{s}})^{W_{\mathfrak{s}}} = \mathcal{O}(T_{\mathfrak{s}}/W_{\mathfrak{s}})$.

It will be convenient to lift everything from $T_{\mathfrak{s}}$ to $X_{\mathrm{nr}}(L)$. However, $W_{\mathfrak{s}}$ does not act naturally on $X_{\mathrm{nr}}(L)$. To overcome this and similar issues, we need the following lemma.

Lemma 3.3 *Let $p : T' \to T$ be a surjection between complex tori, with finite kernel $K = \ker p$. Let Γ be a finite group acting on T by automorphisms of algebraic varieties (so Γ need not fix $1 \in T$). Then there exists a canonical short exact sequence*

$$1 \to K \to \Gamma' \to \Gamma \to 1$$

and a canonical action of Γ' on T' which extends the multiplication action of K on T' and lifts the action of Γ on T.

Proof. Let X be the character lattice of T. Then $\mathcal{O}(T) \cong \mathbb{C}[X]$ and Γ acts on $\mathcal{O}(T)$ by $(\gamma \cdot f)(t) = f(\gamma^{-1}t)$. Since

$$\mathcal{O}(T)^\times = \{z\theta_x : z \in \mathbb{C}^\times, x \in X\} \cong \mathbb{C}^\times \times X,$$

Γ also acts naturally on $X \cong \mathcal{O}(T)^\times/\mathbb{C}^\times$. For $\gamma \in \Gamma$, we denote this action by $l_\gamma : X \to X$. Notice that it defines an action of Γ on $T = \mathrm{Hom}_{\mathbb{Z}}(X, \mathbb{C}^\times)$ by algebraic group automorphisms. The given action on $\mathcal{O}(T)$ can now be written as

$$\gamma(z\theta_x) = zz_\gamma^{-1}(l_\gamma(x))\theta_{l_\gamma(x)},$$

for a unique $z_\gamma \in T$. Consequently the original action of Γ on T can be expressed as

$$\gamma(t) = z_\gamma l_\gamma(t). \tag{50}$$

The character lattice X' of T' contains X with finite index (namely $|K|$). Thus l_γ induces a canonical linear action of Γ on X', which we also denote by l_γ. For every $\gamma \in \Gamma$ we choose a $z'_\gamma \in p^{-1}(z_\gamma)$, and we define

$$\phi_\gamma : T' \to T', \quad \phi_\gamma(t') = z'_\gamma l_\gamma(t').$$

Clearly, ϕ_γ is a lift of (50), so for every $\gamma, \gamma' \in \Gamma$ there exists a unique $z'_{\gamma,\gamma'} \in K$ with

$$\phi_\gamma \circ \phi_{\gamma'} \circ \phi_{\gamma\gamma'}^{-1}(t') = z'_{\gamma,\gamma'} t' \quad \forall t' \in T'. \tag{51}$$

Let Γ' be the subgroup of $\mathrm{Aut}(T')$ generated by the ϕ_γ ($\gamma \in \Gamma$) and K. Then (51) gives a canonical isomorphism $\Gamma'/K \cong \Gamma$.

The only unnatural steps in the above argument are the choices of the z'_γ. Different choices would lead to different $z'_{\gamma,\gamma'}$ in (51), but to the same group Γ'. Hence Γ' is canonically determined by the data T, T' and Γ.

Next we recall the Plancherel isomorphism for $\mathcal{S}(G)^{\mathfrak{s}}$, as discovered by Harish-Chandra and worked out by Waldspurger. As induction data for G we take quadruples (P, M, ω, χ), where

- P is a parabolic subgroup of G with a Levi factor M;
- $(\omega, V_\omega) \in \mathrm{Irr}_{L^2}(M)$, the set of (isomorphism classes of) irreducible smooth square-integrable modulo centre representations of M;
- $\chi \in X_{\mathrm{nr}}(M)$.

To such a datum, we associate the smooth G-representation $I_P^G(\omega \otimes \chi)$. When χ is unitary, the M-invariant inner product on $(\omega \otimes \chi, V_\omega)$ induces a G-invariant inner product on $I_P^G(V_\omega)$, so $I_P^G(\omega \otimes \chi)$ is pre-unitary [10, Proposition 3.1.4]. However, $I_P^G(V_\omega)$ is only complete with respect to the associated metric if $P = G$ and $\dim(V_\omega)$ is finite.

Let $(\check{\omega}, \check{V}_\omega)$ be the smooth contragredient of ω and put

$$\mathfrak{L}(\omega, P) = I_{P\times P}^{G\times G}(\omega \otimes \check{\omega}) = I_P^G(\omega) \otimes I_P^G(\check{\omega}).$$

Since $I_P^G(\check{\omega})$ can be identified with the smooth contragredient of $I_P^G(\omega)$ [10, Proposition 3.1.2], $\mathfrak{L}(\omega, P)$ can be regarded as the algebra of finite rank linear operators on $I_P^G(V_\omega)$. Notice that for every $\chi \in X_{\mathrm{nr}}(M)$, we can identify $\mathfrak{L}(\omega \otimes \chi, P)$ with $\mathfrak{L}(\omega, P)$ as algebras. The inner product on $I_P^G(V_\omega)$ induces a *-operation on this algebra. That makes $\mathcal{O}(X_{\mathrm{nr}}(M)) \otimes \mathfrak{L}(\omega, P)$ to a *-algebra with

$$f^*(\chi) = f(\check{\chi})^*.$$

There is a natural *-homomorphism

$$\begin{array}{rcl} \mathcal{H}(G) & \to & \mathcal{O}(X_{\mathrm{nr}}(M)) \otimes \mathfrak{L}(\omega, P), \\ f & \mapsto & \big(\chi \mapsto I_P^G(\omega \otimes \chi)(f)\big). \end{array} \tag{52}$$

We put $T_\omega = \{\omega \otimes \chi : \chi \in X_{\mathrm{nr}}(M)\}$ and we record the covering map

$$X_{\mathrm{nr}}(M) \to T_\omega : \chi \mapsto \omega \otimes \chi. \tag{53}$$

The group

$$X_{\mathrm{nr}}(M, \omega) = \{\chi \in X_{\mathrm{nr}}(M) : \omega \otimes \chi \cong \omega\}$$

is finite, because all its elements must be trivial on $Z(M)$. All the fibres of (53) are cosets of $X_{\mathrm{nr}}(M, \omega)$ in $X_{\mathrm{nr}}(M)$.

For every $k \in X_{\mathrm{nr}}(M, \omega)$, there exists a unitary M-intertwiner $\omega \to \omega \otimes k$, unique up to scalars. The same map $V_\omega \to V_\omega$ also intertwines $\omega \otimes \chi$ with $\omega \otimes \chi k$, for any $\chi \in X_{\mathrm{nr}}(M)$. Applying I_P^G, we get a family a G-intertwiners

$$\pi(k, \omega, \chi) : I_P^G(\omega \otimes \chi) \to I_P^G(\omega \otimes \chi k), \tag{54}$$

independent of χ and unitary when $\chi \in X_{\mathrm{unr}}(M)$. Let

$$\check{\pi}(k, \omega, \chi) : I_P^G(\check{\omega} \otimes \check{\chi})) \to I_P^G(\check{\omega} \otimes \check{\chi k})$$

be the inverse transpose $\pi(k, \omega, \chi)$. Since $\pi(k, \omega, \chi)$ is unique up to scalars,

$$I(k, \omega, \chi) := \pi(k, \omega, \chi) \otimes \check{\pi}(k, \omega, \chi) \in \mathrm{Hom}_{G \times G}\big(\mathfrak{L}(\omega \otimes \chi, P), \mathfrak{L}(\omega \otimes \chi k, P)\big) \tag{55}$$

is canonical. Moreover, it is unitary for $\chi \in X_{\mathrm{unr}}(M)$ and independent of χ as map between vector spaces.

Let $W(T_\omega)$ be the stabilizer of T_ω in $W(G, M) = N_G(M)/M$, with respect to the action on $\mathrm{Irr}(M)$ as in (49). Then $W(T_\omega)$ acts naturally on T_ω. From Lemma 3.3, we get a group extension

$$1 \to X_{\mathrm{nr}}(M, \omega) \to W'(T_\omega) \to W(T_\omega) \to 1 \tag{56}$$

and an action of $W'(T_\omega)$ on $X_{\mathrm{nr}}(M)$ compatible with the covering (53). In [43], the representations $\omega \otimes \chi$ and $\omega \otimes \chi k$ are often not distinguished. The introduction of $W'(T_\omega)$ and of the $I(k, \omega, \chi)$ allows us to compare $I_P^G(\omega \otimes \chi)$ and $I_P^G(\omega \otimes \chi k)$ in a systematic way. From [43, §VI.1], one can see that actually our setup is just another way to keep track of all the ingredients of [43].

The following results are proven in [43, Paragraphe V]. For $w' \in W'(T_\omega)$, there exist unitary G-intertwining operators

$$\pi(w', \omega, \chi) : I_P^G(\omega \otimes \chi) \to I_P^G(\omega \otimes w'(\chi)) \quad \chi \in X_{\mathrm{unr}}(M), \tag{57}$$

unique up to scalars. These give canonical unitary intertwiners

$$I(w', \omega, \chi) = \pi(w', \omega, \chi) \otimes \check{\pi}(w', \omega, \chi) \in \mathrm{Hom}_{G \times G}\big(\mathfrak{L}(\omega \otimes \chi, P), \mathfrak{L}(\omega \otimes w'(\chi), P)\big) \tag{58}$$

with the following properties [43, Lemme V.3.1]:

- as functions of χ, $\pi(w', \omega, \chi)$ and $I(w', \omega, \chi)$ are continuous with respect to the Zariski topology on the real algebraic variety $X_{\mathrm{unr}}(M)$;
- $I(w_2', \omega, w_1'(\chi)) \circ I(w_1', \omega, \chi) = I(w_2' w_1', \omega, \chi)$ for $w_1', w_2' \in W'(T_\omega)$.

The properties of the intertwining operators (57) imply that, for every $g \in G$, $\omega \in \mathrm{Irr}_{L^2}(M)$, $\chi \in X_{\mathrm{nr}}(M)$ and every parabolic subgroup $P' \subset G$ with Levi factor gMg^{-1}, the representations $I_P^G(\omega \otimes \chi)$ and $I_{P'}^G(g \cdot \omega \otimes g \cdot \chi)$ have the same irreducible subquotients, counted with multiplicity [37, Corollary 2.7].

We remark that $I(w', \omega, \chi)$ is called ${}^\circ c_{P|P}(w', \omega \otimes \chi)$ in [43]. The intertwining operators (58) give rise to an action of $W'(T_\omega)$ on the algebra

$$C^\infty(X_{\mathrm{unr}}(M)) \otimes \mathfrak{L}(\omega, P) \quad \text{by} \quad (w' \cdot f)(w'\chi) = I(w', \omega, \chi) f(\chi).$$

We fix a parabolic subgroup P_L with Levi factor L, and we recall that $\mathfrak{s} = [L, \sigma]_G$. To study representations in the Bernstein block $\mathrm{Rep}(G)^{\mathfrak{s}}$, it suffices to consider induction data such that $P \supset P_L$, $M \supset L$ and the cuspidal support of ω lies in $[L, \sigma]_M$. Then $W(T_\omega)$ can be regarded as a subgroup of $W_{\mathfrak{s}}$.

Choose representatives for the G-association classes of parabolic subgroups P containing P_L. Notice that every such P has a unique Levi factor M containing L. We also choose representatives ω for the action of $W_{\mathfrak{s}} \ltimes X_{\mathrm{unr}}(M)$ on $\mathrm{Irr}_{L^2}(M) \cap \mathrm{Irr}(M)^{\mathfrak{s}_M}$, where $\mathfrak{s}_M = [L, \sigma]_M$. We denote the resulting set of representative triples by $(P, M, \omega)/\sim$. Harish-Chandra established the following Plancherel isomorphism (see [33, Theorem 8.9] for an alternative proof).

Theorem 3.4 [43, Théorème VII.2.5]
The maps (52) *induces isomorphisms of topological *-algebras*

$$\begin{aligned} \mathcal{S}(G)^{\mathfrak{s}} \quad &\to \bigoplus\nolimits_{(P,M,\omega)/\sim} \big(C^\infty(X_{\mathrm{unr}}(M)) \otimes \mathfrak{L}(\omega, P)\big)^{W'(T_\omega)}, \\ \mathcal{S}(G, K_{\mathfrak{s}})^{\mathfrak{s}} &\to \bigoplus\nolimits_{(P,M,\omega)/\sim} \Big(C^\infty(X_{\mathrm{unr}}(M)) \otimes \mathrm{End}_{\mathbb{C}}\big(I_P^G(V_\omega)^{K_{\mathfrak{s}}}\big)\Big)^{W'(T_\omega)}. \end{aligned}$$

Plymen [26] showed that Theorem 3.4 has a natural extension to C^*-algebras. Let $\mathfrak{H}(\omega \otimes \chi, P)$ be the Hilbert space completion of $I_P^G(V_{\omega\otimes\chi}) = I_P^G(V_\omega)$ and let $\mathfrak{K}(\omega \otimes \chi, P)$ be the C^*-algebra of compact operators on $\mathfrak{H}(\omega \otimes \chi, P)$.

Theorem 3.5 *The maps* (52) *induces isomorphisms of C^*-algebras*

$$\begin{aligned} C_r^*(G)^{\mathfrak{s}} \quad &\to \bigoplus\nolimits_{(P,M,\omega)/\sim} C\big(X_{\mathrm{unr}}(M); \mathfrak{K}(\omega, P)\big)^{W'(T_\omega)}, \\ C_r^*(G, K_{\mathfrak{s}})^{\mathfrak{s}} &\to \bigoplus\nolimits_{(P,M,\omega)/\sim} \Big(C(X_{\mathrm{unr}}(M)) \otimes \mathrm{End}_{\mathbb{C}}\big(I_P^G(V_\omega)^{K_{\mathfrak{s}}}\big)\Big)^{W'(T_\omega)}. \end{aligned}$$

Proof. First we note that we have intertwining operators associated with the group $W'(T_\omega)$, instead of $W(T_\omega)$ in [26, 43]. The reason for this is explained after (56). In view of Theorem 3.4, it only remains to prove that completing with respect to the operator norm of $C_r^*(G)$ boils down to replacing $C^\infty(X_{\mathrm{unr}}(M)) \otimes \mathfrak{L}(\omega, P)$ by $C\big(X_{\mathrm{unr}}(M); \mathfrak{K}(\omega, P)\big)$. This is shown in [26, Theorem 2.5]. □

3.2 The Space of Irreducible Representations

As in Paragraph 2.2, we need more information about the space of all irreducible smooth G-representations (tempered or not). Suppose that $\pi \in \mathrm{Irr}(G)$ has supercuspidal support $\sigma \otimes \chi$, where $\sigma \in \mathrm{Irr}_{L^2}(M)$ and $\chi \in X_{\mathrm{nr}}(M)$. Then $\log |\chi| \in \mathfrak{a}_M$, and its image in $\mathfrak{a}_0$ is uniquely determined, up to $W(G, M_0)$, by π. In other words,

$$cc(\pi) := W(G, M_0) \log |\chi| \tag{59}$$

is an invariant of π. Since the norm on $\mathfrak{a}_0$ comes from a $W(G, M_0)$-invariant inner product, $\|cc(\pi)\| := \|\log |\chi|\|$ is well defined.

Theorem 3.6 *(a) For every $\pi \in \mathrm{Irr}(G)$ there exists an induction datum (P, M, ω, χ), unique up to conjugation, such that π is a constituent of $I_P^G(\omega \otimes \chi)$ and $\|cc(\omega)\|$ is maximal for this property. In this case we call π a* Langlands constituent *of $I_P^G(\omega \otimes \chi)$.*

(b) π is tempered if and only if $\chi \in X_{\mathrm{unr}}(M)$.

(c) For any induction datum (P, M, ω, χ), every constituent of $I_P^G(\omega \otimes \chi)$ is either a Langlands constituent or a constituent of some $I_{P'}^G(\omega' \otimes \chi')$ with $\|cc(\omega')\| > \|cc(\omega)\|$.

(d) Suppose that L is a standard Levi subgroup and that $\pi \in \mathrm{Irr}(G)^{\mathfrak{s}}$, where $\mathfrak{s} = [L, \sigma]_G$. Then we can choose (P, M, ω, χ) from part (a) such that $P \supset P_0$, $M \supset L$ and $\omega \in \mathrm{Irr}(M)^{[L,\sigma]_M}$.

Proof. (a) See [37, Theorem 2.15(b)].
(b) This is a direct consequence of [37, Proposition 2.14(b) and Theorem 2.15(a)].
(c) By [37, Lemma 2.13] (P, M, ω, χ) is equivalent to an induction datum ξ^+ in positive position. By [37, Corollary 2.7], $I_P^G(\omega \otimes \chi)$ has the same irreducible subquotients, counted with multiplicity, as the parabolically induced representation associated with ξ^+. Therefore we may assume that (P, M, ω, χ) is in positive position, that is, $P \supset P_0$ and $\log |\chi|$ lies in the closed positive cone in $\mathfrak{a}_0$ (determined by P_0).

Then [37, Theorem 2.15(a)] says that the Langlands constituents of $I_P^G(\omega \otimes \chi)$ are precisely its irreducible quotients. Furthermore, by [37, Proposition 2.15(a)] $I_P^G(\omega \otimes \chi)$ is a direct sum of representations of the form $I_Q^G(\tau \otimes |\chi|)$, where $(Q, \tau, \log |\chi|)$ is a datum for the Langlands classification of $\mathrm{Irr}(G)$. Suppose that π' is a constituent of $I_Q^G(\tau \otimes |\chi|)$, but not a quotient. By [37, Lemma 2.11(a) and Lemma 2.12], π' is the Langlands quotient of $I_{Q'}^G(\tau' \otimes \nu')$, for a Langlands datum $(Q', \tau', \log \nu')$ with $Q' \supset Q$ and $\|cc(\tau')\| > \|cc(\tau)\|$. By [37, Proposition 2.15(a)] π' is a Langlands constituent of $I_{P'}^G(\omega' \otimes \chi')$, for some induction datum (P', M', ω', χ') with

$$\|cc(\omega')\| = \|cc(\tau')\| > \|cc(\tau)\| = \|cc(\omega)\| .$$

(d) Let $P_i = M_i U_i$ be a standard parabolic subgroup and let $\overline{P_i}$ be the unique parabolic subgroup with Levi factor M_i that is opposite to P_i. Let $J_{\overline{P_i}}^G : \mathrm{Rep}(G) \to \mathrm{Rep}(M_i)$ be the normalized Jacquet restriction functor.

From [28, §VII.4.2], we recall how π can be realized as a Langlands quotient. Namely, we take $\overline{P_1}$ such that $J^G_{\overline{P_1}}(\pi)$ contains a representation of the form $\tau \otimes \nu$, where $(P_1, \tau, \log \nu)$ is a Langlands datum. By [43, Proposition III.4.1], there exists a parabolic subgroup P_2 with $P_0 \subset P_2 \subset P_1$, and a $\omega' \in \mathrm{Irr}_{L^2}(M_2)$, such that $\tau \otimes \nu$ is a direct summand of $I^{M_1}_{M_1 \cap P_2}(\omega' \otimes \nu)$. From the proof of part (c), we see that π is a Langlands constituent of $I^G_{P_2}(\omega' \otimes \nu)$. By the second adjointness theorem

$$0 \neq \mathrm{Hom}_G(I^G_{P_2}(\omega' \otimes \nu), \pi) \cong \mathrm{Hom}_{M_2}(\omega' \otimes \nu, J^G_{\overline{P_2}}(\pi)). \tag{60}$$

The cuspidal support of $J^G_{\overline{P_2}}(\pi)$ equals that of π, so $\omega' \otimes \nu$ also has cuspidal support in $[L, \sigma]_G$. More precisely, the cuspidal support of $\omega' \otimes \nu$ is of the form $[L', \sigma']_{M_2}$, where L' is a standard Levi subgroup of G conjugate to L. Since every Levi subgroup containing L is G-conjugate to a standard Levi subgroup of G containing L, we may replace (P_2, M_2, ω', ν) by a G-conjugate (P, M, ω, χ) with M standard. Thus, we can arrange that the cuspidal support becomes $[L, \sigma'']_M$, for some cuspidal $\sigma'' \in \mathrm{Irr}(L)$. Then (60) is also valid for $I^G_P(\omega \otimes \chi)$, since $I^G_{P_2}(\omega' \otimes \nu)$ is not affected by G-conjugation of (P_2, M_2, ω', ν). Second adjointness tells us that $J^G_{\overline{P}}(\pi) \in \mathrm{Irr}(M)^{[L,\sigma]_M}$, so also $\omega \otimes \chi \in \mathrm{Irr}(M)^{[L,\sigma]_M}$. Finally, we may replace P by a standard parabolic subgroup with Levi factor M, for this does not change the collection of constituents of $I^G_P(\omega \otimes \chi)$. □

In [37], Theorem 3.6 was used to study the geometry of $\mathrm{Irr}(G)$, and the relation with the subspace of tempered irreducible representations. For our purposes, we need some aspects of that, and we need to know that for almost all induction data every constituent is a Langlands constituent.

Proposition 3.7 *Let (P, M, ω, χ) be an induction datum for G.*

(a) For $r \in \mathbb{R}_{>-1}$, the number of inequivalent Langlands constituents of $I^G_P(\omega \otimes \chi\,|\chi|^r)$ does not depend on r.
(b) For all but finitely many $r \in \mathbb{R}_{>-1}$, $I^G_P(\omega \otimes \chi\,|\chi|^r)$ is completely reducible. Then all its irreducible subquotients are Langlands constituents.

Proof. (a) All the induction data under consideration have the same stabilizer in $W'(T_\omega)$. As $W'(T_\omega)$ is by construction the stabilizer of T_ω in the $\mathcal{W}$ from [37], the statement is a special case of [37, Lemma 2.16].
(b) As noted in the proof of Theorem 3.6(c), $I^G_P(\omega \otimes \chi\,|\chi|^r)$ is a direct sum of representations of the form $I^G_Q(\tau \otimes |\chi|^{r+1})$, where $(Q = LU_Q, \tau, \log|\chi|^{r+1})$ is a Langlands datum. Hence it suffices to show that $I^G_Q(\tau \otimes |\chi|^{r+1})$ is irreducible for almost all $r \in \mathbb{R}_{>-1}$. The conditions of a Langlands datum say that $\tau \in \mathrm{Irr}(L)$ is tempered and that $\log|\chi|^{r+1} \in \mathfrak{a}_L$ is strictly positive with respect to the roots for (Q, L). This implies that, for every $r \in \mathbb{R}_{>-1}$ and every root α for (G, L), $\langle \alpha\,,\ \log|\chi|^{r+1} \rangle \neq 0$. Now [31, Théorème 3.2] says that, for $r \in \mathbb{R}_{>-1}$ close enough to -1, $I^G_Q(\tau \otimes |\chi|^{r+1})$ is irreducible. On the algebraic family of finite length representations $I^G_Q(\tau \otimes |\chi|^{r+1})$ with $r \in \mathbb{R}$, irreducibility is an Zariski-open condition [28, Proposition VI.8.4]. Hence the locus of r's for which this representation is reducible is a finite set. □

4 Morita Equivalences

In this section, we will first formulate a long list of conditions for the objects we want to compare. Assuming these conditions, we will prove a comparison theorem. In the next sections, we will check that these conditions are fulfilled in cases of interest.

4.1 Conditions and First Consequences

We keep the notations from the previous section.

Condition 4.1 For every parabolic subgroup P with $P_L \subset P \subset G$ and Levi factor $M \supset L$, an algebra $\mathcal{H}^M$ and a Morita equivalence

$$\Phi_M : \mathrm{Rep}(M)^{\mathfrak{s}_M} \to \mathrm{Mod}(\mathcal{H}^M)$$

are given. When $P' \supset P$ is another such parabolic subgroup, an algebra injection $\lambda_{MM'} : \mathcal{H}^M \to \mathcal{H}^{M'}$ is given, with the below properties.

(i) The following diagram commutes:

$$\begin{array}{ccc} \mathrm{Rep}(M')^{\mathfrak{s}_{M'}} & \xrightarrow{\Phi_{M'}} & \mathrm{Mod}(\mathcal{H}^{M'}) \\ \uparrow {\scriptstyle I^{M'}_{P\cap M'}} & & \uparrow {\scriptstyle \mathrm{ind}^{\mathcal{H}^{M'}}_{\lambda_{MM'}(\mathcal{H}^M)}} \\ \mathrm{Rep}(M)^{\mathfrak{s}_M} & \xrightarrow{\Phi_M} & \mathrm{Mod}(\mathcal{H}^M) \end{array}$$

(ii) Let $\overline{P}$ be the parabolic subgroup of G which has Levi factor M and is opposite to P. Let $\mathrm{pr}_{\mathfrak{s}_M} : \mathrm{Rep}(M) \to \mathrm{Rep}(M)^{\mathfrak{s}_M}$ be the projection coming from the Bernstein decomposition for M. The following diagram commutes:

$$\begin{array}{ccc} \mathrm{Rep}(M')^{\mathfrak{s}_{M'}} & \xrightarrow{\Phi_{M'}} & \mathrm{Mod}(\mathcal{H}^{M'}) \\ \downarrow {\scriptstyle \mathrm{pr}_{\mathfrak{s}_M} \circ J^{M'}_{\overline{P}\cap M'}} & & \downarrow {\scriptstyle \mathrm{Res}^{\mathcal{H}^{M'}}_{\lambda_{MM'}(\mathcal{H}^M)}} \\ \mathrm{Rep}(M)^{\mathfrak{s}_M} & \xrightarrow{\Phi_M} & \mathrm{Mod}(\mathcal{H}^M) \end{array}$$

(iii) If $P \subset P' \subset P'' \subset G$, then $\lambda_{MM''} = \lambda_{M'M''} \circ \lambda_{MM'}$.

The Conditions 4.1 are quite general, in the sense that they do not involve the structure of the algebras $\mathcal{H}^M$. We will see later that in many cases these conditions hold already by abstract functoriality principles.

The next series of conditions is more specific. For $P = MU$, let $R(M, L)$ be the set of roots of M with respect to the maximal F-split torus A_L in the centre of L. This is a root system when L is a minimal F-Levi subgroup of G. In general, it is only an orthogonal projection of such a root system (in many cases encountered in the literature it is nevertheless a root system). For $P \supset P_L$, we define the set of positive roots as $R^+(M, L) = R(M \cap P_L, L)$, and we call the minimal elements of this set the simple roots of (M, L).

Condition 4.2 Assume Condition 4.1.

(i) $\mathcal{H}^G$ (or $(\mathcal{H}^G)^{\text{op}}$) is an extended affine Hecke algebra $\mathcal{H}(\mathcal{R}, q) \rtimes \Gamma$.
(ii) All the $\mathcal{H}^M$ (or all the $(\mathcal{H}^M)^{\text{op}}$) are parabolic subalgebras and the $\lambda_{MM'}$ are inclusions of parabolic subalgebras.
(iii) Consider the bijection

$$\Phi_L : X_{\text{nr}}(L)/X_{\text{nr}}(L, \sigma) \cong \text{Irr}(L)^{\mathfrak{s}_L} \to \text{Irr}(\mathcal{H}^L) \cong T$$

and its differential $\text{d}\Phi_L : X^*(L) \otimes_{\mathbb{Z}} \mathbb{C} \to Y \otimes_{\mathbb{Z}} \mathbb{C}$.
Then $\text{d}\Phi_L^{-1}$ maps the positive coroots $R^{\vee +}$ for $\mathcal{H}(\mathcal{R}, q)$ to $R^+(G, L)$, and $\text{d}\Phi_L^{-1}(\mathbb{Q}R^\vee)$ has a $\mathbb{Q}$-basis consisting simple roots of (G, L).
(iv) Suppose that $Q \subset \Delta$ and $\text{d}\Phi_L\big(\mathbb{Q}R(M, L)\big) \cap R^\vee = R_Q^\vee$. Then $\mathcal{H}^M = \mathcal{H}^Q \rtimes \Gamma_M$ for some $\Gamma_M \subset \Gamma(Q, Q)$. If, moreover, $\text{d}\Phi_L\big(\mathbb{Q}R(M, L)\big) = \mathbb{Q}Q^\vee$, then Γ_M satisfies Condition 2.1 for Q.

In practice, the positivity part of Condition 4.2(iii) is innocent. Namely, usually one starts by fixing a minimal parabolic subgroup and proves statements with that parabolic as the standard one. Suppose that $R(G, L)$ is a root system and that all the above conditions hold, except the positivity part of Condition 4.2(iii). Then $\text{d}\Phi_L^{-1}(\mathbb{Q}R^\vee) \cap R(G, L)$ is a parabolic root subsystem of $R(G, L)$, so it is conjugate under $W(G, L)$ to a standard parabolic root subsystem, say $R(M, L)$. Applying an element of $W(M, L)$, we can, moreover, arrange that the image of $R^{\vee +}$ consists of positive roots. Equivalently, with respect to a different parabolic subgroup P'_L of G with Levi factor L, Condition 4.2(iii) is fulfilled.

Then we restart the whole procedure with P'_L instead of P_L, and the same arguments as before will also prove the required positivity statements. This applies to all the examples discussed in Sects. 5 and 6.

We draw some first consequences from the above conditions.

Lemma 4.3 *Assume Conditions 4.1 and 4.2.*

(a) There exists a canonical surjective homomorphism of complex tori $\Phi_{\text{nr}} : X_{\text{nr}}(L) \to T$, *with finite kernel* $X_{\text{nr}}(L, \sigma)$.
(b) When $d\Phi_L(\mathbb{Q}R(M, L)) \cap R^\vee = R_Q^\vee$, *the image of* $X_{\text{nr}}(M)$ *under the map from part (a) is contained in* T^Q. *When, moreover,* $d\Phi_L(\mathbb{Q}R(M, L)) = \mathbb{Q}Q^\vee$, *the image of* $X_{\text{nr}}(M)$ *equals* T^Q.
(c) For all $\omega \in \text{Rep}(M)^{\mathfrak{s}_M}$ *and* $\chi \in X_{\text{nr}}(M)$: $\Phi_M(\omega \otimes \chi) = \Phi_M(\omega) \otimes \Phi_{\text{nr}}(\chi)$.

Proof. (a) The map $\chi \mapsto \sigma \otimes \chi$ induces an isomorphism of algebraic varieties

$$X_{nr}(L)/X_{nr}(L,\sigma) \to \mathrm{Irr}(L)^{\mathfrak{s}_L} = \{\sigma \otimes \chi : \chi \in X_{nr}(L)\}.$$

By Condition 4.2(ii), Φ_L gives a bijection $\mathrm{Irr}(L)^{\mathfrak{s}_L} \to \mathrm{Irr}(\mathcal{H}^L) = T$. This lifts to a surjective group homomorphism

$$\Phi_{nr} : X_{nr}(L) \to T \quad \text{with} \quad \Phi_L(\sigma \otimes \chi) = \Phi_L(\sigma) \otimes \Phi_{nr}(\chi). \tag{61}$$

(b) For every $Q \subset \Delta$, we defined the subtorus

$$T^Q = \{t \in T : t(x) = 1 \ \forall x \in \mathbb{Q}Q \cap X\}$$

of T. Using Condition 4.2(iv), we can write

$$X_{nr}(M) = \{\chi \in X_{nr}(L) : \chi = 1 \text{ on } \mathbb{Q}R(M,L)^\vee \cap X^*(X_{nr}(L))\}.$$

The relation between M and Q shows that the preimage of $\Phi_{nr}^{-1}(T^Q)$ contains $X_{nr}(M)$. When $d\Phi_L(\mathbb{Q}R(M,L)) = \mathbb{Q}Q^\vee$, Φ_L also induces a bijection between $\mathbb{Q}R(M,L)^\vee$ and $\mathbb{Q}Q$, and $X_{nr}(M)$ is the full preimage of T^Q.
(c) The kernel of $\Phi_{nr} : X_{nr}(M) \to T^Q$ is

$$X_{nr}(M,\sigma) := X_{nr}(L,\sigma) \cap X_{nr}(M).$$

Then $X_{nr}(M,\sigma)$ acts on $X_{nr}(M)$ by translations and $\mathcal{G}_{Q,\sigma}$ acts on $T^Q \cong X_{nr}(M)/X_{nr}(M,\sigma)$. By Lemma 3.3 there exists a canonical short exact sequence

$$1 \to X_{nr}(M,\sigma) \to \mathcal{G}'_{Q,\sigma} \to \mathcal{G}_{Q,\sigma} \to 1, \tag{62}$$

such that the action of $\mathcal{G}'_{Q,\sigma}$ on $X_{nr}(M)$ lifts that of $\mathcal{G}_{Q,\sigma}$ on T^Q. For $\omega \in \mathrm{Rep}(M)^{\mathfrak{s}_M}$ Condition 4.1(ii) and (61) imply that

$$\begin{aligned}
&\mathrm{Res}^{\mathcal{H}^M}_{\lambda_{LM}(\mathcal{H}^L)}(\Phi_M(\omega \otimes \chi)) = \Phi_L(J^M_{\overline{P_L} \cap M}(\omega \otimes \chi)) \\
&= \Phi_L(J^M_{\overline{P_L} \cap M}(\omega) \otimes \chi) = \Phi_L(J^M_{\overline{P_L} \cap M}(\omega)) \otimes \Phi_L(\chi) \\
&= \mathrm{Res}^{\mathcal{H}^M}_{\lambda_{LM}(\mathcal{H}^L)}(\Phi_M(\omega)) \otimes \Phi_{nr}(\chi) = \mathrm{Res}^{\mathcal{H}^M}_{\lambda_{LM}(\mathcal{H}^L)}(\Phi_M(\omega) \otimes \Phi_{nr}(\chi)).
\end{aligned} \tag{63}$$

When ω is irreducible, $\Phi_M(\omega \otimes \chi)$ lies in the same connected component of $\mathrm{Irr}(\mathcal{H}^M)$ as $\Phi_M(\omega)$, so (63) shows that it is an unramified twist of $\Phi_M(\omega)$. Hence

$$\Phi_M(\omega \otimes \chi) = \Phi_M(\omega) \otimes \Phi_{nr}(\chi) \quad \text{when } \omega \text{ is irreducible.} \tag{64}$$

Using the invertibility of Φ_M, both sides of (64) define a group action of $X_{nr}(M)$ on $\mathrm{Mod}(\mathcal{H}^M)$, by exact functors which commute with inductive (and projective) limits. Since these actions agree on irreducible representations, they agree on all representations. □

Lemma 4.4 *Assume Conditions 4.1 and 4.2 and suppose that* $d\Phi_L\big(\mathbb{Q}R(M,L)\big) \cap R^\vee = R_Q^\vee$.

(a) *The map* Φ_L *induces a group isomorphism* $W_{\mathfrak{s}_M} \to W(R_Q)\Gamma_M$.
(b) $W_{\mathfrak{s}_M}$ *fixes* $X_{\mathrm{nr}}(M)$ *pointwise and, when, moreover,* $\mathrm{d}\Phi_L\big(\mathbb{Q}R(M,L)\big) = \mathbb{Q}Q^\vee$, $W(R_Q)\Gamma_M$ *fixes* T^Q *pointwise.*

Proof. (a) It suffices to prove this when $M = G$. By [5, Théorème 2.13], the centre of the category $\mathrm{Rep}(G)^{\mathfrak{s}}$ is

$$\mathcal{O}(\mathrm{Irr}(L)^{\mathfrak{s}_L})^{W_{\mathfrak{s}}} = \mathcal{O}(\mathrm{Irr}(L)^{\mathfrak{s}_L}/W_{\mathfrak{s}}).$$

The pointwise fixator of $X_{\mathrm{nr}}(L)$ in $N_G(L)$ is $Z_G(A_L) = Z_G(Z(L)^\circ) = L$. Since $W_{\mathfrak{s}} \subset N_G(L)/L$, it acts faithfully on $X_{\mathrm{nr}}(L)$ by algebraic group automorphisms. Hence $W_{\mathfrak{s}}$ also acts faithfully on $\mathrm{Irr}(L)^{\mathfrak{s}_L}$.

By (13) the centre of $\mathrm{Mod}(\mathcal{H} \rtimes \Gamma)$ is

$$Z(\mathcal{H} \rtimes \Gamma) = \mathcal{O}(T)^{W\Gamma} = \mathcal{O}(T/W\Gamma), \tag{65}$$

provided that $W\Gamma$ acts faithfully on T. Clearly, W acts faithfully on T. By assumption, every $\gamma \in \Gamma$ acts on R by a diagram automorphism, so it cannot act on T as any nontrivial element of W. Hence to check that $W\Gamma$ acts faithfully on T, it suffices to do so for Γ.

In view of (15), the isomorphism $\Phi_L : \mathrm{Irr}(L)^{\mathfrak{s}_L} \to T$ implies that $\Gamma_\emptyset$ is trivial. We recall from [31, Théorème 3.2] that $I_{P_L}^G(\sigma \otimes \chi)$ is irreducible for χ in a Zariski-open nonempty subset of $X_{\mathrm{nr}}(L)$. If $\gamma \in \Gamma \setminus \{1\}$ would act trivially on T, then so would the cyclic group $\langle\gamma\rangle$ generated by it. In that case

$$\mathrm{ind}_{\mathcal{H}^\emptyset}^{\mathcal{H}\rtimes\Gamma} \mathbb{C}_t = \mathrm{ind}_{\mathcal{H}^\emptyset \rtimes \langle\gamma\rangle}^{\mathcal{H}\rtimes\Gamma} (\mathbb{C}_t \otimes_{\mathbb{C}} \mathbb{C}\langle\gamma\rangle)$$

would be reducible for all $t \in T$ (as $\mathbb{C}\langle\gamma\rangle$ is reducible). That would contradict Condition 4.1(i). So $W\Gamma$ acts faithfully on T and (65) holds.

Now Condition 4.1 says that

$$Z(\mathcal{H} \rtimes \Gamma) = \mathcal{O}(T/W\Gamma) \cong \mathcal{O}(\mathrm{Irr}(L)^{\mathfrak{s}_L}/W_{\mathfrak{s}}).$$

From this and Condition 4.1(i), we deduce that $\Phi_L : \mathrm{Irr}(L)^{\mathfrak{s}_L} \to T$ induces a bijection

$$\mathrm{Irr}(L)^{\mathfrak{s}_L}/W_{\mathfrak{s}} \to T/W\Gamma.$$

On both sides, the finite groups act faithfully by automorphisms of complex algebraic varieties. Consider the open subvariety of T (resp. of $\mathrm{Irr}(L)^{\mathfrak{s}_L}$) where the stabilizers in $W\Gamma$ (resp. in $W_{\mathfrak{s}}$) are trivial. For such a $t \in T$ and $\gamma \in W\Gamma$, the equation $\Phi_L(w\Phi_L^{-1}t) = \gamma t$ holds for a unique $w \in W_{\mathfrak{s}}$. This defines the group isomorphism $W\Gamma \to W_{\mathfrak{s}}$.

(b) The first claim is trivial, because $W_{\mathfrak{s}_M} \subset N_M(L)/L$. The second claim follows directly from the first claim, part (a) and Lemma 4.3(b). □

Lemma 4.5 *Assume*

$$M \mapsto \mathrm{d}\Phi_L\big(\mathbb{Q}R(M, L)\big) \cap \Delta^\vee$$

induces a bijection between:

- *$W_\mathfrak{s}$-association classes of Levi subgroups $M \subset G$ such that $L \subset M$ and $d\Phi_L\big(\mathbb{Q}R(M, L)\big)$ equals the $\mathbb{Q}$-span of a subset of $\Delta^\vee$;*
- *subsets of $\Delta^\vee$, modulo ΓW-association.*

Proof. Suppose that M and M' are such Levi subgroups, and that they are conjugate by an element $w \in W_\mathfrak{s}$. Then the functors $I_P^G : \mathrm{Rep}(M)^{\mathfrak{s}_M} \to \mathrm{Rep}(G)^\mathfrak{s}$ and $I_{P'}^G :$ $\mathrm{Rep}(M')^{\mathfrak{s}_{M'}} \to \mathrm{Rep}(G)^\mathfrak{s}$ have the same image, for

$$I_{P'}^G \circ \mathrm{Ad}(w)^* = \mathrm{Ad}(w)^* \circ I_P^G \cong I_P^G .$$

With Condition 4.1(i), this implies that $\mathrm{ind}_{\lambda_{MG}(\mathcal{H}^M)}^{\mathcal{H}^G}$ and $\mathrm{ind}_{\lambda_{M'G}(\mathcal{H}^{M'})}^{\mathcal{H}^G}$ have the same image. Condition 4.2(ii) says that $\mathcal{H}^M = \mathcal{H}^Q \rtimes \Gamma_M$ and $\mathcal{H}^{M'} = \mathcal{H}^{Q'} \rtimes \Gamma_{M'}$ are parabolic subalgebras of $\mathcal{H}^G$, and then Proposition 2.6 shows that they must be $\mathcal{G}$-associate. Condition 4.2(iv) implies that

$$Q^\vee = \mathrm{d}\Phi_L\big(\mathbb{Q}R(M, L)\big) \cap \Delta^\vee \quad \text{and} \quad Q'^\vee = \mathrm{d}\Phi_L\big(\mathbb{Q}R(M', L)\big) \cap \Delta^\vee$$

are $W\Gamma$-associate. Hence the map of the lemma is well defined on the given equivalence classes.

Using the same notations as above, suppose that $Q^\vee$ and $Q'^\vee$ are $W\Gamma$-associate. Then $\mathrm{d}\Phi_L\big(\mathbb{Q}R(M, L)\big) = \mathbb{Q}Q^\vee$ is $W\Gamma$-associate with $\mathrm{d}\Phi_L\big(\mathbb{Q}R(M', L)\big) = \mathbb{Q}Q'^\vee$. By Lemma 4.4, $\mathbb{Q}R(M, L)$ is $W_\mathfrak{s}$-associate with $\mathbb{Q}R(M', L)$. Hence M and M' are conjugate by an element of $W_\mathfrak{s}$, showing that the map of the lemma is injective.

By Condition 4.2(iii), the subgroup $P_\mathfrak{s} \subset G$ generated by P_L and the root subgroups for roots in $\mathrm{d}\Phi_L^{-1}(\mathbb{Q}R^\vee)$ is a parabolic subgroup of G. The map of the statement sends the standard Levi factor $M_\mathfrak{s}$ of $P_\mathfrak{s}$ to $\Delta^\vee$.

Suppose that the map is not surjective. Choose $\tilde{Q}^\vee \subset \Delta^\vee$ which does not lie in the image and is maximal for that property. Since $\Delta^\vee$ belongs to the image, $\tilde{Q} \neq \Delta$ and we can find $\alpha \in \Delta \setminus \tilde{Q}$ such that $Q^\vee := \tilde{Q}^\vee \cup \{\alpha^\vee\}$ does lie in the image. We write $\Gamma_{\tilde{Q}} = \Gamma_Q \cap \Gamma(\tilde{Q}, \tilde{Q})$.

For every Levi subgroup $M' \subset M$, we choose finitely many representations $\pi_{M',i} \in \mathrm{Irr}(M')^{\mathfrak{s}_{M'}}$ as in Theorem 3.2. Then the representations

$$I_{P' \cap M}^M (\pi_{M',i} \otimes \chi_{M'}) \quad \text{with } \chi_{M'} \in X_{\mathrm{nr}}(M'),$$

for all M' and all possible i, span the Grothendieck group of $\mathrm{Rep}(M)^{\mathfrak{s}_M}$. Applying Condition 4.1, we find that the representations

$$\operatorname{ind}_{\lambda_{M'M}(\mathcal{H}^{M'})}^{\mathcal{H}^M} \Phi_{M'}(\pi_{M',i} \otimes \chi_{M'}) \tag{66}$$

span the Grothendieck group $\operatorname{Gr}(\mathcal{H}^M)$ of $\operatorname{Mod}(\mathcal{H}^M)$. By Lemma 4.3

$$\Phi_{M'}(\pi_{M',i} \otimes \chi_{M'}) = \Phi_{M'}(\pi_{M',i}) \otimes \Phi_{\text{nr}}(\chi_{M'})$$

and $\Phi_{\text{nr}}(\chi_{M'}) \in T^{Q'}$. For $M' \neq M$ and $t \in T^{Q'}$, $\Phi_{M'}(\pi_{M',i}) \otimes t$ is a representation of $\mathcal{H}^{Q'} \rtimes \Gamma_{M'}$ with Q' not $W\Gamma$-associate with $\tilde{Q}$. Hence the collection of representations

$$\Phi_M(\pi_{M,i}) \otimes t \quad \text{with } t \in T^Q \tag{67}$$

spans the quotient

$$\operatorname{Gr}(\mathcal{H}^Q \rtimes \Gamma_M) / \sum_{Q' \subsetneq Q,\, Q' \text{ not associate} \tilde{Q}} \operatorname{ind}_{\lambda_{M'M}(\mathcal{H}^{Q'} \rtimes \Gamma_{M'})}^{\mathcal{H}^Q \rtimes \Gamma_M} \operatorname{Gr}(\mathcal{H}^{Q'} \rtimes \Gamma_{M'}). \tag{68}$$

By Theorem 2.8(a), $\operatorname{ind}_{\lambda_{\tilde{Q}M}(\mathcal{H}^{\tilde{Q}} \rtimes \Gamma_{\tilde{Q}})}^{\mathcal{H}^Q \rtimes \Gamma_M} \operatorname{Gr}(\mathcal{H}^{\tilde{Q}} \rtimes \Gamma_{\tilde{Q}})$ contributes an entire $T^{W(R_{\tilde{Q}})\Gamma_{\tilde{Q}}}$-orbit of representations to (68).

By Lemma 4.4 $T^Q \subset T^{W(R_Q)\Gamma_Q}$, which shows in particular that the translation part z_γ of γ is trivial for all $\gamma \in \Gamma_Q$. As $W(R_{\tilde{Q}})\Gamma_{\tilde{Q}} \subsetneq W(R_Q)\Gamma_Q$, we have

$$T^{W(R_{\tilde{Q}})\Gamma_{\tilde{Q}}} \supset T^Q.$$

We want to see that the left-hand side has higher dimension than the right-hand side. By construction, $W(R_{\tilde{Q}})$ fixes $T^{\tilde{Q}}$ pointwise. The torus $T_1 = (T_Q \cap T^{\tilde{Q}})^\circ$ is one-dimensional, because $|Q \setminus \tilde{Q}| = 1$. For the same reason $\alpha : T_1 \to \mathbb{C}^\times$ is a surjection with finite kernel. The group $\Gamma_{\tilde{Q}} = \Gamma_Q \cap \Gamma(\tilde{Q}, \tilde{Q})$ stabilizes both Q and $\tilde{Q}$, so fixes α. Therefore $\Gamma_{\tilde{Q}}$ fixes T_1 pointwise. It follows that $W(R_{\tilde{Q}})\Gamma_{\tilde{Q}}$ fixes the torus $T^{\tilde{Q}} = T^Q T_1$ pointwise.

Returning to (67) and (68), we see now that the contribution from $\mathcal{H}^{\tilde{Q}} \rtimes \Gamma_{\tilde{Q}}$ encompasses at least one $T^{\tilde{Q}}$-orbit. But that is impossible, because the i's in (67) belong to a finite set and $\dim_{\mathbb{C}}(T^{Q'}) > \dim_{\mathbb{C}}(T^Q)$. This contradiction entails that the map from the statement is surjective. □

4.2 Preservation of Temperedness

We will show that the above conditions imply that the Morita equivalences preserve temperedness and (under an extra condition) discrete series. For Φ_M^{-1} this is relatively easy.

Lemma 4.6 *Assume Conditions 4.1 and 4.2, and let $P = MU$ be a parabolic subgroup containing P_L such that $d\Phi_L(\mathbb{Q}R(M, L)) = \mathbb{Q}Q^\vee$.*

(a) Φ_M^{-1} preserves temperedness of finite length representations.
(b) Φ_M^{-1} sends finite dimensional essentially discrete series $\mathcal{H}^M$-representations to essentially square-integrable M-representations.
(c) Suppose that $\pi' \in \mathrm{Mod}(\mathcal{H}^M)$ has finite dimension, is tempered, essentially discrete series and factors through $\psi_t : \mathcal{H}^Q \rtimes \Gamma_M \to \mathcal{H}_Q \rtimes \Gamma_M$ for some $t \in T^Q$. Then the M-representation $\Phi_M^{-1}(\pi')$ is square-integrable modulo centre.

Proof. (a) Since every irreducible $\mathcal{H}^M$-module has finite dimension, Φ_M restricts to an equivalence between finite length representations on the one hand, and the finite dimensional modules on the other hand.

Let $\pi \in \mathrm{Rep}(M)^{\mathfrak{s}_M}$ be of finite length, and recall the criterion for temperedness from [43, Proposition III.2.2] and (43). As the supercuspidal support of π is contained in $[L, \sigma]_G$, it is equivalent to impose these conditions only with respect to the parabolic subgroup $P_L = LU_L$ [16, Proposition 1.2(i)]. Let $\overline{P_L}$ be the parabolic subgroup opposite to P_L. Then $\overline{P_L} \cap M$ is opposite to $P_L \cap M$. The above condition (for P_L) is equivalent to:

$$\log|\chi| \in \overline{{}^+\mathfrak{a}_{P_L \cap M}^M} = \Big\{ \sum\nolimits_{\alpha \in R(P_L \cap M, L)} c_\alpha \alpha|_{A_L} : c_\alpha \leq 0 \Big\} \tag{69}$$

for every A_L-weight χ of $J^M_{M \cap \overline{P_L}} \pi$.

By the assumption on M and Condition 4.2(iii), $\mathrm{d}\Phi_L^{-1} : \mathbb{Q}Q^\vee \to \mathbb{Q}R(M, L)$ is a linear bijection which sends

$$\mathfrak{a}^{-Q} = \Big\{ \sum\nolimits_{\alpha \in Q} \lambda_\alpha \alpha^\vee : \lambda_\alpha \leq 0 \Big\} \quad \text{to} \quad \overline{{}^+\mathfrak{a}_{P_L \cap M}^M}. \tag{70}$$

Suppose that $\Phi_M(\pi)$ is tempered. By definition, this means that all $\mathcal{H}^L$-weights t of $\mathrm{Res}^{\mathcal{H}^M}_{\lambda_{LM}(\mathcal{H}^L)}(\Phi_M(\pi))$ satisfy $\log|t| \in \mathfrak{a}^{-Q}$. By Condition 4.1(ii) and (70), all A_L-weights χ of $J^M_{M \cap \overline{P_L}} \pi$ have $\log|\chi| \in \overline{{}^+\mathfrak{a}_{P_L \cap M}^M}$. Thus (69) says that π is tempered.
(b) By [43, Proposition III.1.1] and arguments analogous to the above, π is square-integrable modulo centre if and only if

$$\log|\chi| \in {}^+\mathfrak{a}_{\overline{P_L} \cap M}^M = \Big\{ \sum\nolimits_{\alpha \in R(P_L \cap M, L)} c_\alpha \alpha|_{A_L} : c_\alpha < 0 \Big\} \tag{71}$$

for every A_L-weight χ of $J^M_{M \cap \overline{P_L}} \pi$. The criterium for essential square-integrability then becomes

$$\log|\chi| \in {}^+\mathfrak{a}_{\overline{P_L} \cap M}^M + \mathfrak{a}_M. \tag{72}$$

for every A_L-weight χ of $J^M_{M \cap \overline{P_L}} \pi$. Since the rank $|Q|$ of R_Q equals the rank of $R(M, L)$ and $\mathrm{d}\Phi_L^{-1}$ preserves positivity, it maps $\mathfrak{a}^{--Q}$ (the interior of $\mathfrak{a}^{-Q}$) to $\mathfrak{a}_{\overline{P_L} \cap M}^M$. By Lemma 4.3(b) $\Phi_L^{-1}(T^Q) = X_{\mathrm{nr}}(M)$.

Suppose that $\Phi_M(\pi)$ is essentially discrete series. By definition, this means that all $\mathcal{H}^L$-weights t of $\mathrm{Res}^{\mathcal{H}^M}_{\lambda_{LM}(\mathcal{H}^L)}(\Phi_M(\pi))$ satisfy $|t| \in \exp(\mathfrak{a}^{--Q})T^Q$. By the above and Condition 4.1(ii), (72) holds for all A_L-weights χ of $J^M_{M\cap \overline{P_L}}\pi$. Hence π is essentially square-integrable.

(c) Recall that a M-representation is essentially square-integrable if its restriction to the derived subgroup of M is square-integrable. If a M-representation with a central character is tempered, then $Z(M)$ acts on it by a unitary character. Hence all tempered essentially square-integrable representations with a central character are square-integrable modulo centre.

By parts (a) and (b), $\Phi_M^{-1}(\pi')$ is tempered and essentially square-integrable. Since $\Phi_M^{-1}(\pi')$ lies in one Bernstein component $\mathrm{Rep}(M)^{\mathfrak{s}_M}$, the maximal compact subgroup of $Z(M)$ acts on it by a single character χ_0, which is automatically unitary. Then $X_{\mathrm{nr}}(M)$ (resp. $X_{\mathrm{unr}}(M)$) parametrizes the extensions of χ_0 to a character (resp. unitary character) of $Z(M)$. Lemma 4.3(c) says that $Z(M)$ acts on $\Phi_M^{-1}(\pi')$ by the character determined by χ_0 and $\Phi_{\mathrm{nr}}^{-1}(t)$. Lemma 2.3(b) shows that $t \in T^Q_{\mathrm{un}}$ and $\Phi_{\mathrm{nr}}^{-1}(T_{\mathrm{un}}) = X_{\mathrm{unr}}(L)$, so $Z(M)$ acts by a unitary character. □

Part (a) of Lemma 4.6 admits a quick generalization to all Levi subgroups that we encounter. On the other hand, that is not possible for parts (b) and (c). In fact, for Levi subgroups $M \subset G$ containing L but not of the form as in Lemma 4.6, $\mathrm{Rep}(M)^{\mathfrak{s}_M}$ contains no essentially square-integrable representations. We delay the proof of that claim to Proposition 4.10. We note that in those cases $\mathrm{Irr}_{L^2}(\mathcal{H}^M)$ can still be nonempty.

Lemma 4.7 *Assume Conditions 4.1 and 4.2, and let $P' = M'U'$ be a parabolic subgroup containing P_L such that $\mathbb{Q}R(M', L) \cap \mathrm{d}\Phi_L^{-1}(R^\vee)$ does not span $\mathbb{Q}R(M', L)$. Then $\Phi_{M'}^{-1}$ preserves temperedness of finite length representations.*

Proof. By Condition 4.2(iii) $\mathrm{d}\Phi_L(\mathbb{Q}R(M', L)) \cap R^\vee$ is a standard parabolic root subsystem of $R^\vee$, that is, of the form $R_Q^\vee$ for a unique $Q \subset \Delta$. By Lemma 4.5, there exists a Levi subgroup $M \subset M'$ such that $L \subset M$ and $\mathrm{d}\Phi_L(\mathbb{Q}R(M, L)) = \mathbb{Q}Q^\vee$.

By Condition 4.2(iv) $\mathcal{H}^{M'} = \mathcal{H}^Q \rtimes \Gamma_{M'}$ and $\mathcal{H}^M = \mathcal{H}^Q \rtimes \Gamma_M$. Further, by Condition 4.2(ii) $\Gamma_{M'} \supset \Gamma_M$ and $\lambda_{MM'}$ is just the inclusion. The cone $T^{-Q} \subset T_{\mathrm{rs}}$ is the same for $\mathcal{H}^M$, $\mathcal{H}^Q$ and $\mathcal{H}^{M'}$. For any finite dimensional $\mathcal{H}^M$-module V:

$$\mathrm{Wt}\big(\mathrm{ind}_{\mathcal{H}^M}^{\mathcal{H}^{M'}}(V)\big) = \{\gamma(t) : t \in \mathrm{Wt}(V), \gamma \in \Gamma_{M'}\}. \tag{73}$$

Since $\Gamma_{M'}$ preserves T^{-Q}, $\mathrm{ind}_{\mathcal{H}^M}^{\mathcal{H}^{M'}}(V)$ is tempered if and only if V is tempered. Similarly, (73) shows that a finite dimensional $\mathcal{H}^{M'}$-module V' is tempered if and only if $\mathrm{Res}^{\mathcal{H}^{M'}}_{\mathcal{H}^M}(V')$ is tempered. We note also that

$$\mathrm{ind}_{\mathcal{H}^M}^{\mathcal{H}^{M'}} \mathrm{Res}^{\mathcal{H}^{M'}}_{\mathcal{H}^M}(V') \cong \mathbb{C}[\Gamma_{M'}] \otimes_{\mathbb{C}[\Gamma_M]} V' \cong \mathbb{C}[\Gamma_{M'}/\Gamma_M] \otimes_{\mathbb{C}} V', \tag{74}$$

a $\mathcal{H}^Q \rtimes \Gamma_{M'}$-module for the diagonal action. Then (74) contains V' as the direct summand $\mathbb{C}[\Gamma_{M'}/\Gamma_{M'}] \otimes_{\mathbb{C}} V'$, and the restriction of (74) to $\mathcal{H}^M$ is a direct sum of copies of $\mathrm{Res}^{\mathcal{H}^{M'}}_{\mathcal{H}^M}(V')$.

We recall from [28, Lemme VII.2.2] that $I_{P\cap M'}^{M'}$ always preserves temperedness. Consider a finite dimensional tempered module $V' \in \mathrm{Mod}(\mathcal{H}^{M'})$. By Lemma 4.6(a) the M'-representation $I_{P\cap M'}^{M'} \circ \Phi_M^{-1} \circ \mathrm{Res}_{\mathcal{H}^M}^{\mathcal{H}^{M'}}(V')$ is tempered. By Condition 4.1 it is isomorphic to

$$\Phi_{M'}^{-1} \circ \mathrm{ind}_{\mathcal{H}^M}^{\mathcal{H}^{M'}} \circ \mathrm{Res}_{\mathcal{H}^M}^{\mathcal{H}^{M'}}(V') \cong I_{P\cap M'}^{M'} \circ \mathrm{pr}_{\mathfrak{s}_M} \circ J_{\overline{P}\cap M'}^{M'} \circ \Phi_{M'}^{-1}(V'), \tag{75}$$

Since $\Phi_{M'}^{-1}$ is an equivalence, (74) shows that (75) contains $\Phi_{M'}^{-1}(V')$ as a direct summand. So the latter is tempered as well. □

Let $M \supset L$ be a Levi subgroup of G and write $d\Phi_L(\mathbb{Q}R(M, L)) \cap R^\vee = R_Q^\vee$. As family of parabolic subalgebras of $\mathcal{H}^M$ we take $\mathcal{H}^M$ and the $\mathcal{H}^{M_1}$ where $M \supsetneq M_1 \supset L$ and $d\Phi_L(\mathbb{Q}R(M_1, L)) = \mathbb{Q}Q_1$ for some $Q_1 \subsetneq Q$. By Lemma 4.5 every (proper) subset of Q is obtained in this way. Recall the notion of Langlands constituents from Theorems 2.10 and 3.6.

Lemma 4.8 *Let M' be a Levi subgroup of M containing L, such that $d\Phi_L(\mathbb{Q}R(M', L)) = \mathbb{Q}Q'^\vee$ for a subset $Q' \subset Q$. Let $\sigma \in \mathrm{Irr}_{L^2}(\mathcal{H}_{Q'} \rtimes \Gamma_{M'})$ and $t \in T^{Q'}$. Then*

$$\Phi_{M'}^{-1}(\sigma \otimes t) = \Phi_{M'}^{-1}(\sigma \otimes t\,|t|^{-1}) \otimes \Phi_{\mathrm{nr}}^{-1}(|t|) \quad \textit{with} \quad \Phi_{M'}^{-1}(\sigma \otimes t\,|t|^{-1}) \in \mathrm{Irr}_{L^2}(M')^{\mathfrak{s}_{M'}}.$$

The Morita equivalence Φ_M^{-1} restricts to a bijection between the Langlands constituents of $\mathrm{ind}_{\mathcal{H}^{M'}}^{\mathcal{H}^M}(\sigma \otimes t)$ and those of

$$\Phi_M^{-1}\big(\mathrm{ind}_{\mathcal{H}^{M'}}^{\mathcal{H}^M}(\sigma \otimes t)\big) \cong I_{(M\cap P_L)M'}^{M}\big(\Phi_{M'}^{-1}(\sigma \otimes t\,|t|^{-1}) \otimes \Phi_{\mathrm{nr}}^{-1}(|t|)\big).$$

Remarks By Lemma 4.3(a), Φ_{nr} becomes injective when restricted to unramified characters with values in $\mathbb{R}_{>0}$. Therefore $\Phi_{\mathrm{nr}}^{-1}(|t|) \in X_{\mathrm{nr}}(M')$ is well defined.

When $Q' = Q$, Langlands constituents are not defined on the Hecke algebra side. In that case, the lemma must be interpreted differently. The functor $\mathrm{ind}_{\mathcal{H}^{M'}}^{\mathcal{H}^M} = \mathrm{ind}_{\mathcal{H}^Q \rtimes \Gamma_{M'}}^{\mathcal{H}^Q \rtimes \Gamma_M}$ preserves complete reducibility (by Clifford theory, as Γ_M is finite). By Condition 4.1(i), so does $I_{(P_L\cap M)M'}^M$. In view of Proposition 3.7(b), the lemma becomes true in the case $Q' = Q$, provided we declare that all irreducible subquotients of $\mathrm{ind}_{\mathcal{H}^{M'}}^{\mathcal{H}^M}(\sigma \otimes t)$ are Langlands constituents.

Proof. The alternative expression for $\Phi_{M'}^{-1}(\sigma \otimes t)$ comes from Lemma 4.3(c). By Lemma 2.3 $\sigma \otimes t\,|t|^{-1} \in \mathrm{Irr}_{L^2}(\mathcal{H}^{Q'} \rtimes \Gamma_{M'})$, and by Lemma 4.6(c) $\Phi_{M'}^{-1}(\sigma \otimes t\,|t|^{-1}) \in \mathrm{Irr}_{L^2}(M')$.

Case I. Suppose that $\mathrm{ind}_{\mathcal{H}^{M'}}^{\mathcal{H}^M}(\sigma \otimes t)$ is completely reducible. By Proposition 2.11(b), all its irreducible subquotients are Langlands constituents. Since Φ_M is an equivalence, $\Phi_M^{-1}\big(\mathrm{ind}_{\mathcal{H}^{M'}}^{\mathcal{H}^M}(\sigma \otimes t)\big)$ is also completely reducible. By Proposition 3.7(b) all its irreducible subquotients are Langlands constituents. Hence Φ_M^{-1} provides a bijection between these two collections of Langlands constituents.

Case II. Suppose that $\mathrm{ind}_{\mathcal{H}^{M'}}^{\mathcal{H}^M}(\sigma \otimes t)$ is not completely reducible. By Proposition 2.11(b), there exists an $r \in \mathbb{R}_{>-1}$ such that $\mathrm{ind}_{\mathcal{H}^{M'}}^{\mathcal{H}^M}(\sigma \otimes t\,|t|^r)$ is completely reducible. By Proposition 2.11(a), Case I and Proposition 3.7(a), the four representations

$$\mathrm{ind}_{\mathcal{H}^{M'}}^{\mathcal{H}^M}(\sigma \otimes t),\ \mathrm{ind}_{\mathcal{H}^{M'}}^{\mathcal{H}^M}(\sigma \otimes t\,|t|^r),\ \Phi_M^{-1}\big(\mathrm{ind}_{\mathcal{H}^{M'}}^{\mathcal{H}^M}(\sigma \otimes t\,|t|^r)\big) \text{ and } \Phi_M^{-1}\big(\mathrm{ind}_{\mathcal{H}^{M'}}^{\mathcal{H}^M}(\sigma \otimes t)\big)$$

have the same number of inequivalent Langlands constituents.

Let π' be a non-Langlands constituent of $\mathrm{ind}_{\mathcal{H}^{M'}}^{\mathcal{H}^M}(\sigma \otimes t)$. By Theorem 2.10(c), there exists a Levi subgroup $M_1 \subset M$, with $M_1 = M$ or $\mathrm{d}\Phi_L(\mathbb{Q}R(M_1, L)) = \mathbb{Q}Q_1^\vee \subsetneq Q^\vee$, such that π' is a constituent of $\mathrm{ind}_{\mathcal{H}^{M_1}}^{\mathcal{H}^M}(\sigma_1 \otimes t_1)$ for some $t_1 \in T^{Q_1}$ and $\sigma_1 \in \mathrm{Irr}_{L^2}(\mathcal{H}_{Q_1} \rtimes \Gamma_{M_1})$ with $\|cc(\sigma_1)\| > \|cc(\sigma)\|$. When $M_1 = M$, Lemma 2.3 shows that the same condition on π' is also fulfilled for the unique Levi subgroup $M_2 \subset M$ with $\mathrm{d}\Phi_L(\mathbb{Q}R(M_2, L)) = \mathbb{Q}Q^\vee$. In that case, we replace M_1 by M_2.

Now $\Phi_{M_1}^{-1}(\sigma_1 \otimes t_1|t_1|^{-1}) \in \mathrm{Irr}_{L^2}(M_1)$ and π' is a constituent of

$$I_{(M\cap P_L)M_1}^M\big(\Phi_{M'}^{-1}(\sigma_1 \otimes t_1|t_1|^{-1}) \otimes \Phi_{\mathrm{nr}}^{-1}(|t_1|)\big). \tag{76}$$

Recall that the invariant $\|cc_{M_1}\|$ on $\mathrm{Irr}(M_1)^{\mathfrak{s}_{M_1}}$ is defined via a $W_{\mathfrak{s}}$-invariant inner product on $\mathfrak{a}_L = X^*(A_L) \otimes_{\mathbb{Z}} \mathbb{R}$. Via Lemmas 4.3(a) and 4.4(a) this can be transferred (canonically) to a $W\Gamma$-invariant inner product on $\mathfrak{a}$. The supercuspidal support (which is involved in cc_{M_1}) on $\mathrm{Irr}(M_1)^{\mathfrak{s}_{M_1}}$ is (up to conjugation) given by $J_{\overline{P_L}\cap M_1}^{M_1}$. Then Condition 4.1(ii) shows that

$$\begin{aligned}\big\|cc_{M_1}\big(\Phi_{M_1}^{-1}(\sigma_1 \otimes t_1|t_1|^{-1})\big)\big\| = \big\|cc(\sigma_1 \otimes t_1|t_1|^{-1})\big\| = \|cc(\sigma_1)\| > \\ \|cc(\sigma)\| = \big\|cc(\sigma \otimes t\,|t|^{-1})\big\| = \big\|cc_{M'}\big(\Phi_{M'}^{-1}(\sigma \otimes t\,|t|^{-1})\big)\big\|.\end{aligned} \tag{77}$$

Now Theorem 3.6(c) says that $\Phi_M^{-1}(\pi')$ is not a Langlands constituent of $\Phi_M^{-1}\big(\mathrm{ind}_{\mathcal{H}^{M'}}^{\mathcal{H}^M}(\sigma \otimes t)\big)$. Summarizing, we know that:

- Φ_M^{-1} provides a bijection between the collections of inequivalent irreducible subquotients of $\mathrm{ind}_{\mathcal{H}^{M'}}^{\mathcal{H}^M}(\sigma \otimes t)$ and of $\Phi_M^{-1}\big(\mathrm{ind}_{\mathcal{H}^{M'}}^{\mathcal{H}^M}(\sigma \otimes t)\big)$;
- these two collections have the same number of Langlands constituents and the same number of non-Langlands constituents;
- Φ_M^{-1} maps non-Langlands constituents of $\mathrm{ind}_{\mathcal{H}^{M'}}^{\mathcal{H}^M}(\sigma \otimes t)$ to non-Langlands constituents of $\Phi_M^{-1}\big(\mathrm{ind}_{\mathcal{H}^{M'}}^{\mathcal{H}^M}(\sigma \otimes t)\big)$.

Consequently, Φ_M^{-1} also provides a bijection between the collections of inequivalent Langlands constituents on both sides. □

Now we are ready for the proof of main result of this paragraph.

Theorem 4.9 *Assume Conditions 4.1 and 4.2, and let $P = MU$ be a parabolic subgroup containing P_L.*

(a) *Φ_M restricts to an equivalence between the category of finite length tempered representations in* $\mathrm{Rep}(M)^{\mathfrak{s}_M}$ *and the category of finite dimensional tempered $\mathcal{H}^M$-modules.*
(b) *Suppose that $d\Phi_L(\mathbb{Q}R(M, L)) = \mathbb{Q}Q^\vee$ for some $Q \subset \Delta$. Then Φ_M sends finite length essentially square-integrable M-representations to essentially discrete series $\mathcal{H}^M$-representations, and Φ_M^{-1} does the converse.*

Proof. (a) In view of Lemmas 4.6(a) and 4.7, it suffices to prove that Φ_M preserves temperedness of finite length representations.

Suppose, on the contrary, that there exists a finite length tempered $\pi \in \mathrm{Rep}(M)^{\mathfrak{s}_M}$ such that $\Phi_M(\pi) \in \mathrm{Mod}(\mathcal{H}^M)$ is not tempered. Since $\Phi_M(\pi)$ has finite length, it has a composition series with finite dimensional irreducible quotients. It follows directly from the definition of temperedness for $\mathcal{H}^M = \mathcal{H}^Q \rtimes \Gamma_M$ that at least one of these irreducible subquotients, say ρ_1, is not tempered. Then we may replace $\Phi_M(\pi)$ by ρ_1 and π by $\Phi_M^{-1}(\rho_1)$. Hence it suffices to prove the claim for irreducible representations.

We take the same family of parabolic subalgebras of $\mathcal{H}^M$ as in Lemma 4.8. By Theorem 2.10(c), there exists a Levi subgroup $M_1 \subset M$, with $M_1 = M$ or $d\Phi_L(\mathbb{Q}R(M_1, L)) = \mathbb{Q}Q_1^\vee \subsetneq Q^\vee$, such that $\Phi_M(\pi)$ is a Langlands constituent of $\mathrm{ind}_{\mathcal{H}^{M_1}}^{\mathcal{H}^M}(\sigma_1 \otimes t_1)$ for some $t_1 \in T^{Q_1}$ and $\sigma_1 \in \mathrm{Irr}_{L^2}(\mathcal{H}_{Q_1} \rtimes \Gamma_{M_1})$. When $M_1 = M$, Lemma 2.3 shows that the same condition on π' is also fulfilled for the unique Levi subgroup $M_2 \subset M$ with $d\Phi_L(\mathbb{Q}R(M_2, L)) = \mathbb{Q}Q$. In that case we replace M_1 by M_2.

By Lemma 4.8, π is a Langlands constituent of

$$\Phi_M^{-1}\big(\mathrm{ind}_{\mathcal{H}^{M_1}}^{\mathcal{H}^M}(\sigma \otimes t)\big) \cong I_{(M \cap P_L)M_1}^M\big(\Phi_{M_1}^{-1}(\sigma \otimes t\,|t|^{-1}) \otimes \Phi_{\mathrm{nr}}^{-1}(|t|)\big). \tag{78}$$

Suppose that $\Phi_M(\pi)$ is not tempered, so $t \notin T_{\mathrm{un}}$ by Theorem 2.10(d). Then $\Phi_{\mathrm{nr}}^{-1}(|t|) \in X_{\mathrm{nr}}(M') \setminus X_{\mathrm{unr}}(M')$, and by Theorem 3.6(b) π is not tempered.

With Lemma 4.6(a), we see that Φ_M^{-1} preserves both temperedness and non-temperedness of irreducible representations. Hence so does Φ_M.

(b) For Φ_M^{-1} this is Lemma 4.6(b), so we only have to consider the claim for Φ_M. Up to (78) we can follow the proof of part (a), only replacing tempered by essentially discrete series everywhere.

Suppose that $\Phi_M(\pi)$ is not essentially discrete series. By the uniqueness in Theorem 2.10(b), Q' is a proper subset of Q. Then M' is a proper Levi subgroup of M, so by the uniqueness in Theorem 3.6(a) π is not essentially square-integrable. With Lemma 4.6(b), we conclude that Φ_M^{-1} sends those irreducible representations which are essentially discrete series to essentially square-integrable representations, and those which are not essentially discrete series to representations that are not essentially square-integrable. Now it is clear that Φ_M also respects these properties. □

For essentially square-integrable representations, we can be more precise than Theorem 4.9. We write $d\Phi_L(\mathbb{Q}R(M, L)) \cap R^\vee = R_Q^\vee$.

Proposition 4.10 *Assume Conditions 4.1 and 4.2, and let $P = MU$ be a parabolic subgroup containing P_L.*

(a) Suppose that $\mathbb{Q}R(M, L) \cap \mathrm{d}\Phi_L^{-1}(R^\vee)$ does not span $\mathbb{Q}R(M, L)$. Then $\mathrm{Rep}(M)^{\mathfrak{s}_M}$ contains no finite length essentially square-integrable representations.

(b) Suppose that $\pi \in \mathrm{Rep}(M)^{\mathfrak{s}_M}$ is square-integrable modulo centre and has finite length. Then $\Phi_M(\pi)$ is tempered, essentially discrete series and factors through $\psi_t : \mathcal{H}^Q \rtimes \Gamma_M \to \mathcal{H}_Q \rtimes \Gamma_M$ for some $t \in T_{\mathrm{un}}^Q$.

(c) Suppose that $d\Phi_L(\mathbb{Q}R(M, L)) = \mathbb{Q}Q^\vee$. Then Φ_M gives a bijection between $\mathrm{Irr}_{L^2}(M)^{\mathfrak{s}_M}$ and $\mathrm{Irr}_{L^2}(\mathcal{H}^M)$.

Proof. (a) Suppose, contrary to what we need to show, that $\mathrm{Rep}(M)^{\mathfrak{s}_M}$ does contain a representation of the indicated kind. Since it has finite length, it has an irreducible subrepresentation, say π. Let ζ be the central character of π, and let $|\zeta| \in X_{\mathrm{nr}}(M)$ be its absolute value. Then $\pi \otimes |\zeta|^{-1} \in \mathrm{Rep}(M)^{\mathfrak{s}_M}$ is an irreducible essentially square-integrable representation with a unitary central character. Hence it is square-integrable modulo centre and in particular tempered. By Theorem 4.9(a) $\Phi_M(\pi \otimes |\zeta|^{-1})$ is also tempered.

Let $P_1 = M_1U_1 \subset G$ be the parabolic subgroup such that $M \supset M_1 \supset L$ and $\mathrm{d}\Phi_L(\mathbb{Q}R(M_1, L))$ is spanned by $\mathbb{Q}R(M, L) \cap \mathrm{d}\Phi_L^{-1}(R^\vee)$. From (73) we know that $\mathrm{Res}^{\mathcal{H}^M}_{\lambda_{MM_1}(\mathcal{H}^{M_1})} \circ \Phi_M(\pi \otimes |\zeta|^{-1})$ is tempered and nonzero. By Lemma 4.6(a)

$$\Phi_{M_1}^{-1} \circ \mathrm{Res}^{\mathcal{H}^M}_{\lambda_{MM_1}(\mathcal{H}^{M_1})} \circ \Phi_M(\pi \otimes |\zeta|^{-1}) = J^M_{P_1 \cap M}(\pi \otimes |\zeta|^{-1})$$

is also tempered and nonzero. But [43, Lemme III.3.2] says that this contradicts the square-integrability (modulo centre) of $\pi \otimes |\zeta|^{-1}$. (b) This follows from Theorem 4.9, in the same way as Lemma 4.6(c) followed from parts (a) and (b) of that lemma. (c) This follows from Theorem 4.9(b), Lemma 4.6(c) and part (b). □

4.3 Comparison of Completions

In this paragraph, we will show that the equivalences Φ_M induce Morita equivalences between the appropriate Schwartz algebras. In Proposition 2.6, we described the Plancherel isomorphism for the Schwartz completion of an affine Hecke algebra, in terms of the following data:

- the set of parabolic subalgebras $\mathcal{H}^Q \rtimes \Gamma_Q$ of $\mathcal{H} \rtimes \Gamma$, up to ΓW-equivalence,
- the tori T_{un}^Q,
- the sets $\mathrm{Irr}_{L^2}(\mathcal{H}^Q \rtimes \Gamma_Q)$, up to the actions of T_{un}^Q and $W\Gamma(Q, Q)$,
- the groupoid $\mathcal{G}$,
- the intertwining operators $I(g, Q, \sigma, t)$ for $g \in \mathcal{G}_{Q,\sigma}$.

These data depend mainly on the categories $\mathrm{Mod}(\mathcal{H}^Q \rtimes \Gamma_Q)$. In Condition 4.2, we included the possibility that not the $\mathcal{H}^M$, but the $(\mathcal{H}^M)^{\mathrm{op}}$ are affine Hecke algebras, so that Φ_M becomes an equivalence between $\mathrm{Rep}(G)^{\mathfrak{s}}$ and $\mathrm{Mod}\big((\mathcal{H}^Q \rtimes \Gamma_Q)^{\mathrm{op}}\big)$. Then we use Lemma 2.7 to describe the Plancherel isomorphism of $\mathcal{S}(\mathcal{R}, q) \rtimes \Gamma$ in terms of right modules of its subalgebras $\mathcal{H}^Q \rtimes \Gamma_Q$, that is in terms of the categories $\mathrm{Mod}(\mathcal{H}^M)$. With this in mind, it suffices to consider the case where each $\mathcal{H}^M$ is an (extended) affine Hecke algebra.

On the other hand, in Theorem 3.4 the Plancherel isomorphism for $\mathcal{S}(G)^{\mathfrak{s}}$ was formulated in terms of:

- the set of parabolic subgroups $P \supset P_L$, up to conjugation by $W_{\mathfrak{s}}$,
- the tori $X_{\mathrm{unr}}(M)$,
- the sets $\mathrm{Irr}_{L^2}(M)^{\mathfrak{s}_M}$, up to the actions of $X_{\mathrm{unr}}(M)$ and $\mathrm{Stab}_{W_{\mathfrak{s}}}(M)$,
- the groups $W'(T_\omega)$,
- the intertwining operators $I(w', \omega \otimes \chi)$ for $w' \in W'(T_\omega)$.

We will compare these two data sets and manipulate them until we get a nice bijection from one side to the other.

By Proposition 4.10(a), only the P with $d\Phi_L(\mathbb{Q}R(M, L))$ of the form $\mathbb{Q}Q^\vee$ occur in the Plancherel isomorphism, since for the other P the set $\mathrm{Irr}_{L^2}(M)^{\mathfrak{s}_M}$ is empty. Given $Q \subset \Delta$, we define Γ_Q as Γ_M, where $\mathbb{Q}R(M, L) = \mathbb{Q}Q^\vee$.

By Condition 4.2(iv), there is a canonical bijection from the set of parabolic subgroups $P = MU_P$, with $P \supset P_L$, $M \supset L$ and $d\Phi_L(\mathbb{Q}R(M, L))$ of the form $\mathbb{Q}Q^\vee$ and modulo conjugation by elements of $W_{\mathfrak{s}}$, to the parabolic subalgebras $\mathcal{H}^Q \rtimes \Gamma_Q$ of $\mathcal{H}^G$, up to association by $W\Gamma$. From Theorem 3.4, one sees that two such Levi subgroups $M \subset G$ are $W_{\mathfrak{s}}$-conjugate if and only if the tempered parts of the two subsets $I_P^G(\mathrm{Rep}(M)^{\mathfrak{s}_M})$ coincide. By Condition 4.1(i) and Theorem 4.9(a), this means precisely that two subsets $\mathrm{ind}_{\lambda_{MG}(\mathcal{H}^M)}^{\mathcal{H}^G}(\mathrm{Mod}(\mathcal{H}^M))$ of $\mathrm{Mod}(\mathcal{H}^G)$ coincide. By Theorem 2.4, that happens if and only if the two $\mathcal{H}^M$ are ΓW-equivalent. Thus we can pick of representatives for such P modulo $W_{\mathfrak{s}}$-conjugacy, and then the corresponding $\mathcal{H}^M$ form representatives for ΓW-equivalence classes of parabolic subalgebras $\mathcal{H}^M = \mathcal{H}^Q \rtimes \Gamma_Q$ of $\mathcal{H}^G$.

By Proposition 4.10,b Φ_M gives a bijection between $\mathrm{Irr}_{L^2}(M)^{\mathfrak{s}_M}$ and $\mathrm{Irr}_{L^2}(\mathcal{H}^M)$. Upon parabolic induction, every $X_{\mathrm{unr}}(M)$-orbit in $\mathrm{Irr}_{L^2}(M)^{\mathfrak{s}_M}$ (resp. every T_{un}^Q-orbit in $\mathrm{Irr}_{L^2}(\mathcal{H}^M)$) gives rise to a family of tempered representations in $\mathrm{Rep}(G)^{\mathfrak{s}}$ (resp. in $\mathrm{Mod}(\mathcal{H}^G)$). From Theorem 3.4, we see that $I_P^G(\omega)$ and $I_P^G(\omega')$ belong to the same such family if and only if $\omega' = w(\omega \otimes \chi)$ for some $w \in \mathrm{Stab}_{W_{\mathfrak{s}}}(M)$ and $\chi \in X_{\mathrm{unr}}(M)$. Similarly, by 4.2(ii) and Proposition 2.6

$$\mathrm{ind}_{\lambda_{MG}(\mathcal{H}^M)}^{\mathcal{H}^G}(\sigma) \quad \text{and} \quad \mathrm{ind}_{\lambda_{MG}(\mathcal{H}^M)}^{\mathcal{H}^G}(\sigma')$$

belong to the same family in $\mathrm{Mod}(\mathcal{H}^G)$ if and only if $\sigma' = g(\sigma \circ \phi_t)$ for some $g \in \mathcal{G}_{QQ}$ and $t \in T_{\mathrm{un}}^Q$. Applying Φ_G and Condition 4.1(i), we see that the respective equivalence relations on $\mathrm{Irr}_{L^2}(M)^{\mathfrak{s}_M}$ and $\mathrm{Irr}_{L^2}(\mathcal{H}^M)$ agree via Φ_M.

Let the set of representatives $(Q, \sigma)/\sim$ be as in (34). Let $(P, M, \omega)/\sim$ be its image under Lemma 4.5 and the Φ_M^{-1}. Then $(P, M, \omega)/\sim$ is a set of representatives as in Theorem 3.4. Lemma 4.3(c) and Condition 4.1(i) guarantee that

$$\Phi_G(I_P^G(\omega \otimes \chi)) = \operatorname{ind}_{\lambda_{MG}(\mathcal{H}^M)}^{\mathcal{H}^G}(\sigma \otimes \Phi_{\mathrm{nr}}(\chi)) = \pi(Q, \sigma, \Phi_{\mathrm{nr}}(\chi)). \tag{79}$$

Hence Φ_G matches the finite length tempered elements of Rep(G) associated with (P, M, ω) (via Theorem 3.4) with the finite dimensional tempered $\mathcal{H}^G$-modules associated with (Q, σ) (via Proposition 2.6). By Theorem 3.4 $I_P^G(\omega \otimes \chi)$ and $I_P^G(\omega \otimes \chi')$ are isomorphic if and only $\chi' = w'\chi$ for some $w' \in W'(T_\omega)$. Analogously, Proposition 2.6 entails that $\pi(Q, \sigma, t)$ and $\pi(Q, \sigma, t')$ are isomorphic if and only if $t' = g(t)$ for some $g \in \mathcal{G}_{Q,\sigma}$. From this and (62), we see that Φ_{nr} (from Lemma 4.3a) induces a bijection

$$X_{\mathrm{unr}}(M)/W'(T_\omega) \to T_{\mathrm{un}}^Q/\mathcal{G}_{Q,\sigma} \cong X_{\mathrm{unr}}(M)/\mathcal{G}'_{Q,\sigma}. \tag{80}$$

In the proof of Lemma 4.4, we checked that $W_{\mathfrak{s}}$ (resp. $W\Gamma$) acts faithfully on $X_{\mathrm{nr}}(L)$ (resp. on T). Then we see from (56) and (21) that the group actions in (80) are faithful. Comparing the outer sides of (80) and using the same method as in the proof of Lemma 4.4, we deduce that $W'(T_\omega) = \mathcal{G}'_{Q,\sigma}$ as subgroups of Aut($X_{\mathrm{unr}}(M)$).

Now we come to the intertwining operators. Recall from (57) and (58) that $I(w', \omega \otimes \chi)$ for $w' \in W'(T_\omega)$ comes from a unitary operator

$$\pi(w', \omega, \chi) : I_P^G(\omega \otimes \chi) \to I_P^G(\omega \otimes w'(\chi)). \tag{81}$$

For bookkeeping purposes, we replace T^Q by $X_{\mathrm{nr}}(M)$ and $\mathcal{G}_{Q,\sigma}$ by $\mathcal{G}'_{Q,\sigma}$, at the same time defining

$$\pi(Q, \sigma, \chi) := \pi(Q, \sigma, \Phi_{\mathrm{nr}}(\chi)) \quad \text{and} \quad \pi(g', Q, \sigma, \chi) = \pi(g, Q, \sigma, \Phi_{\mathrm{nr}}(\chi))$$

when $g' \in \mathcal{G}'_{Q,\sigma}$ is a lift of $g \in \mathcal{G}_{Q,\sigma}$. In particular, for $k \in X_{\mathrm{nr}}(M, \sigma)$, the interwiner $\pi(k, Q, \sigma, \chi)$ is the identity as map on the underlying vector spaces, it only changes χ to $k\chi$. Then (26) says that the action of $\mathcal{G}'_{Q,\sigma}$ in Proposition 2.6 and (34) comes from unitary intertwiners

$$\pi(g', Q, \sigma, \chi) \in \mathrm{Hom}_{\mathcal{H}^G}(\pi(Q, \sigma, \chi), \pi(Q, \sigma, g'(\chi)). \tag{82}$$

Lemma 4.11 *The intertwining operators* (81) *and* (82) *can be normalized so that*

$$\Phi_G(\pi(w', \omega, \chi)) = \pi(g', Q, \sigma, \Phi_{\mathrm{nr}}(\chi))$$

whenever w' corresponds to g' under the identification $W'(T_\omega) = \mathcal{G}'_{Q,\sigma}$ from (80).

Proof. Both (81) and (82) are unique up to scalars, because they depend algebraically on χ and because for generic $\chi \in X_{\mathrm{unr}}(M)$ the involved representations are irreducible. (The latter follows for example from the Plancherel isomorphisms.) Therefore, if $w' = g'$ in the indicated way, $\Phi_G(\pi(w', \omega, \chi))$ equals $\pi(g', Q, \sigma, \Phi_{\mathrm{nr}}(\chi))$ up to a complex number of absolute value 1. To make this scalar 1, we simply replace $\pi(w', \omega, \chi)$ by $\Phi_G^{-1}\big(\pi(g', Q, \sigma, \Phi_{\mathrm{nr}}(\chi))\big)$. □

We remark that the normalization from Lemma 4.11 is harmless, because it does not change $I(w', \omega \otimes \chi)$.

Theorem 4.12 *Under the Conditions 4.1, 4.2 and 2.1,* $\Phi_G : \mathrm{Rep}(G)^{\mathfrak{s}} \to \mathrm{Mod}(\mathcal{H}^G)$ *induces Morita equivalences*

$$\mathcal{S}(G)^{\mathfrak{s}} \sim_M \mathcal{S}(\mathcal{R}, q) \rtimes \Gamma \quad \text{and} \quad C_r^*(G)^{\mathfrak{s}} \sim_M C_r^*(\mathcal{R}, q) \rtimes \Gamma.$$

Proof. In view of Proposition 2.6 and Theorem 3.4, we have to compare the Schwartz algebras

$$\begin{aligned} &\bigoplus\nolimits_{(P,M,\omega)/\sim} \big(C^\infty(X_{\mathrm{unr}}(M)) \otimes \mathrm{End}_{\mathbb{C}}(I_P^G(V_\omega)^{K_{\mathfrak{s}}}\big)^{W'(T_\omega)} \quad \text{and} \\ &\bigoplus_{(Q,\sigma)/\mathcal{G}} C^\infty\big(T_{\mathrm{un}}^Q; \mathrm{End}_{\mathbb{C}}(V_{Q,\sigma})\big)^{\mathcal{G}_{Q,\sigma}} = \bigoplus_{(Q,\sigma)/\mathcal{G}} C^\infty\big(X_{\mathrm{unr}}(M); \mathrm{End}_{\mathbb{C}}(V_{Q,\sigma})\big)^{\mathcal{G}'_{Q,\sigma}}. \end{aligned} \tag{83}$$

To justify the equality in the second line, we note that a section of the algebra bundle over $X_{\mathrm{unr}}(M)$ is $X_{\mathrm{nr}}(M, \sigma)$-invariant if and only if it descends to a section of the analogous algebra bundle over T_{un}^Q.

By the above constructions, the Φ_M provide a bijection between the indexing sets for the sums in (83), so it suffices to compare

$$\begin{aligned} A_1 &:= C^\infty\big(X_{\mathrm{unr}}(M); \mathrm{End}_{\mathbb{C}}(I_P^G(V_\omega)^{K_{\mathfrak{s}}})\big)^{W'(T_\omega)} \quad \text{with} \\ A_2 &:= C^\infty\big(X_{\mathrm{unr}}(M); \mathrm{End}_{\mathbb{C}}(V_{Q,\sigma})\big)^{\mathcal{G}'_{Q,\sigma}} \end{aligned} \tag{84}$$

when (P, M) corresponds to Q via Lemma 4.5 and $\Phi_M(\omega) = \sigma$. The Morita equivalences $\mathcal{S}(G, K_{\mathfrak{s}})^{\mathfrak{s}} \sim_M \mathcal{S}(G)^{\mathfrak{s}}$ and Φ_G send $I_P^G(\omega \otimes \chi)^{K_{\mathfrak{s}}}$ to $\pi(Q, \sigma, \chi)$ and by Lemma 4.11 this is compatible with the intertwining operators. Identifying $W'(T_\omega)$ and $\mathcal{G}'_{Q,\sigma}$ via (80), we consider the following bimodules for A_1 and A_2:

$$\begin{aligned} B_1 &:= C^\infty\big(X_{\mathrm{unr}}(M); \mathrm{Hom}_{\mathbb{C}}(I_P^G(V_\omega)^{K_{\mathfrak{s}}}, V_{Q,\sigma})\big)^{W'(T_\omega)}, \\ B_2 &:= C^\infty\big(X_{\mathrm{unr}}(M); \mathrm{Hom}_{\mathbb{C}}(V_{Q,\sigma}, I_P^G(V_\omega)^{K_{\mathfrak{s}}})\big)^{W'(T_\omega)}. \end{aligned}$$

Here the $W'(T_\omega)$-actions are

$$\begin{aligned} (w' \cdot f_1)(w'\chi) &= \pi(w', \omega \otimes \chi) f_1(\chi) \pi(w', Q, \sigma, \chi)^{-1} \qquad f_1 \in B_1, \\ (w' \cdot f_2)(w'\chi) &= \pi(w', Q, \sigma, \chi) f_2(\chi) \pi(w', \omega \otimes \chi)^{-1} \qquad f_2 \in B_2. \end{aligned}$$

Notice that by Lemma 4.11, these are honest group actions, not just up to some scalars. We claim that

$$B_1 \otimes_{A_1} B_2 \cong A_2 \quad \text{and} \quad B_2 \otimes_{A_2} B_1 \cong A_1 \tag{85}$$

as bimodules over A_2, respectively, A_1. Since all these algebras and modules are of finite rank over $C^\infty(X_{\text{unr}}(M))^{W'(T_\omega)}$, it suffices to check this locally, at any $\chi \in X_{\text{unr}}(M)$. Then the proof of the first half of (85) reduces to checking that

$$\text{Hom}_{\mathbb{C}}(I_P^G(V_\omega)^{K_{\mathfrak{s}}}, V_{Q,\sigma})^{W'(T_\omega)_\chi} \otimes_{\text{End}_{\mathbb{C}}(I_P^G(V_\omega)^{K_{\mathfrak{s}}})^{W'(T_\omega)_\chi}} \text{Hom}_{\mathbb{C}}(V_{Q,\sigma}, I_P^G(V_\omega)^{K_{\mathfrak{s}}})^{W'(T_\omega)_\chi} \cong \text{End}_{\mathbb{C}}(V_{Q,\sigma})^{W'(T_\omega)_\chi}, \tag{86}$$

and the other way round for $B_2 \otimes_{A_2} B_1 \cong A_1$.

By the uniqueness of $\pi(w', Q, \sigma, \chi)$ up to scalars, $w' \mapsto \pi(w', Q, \sigma, \chi)$ defines a projective representation of $W'(T_\omega)_\chi$. Let W'' be a finite central extension of $W'(T_\omega)_\chi$, such that this lifts to a linear representation of W''. By (85), the map $w' \mapsto \pi(w', \omega \otimes \chi)$ also lifts to a linear representation of W''. Then W'' and $W'(T_\omega)$ have the same invariants in the all involved modules, so we can rewrite (86) as

$$\text{Hom}_{\mathbb{C}[W'']}(I_P^G(V_\omega)^{K_{\mathfrak{s}}}, V_{Q,\sigma}) \otimes_{\text{End}_{\mathbb{C}[W'']}(I_P^G(V_\omega)^{K_{\mathfrak{s}}})} \text{Hom}_{\mathbb{C}[W'']}(V_{Q,\sigma}, I_P^G(V_\omega)^{K_{\mathfrak{s}}}) \cong \text{End}_{\mathbb{C}[W'']}(V_{Q,\sigma}). \tag{87}$$

This is a statement about finite dimensional representations of the finite group W''. One can verfiy (87) by reducing it to the case of irreducible W''-representations, where it is obvious.

This also proves (86) and (85), and shows that the algebras in (83) are Morita equivalent. Combining that with Theorem 3.4 and (34), we find the desired Morita equivalences of Schwartz algebras.

To prove that $C_r(G)^{\mathfrak{s}}$ and $C_r(\mathcal{R}, q) \rtimes \Gamma$ are Morita equivalent, we can use exactly the same argument. We only have to replace C^∞ by continuous functions everywhere, and to use Theorem 3.5 instead of Theorem 3.4. □

5 Hecke Algebras from Bushnell–Kutzko Types

Let $L \subset G$ be a Levi subgroup and let $\sigma \in \text{Irr}(L)$ be supercuspidal. Recall from [9, §4] that a type for $\mathfrak{s} = [L, \sigma]_G$ consists of a compact open subgroup $J \subset G$, and a $\lambda \in \text{Irr}(J)$, such that $\text{Rep}(G)^{\mathfrak{s}}$ is precisely the category of smooth G-representations which are generated by their λ-isotypical subspace. To such a type one associates the algebra

$$\mathcal{H}(G, J, \lambda) = \text{End}_G(\text{ind}_J^G \lambda),$$

which (by definition) acts from the right on $\text{ind}_J^G \lambda$. Then there is a Morita equivalence

$$\begin{array}{rcl} \Phi_G : \text{Rep}(G)^{\mathfrak{s}} & \to & \text{Mod}(\mathcal{H}(G, J, \lambda)) \\ \pi & \mapsto & \text{Hom}_J(\lambda, \pi) \cong \text{Hom}_G(\text{ind}_J^G \lambda, \pi). \end{array} \tag{88}$$

For a Levi subgroup $M \subset G$ containing L, Bushnell and Kutzko [9, §8] developed the notion that (J, λ) covers a $[L, \sigma]_M$-type (J_M, λ_M). Roughly speaking, this means that $J_M = J \cap M$, that $\lambda_M = \mathrm{Res}^J_{J_M} \lambda$ and that $\mathcal{H}(G, J, \lambda)$ contains invertible "strongly positive" elements. Under these conditions, writing $\mathfrak{s}_M = [L, \sigma]_M$, there is a Morita equivalence $\Phi_M : \mathrm{Rep}(M)^{\mathfrak{s}_M} \to \mathrm{Mod}(\mathcal{H}(M, J_M, \lambda_M))$ as in (88), which is in several ways compatible with Φ_G.

Lemma 5.1 *Suppose that (J, λ) is a cover of a $[L, \sigma]_L$-type (J_L, λ_L). Then Condition 4.1 is fulfilled, with $\mathcal{H}^M = \mathcal{H}(M, J_M, \lambda_M))$.*

Proof. Let P and P' be as in the condition. By [9, Proposition 8.5] $(J_{M'}, \lambda_{M'})$ is an $\mathfrak{s}_{M'}$-type, (J_M, λ_M) is an $\mathfrak{s}_M$-type and the former covers the latter.

By [9, Corollary 8.4] there exists a unique algebra monomorphism

$$t_{\overline{P} \cap M} : \mathcal{H}(M, J_M, \lambda_M) \to \mathcal{H}(M', J_{M'}, \lambda_{M'})$$

such that

$$\mathrm{Res}^{\mathcal{H}(M', J_{M'}, \lambda_{M'})}_{t_{\overline{P} \cap M}(\mathcal{H}(M, J_M, \lambda_M))} \circ \Phi_{M'} = \Phi_M \circ \mathrm{pr}_{\mathfrak{s}_M} \circ R^{M'}_{\overline{P} \cap M'}.$$

Here $R^{M'}_{\overline{P} \cap M'}$ means the unnormalized parabolic restriction functor. To obtain the version with the normalized Jacquet functor $J^{M'}_{\overline{P} \cap M'}$, we must adjust $t_{\overline{P} \cap M}$ by the square root of a modular character. This yields our $\lambda_{MM'}$. The uniqueness of $\lambda_{MM'}$ and the transitivity of normalized Jacquet restriction entail that

$$\lambda_{M'M''} \circ \lambda_{MM'} = \lambda_{MM''} \quad \text{when} \quad P \subset P' \subset P'' \subset G.$$

On general grounds $\mathrm{ind}^{\mathcal{H}(M', J_{M'}, \lambda_{M'})}_{\lambda_{MM'}(\mathcal{H}(M, J_M, \lambda_M))}$ is the left adjoint of $\mathrm{Res}^{\mathcal{H}(M', J_{M'}, \lambda_{M'})}_{\lambda_{MM'}(\mathcal{H}(M, J_M, \lambda_M))}$. By Bernstein's second adjointness theorem $I^{M'}_{P \cap M'} : \mathrm{Rep}(M) \to \mathrm{Rep}(M')$ is the left adjoint of $J^{M'}_{\overline{P} \cap M'}$. Hence

$$I^{M'}_{P \cap M'} : \mathrm{Rep}(M)^{\mathfrak{s}_M} \to \mathrm{Rep}(M')^{\mathfrak{s}_{M'}}$$

is the left adjoint of $\mathrm{pr}_{\mathfrak{s}_M} \circ J^{M'}_{\overline{P} \cap M'}$. By the uniqueness of adjoints

$$\Phi_{M'} \circ I^{M'}_{P \cap M'} = \mathrm{ind}^{\mathcal{H}(M', J_{M'}, \lambda_{M'})}_{\lambda_{MM'}(\mathcal{H}(M, J_M, \lambda_M))} \circ \Phi_M. \qquad \square$$

Having checked Condition 4.1 in a general framework, we turn to more specific instances where Condition 4.2 holds. In most cases, the intermediate algebras $\mathcal{H}^M$ are not mentioned explicitly in the literature. One can obtain them by applying the same references to the group M instead of G. Using the canonical construction of $\lambda_{MM'}$ as in the proof of Lemma 5.1, Condition 4.2(ii) will be satisfied automatically in those cases. We will check the remaining conditions, mainly by providing relevant references. Recall that to achieve Condition 4.2(iii), we can use the method described on page 239.

Iwahori–spherical representations.
This is the classical case. Let $\mathfrak{o}_F$ be the ring of integers of the non-Archimedean local field F, let $\mathfrak{p}_F$ be its maximal ideal, and let $k_F = \mathfrak{o}_F/\mathfrak{p}_F$ be the residue field. Choose an apartment $\mathbb{A}$ of the Bruhat–Tits building of G and let L be the correponding minimal F-Levi subgroup of G. Let I be an Iwahori subgroup of G associated with a chamber of $\mathbb{A}$. For convenience of the exposition, we define I here as the maximal compact subgroup of the pointwise fixator of that chamber. Let P_L be the parabolic subgroup of G with Levi factor L such that $P_L(k_F)$ is the neutral component (in the sense of k_F-groups) of the reduction of I modulo $\mathfrak{p}_F$.

Borel [7] showed that the trivial representation of I is an $\mathfrak{s}$-type where $\mathfrak{s} = [L, \mathrm{triv}_L]_G$. Borel assumes that G is semisimple, but it is easy to generalize his arguments to reductive G.

By [17, §3], there is a *-algebra isomorphism

$$C_c(I\backslash G/I) \cong \mathcal{H}(G, I, \mathrm{triv}) \cong \mathcal{H}\big(X_*(L), R^\vee(G, L), X^*(L), R(G, L), \Delta, q_I\big), \quad (89)$$

where the basis Δ is determined by P_L and $q_{I,\alpha} = \mathrm{vol}(Is_\alpha I)/\mathrm{vol}(I)$ for a simple reflection s_α. From [7, §3.1], one sees that Conditions 4.2(iii) and (iv) hold. Here $\Gamma_M = 1$ for all M, so Condition 2.1 is vacuous.

Of course Theorem 4.9 was already known for irreducible Iwahori-spherical representations. Indeed, by [19, Sect. 8] and [1, Theorem 15.1.(2) and Proposition 16.6], the bijection $\mathrm{Irr}(G)^{\mathfrak{s}} \to \mathrm{Irr}(\mathcal{H}(G, I, \mathrm{triv}))$ preserves temperedness and essential square-integrability. Moreover, Theorem 4.12 has been proven for Schwartz algebras in [12, Theorem 10.2]: (89) extends to an isomorphism of Fréchet *-algebras

$$\mathcal{S}(I\backslash G/I) \cong \mathcal{S}\big(X_*(L), R^\vee(G, L), X^*(L), R(G, L), \Delta, q_I\big).$$

Principal series representations of split groups.
Suppose that G is F-split and let T be a maximal split torus of G. Fix a smooth character $\chi_{\mathfrak{s}} \in \mathrm{Irr}(T)$ and put $\mathfrak{s} = [T, \chi_{\mathfrak{s}}]_G$, so that

$$X_{\mathrm{nr}}(T) \to T_{\mathfrak{s}} : \chi \mapsto \chi\chi_{\mathfrak{s}}$$

is a homeomorphism. Notice that $\chi_{\mathfrak{s}}$ restricted to the unique maximal compact subgroup T_{cpt} of T is a type for $[T, \chi_{\mathfrak{s}}]_T$. By [29, Lemma 6.2], there exist a root subsystem $R_{\mathfrak{s}} \subset R^\vee(G, T)$ and a subgroup $\mathfrak{R}_{\mathfrak{s}} \subset W_{\mathfrak{s}}$ such that $W_{\mathfrak{s}} = W(R_{\mathfrak{s}}) \rtimes \mathfrak{R}_{\mathfrak{s}}$.

Theorem 5.2 [29, Theorem 6.3]
*There exists a type (J, λ) for $\mathfrak{s}$ and a *-algebra isomorphism*

$$\mathcal{H}(G, J, \lambda) \cong \mathcal{H}(T_{\mathfrak{s}}, R_{\mathfrak{s}}, q) \rtimes \mathfrak{R}_{\mathfrak{s}},$$

where $q_\alpha = |k_F|$ for all $\alpha \in R_{\mathfrak{s}}$. Moreover, (J, λ) is a cover of (T_{cpt}, χ).

Furthermore, Conditions 4.2(iii) and (iv) hold by construction. If $\mathbb{Q}R(M, T) = \mathbb{Q}Q^\vee$, then $X \rtimes W_Q \Gamma_Q \subset X_*(T) \rtimes W(M, T)$, so by Remark 2.2 Condition 2.1 holds as well.

We note that for these Bernstein components Theorem 4.9(b) was already proven in [29, Theorem 10.7], while Theorem 4.9(a) follows from [12, Theorem 10.1], using [29, Sect. 8].

Level-zero representations.
These are G-representations which are generated by non-zero vectors fixed by the pro-unipotent radical of a parahoric subgroup of G. Iwahori-spherical representations constitute the most basic example of this kind. A type (J, λ) for any Bernstein component $\mathfrak{s}$ consisting level-zero representations was exhibited in [23], while it was proven in [24, Theorem 4.9] that it actually is a type. More precisely, by [24, §3.8] (J, λ) is a cover of a type for the underlying supercuspidal Bernstein component of a Levi subgroup L of G.

By [23, Theorem 7.12] (see also [21])

$$\mathcal{H}(G, J, \lambda) \cong \mathcal{H}(\mathcal{R}, q) \rtimes \mathbb{C}[\Gamma, \natural_{\mathfrak{s}}] \tag{90}$$

for suitable $\mathcal{R}$, q and Γ. In all examples of level-zero Bernstein blocks which have been worked out, the 2-cocycle $\natural_{\mathfrak{s}}$ of Γ is trivial. But even if it were non-trivial, we could easily deal with it. There always exists a finite central extension

$$1 \to \Gamma_1 \to \Gamma_2 \xrightarrow{\phi_\Gamma} \Gamma \to 1$$

such that the pullback of $\natural_{\mathfrak{s}}$ to Γ_2 splits. Then $\mathcal{H}(G, J, \lambda)$ can be regarded as the direct summand of $\mathcal{H}(\mathcal{R}, q) \rtimes \Gamma_2$ associated with a minimal central idempotent $p_{\natural_{\mathfrak{s}}} \in \mathbb{C}[\Gamma_1]$. The algebra $\mathcal{H}(\mathcal{R}, q) \rtimes \Gamma_2$, with the parabolic subalgebras $\mathcal{H}^Q \rtimes \phi_\Gamma^{-1}(\Gamma_Q)$, is of the kind studied in Sect. 2. In this situation, the Conditions 4.1 and 4.2 must be adjusted slightly, now each $\mathcal{H}^M$ should be $p_{\natural_{\mathfrak{s}}} \mathcal{H}^Q \rtimes \phi_\Gamma^{-1}(\Gamma_M)$ for some $Q \subset \Delta$. With these minor modifications, all the arguments in Sect. 4 remain valid.

Conditions 4.2(iii) and (iv) follow from the setup in [23, §3.12–3.14] and [24, §1.10], combined with the description of $\mathcal{R}$ in [23, Proposition 7.3]. The groups Γ_Q for $\mathbb{Q}Q^\vee = \mathbb{Q}R(M, L)$ satisfy Condition 2.1 because they are contained in $X \rtimes W(M, S)$, where $W(M, S)$ is the Weyl group of M with respect to a maximal F-split torus $S \subset L$.

As in the above examples, there is previous work on temperedness also. It is claimed in [12, Theorem 10.1] that Theorem 4.9(a) holds here. For this, one needs to know that (90) preserves the traces (maybe up to a positive factor) and the natural *-operations. The former follows from the support of the basis elements T_w of $\mathcal{H}(G, J, \lambda)$ constructed in [23] (only the unit element T_e is supported on J). For a simple (affine) reflection s, both T_s and T_s^* have support JsJ, so they differ only by a scalar factor. They also satisfy the same quadratic relation, so $T_s^* = T_s$. This implies that (90) is an isomorphism of *-algebras.

Inner forms of $\mathrm{GL}_n(F)$.
Let D be a division algebra with centre F. Every Levi subgroup of $G = \mathrm{GL}_m(D)$ is of the form $L = \prod_i \mathrm{GL}_{m_i}(D)^{e_i}$, where $\sum_i m_i e_i = m$. Fix a supercuspidal $\omega \in \mathrm{Irr}(L)$, of the form $\omega = \bigotimes_{i=1}^k \omega_i^{\otimes e_i}$, where $\omega_i \in \mathrm{Irr}(\mathrm{GL}_{m_i}(D))$ is supercuspidal and not inertially equivalent with ω_j if $i \neq j$. Then $T_\mathfrak{s} \cong \prod_{i=1}^k (\mathbb{C}^\times)^{e_i}$, $R_\mathfrak{s}$ is of type $\prod_{i=1}^k A_{e_i-1}$ and the stabilizer of $\mathfrak{s} = [\omega, L]_G$ in $W(G, L)$ is $W(R_\mathfrak{s}) \cong \prod_{i=1}^k S_{e_i}$.

Theorem 5.3 [35, 36]
*There exists a type (J, λ) for $\mathfrak{s}$, which is a cover of a $[\omega, L]_L$-type. There exists a parameter function $q_\mathfrak{s} : R_\mathfrak{s} \to q^{\mathbb{N}}$ such that there is an isomorphism of *-algebras*

$$\mathcal{H}(G, J, \lambda) \cong \mathcal{H}(X^*(T_\mathfrak{s}), R_\mathfrak{s}, X_*(T_\mathfrak{s}), R_\mathfrak{s}^\vee, q_\mathfrak{s}),$$

where the right-hand side is a tensor product of affine Hecke algebras of type GL_e with $e \leq m$. Moreover, this isomorphism sends the natural trace of $\mathcal{H}(G, J, \lambda)$ to a positive multiple of the trace of the right-hand side.

We remark that the claims about the * and the traces are not made explicit in [35, 36]. They can be deduced in the same way as for level-zero representations, see above. With [12, Theorem 10.1] that proves Proposition 4.10 for these groups.

Via the tensor product factorization Condition 4.2(iii) reduces to the case of a supercuspidal representation $\sigma^{\otimes e}$ of $\mathrm{GL}_r(D)^e$. There it is a consequence of the constructions involved in [35, Théorème 4.6], which entail that the same notion of positivity in real tori is used for $(\mathrm{GL}_r(D)^e, \mathrm{GL}_1(D)^{re})$ and for $\mathcal{H}(\mathrm{GL}_e, q)$. Condition 4.2(iv) is irrelevant because all the groups Γ_M are trivial.

For the Schwartz algebras of these groups, Theorem 4.12 can be found in [2, Theorem 6.2]. The proof over there is similar but simpler, because not all complications from Sect. 4 arise.

Inner forms of $\mathrm{SL}_n(F)$.
Let G be the kernel of the reduced norm map $\mathrm{GL}_m(D) \to F^\times$. It is an inner form of $\mathrm{SL}_n(F)$, and every inner form looks like this. It was shown in [1] that for every inertial equivalence class $\mathfrak{s}$, $\mathcal{H}(G)^\mathfrak{s}$ is Morita equivalent with an algebra which is closely related to affine Hecke algebras of type GL_e (yet is of a more general kind). It is not known whether there exists an $\mathfrak{s}$-type for every $\mathfrak{s}$, but in any case the constructions in [1] are derived from the work of Sécherre and Stevens on inner forms of $\mathrm{GL}_n(F)$, so types are not far away. Condition 4.1(i) is [2, Theorem 1.5(b)], the maps $\lambda_{MM'}$ are simply inclusions, and Condition 4.1(ii) follows by uniqueness of adjoints.

Condition 4.2 does not hold precisely for the algebras $\mathcal{H}^M$ obtained in this setting (in fact the Plancherel isomorphism for these $\mathcal{H}^M$ has not been worked out), so we cannot apply Proposition 4.10 or Theorem 4.12. Nevertheless the conclusions of these results hold, see [2].

Let us summarize the conclusions from this section.

Corollary 5.4 *Let $\mathfrak{s} = [L, \sigma]_G$ be an inertial equivalence class of the kind discussed in this paragraph (principal series of split group, level zero, inner form of* $\mathrm{GL}_n(F)$ *or* $\mathrm{SL}_n(F)$*). Then $\mathcal{S}(G)^{\mathfrak{s}}$ is Morita equivalent to the Schwartz completion of an extended affine Hecke algebra and $C_r^*(G)^{\mathfrak{s}}$ is Morita equivalent to the C^*-completion of the same extended affine Hecke algebra.*

Proof. Except for the last case, this follows by applying Theorem 4.12. We just checked that all its assumptions are fulfilled. For the inner forms of $\mathrm{SL}_n(F)$, [2, Theorem 6.4] gives the result in the case of Schwartz algebras. As in the proof of Theorem 4.12, the method in [2, §6.2] also works for the C^*-algebras, with minor modifications.

6 Hecke Algebras from Bernstein's Progenerators

We return to the notations from Sects. 3 and 4. Let $\mathfrak{s} = [L, \sigma]_G$ be any inertial equivalence class for G. Bernstein [6, §3] constructed a projective generator $\Pi_{\mathfrak{s}}$ for the category $\mathrm{Rep}(G)^{\mathfrak{s}}$. By [28, VI.10.1], for any Levi subgroup $M \subset G$ containing L:

$$\Pi_{\mathfrak{s}_M} = I_{P_L \cap M}^M (\Pi_{\mathfrak{s}_L}),$$

and this is a progenerator of $\mathrm{Rep}(M)^{\mathfrak{s}_M}$. In other words, the map

$$\Phi_M : V \mapsto \mathrm{Hom}_M(I_{P_L \cap M}^M (\Pi_{\mathfrak{s}_L}), V)$$

is an equivalence between $\mathrm{Rep}(M)^{\mathfrak{s}_M}$ and the category of right modules of $\mathrm{End}_M(I_{P_L \cap M}^M \Pi_{\mathfrak{s}_L})$. For $P_L \subset P = M U_P \subset G$ we put

$$\mathcal{H}^M = \mathrm{End}_M(I_{P_L \cap M}^M \Pi_{\mathfrak{s}_L})^{\mathrm{op}} = \mathrm{End}_M(\Pi_{\mathfrak{s}_M})^{\mathrm{op}}.$$

Then Φ_M provides an equivalence of categories $\mathrm{Rep}(M)^{\mathfrak{s}_M} \to \mathrm{Mod}(\mathcal{H}^M)$.

Lemma 6.1 *In the above setting Condition 4.1 is fulfilled.*

Proof. The functoriality of normalized parabolic induction gives natural injections

$$\lambda_{MM'} : \mathcal{H}^M \to \mathcal{H}^{M'} \quad \text{for} \quad P \subset P' \subset G.$$

By naturality, the $\lambda_{MM'}$ satisfies Condition 4.1(iii). By Bernstein's second adjointness theorem, for $V' \in \mathrm{Rep}(M')^{\mathfrak{s}_{M'}}$:

$$\begin{aligned}\Phi_M(J^{M'}_{\overline{P}\cap M'}V') &= \mathrm{Hom}_M(I^M_{P_L\cap M}\Pi_{\mathfrak{s}_L}, J^{M'}_{\overline{P}\cap M'}V')\\ &\cong \mathrm{Hom}_{M'}(I^{M'}_{P\cap M'}I^M_{P_L\cap M}\Pi_{\mathfrak{s}_L}, V')\\ &\cong \mathrm{Hom}_{M'}(I^{M'}_{P_L\cap M'}\Pi_{\mathfrak{s}_L}, V') = \Phi_{M'}(V')\end{aligned}$$

as $\mathcal{H}^M$-modules (via $\lambda_{MM'}$). This establishes the first commutative diagram in Condition 4.1. As in the proof of Lemma 5.1, the second commutative diagram follows from that by invoking the uniqueness of left adjoints.

In the remainder of this section, we assume that G is:

- either a symplectic group, not necessarily split,
- or a special orthogonal group, not necessarily split,
- or an inner form of $\mathrm{GL}_n(F)$.

Besides the discussion of inner forms of $\mathrm{GL}_n(F)$ in the previous section, we point out that types for Bernstein components of symplectic or special orthogonal groups have been constructed in [22]. However, as far as we know the Hecke algebras associated with these types are in only few cases known explicitly.

For the groups listed above, Heiermann has subjected $(\mathcal{H}^G)^{\mathrm{op}} = \mathrm{End}_G(I^G_{P_L}\Pi_{\mathfrak{s}_L})$ to a deep study. In [15] he proved that it is an extended affine Hecke agebra with positive parameters. The constructions in [15, §5] are such that every $\mathrm{End}_M(I^M_{P_L\cap M}\Pi_{\mathfrak{s}_L})$ arises as a parabolic subalgebra. For Condition 4.2(iii), see [16, §3]. It served as a step towards Theorem 4.9 for these groups [16, Théorème 5].

By [15, Proposition 1.15], the groups $W_Q\Gamma_Q$ are always contained in $W(\tilde{R}_Q)$ where $\tilde{R}_Q \subset \mathbb{Q}R_Q$ is a larger root system. In view of Remark 2.2, Condition 4.2(iv) holds.

In fact, a more precise description of the root data and the groups Γ_M is available. By [15, 1.13], the root datum underlying the affine Hecke algebra $\mathrm{End}_G(I^G_{P_L}\Pi_{\mathfrak{s}_L})$ is a tensor product of root data of four types: GL_n, Sp_{2n}, SO_{2n+1} and SO_{2n}. The groups Γ_M are described in [15, 1.15], but unfortunately some elements were overlooked (for a complete picture, we refer to [13]). The only nontrivial Γ_M come from the type D factors, it can happen that for a root datum of type $(\mathrm{SO}_{2n})^e$ we have (extended) Weyl groups

$$W_M \cong W(D_n)^e, \qquad W_M\Gamma_M \cong W(D_{ne}) \cap W(B_n)^e. \tag{91}$$

Then $|\Gamma_M| = 2^{e-1}$. In the above setting, Theorem 4.12 says:

Theorem 6.2 *Let G be a symplectic group or a special orthogonal group over F (not necessarily split), or an inner form of $GL_n(F)$. Let $\mathfrak{s}$ be any inertial equivalence class for G.*

Then $\mathcal{S}(G)^{\mathfrak{s}}$ is Morita equivalent with the Schwartz completion of an extended affine Hecke algebra. The underlying root datum is a tensor product of root data of type GL_n, Sp_{2n}, SO_{2n+1} and SO_{2n}, and the group Γ is a direct product of groups Γ_M as in (91). *Furthermore $C_r^*(G)^{\mathfrak{s}}$ is Morita equivalent with the C^*-completion of that extended affine Hecke algebra.*

Theorem 6.2 was one of the motivations for the author to write a paper about the K-theory of C^*-completions of (extended) affine Hecke algebras [41]. It enables us to show that the K-groups of the reduced C^*-algebras of the above groups are torsion-free.

Theorem 6.3 *Let G be as in Theorem 6.2. Then $K_*(C_r^*(G))$ is a free abelian group. It is countably infinite (unless $G = 1$).*

Proof. Recall the Bernstein decomposition from (45):

$$C_r^*(G) \cong \prod\nolimits_{\mathfrak{s}\in\mathfrak{B}(G)} C_r^*(G)^{\mathfrak{s}}.$$

Since topological K-theory is a continuous functor on the category of Banach algebras, it commutes with direct sums. This reduces the theorem to one factor $C_r^*(G)^{\mathfrak{s}}$. By Morita invariance and Theorem 6.2, it suffices to show that the K-theory of the C^*-completion of an extended affine Hecke algebra as in Theorem 6.2 is a finitely generated free abelian group. It was checked in [41, (62)] that the Künneth theorem for topological K-theory [34] applies to such algebras. Thus we only need to prove the result when the underlying root datum is of type GL_n, Sp_{2n} or SO_{2n+1} and Γ is trivial, and when the root datum is of type $(\mathrm{SO}_{2n})^e$ and Γ is as in (91). These K-groups were computed in [41], see respectively (99), Theorem 3.3, (122), and Proposition 3.5. They are free abelian and have finite rank (given explicitly in terms of partitions). □

Acknowledgements We thank the referee for suggestions and a careful reading.

References

1. A.-M. Aubert, P.F. Baum, R.J. Plymen, M. Solleveld, Hecke algebras for inner forms of p-adic special linear groups. J. Inst. Math. Jussieu **16**(2), 351–419 (2017)
2. A.-M. Aubert, P.F. Baum, R.J. Plymen, M. Solleveld, The noncommutative geometry of inner forms of p-adic special linear groups arXiv:1505.04361 (2015)
3. D. Barbasch, D. Ciubotaru, Unitary equivalences for reductive p-adic groups. Amer. J. Math. **135**(6), 1633–1674 (2013)
4. D. Barbasch, A. Moy, A unitarity criterion for p-adic groups. Inv. Math. **98**, 19–37 (1989)
5. J. Bernstein, P. Deligne, Le "centre" de Bernstein, in *Représentations des groupes réductifs sur un corps local* (Travaux en cours, Hermann, Paris, 1984), pp. 1–32
6. J. Bernstein, K.E. Rumelhart, Representations of p-adic groups. Draft (1993)
7. A. Borel, Admissible representations of a semi-simple group over a local field with vectors fixed under an Iwahori subgroup. Inv. Math. **35**, 233–259 (1976)
8. C.J. Bushnell, G. Henniart, P.C. Kutzko, Types and explicit Plancherel formulae for reductive p-adic groups, in *On certain L-functions, Clay Mathematics Proceedings*, vol. 13 (American Mathematical Society, 2011), pp. 55–80
9. C.J. Bushnell, P.C. Kutzko, Smooth representations of reductive p-adic groups: structure theory via types. Proc. London Math. Soc. **77**(3), 582–634 (1998)
10. W. Casselman, Introduction to the theory of admissible representations of p-adic reductive groups. Preprint (1995)

11. D. Ciubotaru, Types and unitary representations of reductive p-adic groups. Invent. Math. **213**, 237–269 (2018)
12. P. Delorme, E.M. Opdam, The Schwartz algebra of an affine Hecke algebra. J. Reine Angew. Math. **625**, 59–114 (2008)
13. D. Goldberg, Reducibility of induced representations for Sp(2N) and SO(N). Amer. J. Math. **116**(5), 1101–1151 (1994)
14. Harish-Chandra, The Plancherel formula for reductive p-adic groups. Collected papers vol. IV (Springer-Verlag, New York, 1984), pp. 353–367
15. V. Heiermann, Opérateurs d'entrelacement et algèbres de Hecke avec paramètres d'un groupe réductif p-adique - le cas des groupes classiques. Selecta Math. **17**(3), 713–756 (2011)
16. V. Heiermann, Algèbres de Hecke avec paramètres et représentations d'un groupe p-adique classique: préservation du spectre tempéré. J. Algebra **371**, 596–608 (2012)
17. N. Iwahori, H. Matsumoto, On some Bruhat decomposition and the structure of the Hecke rings of the p-adic Chevalley groups. Inst. Hautes Études Sci. Publ. Math **25**, 5–48 (1965)
18. D. Kazhdan, Cuspidal geometry of p-adic groups. J. Analyse Math. **47**, 1–36 (1986)
19. D. Kazhdan, G. Lusztig, Proof of the Deligne-Langlands conjecture for Hecke algebras. Invent. Math. **87**, 153–215 (1987)
20. G. Lusztig, Affine Hecke algebras and their graded version. J. Amer. Math. Soc. **2**(3), 599–635 (1989)
21. G. Lusztig, Classification of unipotent representations of simple p-adic groups. Int. Math. Res. Notices **11**, 517–589 (1995)
22. M. Miyauchi, S. Stevens, Semisimple types for p-adic classical groups. Math. Ann. **358**, 257–288 (2014)
23. L. Morris, Tamely ramified intertwining algebras. Invent. Math. **114**(1), 1–54 (1993)
24. L. Morris, Level zero G-types. Compositio Math. **118**(2), 135–157 (1999)
25. E.M. Opdam, On the spectral decomposition of affine Hecke algebras. J. Inst. Math. Jussieu **3**(4), 531–648 (2004)
26. R.J. Plymen, Reduced C^*-algebra for reductive p-adic groups. J. Funct. Anal. **88**(2), 251–266 (1990)
27. A. Ram, J. Ramagge, Affine Hecke algebras, cyclotomic Hecke algebras and Clifford theory. in *A tribute to C.S. Seshadri (Chennai 2002)*, (Trends in Mathematics, Birkhäuser, 2003), pp. 428–466
28. D. Renard, *Représentations des groupes réductifs p-adiques*, Cours spécialisés, **17**, Société Mathématique de France (2010)
29. A. Roche, Types and Hecke algebras for principal series representations of split reductive p-adic groups. Ann. Sci. École Norm. Sup. **31**(3), 361–413 (1998)
30. A. Roche, Parabolic induction and the Bernstein decomposition. Comp. Math. **134**, 113–133 (2002)
31. F. Sauvageot, Principe de densité pour les groupes réductifs. Comp. Math. **108**, 151–184 (1997)
32. P. Schneider, E.-W. Zink, K-types for the tempered components of a p-adic general linear group. With an appendix by Schneider and U. Stuhler. J. Reine Angew. Math. **517**, 161–208 (1999)
33. P. Schneider, E.-W. Zink, The algebraic theory of tempered representations of p-adic groups. Part II: Projective generators. Geom. Func. Anal. **17**(6), 2018–2065 (2008)
34. C. Schochet, Topological methods for C^*-algebras. II. Geometry resolutions and the Künneth formula. Pacific J. Math. **98**(2), 443–458 (1982)
35. V. Sécherre, Représentations lisses de $GL_m(D)$ III: types simples. Ann. Scient. Éc. Norm. Sup. **38**, 951–977 (2005)
36. V. Sécherre, S. Stevens, Smooth representations of $GL(m, D)$ VI: semisimple types. Int. Math. Res. Notices **13**, 2994–3039 (2012)
37. M. Solleveld, Periodic cyclic homology of reductive p-adic groups. J. Noncommutative Geom. **3**(4), 501–558 (2009)
38. M. Solleveld, Parabolically induced representations of graded Hecke algebras. Algebr. Represent. Theory **15**(2), 233–271 (2012)

39. M. Solleveld, On the classification of irreducible representations of affine Hecke algebras with unequal parameters. Represent. Theory **16**, 1–87 (2012)
40. M. Solleveld, Hochschild homology of affine Hecke algebras. J. Algebra **384**, 1–35 (2013)
41. M. Solleveld, Topological K-theory of affine Hecke algebras. Ann. K-theory **3**(3), 395–460 (2018)
42. M.-F. Vignéras, On formal dimensions for reductive p-adic groups, in *Festschrift in honor of I.I. Piatetski-Shapiro on the occasion of his sixtieth birthday, Part I, Israel Mathematical Conference Proceedings* vol. 2 (Weizmann, Jerusalem, 1990), pp. 225–266
43. J.-L. Waldspurger, La formule de Plancherel pour les groupes p-adiques (d'après Harish-Chandra). J. Inst. Math. Jussieu **2**(2), 235–333 (2003)

On Relatively Tempered Representations for p-adic Symmetric Spaces

Shuichiro Takeda

Contents

1	Introduction		264
2	θ-split Parabolic Subgroups and Root Systems		266
	2.1	Root Systems	266
	2.2	θ-split Parabolic Subsets	267
	2.3	Coroots	269
	2.4	Canonical Pairing for $\mathfrak{a}^*$	269
	2.5	Canonical Pairing for $\mathfrak{s}^*$	271
	2.6	Restriction from $\mathfrak{a}^*$ to $\mathfrak{s}^*$	272
	2.7	Positive Cones	273
	2.8	Real Parts of Characters	274
3	The Casselman Criteria for Distinguished Representations		277
	3.1	H-distinguished Representation	277
	3.2	The Map r_P	278
	3.3	Relative Exponents	278
	3.4	H-relatively Tempered Representations	280
	3.5	The Casselman Criteria	282
4	Subrepresentation Theorem		283
	4.1	Subrepresentation Theorems	283
	4.2	Some Lemmas	284
	4.3	Weak Jacquet Module	285
	4.4	Proof of Theorem and Its Corollaries	287
References			289

Abstract We generalize a well known subrepresentation theorem for tempered representations to the context of relatively tempered representations for p-adic symmetric spaces, assuming $p \neq 2$.

S. Takeda (✉)
Mathematics Department, University of Missouri-Columbia, 202 Math Sciences Building, Columbia, MO 65211, USA
e-mail: takedas@missouri.edu

A.-M. Aubert et al. (eds.), *Representations of Reductive p-adic Groups*,
Progress in Mathematics 328, https://doi.org/10.1007/978-981-13-6628-4_9

1 Introduction

Let F be a non-Archimedean local field whose residue characteristic is odd. Let G be a connected reductive group over F with an F-involution θ, and let H be the subgroup of θ-fixed points of G, so that the quotient $H\backslash G$ is a symmetric space.

An admissible (not necessarily irreducible) representation (π, V) of G is said to be H-distinguished if there exists a nonzero H-invariant linear form $\lambda : V \to \mathbb{C}$ such that $\lambda(\pi(h)v) = \lambda(v)$ for all $h \in H$ and $v \in V$, namely $\lambda \in \mathrm{Hom}_H(\pi, \mathbf{1})$. (Note that λ is not necessarily smooth.) In [6], Kato and Takano defined the H-matrix coefficient of an H-distinguished representation (π, V) as follows. Let $v \in V$ and $\lambda \in \mathrm{Hom}_H(\pi, \mathbf{1})$. The H-matrix coefficient $\varphi_{\lambda,v}$ is the function on the symmetric space $H\backslash G$ given by

$$\varphi_{\lambda,v}(g) = \langle \lambda, \pi(g)v \rangle$$

for $g \in G$. Further in [7], they defined an H-relatively square-integrable representation as an H-distinguished representation (π, V) which admits a unitary central character and has the property that $\varphi_{\lambda,v}$ is in $L^2(Z_G H\backslash G)$ for all $\lambda \in \mathrm{Hom}_H(\pi, \mathbf{1})$ and all $v \in V$, namely

$$\int_{Z_G H\backslash G} \left|\varphi_{\lambda,v}(g)\right|^2 \, dg < \infty,$$

where Z_G is the center of G. They also obtained the analogue of the Casselman criterion for square-integrability in terms of exponents.

Now, we can naturally define an H-relatively tempered representation as an H-distinguished representation (π, V) which admits a unitary central character and has the property that for all $\varepsilon > 0$ the H-matrix coefficient $\varphi_{\lambda,v}$ is in $L^{2+\varepsilon}(Z_G H\backslash G)$ for all $\lambda \in \mathrm{Hom}_H(\pi, \mathbf{1})$ and all $v \in V$, namely

$$\int_{Z_G H\backslash G} \left|\varphi_{\lambda,v}(g)\right|^{2+\varepsilon} \, dg < \infty.$$

A minor modification of [7] gives the analogue of the Casselman criterion for H-relative temperedness.

In this paper, we will prove the following subrepresentation theorem, which is a generalization of a well known subrepresentation theorem for tempered representations.

Theorem 1.1 *Let (π, V) be an (H, λ)-relatively tempered representation for some $\lambda \in \mathrm{Hom}_H(\pi, \mathbf{1})$. Then there is a nonzero G-intertwining map*

$$\pi \longrightarrow \mathrm{Ind}_P^G \tau,$$

where $P = MU$ is θ-split and τ is M^θ-relatively square-integrable. Further if π is irreducible, it is a direct summand of $\mathrm{Ind}_P^G \tau$.

This has an immediate corollary.

Corollary 1.2 *If* (π, V) *is irreducible and* (H, λ)*-relatively tempered for some* $\lambda \in \mathrm{Hom}_H(\pi, \mathbf{1})$, *then it is unitary and its contragredient* $\pi^\vee$ *is also* $(H, \lambda^\vee)$*-relatively tempered for some* $\lambda^\vee \in \mathrm{Hom}_H(\pi^\vee, \mathbf{1})$.

Let us emphasize that we can prove our theorem only when the residue characteristic is odd. This is because the important papers of Kato and Takano [6, 7] have this assumption.

The structure of the paper is as follows. In Sect. 2, we will recall the notion of θ-split parabolic subgroup, which one might as well consider as the parabolic subgroup relevant to distinguished representations. We then establish numerous facts on the associated root systems. In Sect. 3, we will introduce the Casselman criterion for relative temperedness. In Sect. 4, we will prove our main theorem.

Notation and Assumptions

Throughout, F is a non-archimedean local field of residue characteristic different from 2, but the characteristic of F can be positive. We let q_F be the number of elements in the residue field and $|\cdot|_F$ the absolute value on F.

Let G be a connected reductive group over F. We often identify G with its F-rational points. For admissible representations π and Π of G, we write $\pi \leq \Pi$ when π is a subquotient of Π. If $P \subseteq G$ is a parabolic subgroup, we denote the normalized Jacquet module of π along P by π_P. Also we often use the notation $P = MU$, which means that M is a Levi factor of P and that the unipotent radical is U. For a representation τ of M, we write $\mathrm{Ind}_P^G \tau$ for the representation of G obtained from τ via normalized parabolic induction. The following form of Frobenius reciprocity holds:

$$\mathrm{Hom}_G(\Pi, \mathrm{Ind}_P^G \tau) = \mathrm{Hom}_M(\Pi_P, \tau).$$

Note that the representations τ and Π need not be irreducible.

Throughout the paper, our reductive group G is equipped with an F-rational involution θ unless otherwise stated. For each subgroup $G_1 \subseteq G$, we write

$$G_1^\theta := \{g_1 \in G_1 \ : \ \theta(g_1) = g_1\}.$$

Also we set $H := G^\theta$ and hence $G_1^\theta = G_1 \cap H$.

By a representation, we always mean an admissible representation which is not necessarily irreducible, need not admit a central character and is not necessarily of finite length, unless stated otherwise. For a representation π of a subgroup $P \subseteq G$ and for a fixed $g \in G$, we define the g-twist ${}^g\pi$ to be the representation of ${}^gP := gPg^{-1}$ whose space is the same as that of π with action ${}^g\pi(gpg^{-1}) = \pi(p)$ for $p \in P$.

2 θ-split Parabolic Subgroups and Root Systems

A parabolic subgroup P of G with Levi decomposition $P = MU$ is said to be θ-split if $P \cap \theta(P) = M$. The θ-split parabolic subgroups are precisely the ones relevant to distinguished representations. In this section, we will recall some results on θ-split parabolic subgroups and related issues, and establish some necessary facts. Though this section is essentially expository, we will be as thorough as possible for convenience of the reader.

2.1 *Root Systems*

An F-split torus $S \subseteq G$ is said to be θ-split if $\theta(s) = s^{-1}$ for all $s \in S$. We let S_0 be a maximal θ-split torus. Further we fix a maximal F-split torus A containing S_0. Note that we have

$$S_0 = \{s \in A : \theta(s) = s^{-1}\} \quad \text{and} \quad A^\theta = \{s \in A : \theta(s) = s\}.$$

All of A, S_0, A^θ are θ-stable, and hence θ naturally acts on the groups of rational characters

$$X^*(A) := \mathrm{Hom}(A, \mathbb{G}_m), \quad X^*(S_0) := \mathrm{Hom}(S, \mathbb{G}_m), \quad X^*(A^\theta) := \mathrm{Hom}(A^\theta, \mathbb{G}_m)$$

of the respective tori.

Define

$$\mathfrak{a}^* := X^*(A) \otimes \mathbb{R}, \qquad \mathfrak{s}^* := X^*(S_0) \otimes \mathbb{R}, \qquad \mathfrak{s}^{*\theta} := X^*(A^\theta) \otimes \mathbb{R}.$$

The involution θ acts on $\mathfrak{a}^*$ with eigenvalues ± 1, and their respective eigenspaces are $\mathfrak{s}^{*\theta}$ and $\mathfrak{s}^*$, so we have the decomposition

$$\mathfrak{a}^* = \mathfrak{s}^* \oplus \mathfrak{s}^{*\theta}. \tag{2.1}$$

In this paper, the projection

$$p : \mathfrak{a}^* \longrightarrow \mathfrak{s}^*$$

will be used frequently, and for each $\nu \in \mathfrak{a}^*$ we often write $\overline{\nu} := p(\nu)$. There is a natural splitting of p given by

$$\mathfrak{s}^* \longrightarrow \mathfrak{a}^*, \quad \overline{\nu} \mapsto \frac{1}{2}(\nu - \theta(\nu)),$$

where ν is any preimage of $\overline{\nu}$. Hence we have

$$\overline{\nu} = \frac{1}{2}(\nu - \theta(\nu)) \quad \text{for all } \nu \in \mathfrak{a}^*.$$

Let $\Phi = \Phi(G, A) \subseteq X^*(A)$ be the set of roots of (G, A), which is θ-stable, and let W be its Weyl group. We choose a set Δ of simple roots so that the corresponding ordering has the property

$$\alpha > 0 \quad \text{and} \quad \theta(\alpha) \neq \alpha \quad \Longrightarrow \quad \theta(\alpha) < 0. \tag{2.2}$$

We let Φ^+ be the set of positive roots with respect to this ordering. We define

$$\Phi^\theta := \{\alpha \in \Phi : \theta(\alpha) = \alpha\} \subseteq X^*(A^\theta) \quad \text{and} \quad \Delta^\theta := \Delta \cap \Phi^\theta. \tag{2.3}$$

Let

$$\overline{\Phi} = p(\Phi)\backslash\{0\} = p(\Phi\backslash\Phi^\theta),$$

where p is the projection as above. It has been proven by Helminck and Wang [5] that $\overline{\Phi}$ is a root system with a basis

$$\overline{\Delta} := p(\Delta)\backslash\{0\} = p(\Delta\backslash\Delta^\theta).$$

We choose the ordering on $\overline{\Phi}$ to be the one determined by $\overline{\Delta}$, and denote by $\overline{\Phi}^+$ the set of positive roots in $\overline{\Phi}$.

Let $\overline{W}$ be the Weyl group of the root system $\overline{\Phi}$. Though we will not use it in this paper, let us give a more explicit description of it. Let W_θ be the subgroup of W generated by root reflections s_α for $\alpha \in \Phi^\theta$, and let $W_1 = \{w \in W \ : \ w(\mathfrak{s}^{*\theta}) = \mathfrak{s}^{*\theta}\}$. Then each $w \in W_1$ acts on $\mathfrak{a}^*/\mathfrak{s}^{*\theta} = \mathfrak{s}^*$, and the kernel of $W_1 \to \mathrm{Aut}(\mathfrak{s}^*)$ is W_θ, so $W_\theta \trianglelefteq W_1$. Then we have

$$\overline{W} = W_\theta\backslash W_1; \tag{2.4}$$

namely $\overline{W}$ is a subquotient of W. (See [3, 2.7].)

2.2 θ-split Parabolic Subsets

We will describe those subsets of roots which correspond to θ-split parabolic subgroups. Recall that each standard parabolic subgroup corresponds to a subset of Δ. Analogously, each θ-split parabolic subgroup corresponds to what we call a θ-split subset of Δ, which in turn corresponds to a subset of $\overline{\Delta}$ as we will describe below.

First, for each subset $\overline{I} \subseteq \overline{\Delta}$, define

$$I = (p^{-1}(\overline{I}) \cap \Delta) \cup \Delta^\theta, \tag{2.5}$$

namely I is the union of Δ^θ and the set of all simple roots in Δ that "lie above" the roots in $\overline{I}$. It is shown in [3, Lemma 2.10] that for each $\overline{\alpha} \in \overline{\Delta}$, there are at most two roots lying above $\overline{\alpha}$. To be more precise, let w_θ be the longest element in W_θ, and define

$$\theta^* := -\theta \circ w_\theta,$$

which has been shown to be an involution. (See [3].) Then we have

$$p^{-1}(\overline{\alpha}) \cap \Delta = \{\alpha, \theta^*(\alpha)\}, \quad \alpha \in \Delta \backslash \Delta^\theta.$$

We call a subset of the form (2.5) a **θ-split subset**. On the other hand, for each θ-split subset $I \subseteq \Delta$, we define

$$\overline{I} := \{\overline{\alpha} : \alpha \in I\} \backslash \{0\} = \{\overline{\alpha} : \alpha \in I \backslash \Delta^\theta\}.$$

The association $I \mapsto \overline{I}$ gives an inclusion preserving one-to-one correspondence between the θ-split subsets of Δ and the subsets of $\overline{\Delta}$. Further, for a subset $I \subseteq \Delta$, the corresponding parabolic P_I is θ-split if and only if I is θ-split. It has been shown that a subset $I \subseteq \Delta$ is θ-split if and only if I contains Δ^θ and is θ^*-stable if and only if the subsystem Φ_I generated by I is θ-stable. Of course the maximum θ-split subset is Δ and $P_\Delta = G$; and the minimum θ-split subset is Δ^θ and P_{Δ^θ} is a minimal θ-split parabolic subgroup.

Let us mention that for each $\alpha \in \Delta \backslash \Delta^\theta$ we have

$$\theta(\alpha) = -\theta^*(\alpha) - \beta \tag{2.6}$$

for some $\beta \in \Phi^\theta \cap \Phi^+$. (See [4, 1.7]. Strictly speaking, there it is stated as $\beta \in \Phi^\theta$, but the positivity is clear because $\theta(\alpha) < 0$ by (2.2) and $\theta^*(\alpha) \notin \Delta^\theta$.)

We call a θ-split parabolic subgroup of the form P_I a **Δ-standard θ-split parabolic subgroup**, or when Δ is understood from the context, simply **standard θ-split parabolic subgroup**.

Recall that the notion of the (usual) standardness is after all completely arbitrary in the sense that any parabolic subgroup can become standard by choosing an appropriate simple roots Δ. The same is true for θ-split parabolic subgroups. Namely

Lemma 2.1 *Any θ-split parabolic subgroup P is standard with respect to an appropriate choice of Δ.*

Proof See [6, p. 11]. □

Of course, once Δ is fixed, non-standard parabolic subgroups are conjugate to standard ones. And for the case of θ-split parabolic subgroups, they are often H-conjugate, but not always. Kato and Takano have shown the following.

Lemma 2.2 (Lemma 2.5 of [6]) *Let $\mathbf{H}$ be the algebraic group for H, and $\mathbf{M}_0$ be the one for the Levi M_0 of the minimum θ-split standard parabolic P_{Δ^θ}, so $\mathbf{H}(F) = H$ and $\mathbf{M}_0(F) = M_0$.*

(a) Any θ-split parabolic subgroup is of the form $\gamma^{-1}P_I\gamma$ for some θ-split subset $I \subseteq \Delta$ and $\gamma \in (\mathbf{M}_0\mathbf{H})(F)$.
(b) If $(\mathbf{M}_0\mathbf{H})(F) = M_0H$, then every θ-split parabolic subgroup is H-conjugate to a standard θ-split one.

It seems to be sometimes useful to have that every θ-split parabolic subgroup is H-conjugate to a standard one. Accordingly, we make the following definition.

Definition 2.3 The involution θ is said to be **standard** if every θ-split parabolic subgroup is H-conjugate to a standard one.

Remark 2.4 We believe that standardness is a fairly mild assumption. For example, all the cases treated in [10] are standard. See [5, §4.12] for a non-standard example.

2.3 Coroots

As usual, we define the set of cocharacters of A by $X_*(A) := \mathrm{Hom}(\mathbb{G}_m, A)$ and set $\mathfrak{a} = X_*(A) \otimes \mathbb{R}$. For each $\alpha \in \Phi$, we denote the corresponding coroot by $\alpha^\vee$ and the set of the coroots by $\Phi^\vee \subseteq X_*(A)$, so $(X^*(A), \Phi, X_*(A), \Phi^\vee)$ is a root datum. Also we denote the set of simple coroots by $\Delta^\vee$. Similarly we have the root datum $(X^*(S_0), \overline{\Phi}, X_*(S_0), \overline{\Phi}^\vee)$ and set $\mathfrak{s} = X_*(S_0) \otimes \mathbb{R}$. For each root $\overline{\alpha} \in \overline{\Phi}$, we denote the corresponding coroot by $\overline{\alpha}^\vee$, and the set of simple coroots by $\overline{\Delta}^\vee$. By analogy with (2.1), one has

$$\mathfrak{a} = \mathfrak{s} \oplus \mathfrak{s}^\theta, \tag{2.7}$$

where $\mathfrak{s}^\theta$ and $\mathfrak{s}$ are, respectively, ± 1 eigenspaces of θ.

2.4 Canonical Pairing for $\mathfrak{a}^*$

Let $P = M_PU_P$ be a (not necessarily) θ-split parabolic subgroup, where M_P is the Levi factor and U_P is the unipotent radical as usual. Let A_P be the split component of the center of M_P. Then there exists a choice of simple roots Δ with respect to which P is standard, so that $P = P_I$ for some $I \subseteq \Delta$, in which case we have

$$A_P = \left(\bigcap_{\alpha \in I} \ker \alpha\right)^\circ \quad \text{and} \quad M_P = Z_G(A_P).$$

We let $X^*(M_P) := \mathrm{Hom}(M_P, \mathbb{G}_m)$ and $X^*(A_P) := \mathrm{Hom}(A_P, \mathbb{G}_m)$ be the groups of F-rational characters of M_P and A_P, respectively, and we let $X_*(A_P) = \mathrm{Hom}(\mathbb{G}_m, A_P)$ be the group of F-rational cocharacters of A_P. Further we define

$$\mathfrak{a}_P^* := X^*(A_P) \otimes \mathbb{R} = X^*(M_P) \otimes \mathbb{R} \quad \text{and} \quad \mathfrak{a}_P := X_*(A_P) \otimes \mathbb{R},$$

where the equality $X^*(A_P) \otimes \mathbb{R} = X^*(M_P) \otimes \mathbb{R}$ is given by the natural homomorphism $X^*(M_P) \to X^*(A_P)$, which is injective with finite cokernel.

We have the canonical pairing

$$X_*(A_P) \times X^*(A_P) \longrightarrow \mathrm{End}(\mathbb{G}_m) \cong \mathbb{Z},$$

which gives rise to a pairing

$$\langle \cdot, \cdot \rangle : \mathfrak{a}_P^* \times \mathfrak{a}_P \longrightarrow \mathbb{R}. \tag{2.8}$$

Let $Q \subseteq P$, so that we have the inclusions

$$A_P \subseteq A_Q \subseteq M_Q \subseteq M_P,$$

which induce

$$X^*(M_P) \subseteq X^*(M_Q) \quad \text{and} \quad X_*(A_P) \subseteq X_*(A_Q).$$

By tensoring with $\mathbb{R}$, we have the natural inclusions

$$\mathfrak{a}_P^* \subseteq \mathfrak{a}_Q^* \quad \text{and} \quad \mathfrak{a}_P \subseteq \mathfrak{a}_Q.$$

Let us denote the annihilators of $\mathfrak{a}_P$ in $\mathfrak{a}_Q^*$ and $\mathfrak{a}_P^*$ in $\mathfrak{a}_Q$, respectively, by $\mathfrak{a}_Q^{P*}$ and $\mathfrak{a}_Q^P$, namely

$$\begin{aligned} \mathfrak{a}_Q^{P*} =&\{\nu \in \mathfrak{a}_Q^* : \langle \nu, x \rangle = 0 \text{ for all } x \in \mathfrak{a}_P\} \\ \mathfrak{a}_Q^{P} =&\{x \in \mathfrak{a}_Q : \langle \nu, x \rangle = 0 \text{ for all } \nu \in \mathfrak{a}_P^*\}. \end{aligned}$$

Then we have the orthogonal direct sum decompositions

$$\mathfrak{a}_Q^* = \mathfrak{a}_P^* \oplus \mathfrak{a}_Q^{P*} \quad \text{and} \quad \mathfrak{a}_Q = \mathfrak{a}_P \oplus \mathfrak{a}_Q^P. \tag{2.9}$$

When Q is minimal, we write the decomposition (2.9) as

$$\mathfrak{a}^* = \mathfrak{a}_P^* \oplus \mathfrak{a}_0^{P*} \quad \text{and} \quad \mathfrak{a} = \mathfrak{a}_P \oplus \mathfrak{a}_0^P, \tag{2.10}$$

and via these decompositions we usually view $\mathfrak{a}_P^*$ and $\mathfrak{a}_P$ as subspaces of $\mathfrak{a}^*$ and $\mathfrak{a}$, respectively. We let

$$\rho_P : \mathfrak{a}^* \longrightarrow \mathfrak{a}_P^* \tag{2.11}$$

be the projection on $\mathfrak{a}_P^*$. Also we obtain

$$\mathfrak{a}_P^* = \mathfrak{a}_G^* \oplus \mathfrak{a}_P^{G*} \quad \text{and} \quad \mathfrak{a}_P = \mathfrak{a}_G \oplus \mathfrak{a}_P^G \tag{2.12}$$

by setting Q to be P and P to be G in (2.9)

If we write $P = P_I$ for some $I \subseteq \Delta$ with an appropriate Δ, it is helpful to keep in mind that

$$\begin{aligned} \mathfrak{a}_G^* &= \{\nu \in \mathfrak{a}^* : \langle \nu, \alpha^\vee \rangle = 0 \text{ for all } \alpha \in \Delta\} \\ \mathfrak{a}_P^{G*} &= \mathrm{span}_{\mathbb{R}}\{\rho_P(\alpha) : \alpha \in \Delta \backslash I\} \\ \mathfrak{a}_0^{P*} &= \mathrm{span}_{\mathbb{R}}(I), \end{aligned}$$

and similarly for $\mathfrak{a}$.

Finally, since the decomposition (2.9) is orthogonal, by taking Q to be minimal, one can see that the pairing for $\mathfrak{a}^* \times \mathfrak{a}$ restricts to the pairing for $\mathfrak{a}_P^* \times \mathfrak{a}_P$, and hence we use the same notation for the pairing as P varies.

We should mention

Lemma 2.5 *The canonical pairing is θ-invariant, θ^*-invariant and W-invariant, namely for $\nu \in \mathfrak{a}^*$ and $x \in \mathfrak{a}$, one has $\langle \theta(\nu), \theta(x) \rangle = \langle \nu, x \rangle$, $\langle \theta^*(\nu), \theta^*(x) \rangle = \langle \nu, x \rangle$, and $\langle w(\nu), w(x) \rangle = \langle \nu, x \rangle$ for all $w \in W$.*

Proof The θ-invariance and W-invariance follow from the definition of how θ and W act. The θ^*-invariance follows from those two. □

Hence we have

Proposition 2.6 *The decomposition* (2.1) *is an orthogonal decomposition with respect to the canonical pairing, namely $(\mathfrak{s}^*)^\perp = \mathfrak{s}^\theta$ and $(\mathfrak{s}^{*\theta})^\perp = \mathfrak{s}$.*

Proof This follows from the θ-invariance of the canonical pairing. □

2.5 Canonical Pairing for $\mathfrak{s}^*$

Next assume $P = M_P U_P$ is θ-split, and let $S_0 \subseteq P$ be the maximum θ-split torus contained in P. We set

$$S_P := S_0 \cap A_P.$$

Let $X^*(S_P) = \mathrm{Hom}(S_P, \mathbb{G}_m)$ and $X_*(S_P) = \mathrm{Hom}(\mathbb{G}_m, S_P)$, and define

$$\mathfrak{s}_P^* := X^*(S_P) \otimes \mathbb{R} \quad \text{and} \quad \mathfrak{s}_P := X_*(S_P) \otimes \mathbb{R}.$$

When P is minimal, we have $\mathfrak{s}_P^* = \mathfrak{s}^*$ and $\mathfrak{s}_P = \mathfrak{s}$. The canonical pairing $X^*(S_P) \times X_*(S_P) \to \mathbb{Z}$ gives rise to a pairing

$$\langle \cdot, \cdot \rangle_{\mathfrak{s}} : \mathfrak{s}_P \times \mathfrak{s}_P^* \longrightarrow \mathbb{R}.$$

Let $Q \subseteq P$ be another θ-split parabolic subgroup. Then we have the orthogonal sum decompositions

$$\mathfrak{s}_Q^* = \mathfrak{s}_P^* \oplus \mathfrak{s}_Q^{P*} \quad \text{and} \quad \mathfrak{s}_Q = \mathfrak{s}_P \oplus \mathfrak{s}_Q^P, \tag{2.13}$$

which are analogous to the decomposition (2.9). So by taking Q to be minimal, one can see that the pairing for $\mathfrak{s}^* \times \mathfrak{s}$ restricts to the pairing for $\mathfrak{s}_P^* \times \mathfrak{s}_P$, and hence we use the same symbol for the pairing as P varies. Also one can see that the pairing $\mathfrak{s}^* \times \mathfrak{s} \to \mathbb{R}$ is invariant under the Weyl group $\overline{W}$ as in (2.4).

As before, when Q is minimal, we write the decomposition (2.13) as

$$\mathfrak{s}^* = \mathfrak{s}_P^* \oplus \mathfrak{s}_0^{P*} \quad \text{and} \quad \mathfrak{s} = \mathfrak{s}_P \oplus \mathfrak{s}_0^P. \tag{2.14}$$

Via these decompositions we usually view $\mathfrak{s}_P^*$ and $\mathfrak{s}_P$ as subspaces of $\mathfrak{s}^*$ and $\mathfrak{s}$, respectively. Also let

$$\rho_P : \mathfrak{s}^* \longrightarrow \mathfrak{s}_P^* \tag{2.15}$$

be the projection on $\mathfrak{s}_P^*$. Here we use the same symbol ρ_P as in (2.11), but this will be justified later. The decomposition (2.13) also gives

$$\mathfrak{s}_P^* = \mathfrak{s}_G^* \oplus \mathfrak{s}_P^{G*} \quad \text{and} \quad \mathfrak{s}_P = \mathfrak{s}_G \oplus \mathfrak{s}_P^G. \tag{2.16}$$

Now for any θ-split P, one can choose Δ in such a way that the corresponding ordering is as in (2.2) and P is standard with respect to Δ, in which case we can write $P = P_I$ for some θ-split $I \subseteq \Delta$. Then again it is helpful to keep in mind that

$$\begin{aligned} \mathfrak{s}_G^* &= \{\nu \in \mathfrak{s}^* : \langle \nu, \overline{\alpha}^\vee \rangle_{\mathfrak{s}} = 0 \text{ for all } \overline{\alpha} \in \overline{\Delta}\} \\ \mathfrak{s}_P^{G*} &= \operatorname{span}_{\mathbb{R}}\{\rho_P(\overline{\alpha}) : \overline{\alpha} \in \overline{\Delta} \backslash \overline{I}\} \\ \mathfrak{s}_0^{P*} &= \operatorname{span}_{\mathbb{R}}(\overline{I}), \end{aligned}$$

and similarly for $\mathfrak{s}$.

2.6 *Restriction from* $\mathfrak{a}^*$ *to* $\mathfrak{s}^*$

Since we have the natural inclusion $X_*(S_0) \subseteq X_*(A)$, the pairing $X^*(A) \times X_*(A) \to \mathbb{Z}$ restricts to $X^*(A) \times X_*(S_0) \to \mathbb{Z}$. Hence we have $\langle \cdot, \cdot \rangle : \mathfrak{a}^* \times \mathfrak{s} \to \mathbb{R}$, which factors through $\langle \cdot, \cdot \rangle_{\mathfrak{s}} : \mathfrak{s}^* \times \mathfrak{s} \longrightarrow \mathbb{R}$, namely for each $\nu \in \mathfrak{a}^*$ and $x \in \mathfrak{s}$, we have

$$\langle \nu, x \rangle = \langle \overline{\nu}, r \rangle_{\mathfrak{s}}. \tag{2.17}$$

With this said, the following is elementary.

Proposition 2.7 *Let $Q \subseteq P$ be both θ-split.*

(a) $\mathfrak{s}_P^*$ and $\mathfrak{s}_P$ are the -1-eigenspaces of $\mathfrak{a}_P^*$ and $\mathfrak{a}_P$ under the involution θ. Namely we can write

$$\mathfrak{a}_P^* = \mathfrak{s}_P^* \oplus \mathfrak{s}_P^{*\theta} \quad \textit{and} \quad \mathfrak{a}_P = \mathfrak{s}_P \oplus \mathfrak{s}_P{}^{\theta},$$

where $\mathfrak{s}_P^{*\theta}$ *and* $\mathfrak{s}_P^{\theta}$ *are, respectively, the* $+1$*-eigenspaces of* $\mathfrak{a}_P^*$ *and* $\mathfrak{a}_P$ *under* θ*. Further this direct sum is an orthogonal direct sum.*

(b) *The projection map* $p : \mathfrak{a}_Q^* \to \mathfrak{s}_Q^*$ *and the injection map* $\mathfrak{s}_Q \to \mathfrak{a}_Q$ *preserve the orthogonal decompositions* (2.9) *and* (2.13)*. Namely the projection*

$$\mathfrak{a}_Q^* = \mathfrak{a}_P^* \oplus \mathfrak{a}_Q^{P*} \longrightarrow \mathfrak{s}_Q^* = \mathfrak{s}_P^* \oplus \mathfrak{s}_Q^{P*}$$

maps $\mathfrak{a}_P^*$ *to* $\mathfrak{s}_P^*$ *and* $\mathfrak{a}_Q^{P*}$ *to* $\mathfrak{s}_Q^{P*}$*, and the injection*

$$\mathfrak{s}_Q = \mathfrak{s}_P \oplus \mathfrak{s}_Q^P \longrightarrow \mathfrak{a}_Q = \mathfrak{a}_P \oplus \mathfrak{a}_Q^P$$

maps $\mathfrak{s}_P$ *to* $\mathfrak{a}_P$ *and* $\mathfrak{s}_Q^P$ *to* $\mathfrak{a}_Q^P$*.*

Proof The proof is elementary linear algebra and left to the reader. □

This immediately implies the following, which justifies the use of the same symbol ρ_P for both (2.11) and (2.15).

Corollary 2.8 *For each* θ*-split* P*, the projection* $\rho_P : \mathfrak{a}^* \to \mathfrak{a}_P^*$ *restricts to the projection* $\rho_P : \mathfrak{s}^* \to \mathfrak{s}_P^*$*. In other words, we have the commutative diagram*

$$\begin{array}{ccc} \mathfrak{a}^* & \xrightarrow{\rho_P} & \mathfrak{a}_P^* \\ \downarrow & & \downarrow \\ \mathfrak{s}^* & \xrightarrow{\rho_P} & \mathfrak{s}_P^* \end{array}$$

where the vertical arrows are the projection maps as in the previous proposition.

Proof Apply the above proposition with Q minimal. □

The following is also immediate.

Corollary 2.9 *The projection* $\mathfrak{a}^* \to \mathfrak{s}^*$ *maps* $\mathfrak{a}_G^*$ *to* $\mathfrak{s}_G^*$*.*

Proof Apply the above proposition with Q minimal and $P = G$. □

2.7 Positive Cones

We define the positive cones for both $\mathfrak{a}_P^*$ and $\mathfrak{s}_P^*$ as usual. For this assume that Δ is so chosen that the corresponding ordering is as in (2.2) and P is standard with respect to Δ, so that $P = P_I$ for some $I \subseteq \Delta$. Then we define

$$
\begin{aligned}
{}^{+}\mathfrak{a}_P^* &:= \{\nu \in \mathfrak{a}_P^* \ : \ \nu = \sum_{\alpha\in\Delta\backslash I} c_\alpha \rho_P(\alpha), \ c_\alpha > 0\} \\
\overline{{}^{+}\mathfrak{a}_P^*} &:= \{\nu \in \mathfrak{a}_P^* \ : \ \nu = \sum_{\alpha\in\Delta\backslash I} c_\alpha \rho_P(\alpha), \ c_\alpha \geq 0\},
\end{aligned}
$$

and if $P = P_I$ is θ-split, we define

$$
\begin{aligned}
{}^{+}\mathfrak{s}_P^* &:= \{\overline{\nu} \in \mathfrak{s}_P^* \ : \ \overline{\nu} = \sum_{\overline{\alpha}\in\overline{\Delta}\backslash\overline{I}} c_{\overline{\alpha}}\, \rho_P(\overline{\alpha}), \ c_{\overline{\alpha}} > 0\} \\
\overline{{}^{+}\mathfrak{s}_P^*} &:= \{\overline{\nu} \in \mathfrak{s}_P^* \ : \ \overline{\nu} = \sum_{\overline{\alpha}\in\overline{\Delta}\backslash\overline{I}} c_{\overline{\alpha}}\, \rho_P(\overline{\alpha}), \ c_{\overline{\alpha}} \geq 0\},
\end{aligned}
$$

where ρ_P is as in Corollary 2.8.

These cones are related to each other as follows.

Lemma 2.10 *Let P be θ-split.*

(a) The projection $p : \mathfrak{a}^* \to \mathfrak{s}^*$ gives rise to the surjections

$$
{}^{+}\mathfrak{a}_P^* \longrightarrow {}^{+}\mathfrak{s}_P^* \quad \textit{and} \quad \overline{{}^{+}\mathfrak{a}_P^*} \longrightarrow \overline{{}^{+}\mathfrak{s}_P^*}.
$$

(b) The injection $\mathfrak{s}^* \to \mathfrak{a}^*$ gives rise to the inclusions

$$
{}^{+}\mathfrak{s}_P^* \longrightarrow {}^{+}\mathfrak{a}_P^* \quad \textit{and} \quad \overline{{}^{+}\mathfrak{s}_P^*} \longrightarrow \overline{{}^{+}\mathfrak{a}_P^*}.
$$

Proof (a) This follows from Corollary 2.8.
(b) By (2.6), for each $\alpha \in \Delta\backslash\Delta^\theta$ we have

$$
\overline{\alpha} = \frac{1}{2}(\alpha - \theta(\alpha)) = \frac{1}{2}(\alpha + \theta^*(\alpha) + \beta)
$$

for some $\beta \in \Phi^\theta$. Note that $\theta^*(\alpha) \in \Delta$, and $\rho_P(\beta) = 0$ because $\beta \in \Phi^\theta$. The lemma follows. □

2.8 Real Parts of Characters

In this subsection, we will review the notion of the real part of a quasi-character of a reductive group with enough details to ensure that our later arguments will be rigorous enough. (See, for example, [11, p. 239].) In this subsection (and only in this subsection), we will let G be an arbitrary connected reductive group over F (not necessarily our fixed G) and $H \subseteq G$ an arbitrary connected reductive subgroup group of G (not necessarily our $H = G^\theta$). (The cases of interest to us are actually when G is a torus or a Levi factor of a parabolic.) We set

$$\mathfrak{a}_G^* := X^*(G) \otimes \mathbb{R} \quad \text{and} \quad \mathfrak{a}_{G,\mathbb{C}}^* := \mathfrak{a}_G^* \otimes \mathbb{C}.$$

Here it should be mentioned that this $\mathfrak{a}_G^*$ is indeed equal to the previously defined $\mathfrak{a}_G^*$ and hence the notation. Further we set

$$\widehat{G} = \mathrm{Hom}_{\mathrm{cont}}(G, \mathbb{C}^\times) \quad \text{and} \quad \widehat{G}^{\mathrm{un}} = \mathrm{Hom}_{\mathrm{cont}}(G/G^1, \mathbb{C}^\times),$$

where $G^1 = \bigcap_{\chi \in X^*(G)} \ker |\chi|_F$. Namely $\widehat{G}$ is the group of quasi-characters of G and $\widehat{G}^{\mathrm{un}}$ is the group of unramified quasi-characters of G. It is well-known that there is a surjective group homomorphism

$$\mathfrak{a}_{G,\mathbb{C}}^* \twoheadrightarrow \widehat{G}^{\mathrm{un}}, \quad \sum_i \chi_i \otimes s_i \mapsto (g \mapsto \prod_i |\chi_i(g)|^{s_i}), \tag{2.18}$$

for $\chi_i \in X^*(G)$ and $s_i \in \mathbb{C}$, and the kernel of this homomorphism is the lattice L of the form $(2\pi i/\log q_F)L_0$ for some lattice $L_0 \subseteq X^*(G) \otimes \mathbb{Q}$, so we have the isomorphism

$$\mathfrak{a}_{G,\mathbb{C}}^*/L \xrightarrow{\sim} \widehat{G}^{\mathrm{un}}.$$

Note that the lattice L is "purely imaginary", and hence we have a well-defined map

$$\mathrm{Re} : \mathfrak{a}_{G,\mathbb{C}}^*/L \longrightarrow \mathfrak{a}_G^*, \quad \nu \mapsto \mathrm{Re}(\nu),$$

where $\mathrm{Re}(\nu)$ is the real part of ν, namely

$$\mathrm{Re}(\chi_1 \otimes s_1 + \cdots + \chi_k \otimes s_k) = \chi_1 \otimes \mathrm{Re}(s_1) + \cdots + \chi_k \otimes \mathrm{Re}(s_k)$$

for $\chi_i \in X^*(G)$ and $s_i \in \mathbb{C}$. Now for each $\chi \in \widehat{G}$, we have $|\chi| \in \widehat{G}^{\mathrm{un}}$. Since it is real valued, there exists a $\nu \in \mathfrak{a}_G^*$ which maps to $|\chi|$ under the map (2.18). Then the real part $\mathrm{Re}(\nu)$ is uniquely determined by χ, and we call it the real part of χ and denote it by $\mathrm{Re}(\chi)$. Namely, $\mathrm{Re}(\chi)$ is the image of the composite

$$\widehat{G} \longrightarrow \widehat{G}^{\mathrm{un}} \xrightarrow{\sim} \mathfrak{a}_{G,\mathbb{C}}^*/L \xrightarrow{\mathrm{Re}} \mathfrak{a}_G^*,$$

where the first map is given by $\chi \mapsto |\chi|$, and the last map is the Re map as above. Hence the map $\chi \mapsto \mathrm{Re}(\chi)$ is a group homomorphism. Moreover, the kernel of the map $\chi \mapsto |\chi|$ is the group of unitary characters of G, which we denote by $\widehat{G}^\circ$. Also we denote its image by $\widehat{G}^{\mathrm{un}+}$, which is the set of positive real valued unramified characters, namely

$$\widehat{G}^{\mathrm{un}+} = \mathrm{Hom}_{\mathrm{cont}}(G/G^1, \mathbb{R}^+).$$

Then the map $\chi \mapsto \mathrm{Re}(\chi)$ gives isomorphisms

$$\widehat{G}/\widehat{G}^{\circ} \xrightarrow{\sim} \widehat{G}^{\mathrm{un}+} \xrightarrow{\sim} \mathfrak{a}_G^*$$

of abelian groups.

The inverse of the isomorphism $\widehat{G}^{\mathrm{un}+} \xrightarrow{\sim} \mathfrak{a}_G^*$, which we denote by exp, is described by the Harish-Chandra homomorphism. Recall that the Harish-Chandra homomorphism for G is the homomorphism defined by

$$H_G : G \longrightarrow \mathrm{Hom}(X^*(G), \mathbb{R}), \quad g \mapsto (\chi \mapsto \log_{q_F} |\chi(g)|).$$

This is characterized by the property that

$$q_F^{\langle \chi, H_G(g) \rangle} = |\chi(g)|$$

for all $\chi \in \mathrm{Hom}(X^*(G), \mathbb{R})$. Then for each $\nu \in \mathfrak{a}_G^*$, we define the map

$$\exp(\nu) : G \to \mathbb{R}^+, \qquad g \mapsto q_F^{\langle \nu, H_G(g) \rangle}, \tag{2.19}$$

so in particular if $\chi \in X^*(G)$, then we have

$$\exp(\chi)(g) = |\chi(g)|,$$

or more in general, if $\nu = \chi_1 \otimes r_1 + \cdots + \chi_k \otimes r_k \in \mathfrak{a}_G^*$, where $\chi_i \in X^*(G)$ and $r_i \in \mathbb{R}$, then we have

$$\exp(\nu)(g) = |\chi_1(g)|^{r_1} \cdots |\chi_k(g)|^{r_k}.$$

One can easily verify that exp is indeed the inverse of $|\chi| \mapsto \mathrm{Re}(|\chi|)$.

Next let $H \subseteq G$ be a reductive subgroup. (As we mentioned at the beginning, this H does not have to be our G^{θ}.) Then we also have the "real part map" $\widehat{H}^{\mathrm{un}+} \xrightarrow{\sim} \mathfrak{a}_H^*$ for H. We have the obvious maps $\widehat{G}^{\mathrm{un}+} \to \widehat{H}^{\mathrm{un}+}$ and $\mathfrak{a}_G^* \to \mathfrak{a}_H^*$ both given by restriction. (To be more precise, one needs to use $H^1 \subseteq G^1$ to show that the restriction indeed gives the map for the unramified characters.) Then by disentangling the definitions, one can see that the diagram

$$\begin{array}{ccccc} \widehat{G} & \twoheadrightarrow & \widehat{G}^{\mathrm{un}+} & \underset{\exp}{\overset{\mathrm{Re}}{\rightleftarrows}} & \mathfrak{a}_G^* \\ \downarrow & & \downarrow & & \downarrow \\ \widehat{H} & \twoheadrightarrow & \widehat{H}^{\mathrm{un}+} & \underset{\exp}{\overset{\mathrm{Re}}{\rightleftarrows}} & \mathfrak{a}_H^* \end{array} \tag{2.20}$$

commutes, where all the vertical arrows are given by "restriction".

3 The Casselman Criteria for Distinguished Representations

In this section, after reviewing some of the basics of distinguished representations, we will elucidate the Casselman type criteria for H-relative square integrability and H-relative temperedness, which are essentially proven by Kato-Takano in [7].

3.1 H-distinguished Representation

A (not necessarily irreducible) representation (π, V) of G is said to be H-distinguished if $\mathrm{Hom}_H(\pi, \mathbf{1}) \neq 0$. If $\lambda \in \mathrm{Hom}_H(\pi, \mathbf{1})$ is nonzero, we sometimes say (π, V) is distinguished by λ. More generally, we make the following definition.

Definition 3.1 Assume (π, V) is an H-distinguished representation of G. Let $\lambda \in \mathrm{Hom}_H(\pi, \mathbf{1})$ be nonzero. We say a (not necessarily irreducible) subquotient $\tau \leq \pi$ is **distinguished by** λ or (H, λ)**-distinguished** if λ naturally gives rise to an H-invariant functional on (τ, V_τ). Namely, there exist G-invariant subspaces $V_1 \subseteq V_0 \subseteq V$ such that $V_1 \backslash V_0 = V_\tau, \lambda|_{V_0} \neq 0$ and $\lambda|_{V_1} = 0$, so that λ gives a natural nonzero H-invariant linear functional on τ.

Definition 3.2 Assume (π, V) is an H-distinguished representation of G. For each nonzero $\lambda \in \mathrm{Hom}_H(\pi, \mathbf{1})$, there is a nonzero G-intertwining map

$$T_{\lambda,\pi} : V \longrightarrow C^\infty(H\backslash G), \quad v \mapsto \varphi_{\lambda,v},$$

where $\varphi_{\lambda,v}(g) = \lambda(\pi(g)v)$. We call $\varphi_{\lambda,v}$ an H**-matrix coefficient** or (H, λ)**-matrix coefficient**.

Let

$$ev : C^\infty(H\backslash G) \longrightarrow \mathbb{C}, \quad \varphi \mapsto \varphi(1),$$

be the "evaluation at 1" map. This map makes every subrepresentation of $C^\infty(H\backslash G)$ H-distinguished. Moreover, for a smooth H-distinguished representation (π, V) and for nonzero $\lambda \in \mathrm{Hom}_H(\pi, \mathbf{1})$, we have

$$ev \circ T_{\lambda,\pi} = \lambda;$$

namely λ factors through the image $T_{\lambda,\pi}(V)$ of V under $T_{\lambda,\pi}$, making $T_{\lambda,\pi}(V)$ distinguished by λ. Note that every subrepresentation of $T_{\lambda,\pi}(V)$ is distinguished by λ, and $\ker(T_{\lambda,\pi})$ is the largest subrepresentation of π on which λ is identically zero. Of course, if π is irreducible, $T_{\lambda,\pi}$ is one-to-one, giving a concrete realization of π and λ.

3.2 The Map r_P

Next, we recall the map r_P defined by Kato-Takano in [6]. (Also see the work [9] of Lagier.) Let (π, V) be an H-distinguished representation, and let $\lambda \in \mathrm{Hom}_H(\pi, \mathbf{1})$ be nonzero. For each θ-split parabolic $P = MU$, there exists a linear map

$$r_P : \mathrm{Hom}_H(\pi, \mathbf{1}) \longrightarrow \mathrm{Hom}_{M^\theta}(\pi_P, \mathbf{1})$$

of invariant linear forms such that if $v \in V$ is a canonical lift (in the sense of Casselman) of $\bar{v} \in V_P$, then

$$\langle r_P(\lambda), \bar{v}\rangle = \langle \lambda, v\rangle$$

for all $\lambda \in \mathrm{Hom}_H(\pi, \mathbf{1})$. Moreover, this equality is independent of the choice of the canonical lift.

We need to mention that the map r_P has the expected transitivity property with respect to Jacquet modules. Namely

Proposition 3.3 (Proposition 5.9 of [6]) *Let $Q \subseteq P$ be θ-split, and (π, V) be an H-distinguished representation of G. Then we have*

$$r_Q = r_{Q\cap M} \circ r_P,$$

where M is the Levi factor of P.

Let us note that the map $r_P : \mathrm{Hom}_H(\pi, \mathbf{1}) \longrightarrow \mathrm{Hom}_{M^\theta}(\pi_P, \mathbf{1})$ might fail to be surjective, and hence even if $\tau \leq \pi_P$ is not distinguished by $r_P(\lambda)$ for any $\lambda \in \mathrm{Hom}_H(\pi, \mathbf{1})$, it might still be distinguished. And of course if $r_P(\lambda)$ is non-zero, at least one of the constituents of π_P is distinguished by $r_P(\lambda)$. (Here "distinguished by $r_P(\lambda)$" is defined as in Definition 3.1; namely "distinguished by $r_P(\lambda)$" means $(M^\theta, r_P(\lambda))$-distinguished.)

3.3 Relative Exponents

In this subsection, we will recall the notion of "relative exponents" introduced by Kato and Takano in [7].

First we need to recall some basic concepts. Let (π, V) be a finitely generated representation of G. Let $Z \subseteq Z_G$ be a closed subgroup of the center Z_G of G. Let $\widehat{Z}$ be the set of all quasi-characters of Z. For each $\chi \in \widehat{Z}$, we define

$$V_\chi := \left\{ v \in V \;:\; \begin{array}{c} \text{There exists } d \in \mathbb{N} \text{ such that} \\ (\pi(z) - \chi(z))^d v = 0 \text{ for all } z \in Z \end{array} \right\}.$$

Namely V_χ is the generalized χ-eigenspace. It is well-known that (π, V) has the direct sum decomposition

$$(\pi, V) = \bigoplus_{\chi \in \widehat{Z}} (\pi_\chi, V_\chi),$$

where $V_\chi = 0$ for almost all $\chi \in \widehat{Z}$. Note that each V_χ is indeed G-invariant, giving rise to a subrepresentation (π_χ, V_χ) of (π, V). Note that if $Z = Z_G$, every subquotient of (π_χ, V_χ) that admits a central character has χ as the central character.

Next, let us first recall the usual notion of exponents. For a finitely generated representation (π, V) of G and a parabolic subgroup P, we define

$$\begin{aligned} \mathcal{E}xp_{A_P}(\pi_P) &= \{\chi \in \widehat{A}_P \ : \ V_{P,\chi} \neq 0\}, \\ \mathrm{Exp}_{A_P}(\pi_P) &= \left\{ \mathrm{Re}(\chi) \in \mathfrak{a}_P^* \ : \ \chi \in \mathcal{E}xp_{A_P}(\pi_P) \right\}, \end{aligned}$$

where $V_{P,\chi}$ is the generalized χ-eigenspace as above, and $\mathrm{Re}(\chi)$ is the real part of χ as defined previously. Note that $\nu \in \mathrm{Exp}_{A_P}(\pi_P)$ if and only if $\exp(\nu) = |\chi|$ for some $\chi \in \mathcal{E}xp_{A_P}(\pi_P)$ if and only if $\exp(\nu) \otimes \tau \leq \pi_P$ for some irreducible representation τ with a unitary central character. We call each $\chi \in \mathcal{E}xp_{A_P}(\pi_P)$ a central exponent of π along P and each $\nu \in \mathrm{Exp}_{A_P}(\pi_P)$ an exponent of π along P.

Let $Q \subseteq P$ be parabolic subgroups. We have maps $\mathcal{E}xp_{A_Q}(\pi_Q) \to \mathcal{E}xp_{A_P}(\pi_P)$ and $\mathrm{Exp}_{A_Q}(\pi_Q) \to \mathrm{Exp}_{A_P}(\pi_P)$ given by restriction. Indeed, by using the diagram (2.20) with $G = A_Q$ and $H = A_P$, one can obtain the commutative diagram of sets

$$\begin{array}{ccc} \mathcal{E}xp_{A_Q}(\pi_Q) & \xrightarrow{\mathrm{Re}} & \mathrm{Exp}_{A_Q}(\pi_Q) \\ \downarrow & & \downarrow \\ \mathcal{E}xp_{A_P}(\pi_P) & \xrightarrow{\mathrm{Re}} & \mathrm{Exp}_{A_P}(\pi_P) \end{array}$$

where the horizontal arrows are surjective.

Next assume (π, V) is H-distinguished and $P = MU$ is a θ-split parabolic with S_P being the θ-split component. By following Kato-Takano [7, Section 4], we first define

$$\begin{aligned} \mathcal{E}xp_{S_P}(\pi_P) &= \{\chi \in \widehat{S}_P \ : \ V_{P,\chi} \neq 0\}, \\ \mathrm{Exp}_{S_P}(\pi_P) &= \left\{ \mathrm{Re}(\chi) \in \mathfrak{s}_P^* \ : \ \chi \in \mathcal{E}xp_{S_P}(\pi_P) \right\}. \end{aligned}$$

By applying the diagram (2.20) with $G = A_P$ and $H = S_P$, one can obtain the commutative diagram of sets

$$\begin{array}{ccc} \mathcal{E}xp_{A_P}(\pi_P) & \xrightarrow{\mathrm{Re}} & \mathrm{Exp}_{A_P}(\pi_P) \\ \downarrow & & \downarrow \\ \mathcal{E}xp_{S_P}(\pi_P) & \xrightarrow{\mathrm{Re}} & \mathrm{Exp}_{S_P}(\pi_P) \end{array}$$

where all the arrows are surjective.

Now let $\lambda \in \mathrm{Hom}_H(\pi, \mathbf{1})$ be fixed. By following [7], we define

$$\mathcal{E}xp_{S_P}(\pi_P, \lambda) = \left\{\chi \in \mathcal{E}xp_{S_P}(\pi_P) \ : \ r_P(\lambda) \neq 0 \text{ on } V_{P,\chi}\right\},$$
$$\mathrm{Exp}_{S_P}(\pi_P, \lambda) = \left\{ \mathrm{Re}(\chi) \in \mathfrak{s}_M^* \ : \ \chi \in \mathcal{E}xp_{S_P}(\pi_P, \lambda) \right\},$$

where of course $P = MU$ is assumed to be θ-split. We call each element in $\mathcal{E}xp_{S_P}(\pi_P, \lambda)$ an (H, λ)-**relative central exponent** of π along P, and each element in $\mathrm{Exp}_{S_P}(\pi_P, \lambda)$ an (H, λ)-**relative exponent** of π along P. So $\chi \in \mathcal{E}xp_{S_P}(\pi_P, \lambda)$ if and only if there exists a constituent $\tau \leq \pi_P$ distinguished by $r_P(\lambda)$ with the central character ω_τ such that $\omega_\tau|_{S_P} = \chi$.

Proposition 3.4 *Let $Q \subseteq P \subseteq G$ be θ-split. Then we have the commutative diagram of sets*

$$\begin{array}{ccc} \mathcal{E}xp_{S_Q}(\pi_Q, \lambda) & \xrightarrow{Re} & \mathrm{Exp}_{S_Q}(\pi_Q, \lambda) \\ \downarrow & & \downarrow \\ \mathcal{E}xp_{S_P}(\pi_P, \lambda) & \xrightarrow{Re} & \mathrm{Exp}_{S_P}(\pi_P, \lambda) \end{array}$$

where the horizontal arrows are surjective, and the vertical arrows are given by restriction.

Proof It has been shown in [7, Lemma 4.5] that the restriction map $\mathcal{E}xp_{S_Q}(\pi_Q) \to \mathcal{E}xp_{S_P}(\pi_P)$ gives rise to $\mathcal{E}xp_{S_Q}(\pi_Q, \lambda) \to \mathcal{E}xp_{S_P}(\pi_P, \lambda)$. This gives the vertical arrows of the diagram. The commutativity of the diagram can be shown by using (2.20) with $G = S_Q$ and $H = S_P$. □

3.4 *H-relatively Tempered Representations*

Though we already briefly recalled the notion of H-relative temperedness in the introduction, let us define it here more precisely.

Definition 3.5 Let (π, V) be an H-distinguished representation of G which admits a unitary central character.

(a) For each nonzero $\lambda \in \mathrm{Hom}_H(\pi, \mathbf{1})$, we say (π, V) is (H, λ)-**relatively tempered** if for all $\epsilon > 0$, the (H, λ)-matrix coefficients $\varphi_{\lambda,v}$ are in $L^{2+\epsilon}(Z_G H\backslash G)$ for all $v \in V$.
(b) We say (π, V) is H-**relatively tempered** if it is (H, λ)-relatively tempered for all $\lambda \in \mathrm{Hom}_H(\pi, \mathbf{1})$.

Remark 3.6 Assume π is irreducible. One can see that if φ_{λ,v_0} is in $L^{2+\epsilon}(Z_G H\backslash G)$ for one nonzero $v_0 \in V$, then $\varphi_{\lambda,v}$ are in $L^{2+\epsilon}(Z_G H\backslash G)$ for all $v \in V$, because any $v \in V$ is a finite linear combination of vectors of the form $\pi(g)v_0$ for some g.

This definition is motivated by the following definition introduced by Kato-Takano in [6, 7].

Definition 3.7 Let (π, V) be an H-distinguished representation of G.

1. (a) For each nonzero $\lambda \in \mathrm{Hom}_H(\pi, \mathbf{1})$, (π, V) is said to be **(H, λ)-relatively cuspidal** if the (H, λ)-matrix coefficients $\varphi_{\lambda,v}$ are compactly supported modulo $Z_G H$ for all $v \in V$.
   (b) (π, V) is called **H-relatively cuspidal** if it is (H, λ)-relatively cuspidal for all $\lambda \in \mathrm{Hom}_H(\pi, \mathbf{1})$.
2. Assume π admits a unitary central character.

   (a) For each nonzero $\lambda \in \mathrm{Hom}_H(\pi, \mathbf{1})$, (π, V) is said to be **(H, λ)-relatively square integrable** if the (H, λ)-matrix coefficients $\varphi_{\lambda,v}$ are square integrable modulo $Z_G H$ for all $v \in V$.
   (b) (π, V) is called **H-relatively square integrable**, if it is (H, λ)-relatively square integrable for all $\lambda \in \mathrm{Hom}_H(\pi, \mathbf{1})$.

Remark 3.8 For the definition of the relative cuspidality, one does not need to assume π has a central character, though for relative square integrability one does. Of course, a relatively cuspidal representation with a unitary central character is relatively square integrable.

Remark 3.9 As with relative-temperedness, if π is irreducible, it is (H, λ)-relatively cuspidal (resp. (H, λ)-relatively square integrable) if and only if $\varphi_{\lambda,v}$ is compactly supported (resp. square integrable) modulo $Z_G H$ for one nonzero $v \in V$.

Remark 3.10 Strictly speaking, for the square integrability, Kato-Takano used the word "(H, λ)-square integrable" instead of "(H, λ)-relatively square integrable". Yet, it seems to the author that it is more appropriate and standard to include the word "relatively".

Let us mention that the major achievement of [6] is the following.

Theorem 3.11 (Theorem 6.2 of [6]) *Let (π, V) be an H-distinguished representation of G and $\lambda \in \mathrm{Hom}_H(\pi, \mathbf{1})$. Then it is (H, λ)-relatively cuspidal if and only if $r_P(\lambda) = 0$ for all proper θ-split parabolic subgroups P of G. Further, if we assume the involution θ is standard, then the condition has to be checked only for the standard θ-split parabolic subgroups of G.*

3.5 The Casselman Criteria

Now we have the following "Casselman criteria" for H-relative square integrability and H-relative temperedness.

Theorem 3.12 *Let (π, V) be a finitely generated H-distinguished representation which admits a unitary central character. Fix $\lambda \in \mathrm{Hom}_H(\pi, \mathbf{1})$.*

(a) (π, V) is (H, λ)-relatively square integrable if and only if for all θ-split parabolic P, we have

$$\nu \in +\mathfrak{s}_P^*$$

for all $\nu \in \mathrm{Exp}_{S_P}(\pi_P, \lambda)$.

(b) (π, V) is (H, λ)-relatively tempered if and only if for all θ-split parabolic P, we have

$$\nu \in \overline{+\mathfrak{s}_P^*}$$

for all $\nu \in \mathrm{Exp}_{S_P}(\pi_P, \lambda)$.

Further, if we assume that the involution θ is standard, then the conditions have to be checked only for the standard θ-split parabolic subgroups P_I.

Proof (a) This is the main theorem of [7] (i.e. [7, Theorem 4.7]), though their theorem is stated by using central exponents instead of exponents. To translate their theorem to our "exponent version" is straightforward.

(b) The proof for this is essentially the same as (a), with a very small modification at the end of the proof, and left to the reader. (Details are also found in [8].)

The case of standard θ can be easily derived from the previous part. See [10, Proposition 4.21] for the detail. □

Let us mention a couple of corollaries of this theorem.

Corollary 3.13 *Every (H, λ)-relatively square integrable representation is (H, λ)-relatively tempered.*

Proof This follows because $+\mathfrak{s}_P^* \subseteq \overline{+\mathfrak{s}_P^*}$. □

Corollary 3.14 *Let (π, V) be a finitely generated H-distinguished representation of G, which is tempered and admits a unitary central character. Then π is H-relatively tempered.*

Proof By the Casselman criterion of (usual) temperedness, for all parabolics (hence all θ-split parabolics) $P \subseteq G$ and all $\chi \in \mathcal{E}xp_{A_P}(\pi_P)$, we have $\mathrm{Re}(\chi) \in \overline{+\mathfrak{a}_P^*}$. As in Lemma 2.10 the projection $\mathfrak{a}_P^* \to \mathfrak{s}_P^*$ maps $\overline{+\mathfrak{a}_P^*}$ into $\overline{+\mathfrak{s}_P^*}$. The corollary follows by the surjectivity of $\mathcal{E}xp_{A_P}(\pi_P) \to \mathcal{E}xp_{S_P}(\pi_P)$ and the inclusion $\bigcup_{\lambda \in \mathrm{Hom}_H(\pi, \mathbf{1})} \mathcal{E}xp_{S_P}(\pi_P, \lambda) \subseteq \mathcal{E}xp_{S_P}(\pi_P)$. □

It is often necessary to consider representations which do not admit a central character. For this reason, let us now broaden our definition of relatively tempered representation as follows.

Definition 3.15 Let (π, V) be a finitely generated H-distinguished representation of G which does not admit a unitary central character.

(a) For each nonzero $\lambda \in \mathrm{Hom}_H(\pi, \mathbf{1})$, we say (π, V) is (H, λ)**-relatively tempered** if for all θ-split parabolic P and for all $\nu \in \mathrm{Exp}_{S_P}(\pi_P, \lambda)$, we have $\nu \in \overline{+\mathfrak{s}_P^*}$.
(b) We say (π, V) is H**-relatively tempered** if it is (H, λ)-relatively tempered for all $\lambda \in \mathrm{Hom}_H(\pi, \mathbf{1})$.

By the Casselman criterion (Theorem 3.12), this definition coincides with the previous one if π admits a unitary central character.

Remark 3.16 This definition is also used in [1]. Also it is shown in [2] that this is equivalent to saying that $\varphi_{\lambda,v}$ is a tempered function, as in the non-relative case.

4 Subrepresentation Theorem

In this section, we will prove the subrepresentation theorem discussed in the introduction.

4.1 Subrepresentation Theorems

First let us recall the following theorem by Kato-Takano, which is a relative version of the well-known Jacquet's subrepresentation theorem.

Theorem 4.1 (Theorem 7.1 of [6]) *Let (π, V) be an irreducible H-relatively square integrable representation. Then π is a subrepresentation of an induced representation $\mathrm{Ind}_P^G \tau$, where $P = MU$ is a θ-split parabolic and τ is an irreducible M^θ-relatively cuspidal representation. Further if the involution θ is standard, then for each choice of Δ, one can always choose P to be standard.*

Proof The first part is [6, Theorem 7.1]. Now assume θ is standard and fix Δ. By the first part, we already have $\pi \subseteq \mathrm{Ind}_P^G \tau$ for some θ-split $P = MU$. One can always choose $h \in H$ such that ${}^hP$ is standard. Then one can see $\mathrm{Ind}_P^G \tau \cong \mathrm{Ind}_{{}^hP}^G {}^h\tau$, and ${}^h\tau$ is $({}^hM)^\theta$-relatively cuspidal since $({}^hM)^\theta = {}^h(M^\theta)$. □

Remark 4.2 In this theorem the inducing representation τ is (M^θ, λ)-relatively cuspidal for all $\lambda \in \mathrm{Hom}_{M^\theta}(\tau, \mathbf{1})$.

Now the following is our main theorem, which is a relative version of [11, Proposition III.4.1] which asserts that every tempered representation can be embedded into an induced representation whose inducing data is a square integrable representation. Namely we will prove

Theorem 4.3 *Let (π, V) be an (H, λ)-relatively tempered representation for some $\lambda \in \mathrm{Hom}_H(\pi, \mathbf{1})$ which admits a unitary central character. Then there exists a nonzero G-intertwining map*

$$\pi \longrightarrow \mathrm{Ind}_P^G \tau,$$

where $P = MU$ is θ-split and τ is an irreducible M^θ-relatively square integrable representation of M. In particular, if π is irreducible, it is a direct summand of $\mathrm{Ind}_P^G \tau$. Further if the involution θ is standard, then for each choice of Δ one can always choose P to be standard.

4.2 Some Lemmas

To prove this theorem, we need some preparations. First we need

Lemma 4.4 *Let π be a representation of G with a unitary central character, and let $\nu \in \mathfrak{a}_G^*$. Then $\exp(\nu) \otimes \pi$ is H-distinguished if and only if $\exp(\nu)|_H = 1$ and π is H-distinguished. Moreover, we have*

$$\mathrm{Hom}_H(\exp(\nu) \otimes \pi, \mathbf{1}) = \mathrm{Hom}_H(\pi, \mathbf{1}),$$

where the equality is literally given by the equality on the spaces of linear functionals on the space $V_{\exp(\nu)\otimes\tau} = V_\tau$.

Proof The if-part is obvious. To show the only-if-part, note that if $\exp(\nu) \otimes \tau$ is H-distinguished, then its central character is trivial on $Z_G^\theta = Z_G \cap H$, and hence its absolute value is trivial on Z_G^θ. Namely $\exp(\nu)|_{Z_G^\theta} = 1$. Thus $\exp(\nu) = 1$ by [6, Lemma 1.10 (2)]. Once we know this, it is immediate that π is H-distinguished, and the equality of the hom spaces is clear. □

Next, we need

Lemma 4.5 *Let (π, V) be an H-distinguished representation of G which admits a unitary central character. Assume π is (H, λ)-relatively square integrable for $\lambda \in \mathrm{Hom}_H(\pi, \mathbf{1})$. Then λ factors through a nonzero unitary semi-simple quotient (π^λ, V^λ) of (π, V). In particular, (π, V) has an irreducible unitary quotient distinguished by λ.*

Proof As discussed in Sect. 3.1, λ always factors through the image of the map $T_{\lambda,\pi} : V \to C^\infty(H\backslash G)$. But since π is relatively square integrable, the image of $T_{\lambda,\pi}$ is actually in $L^2(Z_G H\backslash G)$, which is unitary. Let us denote this image by (π^λ, V^λ).

Since π is admissible, so is (π^λ, V^λ). So (π^λ, V^λ) is semi-simple. (Recall that a unitary admissible representation is semi-simple.) □

This implies

Lemma 4.6 *Let (π, V) be a finitely generated H-distinguished representation of G, and let $\lambda \in \mathrm{Hom}_H(\pi, \mathbf{1})$ be fixed. Assume that for every θ-split P one has $\nu \in +\mathfrak{s}_P^*$ for all $\nu \in \mathrm{Exp}_{S_P}(\pi_P, \lambda)$. (Namely π satisfies the Casselman criterion for (H, λ)-relative square integrability.) Then π has an irreducible distinguished quotient σ which is (H, λ')-relatively square integrable for some $\lambda' \in \mathrm{Hom}_H(\sigma, \mathbf{1})$. (Note that if π admits a unitary central character, this is immediate from the previous lemma. But here we do not assume it.)*

Proof As in the proof of the previous lemma, λ always factors through the image of the map $T_{\lambda,\pi} : V \to C^\infty(H\backslash G)$. Let us call the image Σ, so that $\Sigma \subseteq C^\infty(H\backslash G)$. Note that every subrepresentation of Σ is distinguished by λ. Since Σ is a quotient of π, it also satisfies the Casselman criterion, and thus every subrepresentation of Σ that admits a central character is (H, λ)-relatively square integrable. (Note that since Σ satisfies the Casselman criterion, the central character has to be unitary.) Now by [6, Lemma 1.7], Σ has a quotient representation σ' isomorphic to a subrepresentation $\sigma'' \subseteq \Sigma$ which admits a central character. Note that, being an (H, λ)-distinguished subrepresentation of Σ, σ'' is (H, λ)-relatively square integrable, and hence by the previous lemma $\lambda|_{V_{\sigma''}}$ factors through a semisimple quotient. But being a subrepresentation of the image Σ of $T_{\lambda,\pi}$, we know that $\lambda|_{V_{\sigma''}}$ does not factor through any proper subquotient of σ'', which implies σ'' is already semisimple. Hence σ' is semi-simple, and is (H, λ')-relatively square integrable, where λ' is the composite $\sigma' \xrightarrow{\sim} \sigma'' \xrightarrow{\lambda} \mathbb{C}$. So we can take σ to be any irreducible constituent of σ'. □

4.3 Weak Jacquet Module

Our proof of the subrepresentation theorem is a relative version of [11, Proposition III.4.1], in which one considers "weak Jacquet modules". To introduce this notion, let (π, V) be a finitely generated admissible H-distinguished representation of our reductive group G and $P = MU$ be a θ-split parabolic subgroup. Recall that the Jacquet module (π_P, V_P) is written as

$$(\pi_P, V_P) = \bigoplus_{\chi \in \mathcal{E}xp_{S_P}(\pi_P)}{}^{\cdot} \; (\pi_{P,\chi}, V_{P,\chi}),$$

where $V_{P,\chi}$ is the generalized χ-eigenspace, and each $V_{P,\chi}$ is invariant under M. For a fixed $\lambda \in \mathrm{Hom}_H(\pi, \mathbf{1})$ one can further write

$$(\pi_P, V_P) = (\pi_P^\lambda, V_P^\lambda) \oplus (\pi_P^0, V_P^0),$$

where

$$V_P^\lambda := \bigoplus_{\chi \in \mathcal{E}xp_{S_P}(\pi_P,\lambda)} V_{P,\chi} \quad \text{and} \quad V_P^0 := \bigoplus_{\chi \notin \mathcal{E}xp_{S_P}(\pi_P,\lambda)} V_{P,\chi}.$$

Namely, $r_P(\lambda)$ is nonzero on each summand $V_{P,\chi}$ of the space V_P^λ, but identically zero on V_P^0. By following [11, p. 267], let us write

$$(\pi_P^\lambda, V_P^\lambda) = (\pi_P^{\lambda,w}, V_P^{\lambda,w}) \oplus (\pi_P^{\lambda,+}, V_P^{\lambda,+}),$$

where

$$V_P^{\lambda,w} := \bigoplus_{\substack{\chi \in \mathcal{E}xp_{S_P}(\pi_P,\lambda) \\ \mathrm{Re}(\chi)=0}} V_{P,\chi} \quad \text{and} \quad V_P^{\lambda,+} := \bigoplus_{\substack{\chi \in \mathcal{E}xp_{S_P}(\pi_P,\lambda) \\ \mathrm{Re}(\chi)\neq 0}} V_{P,\chi}.$$

We call $V_P^{\lambda,w}$ the weak Jacquet module with respect to λ, and hence the superscript ^{w}. We then have

Lemma 4.7 *Let (π, V) be a finitely generated (H, λ)-relatively tempered representation which does not necessarily admit a central character. The representation $(\pi_P^{\lambda,w}, V_P^{\lambda,w})$ (if nonzero) is $(M^\theta, r_P(\lambda))$-relatively tempered for all θ-split P.*

Proof We will apply the Casselman criterion for relative temperedness to $(\pi_P^{\lambda,w}, V_P^{\lambda,w})$. Let $\chi' \in \mathcal{E}xp_{S_{P'}}(\pi_{P'}, (\lambda))$, where P' is a θ-split parabolic subgroup such that $P' \subseteq P$. But since for all $\chi \in \mathcal{E}xp_{S_P}(\pi_P, \lambda)$ such that $V_{P,\chi} \subseteq V_P^{\lambda,w}$ we have $\mathrm{Re}(\chi) = 0$, the projection of $\mathrm{Re}(\chi')$ onto $\mathfrak{s}_P^*$ is zero, and hence $\mathrm{Re}(\chi') \in \overline{+\mathfrak{s}_P^{P'*}} := \overline{+\mathfrak{s}_{P'}^*} \cap \mathfrak{s}_P^{P'*}$, which shows $(\pi_P^{\lambda,w}, V_P^{\lambda,w})$ is relatively tempered. (Note that the representation $\pi_P^{\lambda,w}$ is a representation of M_P and hence the Casselman criterion has to be with respect to $\overline{+\mathfrak{s}_P^{P'*}}$.) □

This implies

Lemma 4.8 *Let (π, V) be a finitely generated (H, λ)-relatively tempered representation which does not necessarily admit a central character. Then the following conditions are equivalent.*

(a) For all proper θ-split parabolic subgroups P, one has $\nu \in +\mathfrak{s}_P^*$ for all $\nu \in \mathrm{Exp}_{S_P}(\pi_P, \lambda)$.
(b) For all proper θ-split parabolic subgroups P, one has $V_P^{\lambda,w} = 0$.
(c) For all proper maximal θ-split parabolic subgroups P, one has $V_P^{\lambda,w} = 0$.

Further if (π, V) admits a unitary central character, (b), (c) and the following (d) are all equivalent.

(d) (π, V) is (H, λ)-relatively square integrable.

Finally, if the involution θ is standard, then all the above P's may be taken to be standard.

Proof $(a) \Rightarrow (b)$: This is immediate.

$(b) \Rightarrow (c)$: This is immediate, too.

$(c) \Rightarrow (a)$: Let $\nu \in \mathrm{Exp}_{S_P}(\pi_P, \lambda)$, where P is a proper θ-split parabolic subgroup. Further let $Q \supseteq P$ be a proper maximal θ-split parabolic subgroup and let $\nu_Q \in \mathrm{Exp}_{S_Q}(\pi_Q, \lambda)$ be the image of ν under the surjection $\mathrm{Exp}_{S_P}(\pi_P, \lambda) \to \mathrm{Exp}_{S_Q}(\pi_Q, \lambda)$, which is the vertical arrow in Proposition 3.4 with P and Q switched. Since π is (H, λ)-relatively tempered, by definition we have $\nu_Q \in \overline{+\mathfrak{s}_Q^*}$. Also since we have $V_Q^{\lambda,w} = 0$ by our assumption (c), we must have $\nu_Q \neq 0$. Now if $\nu_Q \in \overline{+\mathfrak{s}_Q^*}$ and $\nu_Q \neq 0$, then $\nu_Q \in +\mathfrak{s}_Q^*$ because $\overline{+\mathfrak{s}_Q^*}$ is a "one-dimensional ray". Since Q is arbitrary, we have $\nu_Q \in +\mathfrak{s}_Q^*$ for all maximal θ-split $Q \supseteq P$, which implies $\nu \in +\mathfrak{s}_P^*$.

The equivalences of (b),(c) and (d) are immediate from the Casselman criterion for relative square integrability.

Finally in the case of standard θ, the equivalences follow because the Casselman criterion has to be checked only for standard P. □

4.4 Proof of Theorem and Its Corollaries

Now finally we are ready to prove the subrepresentation theorem.

Proof of Theorem 4.3. By following the argument adapted by [7, proof of Theorem 7.1], we will argue by induction on the rank of the maximal θ-split torus of $A_G \backslash G$. The base step is when the rank is 0, in which case $HA_G \backslash G$ is compact by [5, §4.3], and hence π is already relatively square integrable, and there is nothing to show because one can just take $P = G$.

Assume the rank is > 0. If (π, V) is already (H, λ)-relatively square integrable, then again there is nothing to show. So assume it is not (H, λ)-relatively square integrable. Then by Lemma 4.8, $V_P^{\lambda,w} \neq 0$ for some proper θ-split P. Let us take $P = MU$ to be minimal with respect to this property. Since π is (H, λ)-relatively tempered, $V_P^{\lambda,w}$ is $(M^\theta, r_P(\lambda))$-relatively tempered by Lemma 4.7. But since $(V_P^{\lambda,w})_Q^{r_P(\lambda),w} = 0$ for any smaller parabolic $Q \subsetneq P$, we know by Lemma 4.8 that all the relative exponents of $V_P^{\lambda,w}$ with respect to $r_p(\lambda)$ satisfy the Casselman criterion for relative square integrability. Hence by Lemma 4.6, $V_P^{\lambda,w}$ has an irreducible distinguished quotient σ which is (M^θ, λ')-relatively square integrable for some λ'. By Frobenius reciprocity, we have nonzero $\pi \to \mathrm{Ind}_P^G \sigma$.

Here since σ is (M^θ, λ')-relatively square integrable and hence is (M^θ, λ')-relatively tempered, and the rank of the maximal θ-split torus of $A_M \backslash M$ is strictly smaller, one can apply the induction hypothesis to σ, and obtain $\sigma \hookrightarrow \mathrm{Ind}_{Q \cap M}^M \tau$ for some θ-split $Q \cap M = M_Q U_Q$ and M_Q^θ-relatively square integrable τ. By inducing in stages, we have the desired $\pi \to \mathrm{Ind}_Q^G \tau$.

By Lemma 4.5, we know that τ is unitary. Hence $\mathrm{Ind}_Q^G \tau$ is unitary, which implies that $\mathrm{Ind}_P^G \tau$ is semi-simple. So if π is irreducible, then π is indeed a direct summand.

Finally, if the involution θ is standard, one can use Lemma 4.8 only for standard parabolic subgroups, and hence both P and Q in the above argument can be chosen to be standard. □

The following is an immediate corollary.

Corollary 4.9 *Let (π, V) be an irreducible H-distinguished representation which is (H, λ)-relatively tempered for some $\lambda \in \mathrm{Hom}_H(\pi, \mathbf{1})$. Then π is unitary.*

Proof By Theorem 4.3, we have $\pi \hookrightarrow \mathrm{Ind}_P^G \tau$ for some relatively square integrable τ, which is unitary by Lemma 4.5. Hence π is unitary because $\mathrm{Ind}_P^G \tau$ is unitary. □

Now recall that for any representation (π, V) of G, the complex conjugate representation $(\overline{\pi}, \overline{V})$ is defined as follows: The space $\overline{V}$ is the same as V as an abelian group but the scalar multiplication is via complex conjugate, and $\overline{\pi}$ is defined by $\overline{\pi}(g)v = \pi(g)v$ for all $v \in V = \overline{V}$ and $g \in G$. Then there is a natural isomorphism

$$\mathrm{Hom}_H(\pi, \mathbf{1}) \xrightarrow{\sim} \mathrm{Hom}_H(\overline{\pi}, \mathbf{1}), \quad \lambda \mapsto \overline{\lambda},$$

where $\overline{\lambda}$ is the complex conjugate of λ, namely $\overline{\lambda}(v) = \overline{\lambda(v)}$ for $v \in V$.

If (π, V) is an irreducible unitary representation with a unitary structure $\langle -, - \rangle_\pi$, its contragredient $(\pi^\vee, V^\vee)$ is identified with $(\overline{\pi}, \overline{V})$ via the map

$$(\overline{\pi}, \overline{V}) \xrightarrow{\sim} (\pi^\vee, V^\vee), \quad v \mapsto \langle -, v \rangle_\pi,$$

which induces the isomorphism

$$\mathrm{Hom}_H(\pi, \mathbf{1}) \xrightarrow{\sim} \mathrm{Hom}_H(\pi^\vee, \mathbf{1}), \quad \lambda \mapsto \lambda^\vee,$$

where we define $\lambda^\vee$ to be the composite

$$\lambda^\vee : \pi^\vee \xrightarrow{\sim} \overline{\pi} \xrightarrow{\overline{\lambda}} \mathbb{C}.$$

So in particular, if (π, V) is unitary, then (π, V) is H-distinguished if and only if $(\pi^\vee, V^\vee)$ is H-distinguished.

With this said, we have

Corollary 4.10 *Let (π, V) be irreducible. Then (π, V) is (H, λ)-relatively tempered for some $\lambda \in \mathrm{Hom}_H(\pi, \mathbf{1})$ if and only if $(\pi^\vee, V^\vee)$ is $(H, \lambda^\vee)$-relatively tempered for some $\lambda^\vee \in \mathrm{Hom}_H(\pi^\vee, \mathbf{1})$. Also (π, V) is H-relatively tempered if and only if $(\pi^\vee, V^\vee)$ is H-relatively tempered.*

Proof By symmetry, we have only to show the only-if part. So assume (π, V) is (H, λ)-relatively tempered for some $\lambda \in \mathrm{Hom}_H(\pi, \mathbf{1})$. By the above corollary, π is unitary, and hence we can identify $(\pi^\vee, V^\vee)$ with $(\overline{\pi}, \overline{V})$. Then one can readily see that $(\overline{\pi}, \overline{V})$ is $(H, \overline{\lambda})$-relatively tempered. The first assertion follows.

Since we have the natural isomorphism $\mathrm{Hom}_H(\pi, \mathbf{1}) \xrightarrow{\sim} \mathrm{Hom}_H(\overline{\pi}, \mathbf{1})$, the second assertion follows. □

Acknowledgements The author was partially supported by Simons Foundations Collaboration Grant #35952, NSA Young Investigator Grant H98230-16-1-0312, and an AMS Centennial Fellowship for 2017–2018. Part of the paper was completed while he was visiting the National University of Singapore in fall 2017 and spring 2018, and he would like to express thanks for their hospitality. He would like to thank Wee Teck Gan for useful conversations and Kimball Martin for introducing him to the works of Kato and Takano. Finally he would like to thank Kenji Takano for sending him the preprint [8].

References

1. J. Carmona, P. Delorme, Constant term of Eisenstein integrals on a reductive p-adic symmetric space. Trans. Amer. Math. Soc. **366**, 5323–5377 (2014)
2. P. Delorme, P. Harinck, Wave packets in the Schwartz space of a reductive p-adic symmetric space. J. Lie Theory **24**, 41–75 (2014)
3. A.G. Helminck, Algebraic groups with a commuting pair of involutions and semisimple symmetric spaces. Adv. Math. **71**, 21–91 (1988)
4. A.G. Helminck, G.G. Helminck, A class of parabolic k-subgroups associated with symmetric k-varieties. Trans. Amer. Math. Soc. **350**, 4669–4691 (1998)
5. A.G. Helminck, S.P. Wang, On rationality properties of involutions of reductive groups. Adv. Math. **99**, 26–96 (1993)
6. S. Kato, K. Takano, Subrepresentation theorem for p-adic symmetric spaces. Int. Math. Res. Not. IMRN **11**, Art. ID rnn028, 40 pp (2008)
7. S. Kato, K. Takano, Square integrability of representations on p-adic symmetric spaces. J. Funct. Anal. **258**, 1427–1451 (2010)
8. S. Kato K. Takano, Addendum to the paper "Square integrability of representations on p-adic symmetric spaces", preprint
9. N. Lagier, Terme constant de fonctions sur un espace symétrique réductif p-adique. J. Funct. Anal. **254**, 1088–1145 (2008)
10. J. Smith, Relative discrete series representations for two quotients of p-adic GL_n. Can. J. Math. (to appear)
11. J.-L. La, Formule de Plancherel pour les groupes p-adiques (d'après Harish-Chandra). J. Inst. Math. Jussieu **2**, 235–333 (2003)

Printed by Printforce, the Netherlands